Mechanics of Materials and Interfaces

The Disturbed State Concept

Chandrakant S. Desai

CRC Press
Boca Raton   London   New York   Washington, D.C.

Library of Congress Cataloging-in-Publication Data

Desai, C.S. (Chandrakant S.), 1936–
  Mechanics of materials and interfaces : the disturbed state concept / by Chandrakant S. Desai
    p. cm.
  Includes bibliographical references and index.
  ISBN 0-8493-0248-X (alk. paper)
   1. Strength of materials—Mathematical models. 2. Strains and stresses. 3. Interfaces (Physical sciences) I. Title.

TA405 .D45 2000
620.1′12′015118—dc21   00-052883

This book contains information obtained from authentic and highly regarded sources. Reprinted material is quoted with permission, and sources are indicated. A wide variety of references are listed. Reasonable efforts have been made to publish reliable data and information, but the author and the publisher cannot assume responsibility for the validity of all materials or for the consequences of their use.

Neither this book nor any part may be reproduced or transmitted in any form or by any means, electronic or mechanical, including photocopying, microfilming, and recording, or by any information storage or retrieval system, without prior permission in writing from the publisher.

The consent of CRC Press LLC does not extend to copying for general distribution, for promotion, for creating new works, or for resale. Specific permission must be obtained in writing from CRC Press LLC for such copying.

Direct all inquiries to CRC Press LLC, 2000 N.W. Corporate Blvd., Boca Raton, Florida 33431, or visit our website at www.crcpress.com

Trademark Notice: Product or corporate names may be trademarks or registered trademarks, and are used only for identification and explanation, without intent to infringe.

© 2001 by CRC Press LLC

No claim to original U.S. Government works
International Standard Book Number 0-8493-0248-X
Library of Congress Card Number 00-052883
Printed in the United States of America 1 2 3 4 5 6 7 8 9 0
Printed on acid-free paper

To my father

- who, I believe, is inquisitive and questioning in the space beyond, which is congruent to that of mine.

and

To those giants

- of mechanics, physics, and philosophy, on whose contributions we stand and extend.

Continuum and discontinuum,
Points and spaces,
Exist together, United and Coupled;
Sat and Asat,
Existence and nonexistence;
Exist together, United and Coupled;
Merging in each other.

PREFACE

Understanding and characterizing the mechanical behavior of engineering materials and interfaces or joints play vital roles in the prediction of the behavior, and the analysis and design, of engineering systems. Principles of mechanics and physics are invoked to derive governing equations that allow solutions for the behavior of the systems. Such closed-form or numerical solutions involve the important component of material behavior defined by constitutive laws or equations or models.

Definition of the constitutive laws based on fundamental principles of mechanics, identification of significant parameters, determination of the parameters from appropriate (laboratory and/or field) tests, validation of the models with respect to the test data, implementation of the models in the solution procedures—closed-form or computational—and validation of practical boundary-value problems are all important ingredients in the development and use of realistic material models.

The characterization of the mechanical behavior of engineering materials, called the stress–strain or constitutive models, has been the topic under the general subject of "mechanics of materials". As material behavior is very often nonlinear, the governing equations are also nonlinear. In the early stages, however, it was necessary to linearize the governing differential equations so that the closed-form solution procedures could be used. The advent of the electronic computer, with increasing storage capacity and speed, made it possible to solve nonlinear equations in discretized forms. Hence, the need to assume constant coefficients or material parameters in the linear and closed-form solutions may no longer exist. As a consequence, it is now possible to develop and use models for realistic nonlinear material response.

Almost all materials exhibit nonlinear behavior. In simple words, this implies that the response of the material is not proportional to the input excitation or load. Hence, although the assumption of linearity provided, and still can provide, useful solutions, their validity is highly limited in the nonlinear regimes of the material response. Thus, the fact that modern computers and numerical or computational methods now permit the consideration of nonlinear responses is indeed a highly desirable development.

Among linear constitutive models are Hooke's law that defines linear elastic stress–strain response under mechanical load, Darcy's law that defines the linear velocity-gradient response for fluid flow, and Ohm's law that defines the linear voltage-current relation for electrical flow. It is recognized that the validity of these models is limited. Hooke's law does not apply if the material response involves effect of factors such as state of stress or strain, stress or loading paths, temperature, initial and induced discontinuities, and existence of fluid or gas in the material's porous microstructure. Darcy's law does not apply if the flow is turbulent, and Ohm's law loses validity if the conducting material is nonhomogeneous and thermal effects are present.

For the characterization of the nonlinear behavior of materials, the effects of significant factors such as initial conditions, state of stress, stress or loading path, type of loading, and multiphase nature need to be considered for realistic engineering solutions. The pursuit of the development of models for the nonlinear response has a long history in the subjects of physics and mechanics of materials. Among the models proposed and developed are linear and nonlinear elasticity (e.g., hyper- and hypoelasticity), classical plasticity (von Mises, Tresca, Mohr–Coulomb, Drucker–Prager), continuous hardening or yielding plasticity (critical-state, cap, hierarchical single-surface–HISS), and kinematic and anisotropic hardening in the context of the theory of plasticity.

Viscoelastic, viscoplastic, and elastoviscoplastic models are among those developed to account for time-dependent viscous or creep response. Endochronic theory involving an implicit time scale has been proposed in the context of plasticity and viscoplasticity.

Models based on micromechanical considerations involve the idea that the observed macrolevel response of the material can be obtained by integrating the responses of behavior at the micro- or particle level, often through a process of linear integration. Although this idea is elegant, at this time it suffers from the limitation that the particle-level response is difficult to measure and characterize.

Most of the models are based on the assumption that the material is continuous. As a result, the theories of continuum mechanics have been invoked for their formulation. It is, recognized, however, that discontinuities exist and develop in a deforming material. Thus, the theories based on continuum mechanics may not be strictly valid, and various models based on fracture and continuum damage concepts thereby become relevant.

The classical continuum damage models are based on the idea that a material experiences microcracking and fracturing, which can cause degradation or damage in the material's stiffness and strength. The remaining degraded stiffness (strength) is then defined on the basis of the response of the undamaged part modified by growing damaged parts, which are assumed to act like *voids* and possess no strength at all. As a result, the classical continuum damage models do not allow for the coupling and interaction between the damaged and undamaged parts. This aspect has significant consequences, as the effect of neighboring (damaged) parts is not included in the characterization of the response.

Various nonlocal and microcrack interaction models have been proposed in the context of the classical damage model. An objective here has been to develop constitutive equations that allow for the coupling between the damaged (microcracked) and undamaged parts and the effect of what happens in the neighborhood of a material point. Such enhancements as gradient, Cosserat, and micropolar theories have been proposed to incorporate the nonlocal effects.

The effects of temperature and other environmental factors are incorporated by developing separate theories or by expressing the parameters in the above models as functions of temperature or other environmental factors.

The foregoing models are usually relevant for a specific characteristic of the material behavior such as elastic, plastic, creep, microcracking, and fracture. Each model involves a set of parameters for a specific characteristic that needs to be determined from laboratory tests. There is a growing recognition that development of unified or integrated constitutive descriptions can lead to more efficient, economical, and simplified models with ease of implementation in solution procedures. As a result, a number of efforts have been made toward unified or hierarchical models. The approach presented in this book represents one of these unified concepts: the *disturbed state concept* (DSC).

The DSC is a unified modelling approach that allows, in an integrated manner, for elastic, plastic, and creep strains, microcracking and fracture leading to softening and damage, and stiffening or healing, in a single framework. Its hierarchical nature permits the adoption and use of specialized versions for the foregoing factors. As a result, its development and application are simplified considerably.

The DSC is based on the basic physical consideration that the observed response of a material can be expressed in terms of the responses of its constituents, connected by the coupling or disturbance function. In simple words, the observed material state is considered to represent disturbance or deviation with respect to the behavior of the material for appropriately defined reference states. This approach is consistent with the idea that the current state of a material system, animate or inanimate, can be considered to be the *disturbed* state with respect to its initial and final state(s).

In the case of engineering materials, the DSC stipulates that at any given deformation stage, the material is composed of two (or more) parts. For instance, a dry deforming material is composed of material parts in the original (continuum) state, called the relatively intact (RI) state, and remaining parts in the degraded or stiffened state, called the fully adjusted (FA) state; the meanings of the terms "RI and FA state" will be explained in subsequent chapters. The degraded part can represent effects of relative particle motions and microcracking due to the *natural self-adjustment* (SA) of particles in the material's microstructure and can lead to damage or degradation. Under factors such as chemical, temperature, and fluid effects, the microstructure may experience stiffening or healing. Although the degradation or damage aspect in the DSC is similar to that in the classical damage models, the basic framework of the DSC is general and significantly different from that of the damage concept.

If a material element is composed of more than one material, the DSC can be formulated for the overall observed response (of the composite) by treating the behavior of individual components as reference responses. The behavior of an individual component may be characterized by using a continuum theory or by treating it as a mixture of the RI and FA parts.

Details of the DSC, including formulation of equations, identification of material parameters, determination of parameters from (laboratory) tests, validation at the laboratory test stage, implementation in solution (computer) procedures, and validation and solution of practical boundary-value problems, are presented in this book. Comparisons between the DSC and other available models are discussed, including the advantages the DSC offers. The latter arise due to characteristics such as the compact and unified nature of the DSC, physical meanings of material parameters, considerable reduction in the regression and curve-fitting required in many other models, ease of determination of parameters, and ease of implementation in solution procedures.

One of the DSC's advantages is that it can be used for "solid" materials and for interfaces and joints. The latter play an important role in the behavior of many engineering systems involving combinations of two or more materials. They include contacts in metals, interfaces in soil-medium (structure) problems, joints in rock, and joints in electronic packaging systems. It is shown that the mathematical framework of the DSC for three-dimensional solids can be specialized for the behavior of material contacts idealized as *thin-layer* zones or elements.

The fact that the DSC allows for interaction and coupling between the RI and FA parts offers a number of advantages in that the nonlocal effects are included in the model, hence also the characteristic dimension.

The DSC does not require constitutive description of particle-level processes as the micromechanical models do. The interacting behavior of the material composed of millions of particles is expressed in terms of the coupled responses of the material parts (clusters) in the RI and FA parts. The response of the RI and FA parts can be defined from laboratory tests. Thus, the DSC eliminates the need for defining particle-level behavior, which is difficult to measure at this time. At the same time, it allows for the coupled microlevel processes.

The behavior of material parts in the reference states in the DSC can be defined on the basis of any suitable model(s). Often, such available continuum theories as elasticity and plasticity, and the critical-state concept, are invoked for the characterizations.

The DSC represents a continuous evolution in the pursuit of the development of constitutive models by the author and his co-workers. Although it involves a number of new and innovative ideas, the DSC also relies on the available theories of mechanics and the contributions of many people who have been the giants in this field. For instance, the DSC includes ideas and concepts from the available elasticity, plasticity, viscoplasticity, damage, fracture, and critical-state theories. As the DSC allows adoption of these available models as special cases, they are presented individually in separate chapters, with identification of their use in the DSC.

In summary, the DSC is considered to represent a unique and powerful modelling procedure to characterize the behavior of a wide range of materials and interfaces. Its capabilities go beyond available material models, and it simultaneously leads to significant simplification toward practical applications.

The DSC permits approximation of material systems as *discontinuous* and includes their *continuum* attribute as a special case. Thus, it can provide a generalized basis for the introduction of the subject of mechanics of materials. It is therefore possible that the DSC can be introduced first as the general and basic approach in undergraduate courses on the strength or mechanics of materials in the first few lectures, and then the traditional mechanics of materials can be taught as before, by assuming the material systems to be continuous. The DSC can later be brought into the upper-level undergraduate courses. Hence, the material in this book can be introduced in undergraduate courses.

The advanced topics in the book can be taught in graduate-level courses with prerequisites of continuum theories such as elasticity and plasticity. The book can be useful to the researcher who wants to employ up-to-date, unified and simplified models to account for realistic nonlinear behavior of materials and interfaces. It will also be useful to practitioners involved in the solution of problems requiring realistic models and computer procedures.

The objectives of this text are as follows:

(1) to present a philosophical and detailed theoretical treatment of the DSC, including a comparison with other available models;
(2) to identify the physical meanings of the parameters involved and present procedures to determine them from laboratory test data;
(3) to use the DSC to characterize the behavior of materials such as geologic, ceramic, concrete, metal (alloys), silicon, and asphalt concrete, and interfaces and joints;
(4) to validate the DSC models with respect to laboratory tests used to find the parameters, and *independent* tests not used in the calibration;
(5) to implement the DSC models in computer (finite-element) procedures; and
(6) to validate the computer procedures by comparing predictions with observations from simulated and field boundary-value problems.

The basic theme of the text is to show that the DSC can provide a unified and simplified approach for the mathematical characterization of the mechanical response of materials and interfaces. As the final objective of any material model is to solve practical engineering problems, the text attempts to relate the models to practical use through their implementation in solution (computer) procedures. To this end, a number of problems from different disciplines such as civil, mechanical, and electrical engineering are solved using the computer procedures.

I would like to conclude the preface with the following statements:

> Students of mechanics of materials often raise the question, "Is there a constitutive model which is applicable to all materials?" And I respond: "Although our understanding of the material's response is growing, there is no model available that can characterize *all* materials in all respects. To understand and characterize matter (materials) completely, one may need to become the matter itself! When that happens, there is no difference left, and a full understanding may follow."
>
> This realization is important because the pursuit toward increased comprehension and improved characterization of materials must continue!

A number of my students and co-workers have participated in the development and application of the concepts and models presented in this book; their contributions are cited through references in various chapters. I have learned from them more than I could have from books. I wish to express special gratitude to Professor Antonio Gens, who read the manuscript and offered valuable suggestions. His remarks on mechanics, physics, and philosophy have enlightened and encouraged me. I wish to express my thanks to Professor K. G. Sharma, Professor Giancarlo Gioda, Dr. Marta Dolezalova, Dr. Nasser Khalili, and Dr. Hans Mühlhaus, who read parts of the manuscript and provided helpful comments. Mr. M. Dube, Mr. R. Whitenack, Mr. Z. Wang, and Mr. S. Pradhan provided useful suggestions and assistance. Thanks are due to Mrs. Rachèle Logan for her continued assistance. My mother Kamala, wife Patricia, daughter Maya, and son Sanjay have been sources of constant support and inspiration.

Chandrakant S. Desai
Tucson, Arizona

THE AUTHOR

Chandrakant S. Desai is a Regents' Professor and Director of the Material Modelling and Computational Mechanics Center, Department of Civil Engineering and Engineering Mechanics, University of Arizona, Tucson. He was a Professor in the Department of Civil Engineering, Virginia Polytechnic Institute and State University, Blacksburg, Virginia from 1974 to 1981, and a Research Civil Engineer at the U.S. Army Engineer Waterways Experiment Station, Vicksburg, Mississippi from 1968 to 1974.

Dr. Desai has made original and significant contributions in basic and applied research in material modeling and testing, and computational methods for a wide range of problems in civil engineering, mechanics, mechanical engineering, and electronic packaging. He has authored/edited 20 books and 18 book chapters, and has been the author/co-author of over 270 technical papers. He was the founder and General Editor of the *International Journal for Numerical and Analytical Methods in Geomechanics* from 1977 to 2000, and he has served as a member of the editorial boards of 12 journals. Dr. Desai has also been a chair/member of a number of committees of various national and international societies. He is the President of the International Association for Computer Methods and Advances in Geomechanics. Dr. Desai has also received a number of recognitions: Meritorious Civilian Service Award by the U.S. Corps of Engineers, Alexander von Humboldt Stiftung Prize by the German Government, Outstanding Contributions Medal in Mechanics by the International Association for Computer Methods and Advances in Geomechanics, Distinguished Contributions Medal by the Czech Academy of Sciences, Clock Award by ASME (Electrical and Electronic Packaging Division), Five Star Faculty Teaching Finalist Award, and the El Paso Natural Gas Foundation Faculty Achievement Award at the University of Arizona, Tucson.

CONTENTS

Chapter 1	Introduction	1
Chapter 2	The Disturbed State Concept: Preliminaries	17
Chapter 3	Relative Intact and Fully Adjusted States, and Disturbance	63
Chapter 4	DSC Equations and Specializations	93
Chapter 5	Theory of Elasticity in the DSC	115
Chapter 6	Theory of Plasticity in the DSC	149
Chapter 7	Hierarchical Single-Surface Plasticity Models in the DSC	179
Chapter 8	Creep Behavior: Viscoelastic and Viscoplastic Models in the DSC	273
Chapter 9	The DSC for Saturated and Unsaturated Materials	339
Chapter 10	The DSC for Structured and Stiffened Materials	391
Chapter 11	The DSC for Interfaces and Joints	421
Chapter 12	Microstructure: Localization and Instability	477
Chapter 13	Implementation of the DSC in Computer Procedures	529
Chapter 14	Conclusions and Future Trends	629
Appendix I	Disturbed State, Critical-State, and Self-Organized Criticality Concepts	631
Appendix II	DSC Parameters: Optimization and Sensitivity	663

Index 683

1
Introduction

CONTENTS
1.1 Prelude 1
1.2 Motivation 2
    1.2.1 Explanation of Reference States 3
    1.2.2 Engineering Materials and Matter 4
    1.2.3 Local and Global States 7
1.3 Engineering Materials 8
    1.3.1 Continuous or Discontinuous or Mixture 8
    1.3.2 Transformation and Self-Adjustment 9
    1.3.3 Levels of Understanding 9
    1.3.4 The Role of Material Models in Engineering 10
1.4 Disturbed State Concept 11
    1.4.1 Disturbance and Damage Models 12
    1.4.2 The DSC and Other Models 13
1.5 Scope 14
    1.5.1 Outlines of Chapters 14

1.1 Prelude

Continuity and discontinuity, order and disorder, positive and negative exist simultaneously; they are not separate, and they are contained and culminate in each other. They produce the holistic material world. The material world, *matter*, is a projection or manifestation of the complex and mysterious universe, which we have to deal with and comprehend. "Engineering material" is but a subset of the material world and carries with it the complexity and consciousness of the whole. The metaphysical and physical comprehension of matter entails interconnected phenomena at the macroscopic, microscopic, atomic, and subatomic levels, and beyond.

    The *Vedas*, the ancient scriptures of *knowledge* from India, say that order (or *rta*) is not fully manifested in the physical world or matter, and it exists with disorder, which may contain what remains to be realized. At the same time, there is harmony between *being* (existence, or *sat*), and its external manifestation, which is order (*rta*) (1–3). The manifested state is subject to laws and theories based on measurable quantities (like deformation and failure or collapse), while there always remains the germ of what is to come, which is nonmeasurable and incomprehensible. What is incomprehensible may reside in the space between particles (atoms) as the *life force* (*prana* or *chi*).

No material system under external influences exists in a unique composition. At any stage during deformation, a given material element may be treated as a mixture of a part of its *initial self*, the relatively intact (RI), and another transformed part due to the self-adjustment of the material's microstructure, at the *fully adjusted* (FA) state. A material element may also be composed of more than one material; then each of the components can be treated as a reference. The components exist simultaneously and contribute to the response of the mixture.

For a given material, the fully adjusted state can be described as the *critical state* at which the material approaches the state of invariant properties. For instance, at the critical state, the material approaches a state of constant density or specific volume. The critical state is approached through changes in the microstructure due to microcracking and relative motions of particles. The critical state is asymptotic and cannot be measured precisely or realized, but can be measured and defined *approximately* so as to construct mathematical models. The critical state is like the state Buddhists call *nirvana*, in which all biases, pushes, and pulls, due to *karmic* action (like nonsymmetric forces on materials, say, causing shear stresses), disappear, leading to the equilibrium or isotropic state.

The interpretation of material response that may be governed by factors beyond the mechanistic laws presented here is rather subjective. Its philosophical intent may be of interest to some readers, while to others it may seem not to be relevant from a technological viewpoint. It is presented with the notion that an appreciation of such factors can lead to vistas that may allow improved understanding and characterizations.

1.2 Motivation

The behavior of engineering materials under external forces is similar to the behavior of matter under external influences. It is possible that the behavior of materials at different levels—atomic, microscopic, and macroscopic—is similar; in other words, a collection of atoms may behave the same way as a collection of finite-sized particles. The observed behavior is affected by the components of the material element at a given level. For instance, the microlevel behavior is influenced by the behavior of the particle (skeleton) as well as the *space* between the particles. The particle system may be com-

Introduction

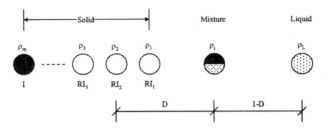

FIGURE 1.1
Relative densities of solid material melting to liquid state.

posed of more than one identifiable component, e.g., solids and liquid. The solid part may be treated as (relatively) intact or continuum, and the part that has experienced progressive microcracking and damage or strengthening by cohesive forces caused by chemicals can be treated as the FA or another reference state.

1.2.1 Explanation of Reference States

The use of the term *relatively* in the relatively intact (RI) state needs an explanation. Consider an example of a material that transforms from its solid state to the liquid state due to melting under a given temperature change. Figure 1.1 shows a symbolic representation of the melting process. The maximum density (which may be unattainable) in the solid state is denoted by ρ_m. However, the material has only a relative existence; in other words, in the solid state, it can exist under different densities, $\rho_1, \rho_2, \ldots$. Consider an intermediate state with density ρ_i during the melting process, which may be composed of parts of solid and liquid states. Then the intermediate state can be considered to be disturbed (D) with respect to its starting density (ρ_1, or $\rho_2 \ldots$). Thus, because a number of relative initial densities are possible for defining the disturbance, we use the term *relatively* to denote the solid reference state. Indeed, if known and measurable, the maximum density state can be used as the reference state; in that case, it may simply be referred to as the intact (I) state. The liquid state with density ρ_L can be adopted as the other reference state in the fully adjusted or fully liquefied condition. Later here and in subsequent chapters, we shall provide further and other explanations of the RI and FA states in the context of deforming materials.

The DSC is based on the fundamental idea that the behavior of an engineering material can be defined by an appropriate connection that characterizes the interaction between the behavior of the components, e.g., at the reference states RI and FA.

A deforming material exhibits the *manifested* and *unmanifested* responses; see Fig. 1.2. The manifested response is what we can measure in the laboratory or field tests on the material, and it can be quantified and defined using physical laws. The unmanifested response cannot be measured and may not

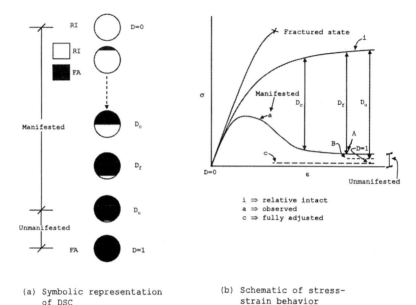

(a) Symbolic representation of DSC

(b) Schematic of stress–strain behavior

FIGURE 1.2
Representations of the DSC.

be amenable to known physical laws. The limitations of the measurement devices do not allow the measurement of the unmanifested response during which the material tends toward the FA or critical state (c), which may represent the fully disintegrated state of the material as a collection of particles or the fractured state involving a multitude of separations. The inclusion of the unmanifested response in the material characterization can indeed provide more realistic models. However, as the unmanifested response cannot be measured, it becomes necessary to quantify and define the FA response approximately, by using the stages *failure* (D_f) or *asymptotic* (D_u) (Fig. 1.2). The inclusion of even such an approximate definition of the FA state can lead to improved models. The DSC is based on the use of the approximate definition of the response of the material in the FA state. Such *approximate* constitutive models are considered to be *incomplete* because a part of the response cannot be measured, defined, or understood fully.

1.2.2 Engineering Materials and Matter

An engineering material is a special manifestation of the matter in the universe, and its response can be considered a subset of the general behavior of matter. We usually characterize the behavior of the material based essentially on the mechanistic considerations that treat it as composed of *inert* or *dead* particles; in other words, the material skeleton is assumed to behave like a "machine."

Figure 1.3 shows a schematic of the different levels at which matter can exist. Its original, most condensed *cosmic* state is marked as "O." Under various forces, it disintegrates and forms local material manifestations, one of them being the engineering material. The disintegrated matter, under various forces, tries to return to the 'O' state; perhaps the states "O" and 'O' are the same! The engineering material, at the local level, also starts from a given state (RI) and, under local engineering forces, tends toward the local (FA) state; $(FA)_\infty$ is the ultimate nonmeasurable state. This may be treated analogously to the seed, which germinates into a tree (a mixture of order and disorder) and then coalesces into the seed. The "morning star" and "evening star" were thought to be different, but it was found that both are the planet Venus! Thus, although we deal with the transformed material state in the engineering sense, in a philosophical sense, the initial and final material states are probably the same.

> Time present and time past are both perhaps present in time future, and time future contained in time past.... What might have been and what has been point to one end, which is always present.
>
> T.S. Eliot (*Burnt Norton*)

Perhaps, we can replace *time* by *matter*.

> "We know now that we live in a historical universe, one in which, not only living organisms, but stars and galaxies are born, mature, grow old and die. There is good reason to believe it to be a universe permeated with life, in which life arises, given enough time, wherever the conditions exist that make it possible," said Nobel laureate Wald (4).

Many quantum physicists have related the understanding of matter with cosmological concepts from the Eastern theological and mystical traditions (5–8). The central role of consciousness in the comprehension of matter in the Vedic tradition has been found to compare with the conclusions of modern physical thoughts (8, 9). Erwin Schrödinger (5), one of the pioneers of quantum mechanics, believed that the issues of determinism can be understood essentially through the Vedic concept of unique and all-pervading consciousness, which is composed of the consciousness of individual components of matter. These and other (recent) thoughts make us aware that scientific and religious (mystical) concepts can essentially be the same; both can lead to the understanding of matter as it exists and to the reality ("truth" or "sat") of the existence.

There is only one consciousness, and all manifestations (matter, living and nonliving) are that (or parts) of that consciousness. Goswami et al. (8) propose the concept of monistic idealism. Here the dualism of mind and matter does not exist, but they interact and exchange energy, and consciousness is considered to be the basic element of reality. This concept states that everything including matter exists in and is manipulated from consciousness.

6 *Mechanics of Materials and Interfaces*

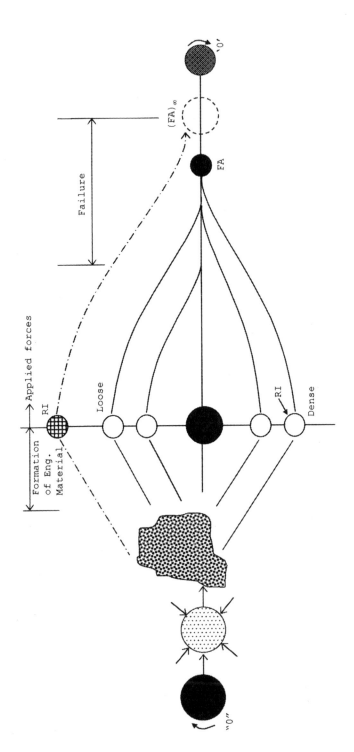

FIGURE 1.3
Levels of material's existence.

The well-known Indian scientist, J.C. Bose, found that apparently nonliving matter (e.g., plants) possesses properties similar to those in animate matter; he developed the Crescograph to measure them experimentally. His findings implied that the boundary between the animate and inanimate vanishes, and points of contact emerge between the domains of the living and nonliving.

The manifested matter, which derives from the same origin (premordial matter), whether animate or living, inanimate, metal, plant, and animal, may follow the same universal law of causality involving action and reaction. They all may exhibit essentially similar phenomena of stress, degradation and fatigue (depression), growth or stiffening (exhilaration), and potential for recovery, as well as permanent unresponsiveness (failure or death at the local level).

Response of the matter, then, is governed by metaphysical laws, which include the mechanistic laws as a subset. Our modelling is based on the mechanistic laws and does not include what exists between the mechanistic and the metaphysical, which is most probably governed by the nonidentifiable and inconceivable property of consciousness that resides in the life force (*prana, chi*) between the material particles. It is this property that may be a cause of the *interaction* between (clusters of) material particles, which at the mechanistic level can be considered to be defined through the *characteristic dimension*. Hence, a model to describe the response of the material is required to include the characteristic dimension. The DSC includes this property implicitly in its formulation.

The motion of particles in any physical system leads to a transition from one state at an instant of time to the next state at the next instant of time. When the next state occurs, the existence of the previous state ceases, but its influence does not. There is always a gap, however small, between the two states. In mechanics, we try to characterize the physical motion from one state to the next by treating the material particle as an inert entity. However, the influence of the gap (which is not known) and of what is contained in it can be profound on the motion from one state to the next; "the things that we see are temporal, but things that are unseen are eternal" (II Corinthians 4:18). It is this influence that may govern the capability of the physical entities to self-adjust or self-organize under the influence of external forces. The issue then is the transformation from one material state to the next. Indeed, the laws of physics and mechanics can be invoked to characterize the transformation as it refers to the skeleton made of inert particles. However, the material does exhibit the attribute of natural self-adjustment to organize such that it responds to the external forces in the optimum way.

1.2.3 Local and Global States

It is apparent that what we have discussed above refers to "local" material states, for "finite" physical systems such as engineering structures. However, in the global or universal sense, similar manifestations occur in which the initial *premordial* or *pristine* material is transformed continuously under cosmic forces and approaches in the limit the ultimate state. It is possible that what happens at the local level is perhaps the reflection of what happens at the global level. Our interest is the local behavior.

1.3 Engineering Materials

Engineering materials are difficult to characterize in their initial natural or artificially manufactured states. Characterization of their behavior under a variety of possible forces—natural, mechanical, and environmental—also poses a challenging problem.

Human understanding of the behavior of materials, which are a mixture of "continuous" and "discontinuous" particle systems *at the same time*, involves mental (human), physical, and mathematical models; the latter are often used to develop numerical models for solution by the *artificial mind*, which is the modern computer.

1.3.1 Continuous or Discontinuous or Mixture

The long pursuit of the mechanics of engineering materials has grappled with the notion that the materials' systems can be treated as *continuous*, such that particles or clusters at the level of interest do not separate or do not overlap. A moment's mental reflection and probing would reveal that particles at any level are not continuous as there is always a gap, or void ("shunya" or space), between them. At the same time, there is some known and some unknown and mysterious thread or force or synchronous cohesion that connects the particles. Even if all physical and chemical forces that contribute to this connection are identified and quantified, there "appears" to exist a force beyond all quantifiable forces that remains to be identified and quantified. Some would say that when the complete understanding occurs there would be no further need to characterize materials, and all will become (again) one material whole! Also, this makes us aware of the fact that the models we develop to characterize the material behavior are only approximations, as they do not completely characterize the response of the entire, or *holistic*, system.

Thus, the limitation of our understanding of the complex discontinuous system requires us to treat materials as continuous. The reality appears to be that both continuous and discontinuous exist simultaneously, i.e., a particle at a given level is connected and disconnected to others at the same time. Hence, in a general sense, almost all reasonably successful efforts and models, in physics and mechanics, until now, have involved some sort of superposition or imposition of discontinuity on continuity. Then, the available continuum models or theories are very often enhanced or enriched by models or constraints to simulate discontinuity.

It is with the foregoing appreciation of the limitation of our modelling that we will deal with materials that are both continuous and discontinuous at the same time.

1.3.2 Transformation and Self-Adjustment

The local and global transformation of the material world, in its physical manifestation and in its "hidden" metaphysical attributes, interests us. One may say that it is this transformation that makes motion or "life" and that makes our endeavors necessary and possible. If we restricted ourselves to the physical world and the transformation did not occur, there would be no problem to solve. Under the external influences, however, the present state of the material changes, and the material modifies its present state to a new state under the given influences. It is the transformation from the present to the new state, so as to define the new state, a process that involves motion or movement of particles, that is the objective of mechanics of materials.

How and why the transformation occurs are important issues in understanding the transformation. The particles constituting the material "yield," or move, so as to resist optimally the external influences, which, in our case, are the mechanical and environmental forces. The particles may come together, move away from each other, rotate by themselves, and/or slip with respect to each other. These motions result in the changes in the physical state of the material, which is usually manifested as changes in the shape, size, and orientation of the material body that is comprised of the particles. Hence, in order to define the new state of the body, it becomes necessary to evaluate the motions under the loads the body is carrying so that we can say with certainty that the engineering body would not "fail," i.e., break apart in the local sense, and move away unacceptably.

The Oriental (Indian and Chinese) and early Western (Greek) thinkers believed that all matter is "living" (6, 7, 10, 11). The idea that a material responds only mechanistically through physical response (motions), which is the foundation of modern science, arose when the attribute of life or consciousness was eliminated from the part of the material world, which we defined as "dead" or "nonliving." This is tragic, since an appreciation of "life force" in materials can not only help in developing enhanced understanding, but can also lead to the humanization of technology (12).

If the quality of self-adjustment is accepted, the pursuit of the understanding of material behavior may open new vistas. At this time, the issue can be controversial—particularly in the treatment of mechanics of materials in the technological context—but its appreciation may be interesting to those who would like to read further, think it over, analyze, speculate, and accept or reject it.

1.3.3 Levels of Understanding

Engineering materials involving a mixture of continuous and discontinuous parts represent complex and nonlinear systems. Hence, it is usually not possible to treat their behavior as simple linear responses or to treat them as a direct accumulation of responses of individual particles or a cluster of particles; from now on, both will be referred to as *particles*. This is partly because such an accumulation would lose at least a part of the influence due to the interconnectedness of the particles. For instance, consider the motion of a handball that bounces repeatedly on the walls of the court. If the motion of the ball were a collection of linear events, it would theoretically be possible to predict its location at any time. However, as the ball itself is not ideally smooth and the walls and floor of the court are rough and undulated, the motion of the ball is nonlinear, and it is almost impossible to predict its *exact* location with time. In this connection, it is interesting to paraphrase Bak and Chen (13): it is not realistic to predict the behavior of a large interactive system by studying its elements and microscopic mechanisms separately, because the response of such a system is not proportional to the disturbances. This implies that the theories for modelling the material behavior based on the micromechanics approach may not provide a rational means of representing the behavior of the *complex interacting* systems such as engineering materials. Indeed, like many available models, they do provide an approximate simulation of the behavior. In the micromechanics models, the behavior of particles is first defined at the local particle (micro-) level, in terms of, say, its shear and normal responses. Then the local or microlevel (constitutive) responses are accumulated to obtain the overall or global response. And very often, the constitutive response at the microlevel is defined based on tests on finite-sized specimens. This appears to be a contradiction.

It would seem appropriate that approaches to define the behavior at the macro- or global level based on particle *mechanisms* that allow for interacting phenomena at the local or microlevel and changes in the microstructure may lead to more consistent theories for the nonlinear and complex material systems. The DSC presented in this book is one such approach. The self-organized criticality (SOC) concept (13; Appendix I) to define critical or threshold states during microstructural changes is another approach that provides models for instability and collapse by considering the interacting mechanisms rather than particle-level descriptions. As will be discussed later, the DSC provides for the instability, or collapse, as well as the precollapse response. Hence, it is considered to be general and unified; Appendix I presents a review of and comparison between the DSC and SOC.

1.3.4 The Role of Material Models in Engineering

Understanding the behavior of matter or materials is a continuing human pursuit involving qualitative and quantitative considerations. The former is based essentially on intuitive and empirical evidence or experience. Intuitive understanding is often based on philosophical and metaphysical interpretations, whereas the empirical comprehension is based on empirical evidence that leads to simplified models. Although they can describe the response of the material approximately, models based strictly on empirical data may not lead to the fundamental approaches often required for the basic description of physical and engineering systems.

Hence, it becomes necessary to develop models based on a combination of mathematics and mechanics, and empirical data, to lead to the calculation of practical quantities such as deformations and stresses required for analysis and design. This approach leads to mathematical expressions or models that connect the response of materials to the (external) mechanical and environmental forces. This connection depends on the behavior of materials, their constitution, and their characteristics. We call these expressions *constitutive laws*, *constitutive models*, or *constitutive equations*. Constitutive laws play a vital role in the prediction of the response of engineering systems. Their development requires consideration of physical laws as well as observations of their behavior under laboratory and/or field conditions that simulate the factors such as loading, geometry, and constitution of materials.

The behavior of engineering systems composed of materials as influenced by the foregoing factors is usually complex. Hence, it is often not possible to employ solution procedures such as those based on closed-form mathematical solutions of differential equations with simplifying assumptions regarding the material properties, geometry, etc. Hence, modern computational methods are often used to solve such nonlinear problems. As a consequence, it becomes necessary to introduce the advanced and realistic constitutive models in such computational procedures as the finite-element, boundary-element, and finite difference methods. Here, the complexities and nonlinearity require special attention toward the robustness and reliability of the computer predictions.

1.4 Disturbed State Concept

This book deals with the disturbed state concept (DSC), which is based on the well-recognized idea that a mixture's response can be expressed in terms of the responses of its interacting components. In the case of the same engineering material, the components are considered to be material parts in the relatively intact (RI) or "continuum" state and the fully adjusted (FA) state, which is the consequence of the self-adjustment of the material's microstructure and can involve decay (damage) or growth (healing). Before the load is applied, the material can be in the continuum state without any disturbance such as anisotropy, microcracking, and flaws; in other words, initially the disturbance is zero. Alternatively, the material may have initial anisotropy, microcracking, and flaws; in that case, there is nonzero initial disturbance.

As loading progresses, the material transforms progressively from the RI state to the FA state through a process of internal *self-adjustment* of its microstructure. This process can involve local (microlevel) unstable or disordered motions of particles tending toward the FA state, in which there may occur "isotropic" particle orientation. A special case of such an orientation is the development of distinct cracks, which can be considered to be the *null* isotropic state, as in the case of the classical continuum damage models. It is recognized that the material experiences growth and coalescence of microcracks, which may lead to distinct cracks. However, the material may often "fail" before the formation of distinct cracks. Hence, the assumption that the FA is the cracked state and acts like a "void" may not be realistic because as the material parts in the FA state are surrounded by the RI material, they possess a certain stiffness and strength. As a result, the RI and FA parts involve *interacting mechanisms* that contribute to the response of the mixture. The FA state is asymptotic and cannot be measured in the laboratory because, before it is reached, the material "fails" in the engineering sense. The FA state is usually defined approximately. For example, it can be defined based on the ultimate (asymptotic) disturbance, D_u (Fig. 1.2). The asymptotic value ($D = 1$) is not measurable when the final FA state is reached.

In the DSC, the disturbance that connects the interacting responses of the RI and FA parts in the same material (or of the components as reference materials) denotes the deviation of the observed response from the responses of the reference states (Fig. 1.2). Thus, depending on the material properties, geometry, and loading, it can represent both decay (damage) or growth (healing or stiffening) in the observed response. For instance, in some cases, the microcracks may grow continuously and result in damage, softening, or degradation of the response, while in other cases, healing (of microcracks) may occur and lead to strengthening or stiffening of the response. Thus, the DSC can allow for the characterization of both the damage and stiffening responses.

As the formulation of the DSC involves both the RI (continuum) and FA states, it provides a systematic *hierarchical* basis for a wide range of models to characterize the material behavior. For example, if there is no disturbance, the DSC specializes to continuum models such as elasticity, plasticity, and viscoplasticity. If the material behavior involves microcracking and fracturing, D is nonzero and various models such as damage with microcrack interaction are obtained. Because the DSC involves interaction between the responses of material parts in the reference states, it can allow for nonlocal effects and characteristic dimension without external enrichments such as Cosserat and gradient theories.

1.4.1 Disturbance and Damage Models

There is a basic difference between the DSC and the classical continuum damage approach (14). In the DSC, we start from the idea that the material under load can be considered a mixture involving continuous interaction between its components. Depending on the mechanical and environmental (thermal, fluid, chemical, etc.) loading, the material mixture can undergo degradation in its strength and stiffness, which leads to the decay or damage-type phenomenon. This is similar to the classical damage approach.

Introduction 13

However, the starting point in the damage approach is different; it starts from the assumption that a part of the material is damaged or cracked. The observed behavior is defined based essentially on that of the remaining continuum or undamaged part. Hence, the damaged part involves no interaction with the continuum part. However, the so-called damaged part may usually become a finite crack or void *only* near the end or failure, because in reality the "damaged" part is the result of the continuous coalescence of microcracks and it possesses certain strength. In the DSC, the FA part represents the distributed, coalescent smeared microcracks, with appropriate deformation and strength characteristics. As a result, the RI and FA parts interact continuously, which is absent in the classical damage model. In order to introduce the microcrack interaction, the damage model requires "external" enrichments such as through kinematics and forces in a (large) number of microcracks, which can add significant complexities. Moreover, as the constitutive behavior of two or more microcracks is not readily measurable, inconsistent assumptions are needed to define the behavior. For instance, very often the microcrack behavior is defined based on test data on macro- or finite-sized specimens.

On the other hand, the DSC includes in its formulation the microcrack interaction through the interacting mechanisms between the RI and FA parts. Also, the definition of the behavior of the material parts in the RI and FA parts relies on the observed (laboratory) behavior of macrolevel or finite-sized specimens. Thus, the DSC model is rooted in the microstructural consideration but does not require constitutive definition at the particle or microlevel. This is considered a distinct advantage compared to the damage models with (external) microcracks interaction and the micromechanical models.

The other major difference between the DSC and damage models is that the foregoing viewpoint in the DSC allows for the possibility of growth or healing, leading to strengthening or stiffening, respectively, of the response of the material under mechanical and environmental loading. Such behavior is possible in many situations, including the case when the material undergoing microcracking and degradation up to a certain threshold or critical deformation state may heal due to factors such as unloading, chemical reaction, oxidation, and locking of microcracks or dislocations. Thus, the DSC includes the possibility of both decay and growth processes, whereas the damage model allows mainly for the degradation or softening response.

1.4.2 The DSC and Other Models

Comparisons between the DSC and other models such as the continuum and damage approach, with enrichments like the gradient and Cosserat theories, and the micromechanics approach are presented in other chapters (e.g., Chapter 12). Appendix I presents a review of and comparison between the DSC, critical-state, and SOC concepts.

1.5 Scope

The scope of this book involves the theoretical development, calibration, and validation of the DSC and its specialized versions.

1.5.1 Outlines of Chapters

Brief descriptions of this book's other chapters, including their computational, validation, and mathematical characteristics, follow.

In Chapter 2 we present preliminaries of the DSC, including its unified character, mechanisms of deformation, the derivation of the DSC equations, and specializations such as composite systems and porous materials. Comparisons of the DSC with other models, and with the SOC, are also presented; however, details of such comparisons are given in Chapter 12 and Appendix I.

The details of the RI and FA states and the disturbance are presented in Chapter 3. Chapter 4 gives details of the incremental DSC constitutive equations, their specializations, and the parameter determination for the disturbance function and models for the fully adjusted state.

Chapters 5 to 8 discuss various theories—elasticity, plasticity, hierarchical single-surface plasticity, and viscoplasticity—based on continuum mechanics including thermal effects, for characterizing the RI response. They include derivations and examples of DSC in which elasticity, plasticity, and viscoplasticity are used to characterize the RI response. Chapter 7 describes the hierarchical single-surface (HISS) plasticity models commonly used for characterizing the RI response. These chapters present examples of a number of materials, including the determination of material parameters from laboratory tests and validation of the constitutive models with respect to the laboratory behavior for the test data used for finding the parameters and *independent* tests not used to find the parameters.

Chapter 9 presents the DSC for saturated and partially saturated materials, in which formulations and validation of the DSC for saturated and partially saturated materials including instability (liquefaction) are described. Chapter 10 deals with characterizing the behavior of "structured" materials, such as stiffening or healing.

Chapter 11 describes the development of the DSC for interfaces and joints using the same mathematical framework as for the "solids." It includes parameter determination as well as validation with respect to laboratory tests for a number of interfaces and joints. Microstructure, localization, threshold transitions, instability and liquefaction, and spurious mesh dependence are discussed in Chapter 12.

Chapter 13 gives details of the implementation of the DSC models in computer (finite-element) procedures. It includes mathematical characteristics of the DSC, predictions and validations of the observed behavior of a number of practical boundary-value problems, and descriptions of computer codes. Finally, conclusions and future trends are presented in Chapter 14.

Appendix I offers a review of and comparisons among the DSC, critical-state (CS), and SOC concepts. Computer procedures for the determination and optimization of material parameters, including validations of laboratory test data, are presented in Appendix II.

References

1. Miller, J., *The Vision of Cosmic Order in the Vedas*, Routledge & Kegal Palu, London, 1985.
2. Swami Nikhilananda, *The Upanishads*, Harper Torchbooks, New York, 1963.
3. Griffiths, R.T.H., *The Hymns of the Rigveda*, Motilal Banarasidas, New Delhi, India, 1973.
4. Wald, G., "The Cosmology of Life and Mind," in *Synthesis of Science and Religion*, Singh, T.D. and Gomatam, R. (Editors), The Bhaktivedanta Institute, San Francisco, CA, 1987.
5. Schrödinger, E., *What Is Life?*, MacMillan Publ. Co., New York, 1965.
6. Capra, F., *The Tao of Physics*, Shambhala, Berkeley, CA, 1976.
7. Zukav, G., *The Dancing Wu Li Masters: An Overview of the New Physics*, William Morrow and Co., New York, 1979.
8. Goswami, A., Reed, R.E., and Goswami, M., *The Self-Aware Universe: How Consciousness Creates the Material World*, Penguin Putnam, Inc., New York, 1995.
9. Fuerstein, G., Kak, S., and Frawley, D., *In Search of the Cradle of Civilization*, Quest Books, Wheaton, IL, 1995.
10. Swami Nikhilananda, *The Upanishads*, Harper Torchbooks, New York, 1963.
11. Max Müller, F. (Translator), *The Upanishads*, Oxford University Press, Oxford, 1884.
12. Prigogine, I. and Stengers, I., *Order Out of Chaos: Man's New Dialogue with Nature*, Bantam Books, New York, 1984.
13. Bak, P. and Chen, K., "Self-organized Criticality," *Scientific American*, January 1991.
14. Kachanov, L.M., *Introduction to Continuum Damage Mechanics*, Martinus Nijhoff Publishers, Dordrecht, The Netherlands, 1986.

2

The Disturbed State Concept: Preliminaries

CONTENTS
2.1 Introduction 18
    2.1.1 Engineering Behavior 21
2.2 Mechanism 23
    2.2.1 Fully Adjusted State 25
    2.2.2 Additional Considerations 27
    2.2.3 Characteristic Dimension 32
2.3 Observed Behavior 32
2.4 The Formulation of the Disturbed State Concept 32
2.5 Incremental Equations 33
    2.5.1 Relative Intact State 35
    2.5.2 Fully Adjusted State 36
    2.5.3 Effective or Net Stress 38
2.6 Alternative Formulations of the DSC 38
    2.6.1 Material Element Composed of Two Materials 40
2.7 The Multicomponent DSC System 41
2.8 DSC for Porous Saturated Media 43
    2.8.1 DSC Equations 44
    2.8.2 Disturbance 46
    2.8.3 Terzaghi's Effective Stress Concept 47
    2.8.4 Example and Analysis 49
    2.8.5 Comments 52
2.9 Bonded Materials 53
    2.9.1 Approach 1 53
    2.9.2 Approach 2 53
    2.9.3 Approach 3 55
    2.9.4 Approach 4 56
    2.9.5 Porous Saturated Bonded Materials 57
    2.9.6 Structured Materials 58
2.10 Characteristics of the DSC 58
    2.10.1 Comparisons and Comments 58
    2.10.2 Self-Organized Criticality 59
2.11 Hierarchical Framework of the DSC 59

Matter is continuous and discontinuous, ordered and disordered, finite and infinite at the same time. Each component has asymptotic attributes that cannot be defined exactly. They culminate or dissolve in each other, can undergo decay and growth at the same time, and yield the interconnected composite that can be defined and understood locally.

2.1 Introduction

A deforming material is considered to be a mixture of "continuous" and "discontinuous" parts. The latter can involve relative motions between particles due to microcracking, slippage, rotations, etc. As a result, the conventional definition of stress $(\sigma)^*$ *at a point* given by

$$\sigma = \frac{P}{A} \qquad (2.1a)$$
$$A \to 0$$

where P is the applied load and A is the area normal to P, does not hold; Fig. 2.1(a). The implication of $A \to 0$ is that the stress is defined at a point. In other words, all points in the material elements retain their neighborhoods before and during load. As a result, abrupt changes in the stress at neighboring points cannot exist, as no cracks or overlaps are permitted.

Now, consider a material element that contains discontinuities due to microcracking and fractures, initial or induced voids or flaws. In this case, the definition of stress, Eq. (2.1a), will not hold, as the stress may change—and abruptly—from point to point in the material element. In other words, the so-called local (at a point) relevance of stress loses its meaning when discontinuities exist. As a result, it becomes necessary to define a weighted value of stress, $\tilde{\sigma}$, to represent its *weighted* distribution over the material element:

$$\tilde{\sigma} = \frac{P}{\tilde{A}} \qquad (2.1b)$$

where $\tilde{A}$ is the weighted *nonlocal* area that includes the effect of discontinuities in the "finite" area over which $\tilde{\sigma}$ is now evaluated (Fig. 2.1(b)). Such an approach is consistent with the physical necessity for the stress $\tilde{\sigma}$ to include the effects and attributes of the happenings (deformations) in the neighboring

---

*Sign convention: For materials that are loaded mainly in tension, the (normal) stresses are considered to be positive. The compressive (normal) stresses are considered positive for material loaded mainly in compression.

The Disturbed State Concept: Preliminaries

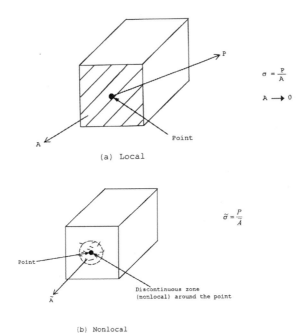

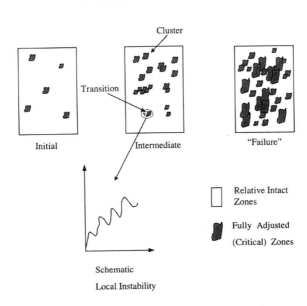

FIGURE 2.1
Definitions of stress.

regions. The DSC allows consideration of the nonlocal effects by defining the stress (Chapters 3 and 12) in a weighted sense, such that the effect of the disturbance (microcracking, etc.) is included in the observed or actual stress.

As introduced in the previous chapter, the *disturbed state concept* (DSC) is based on the basic physical principle that the behavior exhibited through the interacting mechanisms of components in a mixture can be expressed in terms of the responses of the components connected through a coupling function, called the *disturbance function* (D). In the case of the mechanical response of deforming engineering materials, the components are considered to be reference material states. For the element of the *same* material, the reference material states are considered to be its (initial) continuum or relative intact (RI) state, and the fully adjusted (FA) state that results from the transformation of the material in the RI state due to factors such as particle (relative) motions and microcracking. We first consider the DSC for the case of deformations in the *same* material. Then we shall consider the DSC for deforming a material element composed of more than one (different) material.

Analogies for Reference States. If a solid is heated at a certain temperature, it melts or liquefies. The solid and liquid states can then represent two reference states. If the liquid is heated further, it becomes a gas. Then the liquid and gas states can represent the reference states. If a cube of ice melts to water, the ice and water states can be treated as the reference states.

A schematic of the underlying idea in the DSC is shown in Fig. 2.1(c). The material possesses asymptotic (relative) intact and fully adjusted states (Fig. 2.2). The *absolute intact* state may be considered to be the condition of the material, say, at the *theoretical maximum density* (TMD). However, as explained in Chapter 1, the material can exist at other densities, which can be adopted as RI states. Selection of the RI state depends on the characteristics of the material and available test data. For instance, the linear elastic response of a continuum without microcracks can define the RI (e) response with respect to the nonlinear elastic (observed, denoted by a) response affected by microcracking; see Fig. 2.2(a). The elastoplastic (ep) behavior without friction can define the RI response with respect to the elastoplastic behavior with friction; see Fig. 2.2(b). The elastoplastic response can be adopted as the RI response with respect to the behavior affected by microcracks and softening; see Fig. 2.2(c). Figure 2.2(d) shows a schematic of softening and stiffening responses in which the RI response is characterized as elastoplastic.

The asymptotic FA state, $(FA)_\infty$, is the final condition to which the material approaches under external loading [Fig. 2.2(c)]. The behavior of materials at the final state is not measurable in the laboratory but may be defined as the asymptotic value that can be identified approximately. Such a state used in the modelling is considered quasi-FA ($\overline{FA}$), which, for convenience, is referred to simply as the FA state.

The behavior of a material differs when affected by factors such as initial pressure, density, and temperature. Also, there can be more than one RI and FA states. However, the response of the material parts in the RI and FA states

The Disturbed State Concept: Preliminaries

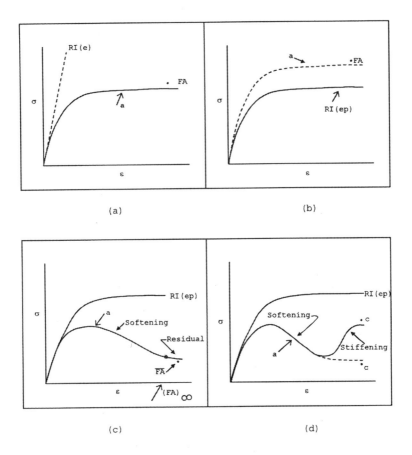

FIGURE 2.2
RI(i), observed (a), and FA (c) responses.

can be expressed in terms of the foregoing factors, which leads to an integrated (DSC) model. This aspect is discussed later in the chapter.

2.1.1 Engineering Behavior

Figure 2.3(a) and (b) show schematics of the response of a material element under the shear stress $\sqrt{J_{2D}}$, the second invariant of the deviatoric stress tensor S_{ij}, and J_1, the first invariant of the total stress tensor, σ_{ij}, which is related to the mean pressure, p, as $p = J_1/3$. It is assumed that the material is initially isotropic and remains isotropic during deformation.

Pure shear stress (with $J_1 = 0$) will cause continuing shear deformations that will lead to an observed engineering "failure" condition (marked 1 in Fig. 2.3(a)) that can be measured in the laboratory. It can be identified as the peak stress or asymptotic or ultimate stress with respect to the behavior in the final range of the stress–strain behavior. Upon further loading, the material may disintegrate

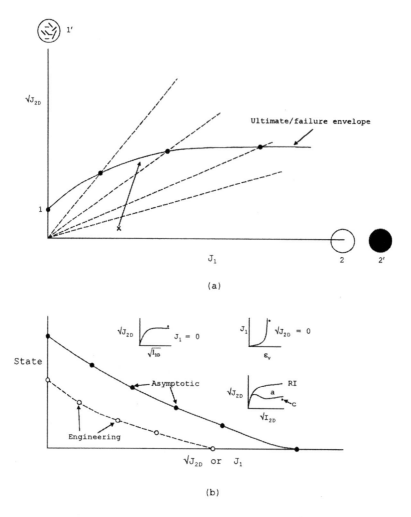

FIGURE 2.3
Material states during loading.

fully and separate into individual particles (1'); this response cannot be measured. Under pure (compressive) mean pressure ($\sqrt{J_{2D}} = 0$), the material compacts and strengthens continuously and will reach the measurable state (2) and nonmeasurable state (2').

A combination of $\sqrt{J_{2D}}$ and J_1 leads to measurable ultimate or failure states defined by the envelope shown in Fig. 2.3(a). The nonmeasurable or asymptotic states may lead to the disintegration of the element under different combinations of $\sqrt{J_{2D}}$ and J_1.

Figure 2.3(b) depicts $\sqrt{J_{2D}}$ vs. $\sqrt{I_{2D}}$ response under pure shear stress ($J_1 = 0$); here $\sqrt{I_{2D}}$ is the second invariant of the deviatoric strain tensor, E_{ij}. For pure mean pressure ($\sqrt{J_{2D}} = 0$), the volume will change (decrease) continuously. With both $\sqrt{J_{2D}}$ and J_1, the stress–strain response affected by both the shear stress and mean pressure will result. Then the RI response can be characterized by using elastic, elastoplastic, or another suitable model (Fig. 2.2). The FA response can be defined at the critical state (c) the material approaches under given mean pressure.

Historical Note. The basic idea underlying the DSC derives from the model for overconsolidated geologic materials proposed in 1974 by Desai (1) in the context of the solution of the problem of slope stability. It was postulated that the behavior of overconsolidated (OC) soil can be decomposed into that of the normally consolidated (NC) and that due to the influence of overconsolidation that entails microcracking and shear planes; see Fig. 2.4(a). Then the observed softening response was expressed in terms of the behavior under the NC state and the effect of overconsolidation. The observed response (stiffness) was then expressed in terms of the stiffness of the two parts. Desai (2) proposed the concept of the residual flow procedure (RFP) for the solution of free surface flow (seepage) in porous media.(3, 4); a review is presented by Bruch (5). Here the response was decomposed into two reference states: the fully saturated with the permeability coefficient, k_s, and the "residual" response given by the difference between the saturated and unsaturated (or partially saturated) conditions due to the difference in permeability coefficients, $k_s - k_{us}$ (Fig. 2.4(b)). Thus, the two reference states were given by the saturated state and the asymptotic unsaturated state at very high negative pressures (p).

The DSC presented in this book can be considered as a generalization of the foregoing two developments in stress analysis and flow through porous media.

2.2 Mechanism

An initially intact material, without any flaws, cracks, or discontinuities, will transform continuously with loading, unloading, and reloading, which is often referred to collectively as loading, from the RI state to the FA state [Fig. 2.1(c)]. If the material before the loading contains initial flaws, cracks, and/or discontinuities (or disturbance), the resulting initial disturbance will influence the subsequent behavior. As the deformation progresses, the extent of the material parts, which may be distributed randomly over the material element depending on factors such as initial conditions and loading, the FA parts can grow or decrease, i.e., lead to degradation or stiffening, respectively [Fig. 2.2(d)]. In the case of the continuous growth of the FA state, the material part in the RI state decreases continuously, during which the microstructural changes can involve the annihilation of particle bonds, leading to a decay process. In the limit, if it is possible to continue the load, the entire material will approach the $(FA)_\infty$ state, in which the material particle may break and separate completely, and then the disturbance approaches the value of unity. As this state is asymptotic, it is not realized in practice—in the field or in the laboratory—because the

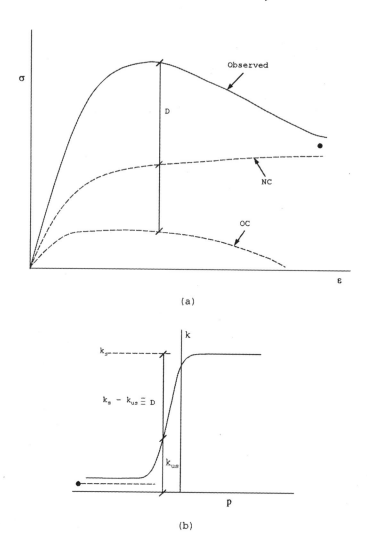

FIGURE 2.4
Disturbance in stiff or structured soil, and flow through a partially saturated medium (1–3).

material "fails," in the engineering sense, in terms of allowable deformation and/or load before the $(FA)_\infty$ state can be reached. Hence, from practical consideration, it often becomes necessary to identify and use approximately the $\overline{FA}$ state when the material enters the ultimate residual state for given initial conditions in the engineering sense when the disturbance = D_u (≤ 1.0). This state is used as the FA state from a practical viewpoint (Fig. 2.2).

In the DSC, the microstructural changes may be such that the material may stiffen due to strengthening of the interparticle bonds [Fig. 2.2(d)]. This may occur due to factors such as predominant hydrostatic stresses that lead to increasing density, the structured nature of material, chemical and thermal effects that lead to increased interparticle bonding and unloading, or rest periods (Chapter 10).

The RI state is often simulated by using such continuum theories as elasticity, plasticity, and viscoplasticity for which well-established formulations are available. These are discussed subsequently in this chapter and in Chapters 5 to 8. Here, we first discuss some aspects of the FA state.

2.2.1 Fully Adjusted State

For engineering materials, the final state at "infinite" or very high loading (Fig. 2.2) may result in a totally disintegrated state in which the separated material particles tend to configure into a "specific" volume. Such a ("loose") material may not have any strength at all unless it is confined. The final disintegrated material state may be considered analogous to the idea of treating the cracked or damaged material part as a "void," as it is assumed in the classical damage mechanics approach (6, 7). One of the main differences between the DSC and the damage concept can be stated here. In the DSC, it is considered that the FA material, in the range of engineering interest, *does* possess certain deformation and strength properties. This is partly because the FA material parts are confined or surrounded by the material parts in the RI state [Fig. 2.1(c)]. Furthermore, in contrast to the damage concept, in which the damaged parts grow continuously, resulting in the continuous loss of strength, in the DSC, the material can also gain strength or stiffen during loading. In other words, under certain loadings and physical conditions, the FA state can entail strengthening. Then, disturbance will be "negative" or have a value greater than unity, indicating strengthening or a growth process. Thus, the DSC allows for the characterization of both degradation (or damage or decay) and stiffening (or healing) in material responses.

The idea of the critical state (CS) in soil mechanics has a connotation similar to the FA state. The material *approaches* the critical state at which there is no further change in volume, i.e., the material assumes an invariant (specific) volume, void ratio, or density under the constant shear stress reached up to that state and given initial mean pressure (8, 9). In practice, however, e.g., in the laboratory, it is usually not possible to measure and identify the *exact* critical state. It is asymptotic, like the FA state. For engineering purposes, we identify and can often adopt the CS as the FA state when the measured volume change is *approximately* zero in the ultimate range of loading. Indeed, there may be instantaneous states of zero volume changes like the point of transition from compactive to dilative volume change in granular materials, where one can identify the point almost exactly; however, the FA state is considered to occur in the ultimate range. To summarize, the definition of material response at the FA state must be approximate because the measurement system would cease to operate when the material specimen "collapsed" from an engineering and a practical viewpoint.

As a further explanation, let us consider two lumps of a material with different initial volumes, and with irregular shapes, as in Fig. 2.5. The irregularities or nonsymmetries of the two lumps are the lumps' initial attributes. Now, let us mold both specimens by applying external pressure to make them "ideally" spherical. After the levels of molding efforts have been increased, both lumps will *tend* toward spherical shapes with different specific, critical, or fully adjusted volumes. It is apparent that it will be (humanly) impossible to achieve the perfect or absolute spherical shapes, with no attributes or irregularities or biases. Hence, we must accept approximate spherical shapes with volumes $\bar{V}_1$ and $\bar{V}_2$ at certain levels of effort (loading) to represent the FA state and use them in our modelling pursuit (Fig. 2.6(a)).

Let us consider the volume of the solid particles as they merge together, when all the particles' attributes have been annihilated, and the volume, V^c, at the FA state is approached. The schematic plots of V/V^c for the two lumps are shown in Fig. 2.6(a). The volume of both approach unity, while $\bar{V}_1$ and $\bar{V}_2$ are the quasi-volumes in the FA state. It is useful to note that the volume V^c (at the FA state) is a unique characteristic for a given set of physical characteristics such as initial density, particle fabric, shape and size, and loading, and can represent the characteristic dimension.

FIGURE 2.5
Molding of clay lumps.

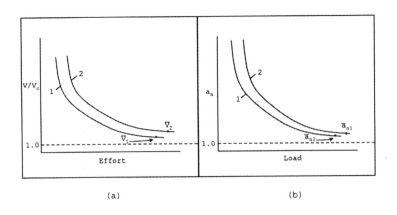

FIGURE 2.6
Behavior of clay lumps.

The behavior of an engineering material (specimen) can be understood similarly. The material has initial attributes, like the nonsymmetries caused by anisotropy, flaws, and defects. Now, consider that such a material is subjected to increasing hydrostatic or isotropic load (pressure) [Fig. 2.6(b)]. As the pressure, like the molding effort for the lumps, is applied, the nonsymmetries or biases will be annihilated and the material will *tend* toward a "condensed" (volume of solids) symmetrical or isotropic state with no attributes of anisotropy. The limiting isotropic state, when the nonsymmetry or anisotropy (a_n), in Fig. 2.6(b), is destroyed completely, will be impossible to achieve in practice and in the laboratory. However, approximate states ($\bar{a}_{n1}, \bar{a}_{n2}$) can be identified, for all practical purposes, to represent the isotropic state, which can represent the FA state. As is indicated in Fig. 2.6(b), the material with initial attributes such as anisotropy, flaws, and different particle sizes will tend toward the isotropic state at different levels of load and effort. Each can have its own or approximate FA state, which can be adopted as the asymptote to each response. Indeed, the final asymptotic state is the perfect isotropic state, i.e., $A_1 \to A_2 \to A_3 \to ... \to A$, where A denotes the final FA state.

2.2.2 Additional Considerations

The disturbance, D, is expressed as the ratio of the material volume, V^c, in the FA state to the total volume, V, of the material element. Hence, for direct evaluation of D, one must evaluate the material parts in the FA states during deformation. Advanced nondestructive techniques based on X-ray computerized tomography and acoustic measurements are being developed, and they can be used to identify the FA states involving evolution of density clusters. However, they are still not fully available. Hence, it is not possible to define D based on such physical measurements. Some limited results are available, however. Figure 2.7 shows the vertical reconstruction of a cylindrical specimen (50.8-cm diameter, 609.6-cm height) of a grout material tested under triaxial loading by using X-ray computerized tomography (10).

FIGURE 2.7
The vertical reconstruction of a failed grout specimen under triaxial compression. (From Ref. 10, with permission.)

The measurements were obtained at different stages of loading; Fig. 2.7 shows results near (before) the peak stress. The dark regions represent material parts that have approached the critical density and can be treated as being at the FA state. The dark zones were integrated to obtain the volume V^c at the FA state. The ratio $D = V^c/V$ was about 0.17, which compared well with the average values of D at that stage of deformation for similar materials (11).

In lieu of direct measurements of the disturbance (V^c), it becomes necessary to define disturbance in terms of internal variables such as accumulated plastic (irreversible) strains and (dissipated) mechanical energy under thermomechanical and other loadings. Such internal variables reflect the microstructural changes that cause disturbance (damage and/or strengthening). Disturbance can also be related to observed laboratory response in terms of such measured quantities as stresses, volume or void ratio change, porosity change, effective stress (pore fluid pressure), saturation degree, and nondestructive properties such as P- and S-wave velocities (11–14). The use of these phenomenological approaches permits the determination of parameters in the expressions for disturbance; details are given in Chapter 3 and other chapters.

In defining the disturbance (damage or strengthening) in a deforming material, we will be able to take advantage of the similarity between decay and/or growth in natural systems. For example, the mechanism and trend of growth or decay of a living organism can be considered to be similar to strengthening or decay in a deforming material. As a result, mathematical functions that express decay and growth can be adopted to describe damage and strengthening.

Consider a collection of granular (spherical) material particles, as shown in Fig. 2.8. The initial microstructure involves air spaces and contacts, whose magnitudes will affect the density. The contacts contribute to the deformation and strength properties through interparticle cohesion and friction. During deformation under compressive pressure accompanied by shear stress, the particles' contacts can deform, slip, rotate, and break. Furthermore, the particles can move in the air spaces, leading to the change in density.

An initially loose material will experience continuing compaction and an increase in the density or a decrease in the volume or void ratio, e (volume of voids/volume of solids). In general, the material will exhibit a nonlinear and continuously hardening response, as seen in Fig. 2.9(a). An initially dense material first compacts as the particles move in the voids, and then experiences an increase in the volume or void ratio or a decrease in the density (Fig. 2.9(b)). The latter is called *dilation*, which is caused predominantly by the upward sliding of the particles.

Depending on the initial nature of the microstructure (distribution of voids, etc.), *density clusters* can form at different locations, which may tend to the critical invariant density or void ratio (see Fig. 2.8). Such an asymptotic state can represent the FA state. As the loading progresses, a greater number of FA state zones develops; in the limit the entire material tends toward the fully adjusted state.

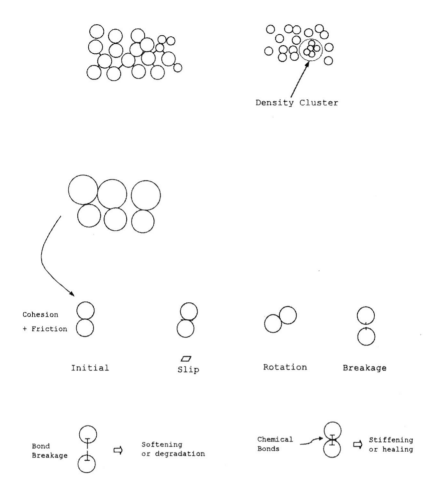

FIGURE 2.8
Particle motions, degradation or softening, and healing.

In the case of the initially dense material, behavior similar to that for the loose material may continue until the transition from compaction to dilation accompanied by microcracking or breakage of particle bonds. This process will continue and grow near and after the peak and in the degradation or softening regime. Increased levels of microcracking, slippage, and rotation of particles will occur in the softening regime (Fig. 2.9), with locally unstable changes in the microstructure. Then the microstructure may experience an intense instability, which can be identified by the critical disturbance (D_c) at the initiation of the residual state, leading to the asymptotic FA state at which there will be no change in volume or density.

It is possible that during deformation, factors such as chemical reactions, increased compaction, or locking of particle contacts or dislocations can cause an increase in the stiffness and strength of the particle bonds. In that case, after

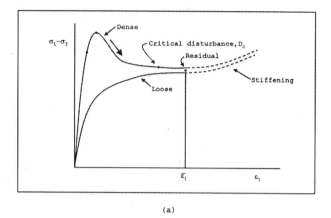

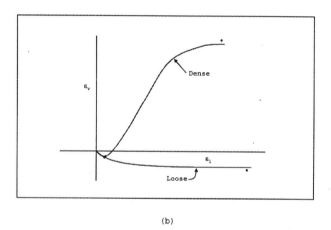

FIGURE 2.9
Schematic of behavior of loose and dense materials.

a threshold value of deformation or strain ($\bar{\varepsilon}_1$), [Fig. 2.9(a)] is reached, the material may stiffen or heal; this can lead to decreased disturbance. The pore-collapse mechanism in porous rocks, the drained post-liquefaction behavior of saturated sands, and the locking of dislocations in silicon with oxygen impurity represent examples of such softening and stiffening behavior (Chapter 10).

Figure 2.10 shows the behavior of a porous chalk under hydrostatic loading (15). Here, a cylindrical specimen of the chalk is subjected to confining or hydrostatic stress, and the resulting volumetric strains ε_v (ε_{ii}) are measured [Fig. 2.10(a)]. The acoustic P- and S-wave velocities are also measured during deformation; see Fig. 2.10(b) and (c).

The initial response of the material is essentially linear elastic (E) with a decrease in the volume; a positive volume indicates a reduction in volume. During this phase, some of the open spaces, initial microcracks, and particle contacts are closed (15); as a result, the acoustic velocities increase. After about

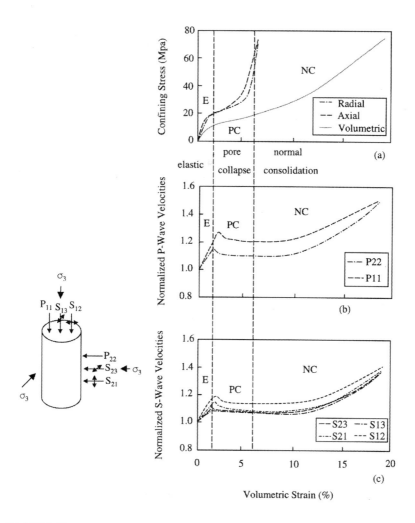

FIGURE 2.10
Stress–strain, compression, and shear wave responses under hydrostatic compression of Ekofisk Chalk (15).

1.0% of the volumetric strain, the particle bonds are broken and the contact cohesion is reduced or lost. As a result, the P- and S-wave velocities decrease during this phase, when the collapse of pores occurs (PC). After critical $\varepsilon_v \approx$ 6.0%, the material exhibits stiffening (or healing). During this consolidation phase, the acoustic velocities increase as a result of the reestablishment of the particle contact and bonds.

It can be said that the disturbance will be essentially zero during the initial phase, will increase during the pore-collapse phase, and then will decrease during the stiffening phase. Details of the formulation of the DSC for this behavior are given in Chapter 10.

2.2.3 Characteristic Dimension

It is appropriate to mention that factors such as the initial mean pressure, the density or void ratio, and the fabric, geometry, and size of particles affect the *zone of influence* around a point in a material element. The zone of influence is considered to be related to the *internal* or *characteristic dimension*. In the DSC, the invariant mass or volume (V^c) which many materials approach during deformation can provide an important measure for the characteristic dimension. These aspects are discussed later, e.g., in Chapter 12.

2.3 Observed Behavior

The deforming material is considered a mixture of the two *interacting* material parts in the RI and FA states. The RI and FA states are termed as *reference* states. Then the observed or actual response of the material is expressed in terms of the responses of material parts in the reference states. The disturbance, D, denotes the deviation of the observed response from those of the reference states. Figure 2.11 shows a symbolic and schematic representation of disturbance in the DSC. The observed or average response (denoted by a) is then expressed in terms of the RI response (denoted by i) and the FA response (denoted by c) by using the disturbance function, D, as an interpolation and coupling mechanism. The behavior of the RI and FA materials, as well as the disturbance function, needs to be defined from laboratory tests.

2.4 The Formulation of the Disturbed State Concept

Let us first consider the case of a (dry) material element (Fig. 2.12). Based on the equilibrium of forces on the material element, composed of the clusters of particles in the RI and FA parts, where F^a = observed or average force, F^c = force in the fully adjusted (FA) part, and F^i = force in the relative intact (RI) part, we have

$$F^a = F^i + F^c \qquad (2.2)$$

Division of both sides by the total area, A (assuming thickness to be unity) leads to

$$\frac{F^a}{A} = \frac{F^i}{A^i} \cdot \frac{A^i}{A} + \frac{F^c}{A^c} \cdot \frac{A^c}{A} \qquad (2.3a)$$

The Disturbed State Concept: Preliminaries

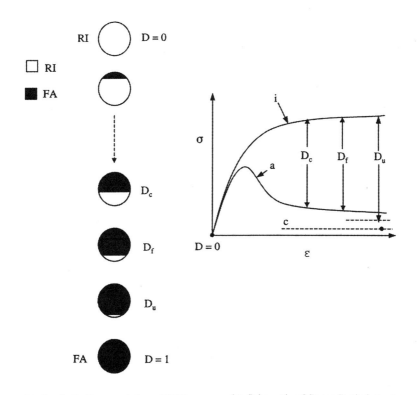

(a) Symbolic Representation of DSC   (b) Schematic of Stress-Strain behavior

FIGURE 2.11
Representations of DSC.

where A^i and A^c are areas corresponding to RI and FA, respectively. Therefore,

$$\sigma^a = \sigma^i \frac{A^i}{A} + \sigma^c \frac{A^c}{A} \qquad (2.3b)$$

where σ^a = observed stress, σ^i = stress in the relative intact part, and σ^c = stress in the FA part.

2.5 Incremental Equations

Equation (2.3b) can be generalized to three dimensions as*

$$\sigma^a_{ij} = (1 - D)\sigma^i_{ij} + D\sigma^c_{ij} \qquad (2.4)$$

where D is the disturbance function, $D = A^c/A$, and $1 - D = A^i/A$. Here D is

---

* Both the tensor and matrix notations are used in this text. The latter can provide for convenient implementation in computer procedures.

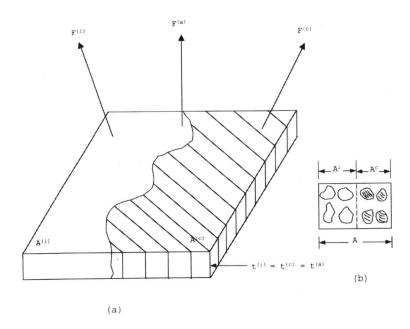

FIGURE 2.12
Schematic of material element and force equilibrium.

assumed to be a scalar in a weighted sense, which implies isotropic condition. However, D can be expressed as a tensor, D_{ijkl}; then

$$\sigma_{ij}^a = (I_{ijk\ell} - D_{ijk\ell})\sigma_{k\ell}^i + D_{ijk\ell} \cdot \sigma_{k\ell}^c \qquad (2.5a)$$

where $I_{ijk\ell}$ is the unit tensor. It is difficult to use Eq. (2.5a) because it is usually not possible to conduct, or to have access to, appropriate laboratory tests for defining the tensor D_{ijkl}. The multiaxial or three-dimensional test with nondestructive (ultrasonic) measurements (12) is one of the possible tests to define D_{ijkl}; this will be discussed later. As a simplification, the tensor of disturbance can be expressed as

$$(D) = \begin{pmatrix} D_1 & 0 & 0 \\ 0 & D_2 & 0 \\ 0 & 0 & D_2 \end{pmatrix} \qquad (2.5b)$$

where D_i ($i = 1, 2, 3$) are disturbance components in the three principal directions, which can be expressed as

$$D_i = \ell_{ij}\left(\frac{A_c}{A}\right)_j \qquad (2.5c)$$

where ℓ_{ij} are the direction cosines and j denotes the (principal) direction.

The Disturbed State Concept: Preliminaries

Since certain devices, such as multiaxial three- and two-dimensional devices, can allow measurements in the principal directions, it can be relatively easier to define D in Eq. (2.5b) compared to the definition full tensor, D_{ijkl}. For practical purposes, it may often be sufficient to treat D as a scalar. A majority of the treatment here uses this assumption.

The incremental form for Eq. (2.4) can be derived as

$$d\sigma_{ij}^a = (1 - D)d\sigma_{ij}^i + Dd\sigma_{ij}^c + dD(d\sigma_{ij}^c - \sigma_{ij}^i) \quad (2.6a)$$

where d denotes increment, dD is the increment or rate of D, and σ_{ij} is the stress tensor. Equation (2.6a) can now be expressed as

$$d\sigma_{ij}^a = (1 - D)C_{ijk\ell}^i d\varepsilon_{k\ell}^i + DC_{ijk\ell}^c d\varepsilon_{k\ell}^c + dD(\sigma_{ij}^c - \sigma_{ij}^i) \quad (2.6b)$$

where $C_{ijk\ell}^i$ and $C_{ijk\ell}^c$ are constitutive tensors for the materials in the RI and FA parts, respectively, and ε_{ij} is the strain tensor.

2.5.1 Relative Intact State

As stated earlier, the RI behavior can be characterized by using an elasticity or plasticity model with an associative response or any other suitable continuum model—viscoplastic, thermoviscoplastic, etc. For example, the δ_0-version in the hierarchical single-surface (HISS) plasticity concept (Chapter 7) can be used to represent the RI state. In that case, irreversible deformations and coupled shear and volumetric responses will be included in the model; however, factors such as nonassociative response (friction), anisotropy, and microcracking (softening) can deviate or disturb the δ_0-response to lead to the observed behavior.

Thus, the RI response is relative in the sense that it excludes the effect of factors that deviate the observed behavior from that of the material in the given RI state. Hence, the choice of the RI state and corresponding constitutive model to define it depend on the material response and available test data under given initial conditions that enable characterization and calibration of parameters for the model for the RI material. As discussed in Chapter 1, the final (densest) material state can be treated as the intact state. However, usually each stress–strain response can have its own *local* RI behavior (Fig. 2.13), e.g., i_1 for a_1, i_2 for a_2, and so on, because the response is affected by factors such as initial conditions, stress, strain, density, pressure, and temperature. When the δ_0-model is adopted, the RI plasticity response can be expressed in terms of these factors so that the entire range of response, expressed through different curves (Fig. 2.13), is simulated.

In the same manner, if a nonlinear elastic response is adopted for the RI state, the dependence of elastic parameters such as elastic modulus, E, and

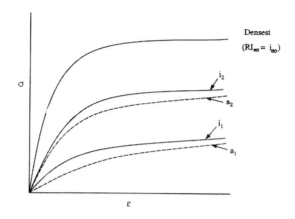

FIGURE 2.13
RI and observed states.

Poisson's ratio, μ, or shear modulus, G, and bulk modulus, K, will define the RI response for given initial conditions, e.g.,

$$E = E(p_0) \tag{2.7a}$$

$$G = G(p_0) \tag{2.7b}$$

where p_0 = initial mean pressure = $(\sigma_1 + \sigma_2 + \sigma_3)/3$. If a single curve is simulated, we use the RI modulus corresponding to the given value of p_0. However, for the model for simulating the response as affected by p_0, the RI modulus can be expressed as a function of p_0, e.g.,

$$E = \bar{K}(p_0)^n \tag{2.7c}$$

where $\bar{K}$ and n are parameters.

2.5.2 Fully Adjusted State

As the loading progresses, the material parts in the RI state transform continuously to the parts in the FA state. Depending on the initial state and induced conditions caused by stress, flaws, discontinuities, and microcracking, the material parts in the FA state can be distributed randomly over the material element [Fig. 2.1(c)]. The transformation occurs due to the self-adjustment of the material's microstructure. It may be surmised that the material particles, while undergoing motions such as translation and rotation, would experience *instantaneous* instability at the local (micro-) level [Fig. 2.1(c)]. However, the measured macro response would appear as the integration of the locally unstable responses at the microlevel and may not indicate instability in the sense of *collapse*. The collapse would occur in the ultimate stages when the instabilities become predominant and critical.

In the FA state, which is asymptotic and cannot be reached or measured in practice (in the laboratory), the material may organize toward the *equilibrated* or "isotropic" state with maximum entropy. The structure and the composition of the material in the FA state are different from that of the material in the RI state. At the same time, both the RI and FA material parts are coupled and interconnected, like water bubbles enclosed at random locations in a matrix of solid material particles. It is evident that the behavior of the mixture of the material element consisting of a skeleton or matrix of solid particles and the bubbles enclosed in it is dependent on the behavior of both. As the bubbles are (totally) enclosed and surrounded by the solid matrix, they are capable of sustaining a part of the applied load. Only when they are broken and are no longer enclosed, that is, they are connected to the atmosphere through cracks in the solid matrix, do they cease to sustain any applied load.

The material part in the FA state is analogous to the enclosed water bubbles in the solid matrix. It can sustain a part of the applied load. Only in the *final* fully adjusted state, denoted by $(FA)_\infty$, may the material experience total breakdown and may the FA part not sustain any load. Such a condition is asymptotic; before it can be reached, the laboratory testing stops, and the material has failed. It is the "engineering" $(\overline{FA})$ state that we can define and use in our model. It occurs before the ultimate $(FA)_\infty$ state, when the material "fails" for all practical purposes. It should be noted that only the engineering FA state can be measured and is adopted to represent the FA state approximately.

There are a number of ways in which we can characterize the behavior of the material in the FA state:

- as a *constrained liquid-solid*. Here a possible representation is the critical state (8, 9, 11, 12), at which the material experiences continuing shear strains under the shear stress ($\sqrt{J_{2D}}$ or τ) reached up to that state with given (initial) pressure ($J_1/3$ or p).
- as a *constrained liquid*. Here we can assume that the material part can carry hydrostatic stress or mean pressure ($J_1/3$), but no shear stress ($\sqrt{J_{2D}}$). That is, as soon as the material part reaches the FA state, the shear stress in it drops to zero (11, 12). This simulation is similar to the yielding response in the case of a perfectly plastic material when the FA stress equals the yield stress; see Examples 6-2 and 6-3 in Chapter 6.
- as a finite *crack* or *void*. This is similar to the assumption in the classical damage model (6, 7). Here the material part in the FA state can carry no stress, that is, $\sqrt{J_{2D}} = J_1/3 = 0$. Then the applied load is carried by the material part in the RI state only, and there is no interaction between the intact and the damaged parts, although the

overall observed behavior is modified because of the damaged parts. This assumption is considered to be unrealistic.
- In some cases, the FA state can be adopted corresponding to the material response at a reference value of a given parameter. For instance, in the case of partially saturated materials, the FA state can be adopted to correspond with the behavior at full saturation or zero suction (Chapter 9).

In Chapter 4 we will present details of various characterizations for the material parts in the RI and FA states.

2.5.3 Effective or Net Stress

In the preceding discussion, we assumed that the material is dry (or drained) and the stresses are effective, that is, they are relevant to the skeleton of material particles. If the material is saturated with a liquid (water) in the pores or partially saturated with a liquid together with a gas (air) in the pores, the formulation requires modifications. Alternative formulations that cover this case are given below.

2.6 Alternative Formulations of the DSC

In the foregoing formulation, Eqs. (2.4) and (2.6), we considered an element of the *same* material composed of parts in the RI state and the remaining parts in the FA state. The latter is assumed to occur from the very beginning of the loading, and its extent changes (increases) with loading. The material element is composed of clusters of material particles in the RI and FA states [Fig. 2.1(c)]. The force equilibrium considers the effect of the interacting particles in the RI and FA states on the observed force (F^a). In other words, the formulation allows for micromechanical response of the clusters of particles in the two reference states. A major advantage of the foregoing approach is that the responses of the material parts in the RI and FA states can be defined based on the observed laboratory behavior of the *same* material element (specimen). For instance, the RI behavior can be characterized as linear or nonlinear elastic or elastoplastic by using the observed response in the early stages of deformation, whereas the FA behavior can be defined from the response near the ultimate stage.

The DSC can also be formulated for a material element composed of two (or more) materials whose responses can be treated as reference responses. Here the responses of the component materials may be characterized based on continuum theories such as elasticity and plasticity. Indeed, it is possible that the *individual* responses of the components may exhibit RI and FA states.

The Disturbed State Concept: Preliminaries

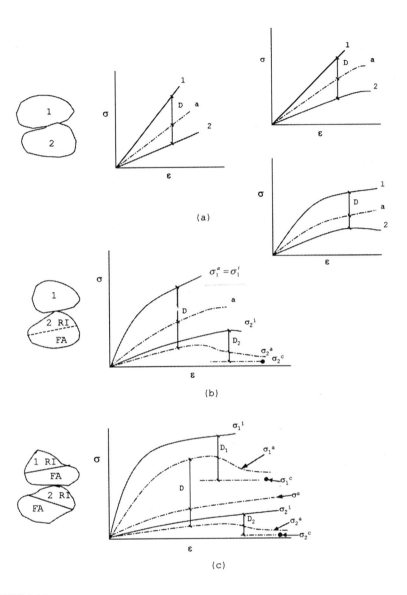

FIGURE 2.14
Material element composed of two or more component-reference materials.

Figure 2.14(a) shows a schematic of two component materials used as reference states. Their behavior can be characterized as elastic or elastoplastic, leading to the observed behavior that can be elastic or elastoplastic. In Fig. 2.14(b), one of the materials can be characterized as elastic or elastoplastic, whereas the other one can involve microcracking, leading to softening with its own RI and FA responses. Figure 2.14(c) shows two materials, each showing softening, and involves RI and FA responses. It is also possible that the material is fully or partially saturated with a liquid (e.g., water). Note that for these situations, it would be necessary to have access to the laboratory behavior of each of the materials composing the element and also to the behavior of the composite σ^a. In the following, we present examples of formulations that allow for the use of individual materials as reference states.

2.6.1 Material Element Composed of Two Materials

Consider the schematic in Fig. 2.14(c). Each of the two reference material states exhibits microcracking and softening behavior and involves FA states. The observed behavior of the mixture of the two materials 1 and 2 can be expressed as

$$\sigma^a = (1 - \bar{D})\sigma_1^a + \bar{D}\sigma_2^a \tag{2.8a}$$

Here the term $\bar{D}$ for the composite element is expressed in terms of the observed responses of materials 1 and 2, e.g.,*

$$\bar{D} = \frac{\sigma_1^a - \sigma^a}{\sigma_1^a - \sigma_2^a} \tag{2.8b}$$

The term $\bar{D}$ denotes a ratio that is analogous to disturbance, and its definition is different from the disturbance in an element of the *same* material, e.g., D_1 and D_2 in Eqs. (2.8c and d) ahead, which are ratios of the volume in the FA state to the total volume of the element, Eq. (2.4). Equation (2.4) can now be used to represent the observed stress, σ^a, for the composite element.

The observed behavior of each of the materials can be written as

$$\sigma_1^a = (1 - D_1)\sigma_1^i + D_1\sigma_1^c \tag{2.8c}$$

and

$$\sigma_2^a = (1 - D_2)\sigma_2^i + D_2\sigma_2^c \tag{2.8d}$$

Hence,

$$\sigma^a = (1 - \bar{D})[(1 - D_1)\sigma_1^i + D_1\sigma_1^c] + \bar{D}[(1 - D_2)\sigma_2^i + D_2\sigma_2^c] \tag{2.9a}$$

or in generalized form,

$$\sigma_{ij}^a = (1 - \bar{D})[(1 - D_1)\sigma_{1ij}^i + D_1\sigma_{1ij}^c] + \bar{D}[(1 - D_2)\sigma_{2ij}^i + D_2\sigma_{2ij}^c] \tag{2.9b}$$

---

* General and alternative definitions of D are presented in chapter 3.

The Disturbed State Concept: Preliminaries

The generalized incremental form of Eq. (2.8a) can be written as

$$d\sigma_{ij}^a = (1 - \bar{D})d\sigma_{1ij}^a + \bar{D}d\sigma_{2ij}^a + d\bar{D}(\sigma_{2ij}^a - \sigma_{2ij}^a) \tag{2.9c}$$

in which the incremental forms for $d\sigma_{1ij}^a$ and $d\sigma_{2ij}^a$ can be substituted by using Eqs. (2.8c and d). Thus, laboratory responses of both materials and the mixture are needed to define Eq. (2.9). The other two situations, Fig. 2.14(a) and (b), can be derived as special cases of Eq. (2.9). For example, for Fig. 2.14(a),

$$\sigma^a = (1 - \bar{D})\sigma_1^a + \bar{D}\sigma_2^a \tag{2.9d}$$

and for Fig. 2.14(b),

$$\sigma^a = (1 - \bar{D})\sigma_1^a + \bar{D}\sigma_2^a$$
$$= (1 - \bar{D})\sigma_1^a + \bar{D}[(1 - D_2)\sigma_2^i + D_2\sigma_2^c] \tag{2.9e}$$

2.7 The Multicomponent DSC System

A material element can be considered to be composed of more than two reference materials. Consider that there are n component materials in a material element as in Fig. 2.15(a). Then equilibrium of forces gives

$$F^a = F_1 + F_2 + \cdots + F_n \tag{2.10a}$$

and

$$\frac{F^a}{A} = \frac{F_1}{A_1} \cdot \frac{A_1}{A} + \frac{F_2}{A_2} \cdot \frac{A_2}{A} + \cdots + \frac{F_n \cdot A_n}{A_n \cdot A} \tag{2.10b}$$

$$= \sigma^1 \bar{D}_1 + \sigma^2 \bar{D}_2 + \cdots + \sigma^n \bar{D}_n \tag{2.10c}$$

where $\Sigma \bar{D}_i = \bar{D}_1 + \bar{D}_2 + \cdots + \bar{D}_n = 1$, which for unit width ($b$) is given by

$$\Sigma D_i = \frac{A_1}{A} + \frac{A_2}{A} + \cdots + \frac{A_n}{A} = \frac{A_1 + A_2 + \cdots + A_n}{A} = 1 \tag{2.10d}$$

Here, $\bar{D}_i$ simply denotes the ratios of areas and do not have the same meaning as the disturbance in the element of the *same* material.

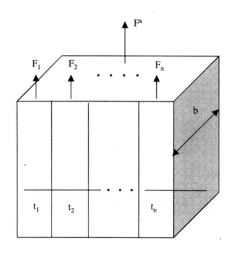
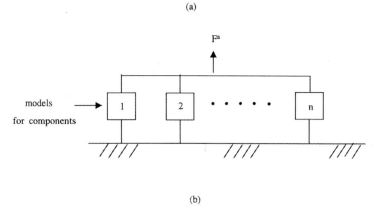

FIGURE 2.15
Multicomponent DSC.

Now Eq. (2.10c) can be written as

$$\underset{\sim}{\sigma}^a = \underset{\sim}{C}_1 \underset{\sim}{\varepsilon}_1 \overline{D}_1 + \underset{\sim}{C}_2 \underset{\sim}{\varepsilon}_2 \overline{D}_2 + \cdots + \underset{\sim}{C}_n \underset{\sim}{\varepsilon}_n \overline{D}_n \quad (2.10e)$$

In Eq. (2.10e), the strains $\varepsilon_i (i = 1, 2, \ldots, n)$ can be different in all the components, and the "disturbances," $\overline{D}_i$ $(i = 1, 2, \ldots, n)$, can be obtained for each component as the ratios of areas, A_i/A. Useful specializations can be obtained by making certain assumptions; if it is assumed that $b = 1$, $\overline{D}_i$ can be expressed in terms of thickness t_i $(i = 1, 2, \ldots, n)$ of each component. Then

$$\underset{\sim}{\sigma}^a = \frac{1}{t}(\underset{\sim}{C}_1 \underset{\sim}{\varepsilon}_1 t_1 + \underset{\sim}{C}_2 \underset{\sim}{\varepsilon}_2 t_2 + \cdots + \underset{\sim}{C}_n \underset{\sim}{\varepsilon}_n t_n) \quad (2.11a)$$

where $\overline{D}_i = t_i/t$ and $\Sigma \overline{D}_i = 1$.

The Disturbed State Concept: Preliminaries

Furthermore, if it is assumed that the strains in all components are the same, i.e., $\varepsilon_1 = \varepsilon_2 = \cdots = \varepsilon_n = \varepsilon^a$, Eq. (2.11a) becomes

$$\underline{\sigma}^a = \frac{1}{t}(\underline{C}_1 t_1 + \underline{C}_2 t_2 + \cdots + \underline{C}_n t_n)\underline{\varepsilon}^a \tag{2.11b}$$

For the one-dimensional case, Eq. (2.11b) gives

$$E^a \cdot \varepsilon^a = \frac{1}{t}(E_1 t_1 + E_2 t_2 + \cdots + E_n t_n)\underline{\varepsilon}^a \tag{2.12a}$$

Therefore, the equivalent composite elastic modulus, E^a, is given by

$$E^a = \frac{1}{t}(E_1 t_1 + E_2 t_2 + \cdots + E_n t_n) \tag{2.12b}$$

In Eqs. (2.11b) and (2.12), the thickness of each unit can be chosen such that $\Sigma t_i = 1$.

Equation (2.11) can lead to the rheological model, shown symbolically in Fig. 2.15(b). Here each of the units, shown as a block in Fig. 2.15(b), can be assigned different material characteristics such as elastic, elastoplastic, viscoelastic, and viscoplastic. Such a model is similar to the *overlay* model (16) for simulating elastoviscoplastic behavior; its details are given in Chapter 8, including example problems. In the multicomponent DSC, it is possible to simulate behavior of a material element with different characterizations for each component, as in the overlay model (Chapter 8).

2.8 DSC for Porous Saturated Media

Terzaghi (17) proposed the effective stress concept for porous saturated materials. The concept, although simple, has proved to be useful for the evaluation of stresses and fluid pressures in saturated media and for the solution of many practical problems. The inherent assumptions in Terzaghi's concept are that the effective stress, a rather fictitious quantity, is carried by the soil skeleton through particle contacts and that the particle contact area, A^s, is negligible throughout the deformation and diffusion process. The latter assumption may be valid approximately for coarse-grained materials (18); however, in general, e.g., for cohesive materials, the contact area may not be small. Hence, it is appropriate and necessary to allow for the effect of the finite magnitudes of particle contacts during deformation on the solid contact and fluid stresses.

It is usually difficult to measure, inside the material, the particle contact areas during deformation. However, it could be possible to define average or weighted values of the changing contact areas, say, as proportional to the measurable void ratio, e (= volume of voids to volume of solids), and thereby define the solid particle stresses and fluid pressures as functions of the contact area. Such a procedure is possible through the use of the DSC and can provide a micromechanical understanding and analysis of the problem.

2.8.1 DSC Equations

Consider an element of porous material (soil) saturated with a fluid (water); see Fig. 2.16(a). The applied or total (compressive) force, F^a, on a material element is carried by the force in the solid skeleton, F^s, transmitted through the contact areas, A^s, the force, F^f, in the fluid is transmitted through the fluid area, A^f [Fig. 2.16(b)]. Then the force equilibrium gives

$$F^a = F^f + F^s \tag{2.13a}$$

Dividing by the nominal area, A, and with rearrangement, Eq. (2.13a) becomes

$$\frac{F^a}{A} = \frac{F^f}{A^f} \cdot \frac{A^f}{A} + \frac{F^s}{A^s} \cdot \frac{A^s}{A} \tag{2.13b}$$

or

$$\sigma^a = p^f \frac{A^f}{A} + \sigma^s \frac{A^s}{A} \tag{2.14a}$$

or

$$\sigma^a = (1 - D)p^f + D\sigma^s \tag{2.14b}$$

where σ^a, σ^s, and p^f are the total stress, solid contact stress, and fluid stress (at a point)[Fig. 12.16(c)], respectively, $A = A^s + A^f$, and D is the disturbance, which can be defined as

$$D = \frac{A^s}{A} \tag{2.15}$$

Initially, as the contact area is negligible, $D \approx 0$ and the contact stress, σ^s, is high, while $p^f = \sigma^a$. As the deformation under a given stress (increment) progresses, the particle-to-particle contacts increase and D approaches unity,

The Disturbed State Concept: Preliminaries

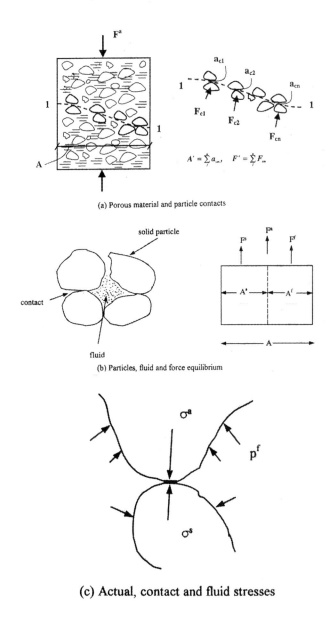

FIGURE 2.16
DSC for porous saturated materials.

corresponding to possible contacts under the applied (compressive) stress. Hence, when $D = 1$, $\sigma^s = \sigma^a$. For $0 < D < 1$, the total stress, σ^a, represents the sum of the changing contact stress, $D\sigma^s$, and the fluid stress, $(1 - D)p^f$. During deformation and flow of water out of the porous material, the contact area, A^s, increases, whereas the fluid area, A^f, decreases. Hence, the stress

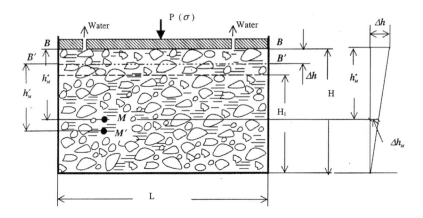

FIGURE 2.17
Flow and deformation in idealized porous material.

($D\sigma^s$) carried by the contacts increases and that $[(1 - D)p^f]$ carried by the fluid decreases.

Assume that all voids are connected during the deformation progress; then the change of fluid pressure produced by applied force, P, or stress ($\sigma = P/A$) (Fig. 2.17) at a given point can be expressed as

$$p^f = \sigma + \gamma^f h \qquad (2.16)$$

where γ^f is unit weight of the fluid and h is the depth of the point from the top. Substitution of Eq. (2.16) into Eq. (2.14b) leads to

$$\sigma^a = D\sigma^s + (1 - D)(\sigma + \gamma^f h) \qquad (2.17)$$

2.8.2 Disturbance

The disturbance can be considered as the deviation of the current deforming state with respect to the initial and final states of the material; Fig. 2.18 shows such states in terms of volume of voids (V^v) or void ratio (e). The initial, current, and final states can be defined by using the void ratio, e; the corresponding void ratios are e^0, e^a, and e^f. Assume that initially the void ratio is higher and that under compressive loading it decreases due to compaction. Then the change (decrease) in the void ratio can be considered to be proportional to the solid-to-solid contact areas. In other words, the contact area increases during deformation and at a given stage [Fig. 2.16(a)]:

$$A^s = \sum_i^n a_{ci} \qquad (2.18)$$

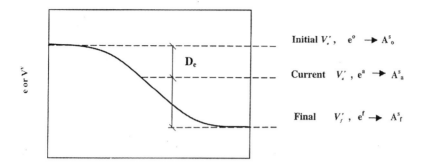

FIGURE 2.18
Disturbance based on the volume of voids or void ratio, e.

where a_{ci} is the contact area of particle group, i, and n is the total number of contacts. Then the disturbance, D, can be defined based on the void ratio as (Fig. 2.18)

$$D_e = \frac{e^0 - e^a}{e^0 - e^f} \tag{2.19a}$$

The disturbance, D_e, in Eq. (2.19a) varies from 0 to 1 for a given stress (increment). In general, the overall or total disturbance can be defined as

$$D_e^t = \frac{e^{max} - e^a}{e^{max} - e^{min}} \tag{2.19b}$$

where e^{max} and e^{min} are the maximum and minimum void ratios in which the material can exist in the loosest and densest states, respectively; D_e^t varies from 0 to 1. Then D_e in Eq. (2.19a), which is a part of D_e^t, will vary from the initial value, D_0 (≥ 0), to the final value, D_f (≤ 1), under the given stress increment. The schematic variation of D is shown in Fig. 2.19. The disturbance, D, can also be defined based on the shear stresses, the pore water pressure, and nondestructive properties (e.g., P- or S-wave velocities) (Chapter 3).

2.8.3 Terzaghi's Effective Stress Concept

Terzaghi's effective stress concept (17) is also based on the force equilibrium [Eq. (2.13)]. As stated earlier, it assumes that the contact area is very small; hence, $A^f \approx A$. Therefore, the contact stress will be very high, as $A^c \approx 0$. Under the assumption that the term $\sigma^s A^s$ must approach a finite limit, the term $(\sigma^s A^s / A)$ in Eq. (2.14) is called the *effective stress*, $\bar{\sigma}$, which represents some measure of average stress carried by the soil skeleton. The term

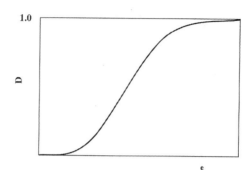

FIGURE 2.19
Schematic of disturbance.

"effective stress" has no accurate physical meaning, however (19). Moreover, the fluid pressure, called the *neutral pressure*, is assumed to be independent of A^s because A^f is assumed to equal the nominal area, A. If the contact area, A^s, can be found (measured), the term A^s/A can be computed and the actual contact stress can be found. However, it is difficult to measure the contact area, A^s, because of many difficulties in such measurements and their interpretation (20, 21). Hence, it was assumed to be negligible in the Terzaghi theory. However, for the calculation of the contact and fluid stress, it is appropriate and necessary to allow for the changing contact area during deformation. In the DSC model, the contact area, A^s, is considered, in an average sense, to be proportional to the void ratio, which represents the ratio of the volume of voids to that of solids. This is achieved by defining the disturbance by using Eq. (2.19).

Terzaghi's effective stress equation is expressed as

$$\sigma^a = \bar{\sigma} + p \tag{2.20}$$

where $\bar{\sigma}$ is the effective stress and p is referred to as neutral pressure or pore water pressure. Since the total stress, σ^a, at a point is the same, the following relations can be assumed based on Eq. (2.14b):

$$\bar{\sigma} = \sigma^s D \tag{2.21a}$$

$$p = p^f(1 - D) \tag{2.21b}$$

It may be noted that $\bar{\sigma}$ and p in Eq. (2.20) are interpreted in the same manner as in the Terzaghi theory; however, their values and variations can be different from those in the Terzaghi theory. This is because σ^s and p^f in Eq. (2.21) depend on the changing area, A^s, the total unit weight, and the height (coordinate) of a point [Eq. (2.15), and Eqs. (2.23) and (2.25) ahead].

2.8.4 Example and Analysis

Consider the problem of flow through a porous medium, as depicted in Fig. 2.17 (22). The porous saturated mass is subjected to a load or stress (increment) σ. The fluid (water) is allowed to flow out uniformly at the top. The following geometrical and material properties are assumed:

> Height, $H = 1.00$ m
> Final height at the end of deformation, $H_1 = 0.80$ m
> Area, $A = 1.0$ m^2
> Applied stress, $\sigma = P/A = 10$ KN/m^2
> Unit weight of water, $\gamma^f = 9.75$ KN/m^3
> Unit weight of solids, $\gamma^s = 20.0$ KN/m^3

The disturbance, D, is defined based on the void ratio according to Eq. (2.19b). Then for the saturated system:

$$D = \frac{V_f^{max} - V_f^a}{V_f^{max} - V_f^{min}} \tag{2.22a}$$

where V_f^{max} and V_f^{min} are the maximum and minimum volumes of the fluid, and V_f^a is the current volume of the fluid during deformation. Here the volume of solids, V^s, is assumed to be constant. For the simple problem in Fig. 2.17, Eq. (2.22a) specializes as

$$D = \frac{\Delta h}{H - H_1} \tag{2.22b}$$

where Δh is the (vertical) deformation from the top at any stage.

Now, consider any point, M, initially at depth $= h_M^0$ (Fig. 2.17). During deformation, the point moves to M', whose depth with respect to the current deformed state is given by

$$h'_M = h_m^0 + \Delta h_m = \frac{h_M^0}{H}(H - \Delta h) \tag{2.23}$$

It is assumed here that there is a uniform decrease in the voids, that is, the variation of deformation is linear (Fig. 2.17).

The total stress at the generic point M is given by

$$\sigma_M^a = \sigma + \gamma^T h'_M \tag{2.24}$$

where γ^T is the varying unit weight of the mixture given by

$$\gamma^T = (\gamma^s + \gamma^f e)\frac{V_0^s}{V} = (\gamma^s + \gamma^f e)\frac{V_0^s}{A(H - \Delta h)} \qquad (2.25)$$

where V_0^s is the initial volume of solids, which remains constant, and is assumed to equal AH_1. The fluid stress at point M is given by

$$p_M^f = \sigma + \gamma^f h_M' \qquad (2.26)$$

Substitution of Eqs. (2.22b), (2.23), and (2.24) into Eq. (2.17) leads to the contact stress as

$$\sigma_M^s = \left[\sigma + \gamma^T h_M' - \left(1 - \frac{\Delta h}{H - H_1}\right)(\sigma + \gamma^f h_M')\right]\frac{H - H_1}{\Delta h} \qquad (2.27)$$

Now, the effective stress, $\bar{\sigma}$ [Eq. (2.20)], can be expressed as

$$\bar{\sigma} = \sigma_M^a - p = \sigma_M^a - (1 - D)p^f = [\gamma^T - (1 - D)\gamma^f] h_m' + D\gamma \qquad (2.28)$$

where the neutral pressure $p = (1 - D)p^f$.

The foregoing equations will be modified for other boundary conditions such as free drainage (22).

Figure 2.20 shows variations of the contact, σ^s [Eq. (2.27)], at different depths with disturbance. It varies from very high values when $D = 0$ to stable

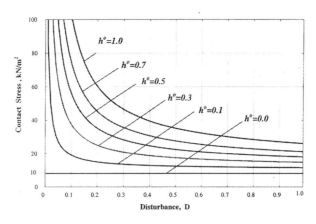

FIGURE 2.20
Variations of contact stress at different depths.

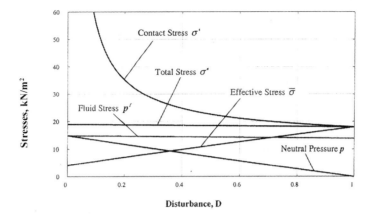

FIGURE 2.21
Variation of σ^a, σ^c, p', $\bar{\sigma}$, and p at depth $h^0 = 0.5$ m.

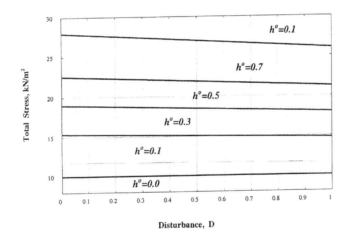

FIGURE 2.22
Total stresses, σ^a, at different depths.

values as $D \to 1$, except that it remains the same ($= \sigma$) at the top. Figure 2.21 shows variations of the total stress, σ^a, contact stress, σ^s, fluid stress, p^f, effective stress, $\bar{\sigma}$, and neutral pressure, p, at depth $h^0 = 0.5$ m. It can be seen that the fluid stress (p^f) variation (reduction) with depth is small. Also, $\bar{\sigma}$ and p increase and decrease with disturbance, respectively. Variations of the total stress, σ^a, at different depths are shown in Fig. 2.22.

Figure 2.23(a) and (b) show variations of the effective stress, $\bar{\sigma}$, and neutral pressure, p (Fig. 2.21). It can be seen that the effective stress increases with disturbance, whereas the neutral pressure decreases with disturbance. For instance, at depths $h^0 = 0.0, 0.50$, and 1.0, and for $D = 0.0$ to 1.0, the increase

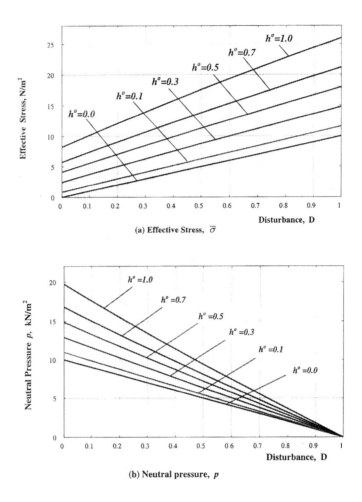

FIGURE 2.23
Variations of effective stress and neutral pressure, p, at different depths.

in the effective stress is 10.00, 13.90, and 17.80 KN/mm², respectively. The corresponding decrease in the neutral pressure is 10.00, 14.88, and 19.75 KN/mm². Only at $h^0 = 0.0$, are the increase in $\bar{\sigma}$ and the decrease in p equal. It may be noted that in contrast to the Terzaghi theory, the DSC model allows for the changing contact area, height, and total unit weight during deformation in the calculations of the effective stress and neutral pressure.

2.8.5 Comments

The foregoing analysis indicates that the DSC model proposed herein can allow evaluation of contact and fluid stresses, depending on the changing contact area, A^s. Also, the model can provide calculation of effective stress and neutral pressure according to the Terzaghi concept, although their

The Disturbed State Concept: Preliminaries

variations are different because the DSC model allows for changing contact area, height, and total unit weight. The advantage of the DSC is that the contact stress and fluid pressure are influenced by the changing contacts during deformation.

2.9 Bonded Materials

There are a number of ways in which the DSC can be formulated to characterize the behavior of bonded materials such as some rocks, aged clay, and artificially cemented or grouted soils. The mechanical behavior of such materials would involve the response (frictional and cohesive) at the particle (grain) contacts and that of the bonding material that cements the particles. Thus, the behavior of bonded materials would involve both cohesive and frictional components.

The behavior of a bonded material represents the response of the matrix of particles in contacts cemented by the bonding material, as Fig. 2.24(a) shows. It will depend on a number of factors such as cohesion, mean effective stress, volume change, microcracking, and fracturing. Microcracking often initiates in the bonded material zones, leading finally to "complete" breakage of bonds, reducing to the original skeleton of particles. The behavior at the particle skeleton with contacts, and the bonding material will involve interaction, and the coupled mechanism will produce the observed response of the mixture or matrix. A number of DSC approaches for bonded materials are described next.

2.9.1 Approach 1

The observed response is expressed in terms of the response of the matrix in its RI state (without microcracking and fracture) and the FA state toward which the observed response approaches asymptotically; see Fig. 2.24(b). Then the incremental equations will be given by Eq. (2.6).

In this approach, the observed test behavior of the matrix can provide the responses for the RI and FA parts. The RI response can be defined as linear elastic or elastoplastic by determining the parameters based on the early part (before peak) of the stress–strain response [Fig. 2.24(b)]. The FA response can be defined by using the residual stress state to approximate the FA state. Here, the constrained liquid or constrained liquid–solid (critical state) simulation can be used. The disturbance can be defined based on the stresses in the RI, observed, and FA states (see Chapter 3).

2.9.2 Approach 2

The observed response can be expressed in terms of the RI response (as in Approach 1) and that of the bonding material (b); see Fig. 2.24(c). Here it would be required to test the specimens of both the bonded matrix and the

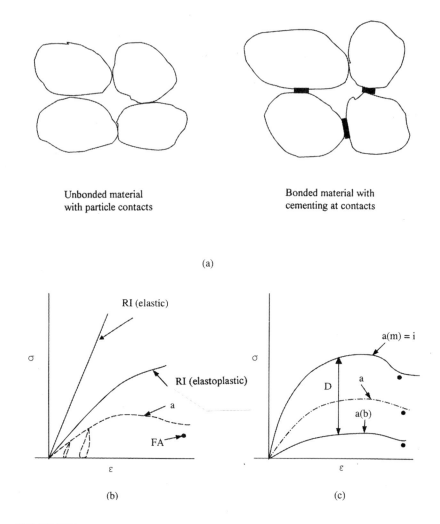

FIGURE 2.24
Behavior of bonded materials: approaches 1 and 2.

bonding material. Then disturbance, D, can be expressed as (see Chapter 3)

$$\bar{D} = \frac{\sigma^i - \sigma^a}{\sigma^i - \sigma^b} \tag{2.29}$$

where i denotes the response of the bonded matrix, $a(m)$ (Fig. 2.24(c)), and the incremental constitutive equations are given by

$$d\underset{\sim}{\sigma}^a = (1 - \bar{D})\underset{\sim}{C}^i d\underset{\sim}{\varepsilon}^i + \bar{D}\underset{\sim}{C}^b d\underset{\sim}{\varepsilon}^b + d\bar{D}(\underset{\sim}{\sigma}^b - \underset{\sim}{\sigma}^i) \tag{2.30a}$$

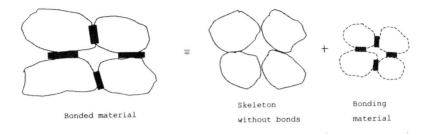

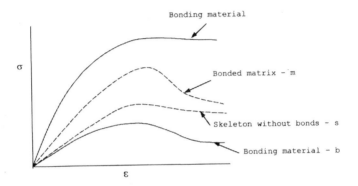

FIGURE 2.25
Response of bonded material: approach 3.

where $\underline{C}^b$ is the constitutive matrix for the bonding material, which can be assumed to be elastic, elastoplastic, elastic with disturbance, or elastoplastic with disturbance, depending on the observed response, $a(b)$. It may be possible to assume that the (average) strains in both the matrix and the bonding material are the same, i.e., $d\underline{\varepsilon}^i = d\underline{\varepsilon}^b = d\underline{\varepsilon}$. Then Eq. (2.30a) reduces to

$$d\underline{\sigma}^a = [(1-\bar{D})\underline{C}^i + \bar{D}\underline{C}^b]d\underline{\varepsilon} + d\bar{D}(\underline{\sigma}^b - \underline{\sigma}^i) \qquad (2.30b)$$

2.9.3 Approach 3

The bonded material (matrix) can be assumed to be composed of the skeleton of particles with contacts but without bonds (s), and the bonding material (b); see Fig. 2.25. Then the observed (incremental) response of the matrix ($d\underline{\sigma}^m$) can be expressed as the sum of the responses, as

$$d\underline{\sigma}^m = d\underline{\sigma}^s \pm d\underline{\sigma}^b \qquad (2.31a)$$

or

$$d\underline{\sigma}^m = \underline{C}^s d\underline{\varepsilon}^s \pm \underline{C}^b d\underline{\varepsilon}^b \qquad (2.31b)$$

where $\underline{C}^s$ and $\underline{C}^b$ are the constitutive matrices for the skeleton (without bonds) and the bonding material, respectively. If the strains in both are equal, Eq. (2.31b) reduces to

$$d\underline{\sigma}^m = (\underline{C}^s \pm \underline{C}^b) d\underline{\varepsilon} \tag{2.31c}$$

This approach is similar to the one proposed by Desai (1) for the behavior of overconsolidated soil, expressed as the sum of the behavior of the soil in the normally consolidated state, and that due to the overconsolidation (bonding) effect. It is also similar to that used recently by Chazallon and Hickes (23). As interaction exists between the solid and bonded phases, it will be necessary to employ an iterative procedure while using Eq. (2.31).

2.9.4 Approach 4

The behavior of the bonding material can be treated as a reference state, while that of the skeleton of soil (without cementation) is treated as the other reference state. The response of the bonding material may be stiffer or softer compared to that of the observed behavior of the matrix (with cementation). The disturbance, $\overline{D}$, can be expressed as

$$\overline{D} = \frac{\sigma^b - \sigma^m}{\sigma^b - \sigma^s} \tag{2.32}$$

where b, m, and s denote bonding material, matrix, and skeleton, respectively. If the response of the bonding material is stiffer, D will vary from 0 to 1; see Fig. 2.26. If it is softer, it will have a value greater than 1; in that case, it may

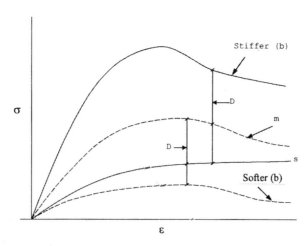

FIGURE 2.26
Response of bonded material: approach 4.

The Disturbed State Concept: Preliminaries

be nondimensionalized with respect to the final value, D_f, corresponding to the residual stress so that it varies from 0 to 1. However, in that case, the value D_f needs to be provided in the analysis and computation.

Figure 2.26 indicates that the behavior of the matrix is affected by progressive microcracking in the bonds. In the limit, the behavior of the bonded matrix may tend toward that of the skeleton of a (dry) granular system without bonds but with particle contacts. Thus, the behavior of the skeleton can be obtained from tests with the granular material (say, sand) in its state without the artificial or natural bonding.

2.9.5 Porous Saturated Bonded Materials

The equilibrium of forces for a material composed of solid particle contacts, bonding material, and fluid, with areas A_s, A_b, and A_f, respectively (Fig. 2.27), gives

$$\frac{F^a}{A} = \frac{F^s}{A_s}\frac{A_s}{A} + \frac{F^b}{A_b}\frac{A_b}{A} + \frac{F_f}{A_f}\frac{A_f}{A} \qquad (2.33a)$$

$$\sigma^a = \sigma^s D_s + \sigma^b D_b + \sigma^f D_f \qquad (2.33b)$$

where D_s, D_b, and D_f are ratios of the areas of the solid contacts, bonding material, and fluid to the total nominal area A. It will be difficult to determine the three areas in a deforming material.

If we assume that the contact area, A_s, is negligible and define effective stress over the nominal area, A, as in the Terzaghi theory [Eq. (2.20)], Eq. (2.33) reduces to

$$\sigma^a = \bar{\sigma} + \sigma^b \frac{A_b}{A} + p\left(1 - \frac{A_b}{A}\right) \qquad (2.34)$$

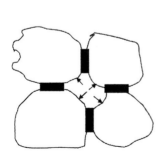
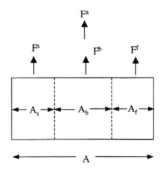

FIGURE 2.27
Saturated bonded material.

where $\bar{\sigma}$ is the effective stress and p is the pore water pressure in the sense of the classical effective stress principle. However, as noted before, the changing contact area is not included in such an approximation. Specialization such as in Eq. (2.34) has been used for defining the behavior of cemented sands (24).

2.9.6 Structured Materials

A *structured* material is defined as a material that possesses a modified microstructure in relation to that of its basic reference state. Thus, many material systems such as reinforced systems (earth, concrete, and metal or ceramic composites), overconsolidated soil, dislocated silicon with impurities, asphalt concrete with enhanced bonding due to chemical effects, and natural soils with respect to their reconstituted or remolded states (25) can be treated as structured materials. The formulation and application of the DSC for structured materials are given in Chapter 10.

2.10 Characteristics of the DSC

It is evident from Eq. (2.6) that the DSC constitutive equations involve

(1) the influence of the deformation characteristics of the interacting material parts in the reference states, and
(2) the possibility of relative motions (translation, rotation) between the material parts in the reference states, which can involve different stresses (and strains) in them.

These two properties are important as they allow for the interaction between the two parts, leading to a diffusion-type process within the mixture in the material element. As a result, Eq. (2.6) can provide a general formulation, compared to that in the classical damage model (6, 7) in which the cracked parts are assumed to carry no stress. Various specializations of the DSC, including the classical damage model, are discussed in Chapter 4.

2.10.1 Comparisons and Comments

One of the distinguishing features of the DSC is that it allows for the interacting mechanisms between the parts in the reference states as affected by the microstructural rearrangement and self-adjustment of particles. However, its formulation does not require particle or microlevel characterization. This is because the responses of the parts in the reference states are defined on the basis of the macrolevel response measured on finite-sized (laboratory) specimens. As a result, it avoids the necessity of defining particle-level constitutive equations, which are required in the micromechanical and microcrack inter-

The Disturbed State Concept: Preliminaries 59

action (with damage) models (26). This is considered to be a distinct advantage, as it is difficult (at this time) to measure constitutive responses at the particle level; therefore, in the micromechanical and microcrack interaction models, it is usually defined on the basis of measured behavior on finite-sized specimens, which is indeed an inconsistency.

Furthermore, a material made up of many (millions of) particles represents a complex and nonlinear system. Hence, it is not possible theoretically to integrate (micro-) particle-level responses to obtain the macrolevel response because a part of the particle interactions may be lost in the process. Indeed, approximations are always possible, as the micromechanical models imply. In the DSC, we depend on the interaction between clusters of material parts in the reference states, which is considered to be more consistent for describing the behavior of the complex system. In Chapters 3 to 12 we discuss how the responses of the material parts in the reference states are determined from laboratory tests.

It is often appropriate to define the RI behavior on the basis of plasticity theory involving a yield surface. However, it is not necessary to invoke yield-surface-based model(s) in the DSC. It can be formulated on the basis of irreversible strains without the yield surface.

2.10.2 Self-Organized Criticality

The self-organized criticality (SOC) concept by Bak and co-workers (27) around the mid-1980s analyzes the *unique* collapse or catastrophic states a material may reach under thermomechanical loading. This concept is similar to the critical state (CS) concept of soil mechanics developed by Roscoe and co-workers in the 1960s (8, 9) in which a soil with given initial confining pressure approaches the critical state when its density or void ratio approaches unique value irrespective of its initial density. At the critical state, the material deforms in shear without a change in its volume. The *synergetics* concept by Haken in 1978 (28) proposed a theory for analyzing the *unique final state*, called the "state of thermal equilibrium" into which a system develops. Here, the original structure of the system disappears, replaced by homogeneous systems.

The DSC allows the characterization of the catastrophic or collapse state as in the foregoing concepts, and also the characterization of the entire behavior of the material including the prepeak and postpeak responses. Hence, the DSC is considered to be general compared to the SOC, CS, and synergetics concepts. Detailed descriptions of the SOC in comparison with the DSC, as well as the generality of the DSC, are presented in Appendix I.

2.11 Hierarchical Framework of the DSC

Figure 2.28 depicts the unified and hierarchical framework of the DSC. It permits, as shown in the subsequent chapters, adoption of constitutive models depending on the specific material properties for a given application's need.

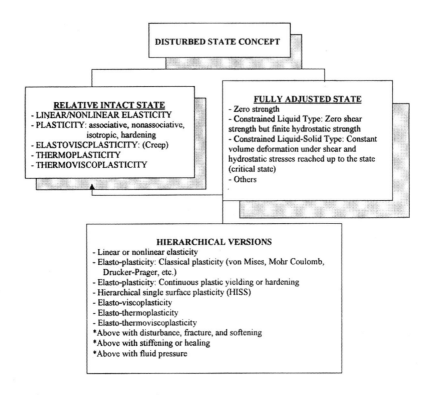

FIGURE 2.28
Hierarchical versions in DSC.

References

1. Desai, C.S., "A Consistent Finite Element Technique for Work-Softening Behavior," *Proc. Int. Conf. On Comp. Meth. in Nonlinear Mech.*, J.T. Oden (editor), Univ. of Texas, Austin, 1974.
2. Desai, C.S., "Finite Element Residual Scheme for Unconfined Flow," *Int. J. Num. Meth. Eng.*, 10, 1976, 1415–1418.
3. Desai, C.S. and Li, G. C., "A Residual Flow Procedure and Application for Free Surface Flow in Porous Media," *Adv. in Water Resources*, 6, 1983, 27–35.
4. Baseghi, B. and Desai, C.S., "Laboratory Verification of the Residual Flow Procedure for Three-Dimensional Free Surface Flow," *Water Resources Research*, 26, 2, 1990, 259–272.
5. Bruch, J.C., Jr., "Fixed Domain Methods for Free and Moving Boundary Flows in Porous Media," *Transport in Porous Media*, 6, 1991, 627–649.
6. Kachanov, L.M., *Theory of Creep,* English translation editor A.J. Kennedy, National Lending Library, Boston, MA, 1958.
7. Kachanov, L.M., *Introduction to Continuum Damage Mechanics,* Martinus Nijhoft Publishers, Dordrecht, The Netherlands, 1986.

8. Roscoe, K.H., Scofield, A. and Wroth, C.P., "On Yielding of Soils," *Geotechnique*, 8, 1958, 22–53.
9. Scofield, A.N. and Wroth, C.P., *Critical State Soil Mechanics*, McGraw-Hill, London, 1968.
10. Felice, C. W., Tester, V. J., and Sharer, J., "A Nondestructive Damage Assessment of PCGC Grout," *Proc. Conf. On Containment of Underground Nuclear Explosions*, Univ. of Nevada, Reno, NV, 1991.
11. Desai, C.S., "Constitutive Modelling Using the Disturbed State as Microstructure Self-Adjustment Concept," Chap. 8 in *Continuum Models for Materials with Microstructure*, H.B. Mühlhaus (Editor), John Wiley, Chichester, U.K., 1995.
12. Desai, C.S. and Toth, J., "Disturbed State Constitutive Modeling Based on Stress-Strain and Nondestructive Behavior," *Int. J. Solids and Structures*, 33, 11, 1619–1650, 1996.
13. Desai, C.S., Park, I. J., and Shao, C., "Fundamental Yet Simplified Model for Liquefaction Instability," *Int. J. Num. Analyt. Meth. in Geomech.*, 22, 721–748, 1998.
14. Desai, C.S., "Liquefaction Using Disturbance and Energy Approaches," *J. of Geotech. and Geoenv. Eng., ASCE*, 126, 7, 2000, 618–631.
15. Scott, T.E., Azeemuddin, M., Zaman, M., and Rogiers, J. C., "Changes in Acoustic Velocity During Pore Collapse of Rocks," News Journal, *Int. Soc. of Rock Mechanics*, 3, 1, 1995, 14–17.
16. Owen, D.R.J. and Hinton, E.. *Finite Elements in Plasticity: Theory and Practice*, Pineridge Press, Swansea, UK, 1980.
17. Terzaghi, K., *Theoretical Soil Mechanics*, John Wiley & Sons, New York, 1943.
18. Bishop, A.W. and Eldin, G., "Undrained Triaxial Tests on Saturated Sands and Their Significance in the General Theory of Shear Strength," *Geotechnique*, 2, 1, 1950.
19. Perloff, W.H. and Baron, W., *Soil Mechanics: Principles and Applications*, The Ronald Press Co., New York, 1976.
20. Lamb, T.W. and Whitman, R. V., "The Role of Effective Stress in the Behavior of Expansive Soils," *Report*, Colorado School of Mines, 54, 4, 1959.
21. Skempton, A., "Effective Stress in Soils, Concrete and Rock," *Proc. Conf. on Pore Pressure and Suction in Soils*, London, 1961.
22. Desai, C.S. and Wang, Z., "Disturbed State Model for Porous Saturated Materials," *Int. J. Geomechanics*, accepted 2000.
23. Chazallon, C. and Hickes, P. Y., 1998, "A Constitutive Model Coupling Elastoplasticity and Damage for Cohesive-Frictional Materials," *J. of Mech. of Cohesive-Frictional Mataerials*, 3(1), 41–63.
24. Abdulla, A.A. and Kiousis, P.D., 1997, "Behavior of Cemented Sands, Parts I and II," *Int. J. Num. Analyt. Meth. Geomech.*, 21, 8, 533–547 and 549–568.
25. Liu, M.D., Carter, J.P., Desai, C.S., and Xu, K. J., "Analysis of the Compression of Structured Soils Using the Disturbed State Concept," *Int. J. Num. Analyt. Meth. Geomech.*, 24, 8, 2000, 723–735.
26. Bazant, Z.P., "Nonlocal Damage Theory Based on Micromechanics of Crack Interactions," *J. Eng. Mech., ASCE*, 120, 1994, 593–617.
27. Bak, P., *How Nature Works*, Copernicus (Springer-Verlag), New York, 1996.
28. Haken, H., *Synergetics: An Introduction*, Springer-Verlag, Berlin, Germany, 1978.

3

Relative Intact and Fully Adjusted States, and Disturbance

CONTENTS
3.1 The Relative Intact and Fully Adjusted States 65
    3.1.1 Characterization of Material at Critical State 68
    3.1.2 Specializations 70
3.2 Disturbance and Function.. 72
    3.2.1 Disturbance Function 73
    3.2.2 Laboratory Tests 73
    3.2.3 Stiffening or Healing 78
    3.2.4 Representation of Disturbance 80
    3.2.5 The Stiffening Effect 82
    3.2.6 Creep Behavior 82
    3.2.7 Rate Dependence 84
    3.2.8 Disturbance Based on Disorder (Entropy) and Free Energy 85
3.3 Material Parameters.. 86
    3.3.1 Fully Adjusted State... 87
    3.3.2 Disturbance Function 88

In Chapter 2, we derived the following incremental DSC equations (1.8) for an element of the *same* material composed of material parts in RI and FA states:

$$d\sigma^a_{ij} = (1 - D)C^i_{ijk\ell}d\varepsilon^i_{k\ell} + DC^c_{ijk\ell}d\varepsilon^c_{k\ell} + dD(\sigma^c_{ij} - \sigma^i_{ij}) \quad (3.1a)$$

or, in matrix notation,*

$$d\underline{\sigma}^a = (1 - D)\underline{C}^i d\underline{\varepsilon}^i + D\underline{C}^c d\underline{\varepsilon}^c + dD(\underline{\sigma}^c - \underline{\sigma}^i) \quad (3.1b)$$

Similar incremental equations can be derived for the alternative formulations for an element composed of more than one material, e.g., Eqs. (2.9c), (2.11), (2.14), and (2.30a) of Chapter 2.

---
*Both the tensor and matrix notations are often used in this text. The latter can provide for convenient implementation in computer procedures.

In this chapter, we present explanations for the RI (i) and FA (c) states and the disturbance function, D, with respect to a given material. We also present procedures for the determination of parameters for the FA state and the disturbance function, D; for the RI models, they are presented in Chapters 5 to 12.

At this time, we consider disturbance that results in damage or degradation; stiffening or healing where the disturbance may decrease is discussed in Chapter 10. In that case, a deforming material element experiences microstructural changes that result in increased disturbance from the initial disturbance, which can be zero or nonzero. The initial disturbance depends on the initial conditions such as anisotropy, flaws, dislocations, and manufacturing defects. During the microstructural modifications and the self-adjustment of material particles, the material parts in the relative intact (RI) state transform to fully adjusted (FA) parts. Figure 3.1(a) shows a schematic of the growth of disturbance with loading and deformation. The shaded zones show the FA parts, while the unshaded

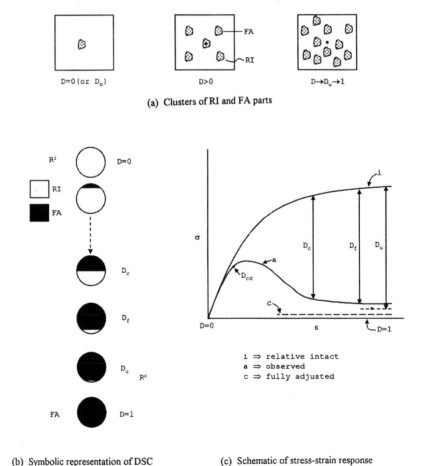

(a) Clusters of RI and FA parts

(b) Symbolic representation of DSC

(c) Schematic of stress-strain response

$i \Rightarrow$ relative intact
$a \Rightarrow$ observed
$c \Rightarrow$ fully adjusted

FIGURE 3.1
Representations of DSC.

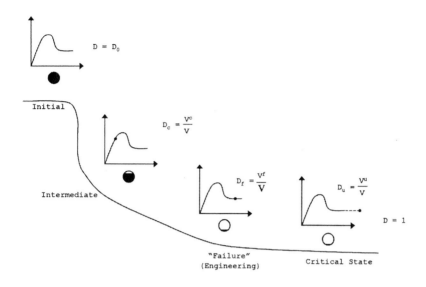

(d) Symbolic representation of disturbance and stress-strain responses

FIGURE 3.1
(Continued.)

parts show clusters of the RI parts. The asymptotic value of disturbance is unity when the FA state is reached over the entire material element. Before the FA state can be reached, the material's microstructure can experience local and global instabilities, denoted by D_c, D_f, D_u, etc. [Fig. 3.1(b)–(d)], which are also discussed later (e.g., Chapter 12). For instance, in the case of a dense granular material, a local instability occurs when the material transits from compactive to dilative volume (D_{cd}), which may not necessarily cause "failure." However, when the critical value of disturbance (D_c) is reached [Fig. 3.1(b)–(d)], failure in the sense of engineering reliability can *initiate*, while at $D = D_f$ the material may be considered to have "failed." For given initial conditions (pressure, density) the behavior may approach a residual (or critical) state (D_u), which may be adopted approximately as the FA state for the evaluation of the disturbance. As shown in Fig. 3.1(c), the final FA state ($D = 1$) is not measurable because the material had "failed" before that state can be realized. However, $D = 1$ can also be used in the formulation of the model, for instance, in the mathematical definition of D ahead.

3.1 The Relative Intact and Fully Adjusted States

Figure 3.2 shows schematics of stress–strain responses of materials whose physical state, defined by density (ρ) or void ratio (e), does not change significantly during deformation. In other words, as in the case of some metals, it

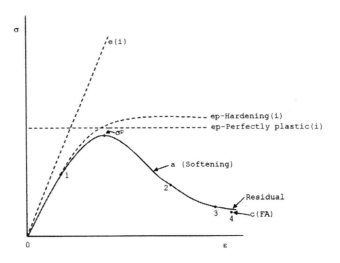

FIGURE 3.2
Stress–strain behavior: no change in physical state.

is assumed that the density or void ratio of the material remains invariant during deformation. The linear elastic response can be considered to be the RI state in relation to the observed elastic–plastic (hardening) behavior. The elastic–plastic hardening response can be treated as an RI state in relation to the elastic perfectly plastic response. The linear elastic or the elastic–plastic hardening responses can be treated as RI in relation to the observed degradation or softening behavior.

In the case of granular and frictional materials like soils (clays, sands), rocks, and concrete, the observed response depends on various factors such as initial mean pressure ($p_0 = \sigma_0 = J_{10}/3$), where $J_1 = \sigma_1 + \sigma_2 + \sigma_3$, density or void ratio (volume of void to volume of solids), and stress or loading path (3, 4, 6, 7). Here the physical state defined by density or void ratio changes during deformation. Figure 3.3(a) shows schematics of the stress–strain behavior of initially loose and dense granular materials. The loose material with initial void ratio e_0^a, and given initial pressure (p_0), compacts continuously, its void ratio decreases, and it approaches a critical value, e^c, (Fig. 3.2(b) and (c)), when the stresses reach critical values J_1^c and $\sqrt{J_{2D}^c}$, where J_{2D} is the second invariant of the deviatoric stress tensor, S_{ij}, and is proportional to the shear stress difference ($\sigma_1 - \sigma_3$) and to octahedral shear stress, τ_{oct}.

The initially dense material with void ratio e_0^a and a given pressure (p_0) may first compact and then experience the transition at point b (Fig. 3.3(b)), corresponding to D_{cd} [Fig. 3.1(c)], when its volume increases or it experiences dilation. That is, the void ratio first decreases, then increases after point b, and upon further deformation, after passing through the peak (point d), it approaches the critical state (e^c, $\sqrt{J_{2D}^c}$, J_1^c); see Fig. 3.3(d). The material's response at the critical state can be used to define the behavior of the material in the FA state.

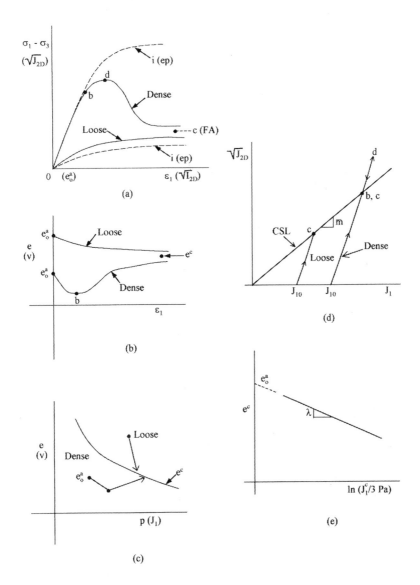

FIGURE 3.3
Behavior of loose and dense granular materials.

Relative Intact Behavior. A loose material with an initial mean pressure (p_0) may compact continuously during shear loading. In that case, the RI response can be characterized by excluding the effect of increased compaction, in which case the observed response can be stiffer than the RI response [Fig. 3.3(a)].

In the case of a dense material, the initial compaction would warrant consideration similar to that for the loose material. However, for practical purposes, and because for the response up to the transition point [b; Fig. 3.3(a)] the disturbance is usually negligible, the RI and the observed responses may be assumed to be approximately the same (up to point b). Then internal "microcracking" and relative particle motions would result in dilation, and the disturbance (damage) after point b would increase continuously. The RI response thereafter can be characterized as nonlinear elastic or elastic–plastic hardening (e.g., δ_0-model, Chapter 7) [Fig. 3.3(a)]. Details of constitutive equations and matrix, $\underline{C}^i$, Eq. (3.1), that characterizes the RI response are given in Chapters 4 to 11.

Fully Adjusted Behavior. The material parts that have reached the FA state during deformation can be considered to be at the critical state, e.g., $(e^c, \sqrt{J_{2D}^c}, J_1^c)$. When the material element follows the same stress path (as in a laboratory test), this (stress) state will be defined uniquely.

Further explanation of the behavior of a material element in the DSC is depicted in Fig. 3.4(a) in the four-dimensional space, $J_1 - \sqrt{J_{2D}} - \sqrt{I_{2D}}, e$, where I_{2D} is the second invariant of the deviatoric strain tensor, E_{ij} (8). Here, as a simplification, the response in which the physical state (e) does not change during deformation is considered; examples of such a behavior are a metal and saturated soil under undrained conditions. The shear loading starts at point A and travels along path A–C, and the observed or average response is denoted by point C. The corresponding RI and FA (critical) states are denoted by points B and D, respectively. Similar depictions can be developed for other loading conditions and stress paths.

In a boundary-value problem, on the other hand, if the stress paths for each material element (or point) can be computed, after a given incremental loading, its stress and void ratio at the critical state can be evaluated. This is shown schematically in Fig. 3.5 for three typical points (1–3) in a half-space loaded incrementally. The loadings of these points start at three different initial pressures (J_{10}). Points 1 and 2 may approach their critical states at 1^c and 2^c, respectively, while point 3 may not reach the critical state at all. However, once a material point has reached the FA (critical) state during deformation, its critical state is defined with respect to the stress path followed by the overall material element (consisting of RI and FA parts); see Fig. 3.1(a).

3.1.1 Characterization of Material at Critical State

When the material is at the critical state, there is no further change in its volume or void ratio, and it deforms in shear under constant value of the shear stress ($\sqrt{J_{2D}^c}$) reached up to that state (9). Based on laboratory observations, it is possible to define the behavior of the material at the critical state in terms of $\sqrt{J_{2D}^c}, J_1^c$, and e^c, as depicted in Fig. 3.3(d) and (e). Then the constitutive response of the material at the critical state can be expressed as

$$\sqrt{J_{2D}^c} = \overline{m} J_1^c \tag{3.2a}$$

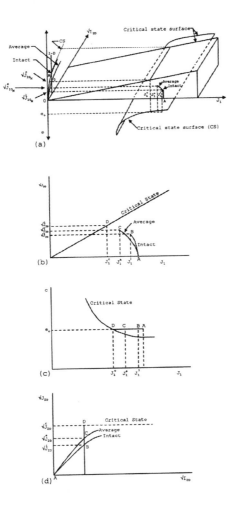

FIGURE 3.4
DSC in four-dimensional space and component spaces. (From Ref. 8, with permission.)

and

$$e^c = e_0^c - \lambda \ln(J_1^c/3p_a) \tag{3.2b}$$

where $\bar{m}$ is the slope of the critical state line (CSL) [Fig. 3.3(d)], λ is the slope of the critical state line [Fig. 3.3(e)], and e_0^c is the value of e^c corresponding to $J_1^c = 3p_a$, where p_a is the atmospheric pressure constant.

Then the behavior of the material, Eq. (3.2), at the critical state can be used to define the response of the material parts in the FA state. The physical interpretation is that the growing clusters of the FA parts (at the critical state), Fig. 3.1(a), as they are surrounded by the RI parts, possess certain strength like a constrained liquid–solid (CLS). That is, they can continue to carry the same shear stress ($\sqrt{J_{2D}^c}$) without change in volume. Thus, the deforming FA parts

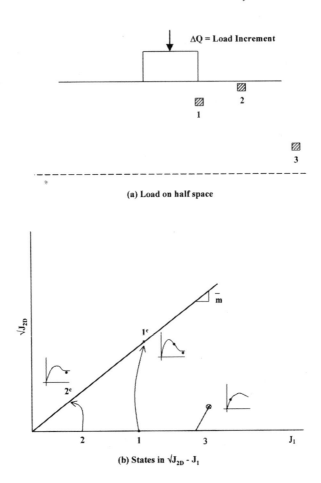

FIGURE 3.5
FA (critical state) in boundary-value problems.

provide interaction with the RI parts and produce the coupled observed response.

3.1.2 Specializations

Now, if it is assumed that the FA part (at the critical state) can carry no shear stress but can continue to carry hydrostatic stress or mean pressure, it behaves like a constrained liquid (CL). In this case, $\bar{m}$ in Eq. (3.2a) is zero. Then the constitutive equation for the FA part will depend only on the volumetric response, Eq. (3.2b).

If the FA material is treated like a "void" or crack, as it is assumed in the classical damage approach (10), it cannot carry any shear or hydrostatic stress. Then $\bar{m} = 0$ and $\lambda = 0$ in Eq. (3.2a,b) and the stress point is at the origin (Fig. 3.6); this idealization does not allow for coupling between the RI and FA parts. Figure 3.6 depicts the above specializations in the $\sqrt{J_{2D}} - J_1$ and $e - J_1$ spaces, in which

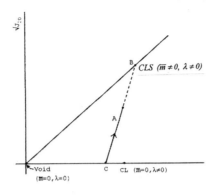

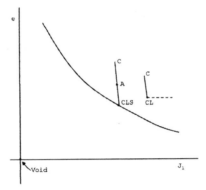

FIGURE 3.6
Representation of FA behavior.

CLS corresponds to the constrained liquid–solid, CL to constrained liquid, and void to no strength, idealizations. Further details on the equations for the FA state and matrix $\underset{\sim}{C}^c$, Eq. (3.1), that define the FA response are given in Chapters 4 to 11.

The behavior of materials such as metals and alloys is not significantly influenced by the hydrostatic stress, and there is no significant change in their density or void ratio during deformation. In other words, their response can be characterized based mainly on the stress–strain response. For these materials, the FA response can be characterized as a *constrained liquid*, which can carry mean pressure but no shear stress. The perfectly plastic response of an elastoplastic material can be considered analogous to such a characterization. Alternatively, the material in the FA state for metallic materials can be assumed to carry no stress at all. Some examples are given in Chapter 6.

The idea of the critical state can be considered a generalization of classical elastoplastic theory (Chapter 6). Just as in the critical state concept, the material, like a metal whose stress–strain behavior is essentially independent of the mean pressure (Fig. 3.2), reaches a state of constant volume at the yield. Then the material deforms under constant yield stress.

3.2 Disturbance and Function

The observed response, R^a, is expressed in terms of the responses, R^i and R^c, of the materials in the reference states, RI and FA, respectively:

$$R^i + R^c \Rightarrow R^a \tag{3.3}$$

Figure 3.1(b) shows a symbolic representation of the R^i, R^a, and R^c states connected through the disturbance, D, which acts as an interpolation mechanism between the interacting materials in the RI and FA states. An approximate analogy can be given here: the above representation is similar to expressing the behavior of a given volume of the mixture of ice and water in terms of the behavior of ice and water. The behavior of ice is analogous to that of the RI material, and the behavior of water is analogous to that of the FA material.

In a deforming material, disturbance, D, can be expressed as the mass (M^c) of material in the FA state to the total mass (M):

$$D = \frac{M^c}{M} \tag{3.4}$$

With no initial disturbance, D varies from 0 to 1, as initially $M^c = 0$, and at higher deformations $M^c \to M$. If we assume that the densities of the materials in various states are the same, D can be expressed approximately as

$$D = \frac{V^c}{V} \tag{3.5a}$$

$$D = \frac{A^c}{A} \tag{3.5b}$$

$$D = \frac{\ell^c}{\ell} \tag{3.5c}$$

where V^c, A^c, and ℓ^c are the volume, area, and length of three-dimensional, two-dimensional, and one-dimensional material elements in the FA state, respectively.

In general, however, D can be expressed as

$$D = \frac{\rho^c}{\rho} \cdot \frac{V^c}{V} = D_1 \cdot D_2 \tag{3.6}$$

where the functions D_1 and D_2 depend on the varying density and volume, respectively.

It may be noted that the form of D in Eq. (3.5) is similar to the damage (ω) in the classical damage approach (10). However, in the DSC, the interpretation of disturbance is different and general; it considers that the FA material has specific strength and that it interacts with the RI material. On the other hand, in the classical damage approach, the FA material is treated as a "void" without interaction with the RI part. Furthermore, in the DSC, D can denote both damage or degradation, and growth or stiffening (strengthening).

3.2.1 Disturbance Function

Once the RI and FA states and disturbance are defined, the next step is to formulate the disturbance function, D. The development of microcracking and damage, as well as compaction and strengthening, can depend on the direction, and hence, it would lead to anisotropic disturbance. As a result, in general, D would be a tensor with directional components (Chapter 2). It is usually difficult to measure the directional properties of deforming materials; some possible ways are discussed in Chapter 2. However, for practical purposes, it may often be sufficient to treat D as a scalar, in an average and weighted sense. Most of the descriptions in this text treat D as a scalar.

There are a number of ways in which we can define D to represent degradation or strengthening. The most direct way is to measure the mass, volume, etc. [Eqs. (3.4) and (3.5)] of the material in the FA state. However, this can be difficult at this time, mainly because of the lack of proper testing devices to penetrate the material and measure these quantities; additional discussion on this aspect is presented in Chapter 2. Hence, the indirect approaches need to be developed based on phenomenological considerations. One such indirect way is to define D from the measurements of stress–strain ($\sigma - \varepsilon$), volumetric (void ratio), effective stress ($\bar{\sigma}$), or pore water pressure (p) and/or nondestructive responses; see Fig. 3.7(a) to (d). The other phenomenological way is to express D in terms of (internal) variables such as plastic strain trajectory, and plastic work or dissipated energy. The definition of D usually relies on the combination of these two phenomenological approaches.

3.2.2 Laboratory Tests

The disturbance that expresses the deviation of the observed behavior with respect to its behavior in the two reference states, RI and FA, can be defined on the basis of the stress–strain, volumetric (void ratio), and nondestructive

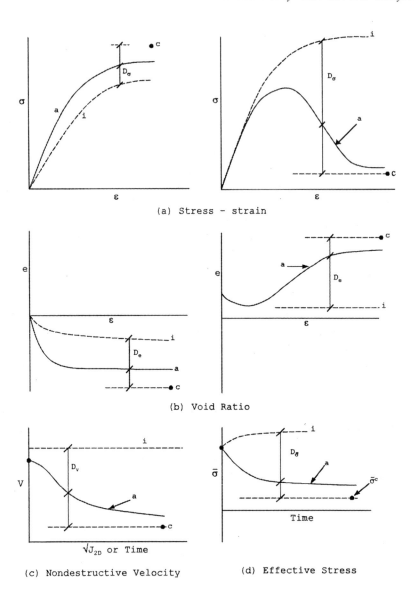

FIGURE 3.7
Disturbance from test data.

(velocity or attenuation) behavior. Figure 3.7(a) and (b) show typical stress–strain and void ratio (volumetric) responses. The disturbance, D, can be expressed as

$$D_\sigma = \frac{\sigma^i - \sigma^a}{\sigma^i - \sigma^c} \tag{3.7a}$$

and

$$D_e = \frac{e^i - e^a}{e^i - e^c} \qquad (3.7b)$$

where σ denotes appropriate stress measures such as axial stress σ_1, shear stress τ, stress difference $(\sigma_1 - \sigma_3)$; $\sqrt{J_{2D}}(J_{2D} = 1/6[(\sigma_1 - \sigma_2)^2 + \sigma_2 - \sigma_3)^2 + \sigma_3 - \sigma_1)^2]$, where σ_i $(i = 1, 2, 3)$ are the principal stresses, or octahedral shear stress, $\tau_{oct} = \sqrt{2/3}\sqrt{J_{2D}}$. It may be noted that values of D_σ and D_e at a given stage during deformation may not be the same. However, in the iterative solution procedures, both the stress equilibrium and volumetric responses can be satisfied simultaneously (Chapter 13).

Often, the disturbance can be defined in terms of $\sqrt{J_{2D}}$ and the measure of hydrostatic or mean stress, J_1, as

$$D' = \frac{\sqrt{J_{2D}^i} - \sqrt{J_{2D}^a}}{\sqrt{J_{2D}^i} - \sqrt{J_{2D}^c}} \qquad (3.8a)$$

$$D'' = \frac{\sqrt{J_1^i} - \sqrt{J_1^a}}{\sqrt{J_1^i} - \sqrt{J_1^c}} \qquad (3.8b)$$

Then they can be used in the constitutive Eqs. (3.1), which are decomposed in deviatoric and hydrostatic components. The use of such a decomposition for the behavior of interfaces is given in Chapter 11.

In the case of the nondestructive behavior, say, ultrasonic or Lamb wave velocity (V), the disturbance can be expressed as [Fig 3.7(c)]

$$D_v = \frac{V^i - V^a}{V^i - V^c} \qquad (3.9)$$

where D_v is disturbance based on velocity, V^i is the velocity in the RI state, which can be adopted as the velocity in the initial state of the material, V^a is the observed velocity, and V^c is the velocity in the FA state, which can be adopted as the asymptotic value to the observed response (5, 11).

In the case of undrained behavior, when the liquid from a saturated porous material does not have sufficient time to drain (under static or cyclic loading), the disturbance can be obtained in various ways; for strain-controlled tests, see Fig. 3.8, or for stress-controlled tests, see Fig. 3.9 (12).

In the case of the strain-controlled test, the disturbance, D_σ, can be found by using Eq. (3.7), in which peak stresses for different cycles are used as the observed response. Here the continuation of the response under the first cycle, not affected by cyclic degradation, can be treated as the RI response

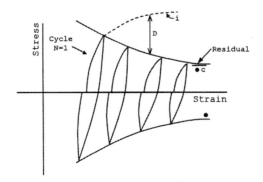

(a) Cyclic stress-strain response

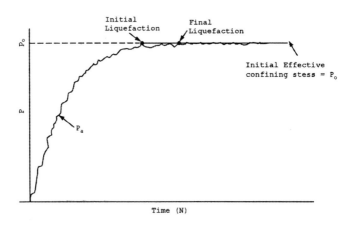

(b) Pore water pressure with cycles (N)

FIGURE 3.8
Cyclic strain-controlled test.

(e.g., using HISS δ_0-plasticity model; Chapter 7). The residual or ultimate stress (σ^c) can be adopted as the stress in the FA state; see Fig. 3.8(a).

For the stress-controlled loading, Fig. 3.9(a), the extension of first-cycle behavior can be treated as the RI response simulated by using a suitable model, e.g., the HISS δ_0-model. Based on the pore water pressure measurements, the observed mean effective stress can be found as

$$(\bar{\sigma})^a = (\bar{\sigma}_1 + \bar{\sigma}_2 + \bar{\sigma}_3)^a/3 = (\sigma_1 + \sigma_2 + \sigma_3 - 3p)^a/3 \quad (3.10)$$

where $\bar{\sigma}_i$ and σ_i ($i = 1, 2, 3$) are the effective and total stresses, respectively, and p is the pore water pressure.

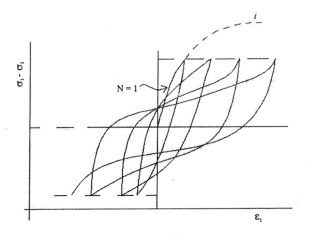

(a) Stress-strain response

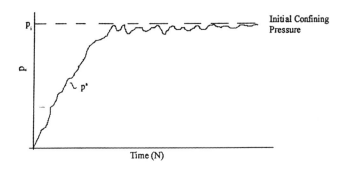

(b) Pore water pressure with cycles (N)

FIGURE 3.9
Cyclic stress-controlled test.

The integration of the incremental plasticity equations for the δ_0-model (Chapter 7)

$$d\underset{\sim}{\sigma}^i = \underset{\sim}{C}^{i(ep)} d\underset{\sim}{\varepsilon}^i \tag{3.11}$$

leads to the computation of the RI stresses and the value of $(\bar{\sigma})^i$ for the RI response. A schematic of $(\bar{\sigma})^i$ and $(\bar{\sigma})^a$ vs. the number of cycles (N) or corresponding deviatoric plastic strain trajectory, ξ_D, from Fig. 3.9(a), is shown in Fig. 3.7(d). Then the disturbance based on the effective stress can be evaluated as (12,13)

$$D_{\bar{\sigma}} = \frac{(\bar{\sigma})^i - (\bar{\sigma})^a}{(\bar{\sigma})^i - (\bar{\sigma})^c} \quad (3.12a)$$

where $(\bar{\sigma})^c$ denotes the measured ultimate (asymptotic) value. Alternatively, the measured excess pore water pressure can be used as [see Fig. 3.9(b)]

$$D_p = \frac{p^a}{p^i} \quad (3.12b)$$

where p^i is the initial effective confining stress $\bar{\sigma}_0$. Further details including instability and liquefaction analysis are given in Chapter 9.

3.2.3 Stiffening or Healing

Some materials, e.g., saturated sands during deformation in the post-liquefaction zone, dislocated silicon crystals with impurities (oxygen, nitrogen, etc.), and asphalt in pavements, may experience a strengthening response during loading and unloading. Some reasons for such "healing" are continuing compaction, chemical rebonding, and locking of dislocations. A schematic of the stress–strain response showing softening and stiffening is shown in Fig. 3.10. Here, disturbance can be expressed using Eq. (3.7). However, it

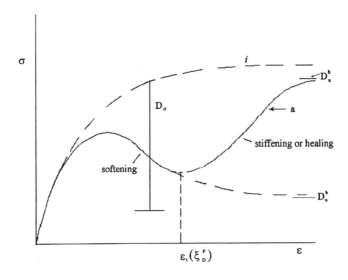

FIGURE 3.10
Softening and stiffening behavior.

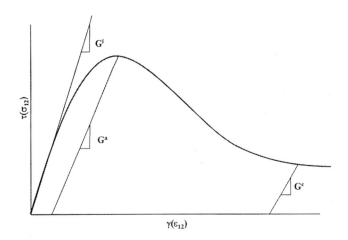

FIGURE 3.11
Disturbance from shear modulus.

may first increase due to microcracking and softening, and then after the threshold plastic deformation (ε_t), it may decrease. Details are given subsequently here and in Chapter 10.

Disturbance from Elastic Moduli. Often, disturbance, D, can be found approximately on the basis of the degrading (secant) elastic (E) or shear (G) moduli. For instance, consider the shear stress (τ or σ_{12}) vs. shear strain (γ or ε_{12}) response in which the changing (degrading) shear modulus, G, is measured with deformation; see Fig. 3.11. The special form of Eq. (3.1), by ignoring the last term, is given by (5)

$$d\sigma_{12}^a = (1 - D)G^i d\varepsilon_{12}^i + DG^c d\varepsilon_{12}^c \qquad (3.13a)$$

where G^i and G^c denote the initial and FA shear moduli, respectively. If it is assumed that $d\varepsilon_{12}^a = d\varepsilon_{12}^i = d\varepsilon_{12}^c$ and that the FA state does not carry any shear stress; that is, $G^c = 0$, Eq. (3.13a) reduces to

$$d\sigma_{12}^a = (1 - D)G^i d\varepsilon_{12}^a \qquad (3.13b)$$

which leads to the expression for the degrading shear modulus, G^a, as

$$G^a = (1 - D)G^i \qquad (3.13c)$$

Similarly, for the case of Young's elastic modulus, E,

$$E^a = (1 - \bar{D})E^i \qquad (3.13d)$$

Then the disturbance can be found as

$$D = \frac{G^i - G^a}{G^i} \tag{3.13e}$$

or

$$\bar{D} = \frac{E^i - E^a}{E^i} \tag{3.13f}$$

Such values of D are considered to be approximate, because they may not include multidimensional effects like in the definitions when they are expressed in terms of total plastic strain trajectory or dissipated work (see below).

3.2.4 Representation of Disturbance

The disturbance function, D, can be expressed in terms of certain internal variables and factors that affect the constitutive behavior. For example, D can be expressed as

$$D = D[\xi, w, S, \phi, t(N), T, \alpha_i] \tag{3.14a}$$

where ξ denotes an internal variable such as the plastic strain trajectory, w is the (dissipated) energy, S is entropy (disorder), ϕ is free energy, t is time or the number of loading cycles (N), T is temperature, and α_i ($i = 1, 2, \ldots$) denotes factors like environmental (chemical) effects and impurities. As a simplification, D can be written as

$$D = D[\xi(t, T, \alpha_i)] \tag{3.14b}$$

or

$$D = D[w(t, T, \alpha_i)] \tag{3.14c}$$

where ξ or w is affected by time, temperature, and α_i. Often, the internal microstructural changes are assumed to be influenced mainly by the shear or deviatoric plastic strain trajectory, ξ_D; hence,

$$D = D[\xi_D, (t, T, \alpha_i)] \tag{3.15}$$

In the functional form, D is often defined by using Weibull (14) functions as

$$D = D_u \left[1 - \left\{ 1 + \left(\frac{\xi_D}{h}\right)^{\bar{w}} \right\}^{-s} \right] \tag{3.16a}$$

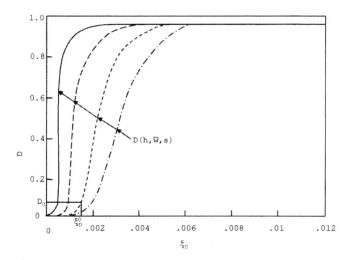

FIGURE 3.12
Schematic of disturbance function (5). (With permission from Elsevier Science.)

where $h, \bar{w}, s$, and D_u are material parameters. This form can provide a general variation of D, to cover materials that may undergo disturbance from the beginning of loading, and also those that involve insignificant disturbance in the beginning; see Fig. 3.12 (5). At the same time, D in Eq. (3.16a) can often be sensitive to small changes in the parameters.

A simpler form of D commonly used is given by

$$D = D_u[1 - \exp(-A\xi_D^Z)] \quad (3.16b)$$

where A, Z, and D_u are material parameters. With only two parameters, h and s (i.e., $w = 0$), Eq. (3.16a) can yield results similar to those from Eq. (3.16b). Note that, as discussed in Chapter 2, the expressions for D, Eqs. (3.16a and b) for the disturbance (damage) are also relevant to describe decay and growth processes in many natural systems.

The parameters, say A, Z, and D_u in Eq. (3.16), can be expressed in terms of such factors as initial confining pressure (σ_0), initial density (ρ_0), size ratio (L/d), where L is the length or height of the test specimen (material element) and d is the mean diameter; temperature (T), and initial dislocation density (N_0) (15–18).

Regarding the size ratio, consider the uniaxial test results in Fig. 3.13 for an artificial rock (15); the initial value of $\sigma_0(=0)$, density ρ_0, and T (room temperature) are assumed to be invariant. It can be seen that the stress–strain response is dependent on L/d. Hence, A, Z, and D_u are dependent on L/d. Thus, the inclusion of L/d in the description of D can allow incorporation of the size effect in the DSC formulation (15, 17).

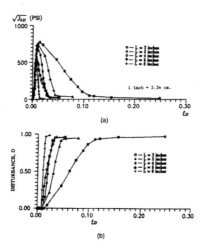

FIGURE 3.13
Stress–strain behavior of soft rock for different L/d ratios, $d = 3.0$ inch:(a) $\sqrt{J_{2D}}$ vs. ξ_D; (b) D vs. ξ_D (15). (©1990, John Wiley & Sons Ltd. Reproduced with permission.)

3.2.5 The Stiffening Effect

The disturbance function, D, Eq. (3.16), can be modified to include the stiffening effect, as (Fig. 3.10) (18,19)

$$D = D_u^b[\{1 - \exp(-A_b \xi_D^{Z_b})\} - \bar{\lambda}\{1 - \exp(-A_h \xi_D^{Z_h})\}] \quad (3.17)$$

where D_u^b, A_b, and Z_b are parameters for the softening behavior to the lower yield or residual stress, and $\bar{\lambda}$, A_h, and Z_h are parameters related to the stiffening response; here, $\bar{\lambda}$ is given by

$$\begin{cases} 0 & \text{if } \xi_D \le \xi_D^p, \\ \dfrac{D_u^h}{D_u^b} & \text{if } \xi_D > \xi_D^p \end{cases} \quad (3.18)$$

where D_u^h is the ultimate disturbance corresponding to the limiting stiffening response, and $\xi_D^p (= \bar{\varepsilon}_t)$ is the threshold value of the deviatoric plastic strain trajectory when stiffening initiates; see Fig. 3.10. Further details are given in Chapter 10.

3.2.6 Creep Behavior

In the case of the viscoplasticity (Chapter 8), the disturbance can be obtained on the basis of the elastoplastic response. In the viscoplastic model, the time-dependent (incremental) viscoplastic response converges to the corresponding

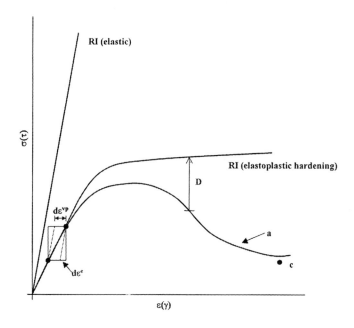

FIGURE 3.14
Viscoplastic response: $d\varepsilon^{vp} = d\varepsilon^{p}$.

plastic strains according to inviscid plasticity (e.g., classical or HISS-δ_0 models; Chapters 6 and 7). Thus, the elastoplastic response can provide the RI response with respect to the observed elastoplastic (softening) behavior; see Fig. 3.14. Then the disturbance, D, can be expressed as

$$D = D_u(1 - e^{A\xi_{vp}^Z}) \qquad (3.19a)$$

where ξ_{vp} is the strajectory of viscoplastic strains, or

$$D = D_{uw}\left(1 - e^{A_w W^{Z_w}}\right) \qquad (3.19b)$$

where w is the dissipated energy

$$w = \int \underline{\sigma}\, d\underline{\varepsilon}^{vp} \qquad (3.20)$$

and D, A, and Z are corresponding parameters. Here, ε^{vp} is the vector of viscoplastic strains. For materials that experience both viscoelastic and viscoplastic deformations, the energy can be expressed as

$$w = \int \underline{\sigma}(d\underline{\varepsilon}^{ve} + d\underline{\varepsilon}^{vp}) \qquad (3.21)$$

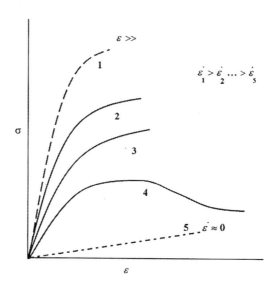

FIGURE 3.15
Responses under different strain rates ($\dot{\varepsilon}$).

where $\underline{\varepsilon}^{ve}$ is the vector of viscoelastic strains. The viscoelastic and viscoplastic strains in Eq. (3.21) and the parameters in Eq. (3.19) can be dependent on temperature and other (environmental) factors. Details of creep or elastoviscoplastic models are given in Chapter 8.

3.2.7 Rate Dependence

In the case of rate-dependent behavior, the disturbance can be defined based on stress–strain responses under different strain ($\dot{\varepsilon}$) or deformation ($\dot{\delta}$) rates. Figure 3.15 shows a schematic of the stress–strain responses at different strain rates at a given temperature. Then the response under the highest strain rate ($\dot{\varepsilon} \gg$) can be assumed to represent one reference (RI) behavior, and that under the lowest strain ($\dot{\varepsilon} \approx 0$) or static condition can be considered as the other reference (FA) state. Then the disturbance can be expressed as

$$D = \frac{\sigma^i - \sigma^a}{\sigma^i - \sigma^c} \qquad (3.22)$$

where σ^i, σ^a, and σ^c are the stresses corresponding to the highest strain rate, the observed, and the lowest strain rate, respectively. In terms of the irreversible (plastic) strains or dissipated energy, D can be expressed as

$$D = D_u(1 - e^{A\xi_D^Z}) \qquad (3.23a)$$

or

$$D = D_{uw}\left(1 - e^{A_w W^{Z_w}}\right) \qquad (3.23b)$$

where ξ_D and w are the temperature-dependent deviatoric plastic or viscoplastic strain trajectory and dissipated energy, respectively, for a given strain rate. Then the parameters D_u, A, and Z can be expressed as functions of the strain rate and temperature. Examples of temperature- and strain-rate-dependent behavior are given in later chapters (e.g., Chapter 8).

3.2.8 Disturbance Based on Disorder (Entropy) and Free Energy

A deforming material can be considered to be composed of *ordered* and *disordered* material parts. In this context, disturbance (D), which connects the two states, can be related to free energy and disorder. Energy can be expressed in terms of Peirel's and Helmholtz free energy, and disorder can be measured in terms of entropy (20–22). Boltzmann (20) proposed the connection between disorder and entropy as

$$\bar{S} = k \ln W \qquad (3.24)$$

where $\bar{S}$ is the entropy, k is Boltzmann's constant, and W is the disorder parameter, which represents the probability that the system would exist in the state relative to all possible states in which it can exist.

Desai et al. (18) and Dishongh and Desai (19) developed a connection between disturbance and rate of dislocation density, $\dot{N}_m$, expressed as (23, 24)

$$\dot{N}_m = \bar{K} N_m k_0 (\sqrt{J_{2D}^a} - \lambda \sqrt{N_m})^{p+r} e^{-Q/KT'} \qquad (3.25a)$$

where k is Boltzmann's constant, $\lambda \sqrt{N_m}$ is the back stress, Q is Peirel's energy, $\bar{K}$, p, and r are material constants, $k_0 = B_0/\tau_0$, B_0 is mobility, T' is the absolute temperature, τ_0 is the (resolved) shear stress, and $\sqrt{J_{2D}^a}$ is the observed shear stress. The expression of D, in Eq. (3.8), by assuming $\sqrt{J_{2D}^c} = 0$, gives

$$\sqrt{J_{2D}^a} = (1 - D)\sqrt{J_{2D}^i} \qquad (3.26)$$

hence, the relation between $\dot{N}_m$ and D is given by (18, 19)

$$\dot{N}_m = \bar{K} N_m k_0 \left[(1 - D)\sqrt{J_{2D}^i} - \lambda \sqrt{N_m}\right]^{p+r} e^{-Q/KT'} \qquad (3.25b)$$

Basaran and Yan (25) proposed an expression for disturbance in terms of disorder and Helmholtz free energy (ϕ) per unit mass given by

$$\phi = e - TS \tag{3.27}$$

where e is the internal energy, T is the absolute temperature, and S is the entropy. The change in disorder, ΔW, at any time is expressed as

$$\Delta W = W_0 - W = e^{\frac{e_0 - \phi_0}{N_0 kT/\bar{m}_s}} - e^{\frac{e - \phi}{N_0 kT/\bar{m}_s}} \tag{3.28}$$

where N_0 is Avogardo's number at reference disorder, W_0, $\bar{m}_s$ is the specific mass (g/mole) (26), and e_0 and ϕ_0 are the internal and Helmholz free energies at the reference state. Then disturbance, D, is expressed as

$$D = \frac{\Delta W}{W_0} = 1 - e^{\left(-\frac{e_0 - \phi_0}{N_0 kT/\bar{m}_s} - \frac{e - \phi}{N_0 kT/\bar{m}_s}\right)} \tag{3.29a}$$

If it is assumed that the temperature change ΔT is small, Eq. (3.29a) can be simplified as

$$D = 1 - e^{\frac{-\Delta e - \Delta \phi}{N_0 kT/\bar{m}_s}} \tag{3.29b}$$

Equation (3.29) is similar to Eq. (3.19), which is expressed in terms of accumulated (deviatoric) plastic strains or dissipated energy. Equation (3.29) can provide an alternative way to express D, while the two in Eq. (3.19) can be relatively simple, particularly from the viewpoint of the definition and determination of parameters from standard laboratory tests on materials.

3.3 Material Parameters

It is necessary to determine material parameters in the RI ($\underline{C}^i$), FA ($\underline{C}^c$), Eq. (3.1), characterizations, and the disturbance function. Appropriate laboratory and/or field tests on the material under various significant factors such as initial conditions, stress paths, volume change response, loading (static, quasistatic, cyclic, repetitive), rate of loading, temperature, and chemicals are required for the determination of the parameters. Detailed descriptions of the test equipment are beyond the scope of this book; however, we shall mention the test devices employed to obtain the observed behavior used for the determination of parameters and validation of the models. Details of the determination of parameters and validations for various models to characterize the RI response are given in Chapters 5 to 12. Here, we provide the procedure for determination of parameters for the FA state and the disturbance function.

3.3.1 Fully Adjusted State

FA as void. If the FA state is treated as a "void" and carries no hydrostatic (mean) and shear stress, the matrix, $\underline{C}^c$ Eq. (3.1), is null. Then it is not necessary to have any parameters for the FA state.

FA as constrained liquid. If the FA state can carry hydrostatic stress and no shear stress and compatibility of strains is assumed, Eq. (3.1) reduces to [see Eq. (4.28)], with the associated assumptions

$$d\sigma_{ij}^a = (1 - D)C_{ijk\ell}^i d\varepsilon_{k\ell}^i + \frac{1}{3}J_1^i \delta_{ij} - dDS_{ij}^i \qquad (3.30)$$

For this specialization, the DSC characterization involves only the RI response, that is, the model used for the RI response can be modified to simulate the FA response. Equation (4.19) in Chapter 4 presents such a specialization of the RI constitutive matrix for the characterization of the FA behavior.

FA as critical state. If the FA behavior is characterized by using the critical state concept, the parameters are those in Eq. (3.2). Their determination is discussed below.

At least three shear tests, triaxial with cylindrical specimens or multiaxial with cubical specimens, with different initial confining pressures ($\sigma_0 = p_0 = J_{10}/3$) are recommended to obtain the plots, such as in Fig. 3.3(a), (b), (d), and (e). Usually, conventional triaxial tests ($\sigma_1 > \sigma_2 = \sigma_3$) are performed under the stress path (CTC); see Fig. 3.16. Sometimes tests with other stress paths (see Fig. 3.16) are available and can be used. The state of stress ($\sigma_1 - \sigma_3$) in the zones where the behavior approaches the critical state provides the *approximate* values of stress at the critical state. Note that the laboratory measurements can provide approximate values because before the final value is reached, the material "fails."

The knowledge of σ_1, $\sigma_2 = \sigma_3$ at the critical state can now be used to evaluate $\sqrt{J_{2D}^c}$ from

$$J_{2D}^c = \frac{1}{6}[(\sigma_1 - \sigma_2)^2 + (\sigma_2 - \sigma_3)^2 + (\sigma_3 - \sigma_1)^2]^c \qquad (3.31a)$$

and J_1^c from

$$J_1^c = (\sigma_1 + \sigma_2 + \sigma_3)^c \qquad (3.31b)$$

Then the slope of the average line on the plot $\sqrt{J_{2D}^c}$ and J_1^c [shown in Fig. 3.3(d)] yields the value of $\overline{m}$.

The values of the void ratio (e^c) (see Fig. 3.3(b)) at the critical stress state (above) are available from the tests under different initial mean pressures.

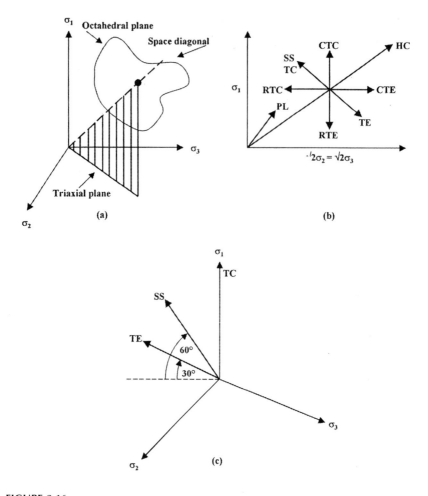

FIGURE 3.16
Representation of stress paths: (a) principal stress space; (b) projections of stress paths on triaxial plane; (c) stress paths in octahedral plane.

The plot of e^c vs. $\ln(J_1^c/3p_a)$ [Fig. 3.3(e)] provides the value of λ as the slope of average line. Here p_a is the atmospheric pressure constant. The value of e_0^c, from Eq. (3.2b), is obtained from the same plot corresponding to $J_1^c/3p_a = 1$.

3.3.2 Disturbance Function

Let us first consider the disturbance function, D, from Eq. (3.16b), which is expressed as follows by taking the logarithm twice:

$$Z\ln(\xi_D) + \ln(A) = \ln\left[-\ln\left(\frac{D_u - D}{D_u}\right)\right] \tag{3.32}$$

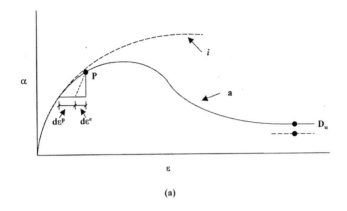

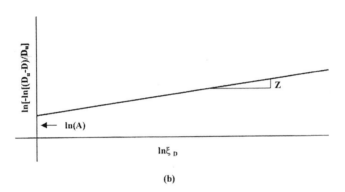

FIGURE 3.17
Determination of disturbance parameters.

The value of ξ_D, say, at any point P (see Fig. 3.17(a)) is obtained from

$$\xi_D = \int (d\underline{E}^p \cdot d\underline{E}^p)^{1/2} \tag{3.33a}$$

where $d\underline{E}^p$ is the vector of the increments of plastic deviatoric strains:

$$d\underline{E}^p = d\underline{\varepsilon}^p - \frac{1}{3}\varepsilon_v^p \underline{I} \tag{3.33b}$$

$$d\underline{\varepsilon}^p = d\underline{\varepsilon} - d\underline{\varepsilon}^e \tag{3.33c}$$

The disturbance, D_σ, at point P is evaluated by using Eq. (3.7), and D_u corresponds to the residual (or critical) stress state, seen in Fig. 3.17. Now, a plot of $\ln[-\ln((D_u - D)/D_u)]$ vs. $\ln \xi_D$ [Fig. 3.17(b)] provides the value of

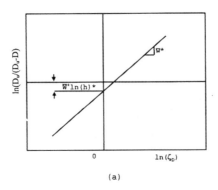

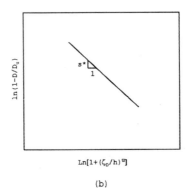

FIGURE 3.18
Parameters for disturbance; Eq (3.15b). (Adapted from Refs. 5 and 28.)

Z as the slope of the average line, and the intercept along the ordinate when $\ln(\xi_D) = 0$ gives the value of A.

The parameters $(h, \bar{w}, s)$ in the general form, as in Eq. (3.16a), can be evaluated by using an iterative procedure in which initial estimates are refined progressively; for example, the Marquard–Levenberg (ML) method, given in Press et al. (27, 28) can be used:

Let us first assume that $s = 1$; then Eq. (3.16a) can be written as

$$\ln\left(\frac{D_u}{D_u - D}\right) = \bar{w}\ln\xi_D - \bar{w}\ln h \tag{3.34}$$

The values of D and ξ_D are evaluated at different points on a given stress–strain curve, and D_u is computed from the residual stress value [Fig. 3.17(a)]. Often, $D_u = 1$ can be adopted. A plot of $\ln(D/(D_u - D))$ vs. $\ln(\xi_D)$ leads to the first estimates of $\bar{w}^*$ and h^* from the slope and the intercept; see Fig. 3.18(a). Now the value of s^* corresponding to the first estimates, $\bar{w}^*$ and h^*, can be found by writing Eq. (3.16a) as

$$\ln\left(1 - \frac{D}{D_u}\right) = -s^* \ln\left[1 + \left(\frac{\xi_D}{h^*}\right)^{\bar{w}^*}\right] \tag{3.35}$$

A plot $\ln(1 - D/D_u)$ vs. $\ln[1 + (\xi_D/h^*)^{\bar{w}^*}]$ gives the first estimate of s^* as the slope of the average line (Fig. 3.18(b)). The above procedure is repeated (in the ML scheme) to obtain progressively refined values of h, w, and s.

References

1. Desai, C.S., "Further on Unified Hierarchical Models Based on Alternative Correction or Disturbance Approach," *Report*, Dept. of Civil Eng. & Eng. Mechs., University of Arizona, Tucson, AZ, USA, 1987.
2. Desai, C.S. and Ma, Y, "Modelling of Joints and Interfaces Using the Disturbed State Concept," *Int. J. Num. and Analyt. Methods in Geomech.*, 16, 9, 1992, 623–653.
3. Desai, C.S., "Constitutive Modelling Using the Disturbed State as Microstructure Self-Adjustment Concept," Chap. 8 in *Continuum Models for Materials with Microstructure*, H.B. Mühlhaus (Editor), John Wiley, Chichester, U.K., 1995.
4. Desai, C.S., "Hierarchical Single Surface and the Disturbed State Constitutive Models with Emphasis on Geotechnical Applications," Chap. 5 in *Geotechnical Engineering*, K.R. Saxena (Editor), Oxford & IBH Pub. Co., New Delhi, India, 1994.
5. Desai, C.S. and Toth, J., "Disturbed State Constitutive Modeling Based on Stress–Strain and Nondestructive Behavior," *Int. J. Solids & Structures*, 33, 11, 1619–1650, 1996.
6. Wathugala, G.W. and Desai, C.S., "Damage Based Constitutive Model for Soils," *Proc. 12th Congress on Appl. Mech.*, Ottawa, Canada, 1989.
7. Armaleh, S.H. and Desai, C.S., "Modelling and Testing of Cohesionless Material Using the Disturbed State Concept," *Int. J. Mech. Behavior of Materials*, 5, 1994, 279–295.
8. Katti, D.R. and Desai, C.S., "Modeling and Testing of Cohesive Soil Using Disturbed State Concept," *J. of Eng. Mech.*, ASCE, 121, 5, 1995, 648–658.
9. Schofield, A.N. and Wroth, C.P., *Critical State Soil Mechanics*, McGraw-Hill, London, UK, 1968.
10. Kachanov, L.M., *Introduction to Continuum Damage Mechanics*, Martinus Nijhoft Publishers, Dordrecht, The Netherlands, 1986.
11. Desai, C.S., "Evaluation of Liquefaction Using Disturbed State and Energy Approaches," *J. of Geotech. and Geoenv. Eng.*, ASCE, 126, 7, 2000, 618–631.
12. Desai, C.S., Park, I.J., and Shao, C., "Fundamental Yet Simplified Model for Liquefaction Instability," *Int. J. Num. Analyt. Meth. Geomech.*, 22, 1998, 721–748.
13. Desai, C.S., Shao, C., and Park, I.J., "Disturbed State Modelling of Cyclic Behavior of Soils and Interfaces in Dynamic Soil-Structure Interfaces," *Proc. 9th Int. Conf. On Computer Methods and Adv. in Geomech.*, Wuhan, China, 1997.

14. Weibull, W.A., "A Statistical Distribution Function of Wide Applicability," *Applied Mechanics,* 18, 1951, 293–297.
15. Desai, C.S., Kundu, T., and Wang, G., "Size Effect on Damage Parameters for Softening in Simulated Rock," *Int. J. Num. Analyt. Methods in Geomech.,* 14, 1990, 509–517.
16. Desai, C.S., Chia, J., Kundu, T., and Prince, J., "Thermomechanical Response of Materials and Interfaces in Electronic Packaging: Parts I and II," *J. of Elect. Packaging, ASME,* 119, 4, 1997, 294–300; 301–309.
17. Desai, C.S., Basaran, C., and Zhang, W., "Numerical Algorithms and Mesh Dependence in the Disturbed State Concept," *Int. J. Num. Methods in Eng.,* 40, 16, 1997, 3059–3083.
18. Desai, C.S., Dishongh, T., and Deneke, P., "Disturbed State Constitutive Model for Thermomechanical Behavior of Dislocated Silicon with Impurities," *J. of Appl. Physics,* 84, 11, 1998.
19. Dishongh, T.J. and Desai, C.S., "Disturbed State Concept for Materials and Interfaces with Application in Electronic Packaging," *Report to NSF,* Dept. of Civil Engng. and Engng. Mechanics, University of Arizona, Tucson, AZ, USA, 1996.
20. Boltzmann, L., *Lectures on Gas Theory,* University of California Press, S. Brush (translator), 1998, 1964.
21. Prigogine, I. and Stengers, I., *Order out of Chaos: Man's New Dialogue with Nature,* Bantam Books, New York, 1984.
22. Halliday, D. and Resnick, R., *Physics,* John Wiley & Sons, New York, 1966.
23. Haasen, P., in *Dislocation Dynamics,* edited by A.R. Rosenfield et al., Battle Institute Materials Science Colloquia, 1967; McGraw-Hill, New York, 1968.
24. Dillon, D.W., Tsai, C.T., and De Angelis, R.J., "Dislocation Dynamics During the Growth of Silicon Ribbon," *J. of Applied Physics,* 60, 5, 1986, 1784–1792.
25. Basaran, C. and Yan, C.Y., "A Thermodynamic Framework for Damage Mechanics of Solder Joints," *J. of Electronic Packaging, ASME,* 120, 4, 1998, 379–384.
26. Dunstan, S., *Principles of Chemistry,* Van Nostrand Reinhold Co., New York, 1968.
27. Press, W.H., Plannery, B.P., Teukolsky, S.A., and Vetterling, W.T., *Numerical Recipes in Pascal,* Cambridge Univ. Press, Cambridge, U.K., 1986.
28. Toth, J. and Desai, C.S., "Development of Lunar Ceramic Composites, Testing and Constitutive Modeling Including Cemented Sand," *Report,* Dept. of Civil Engng. and Engng. Mechanics, University of Arizona, Tucson, AZ, 1994.

4

DSC Equations and Specializations

CONTENTS
4.1 Relative Intact Response 96
    4.1.1 Fully Adjusted Response 97
4.2 Specializations of DSC Equations 98
    4.2.1 Linear Elastic 98
    4.2.2 Elastoplastic 99
    4.2.3 Elastoviscoplastic 99
    4.2.4 Thermal Effects 99
    4.2.5 Disturbance Models 99
    4.2.6 Classical Continuum Damage Model 100
    4.2.7 DSC Model Without Relative Motions 101
    4.2.8 Critical-State Characterization for FA Response 102
    4.2.9 DSC Equations with Critical State 102
4.3 Derivation of Strain Equations 105
    4.3.1 Strain Equations Using Critical State 105
4.4 General Formulation of DSC Equations 108
4.5 Alternative Formulations of DSC 110
4.6 Examples 110

In Chapter 2 we derived the following equations:

$$\sigma_{ij}^a = (1 - D)\sigma_{ij}^i + D\sigma_{ij}^c \tag{4.1}$$

$$\begin{aligned}d\sigma_{ij}^a &= (1 - D)d\sigma_{ij}^i + Dd\sigma_{ij}^c + dD(\sigma_{ij}^c - \sigma_{ij}^i)\\ &= (1 - D)C_{ijk\ell}^i d\varepsilon_{k\ell}^i + DC_{ijk\ell}^c d\varepsilon_{k\ell}^c + dD(\sigma_{ij}^c - \sigma_{ij}^i)\end{aligned} \tag{4.2}$$

As discussed earlier, the DSC involves different stresses (σ_{ij}^i, σ_{ij}^c) and strains (ε_{ij}^i, ε_{ij}^c) in the material parts in the RI and FA states, respectively; see Fig. 4.1. The latter causes relative motions or a diffusion-type process between the material parts in the RI and FA states. As a simplification, we can express a relationship between the strains in the RI and FA parts as [Fig. 4.1(b)]

$$d\varepsilon_{ij}^c = (I_{ijk\ell} + \alpha_{ijk\ell})d\varepsilon_{k\ell}^i \tag{4.3}$$

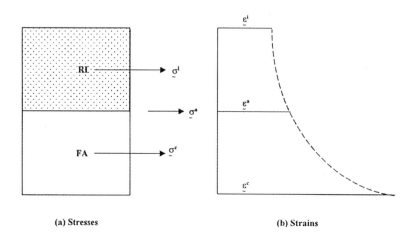

FIGURE 4.1
Stresses and strains in RI, FA, and observed states: the randomly distributed RI and FA parts are shown as collected together.

where $I_{ijk\ell}$ is the unit tensor and $\alpha_{ijk\ell}$ is the relative motion tensor. If the material is assumed isotropic and D is considered to be a weighted (average) value, α can be expressed as a scalar as

$$\alpha = wD \tag{4.4}$$

where w is a parameter. Then Eq. (4.3) becomes

$$d\varepsilon^c_{ij} = (1 + \alpha)d\varepsilon^i_{ij} \tag{4.5}$$

Here $\alpha = wD$, where $0 \leq D \leq 1$ and $w > 0$. With the above definition of the relative motion parameter between the RI and FA strains, the first two terms in Eq. (4.2) can be expressed in terms of the RI strains, $d\varepsilon^i_{ij}$.

Now consider dD in the third term in Eq. (4.2):

$$dD = \frac{dD}{d\xi_D} \cdot d\xi_D \tag{4.6}$$

where D is expressed in terms of ξ_D [see Eq. (4.13a) ahead]. Also

$$\begin{aligned}
d\xi_D &= \frac{d\xi_D}{dE^p_{ij}} \cdot dE^p_{ij} \\
d\xi_D &= (dE^p_{ij} \cdot dE^p_{ij})^{1/2} \\
d\varepsilon^p_{ij} &= \lambda \frac{\partial F}{\partial \sigma_{ij}} \\
d\varepsilon^p_{ii} &= \lambda \frac{\partial F}{\partial \sigma_{kk}} \delta_{ij} \\
dE^p_{ij} &= d\varepsilon^p_{ij} - \frac{1}{3} d\varepsilon^p_{kk} \delta_{ij}
\end{aligned} \tag{4.7}$$

DSC Equations and Specializations

where F is the yield function in the plasticity model, e.g., F in the HISS δ_0-model (Chapter 7). Therefore,

$$dE_{ij}^p = \lambda\left(\frac{\partial F}{\partial \sigma_{ij}} - \frac{1}{3}\frac{\partial F}{\partial \sigma_{kk}}\delta_{ij}\right) \tag{4.8}$$

and

$$d\xi_D = \lambda\left[\left(\frac{\partial F}{\partial \sigma_{ij}} - \frac{1}{3}\frac{\partial F}{\partial \sigma_{kk}}\delta_{ij}\right)\left(\frac{\partial F}{\partial \sigma_{ij}} - \frac{1}{3}\frac{\partial F}{\partial \sigma_{kk}}\delta_{ij}\right)\right]^{1/2} \tag{4.9a}$$

$$= \lambda\left[\frac{\partial F}{\partial \sigma_{ij}}\cdot\frac{\partial F}{\partial \sigma_{ij}} - \frac{2}{3}\frac{\partial F}{\partial \sigma_{ij}}\cdot\frac{\partial F}{\partial \sigma_{kk}}\delta_{ij} + \frac{1}{9}\frac{\partial F}{\partial \sigma_{kk}}\delta_{ij}\frac{\partial F}{\partial \sigma_{kk}}\delta_{ij}\right]^{1/2} \tag{4.9b}$$

Now,

$$-\frac{2}{3}\frac{\partial F}{\partial \sigma_{ij}}\frac{\partial F}{\partial \sigma_{kk}}\delta_{ij} = -\frac{2}{3}\frac{\partial F}{\partial \sigma_{ij}}\frac{\partial F}{\partial \sigma_{kk}}\delta_{ij} \quad \text{for } i = j$$
$$= 0 \quad \text{for } i \neq j \tag{4.10}$$

and

$$\frac{1}{9}\frac{\partial F}{\partial \sigma_{kk}}\delta_{ij}\frac{\partial F}{\partial \sigma_{kk}}\delta_{ij} = \frac{3}{9}\frac{\partial F}{\partial \sigma_{ii}}\frac{\partial F}{\partial \sigma_{kk}}$$

because $\delta_{ij}\cdot\delta_{ij} = \delta_{ii} = 3$. Therefore,

$$d\xi_D = \lambda\left[\frac{\partial F}{\partial \sigma_{ij}}\frac{\partial F}{\partial \sigma_{ij}} - \frac{1}{3}\frac{\partial F}{\partial \sigma_{ii}}\cdot\frac{\partial F}{\partial \sigma_{kk}}\right]^{1/2} \tag{4.11}$$

The parameter λ in Eq. (4.10) is given by (see Chapter 7)

$$\lambda = \frac{\frac{\partial F}{\partial \sigma_{ij}}\cdot C_{ijk\ell}^e d\varepsilon_{k\ell}^i}{\frac{\partial F}{\partial \sigma_{ij}}C_{ijk\ell}^e\frac{\partial F}{\partial \sigma_{k\ell}} - \frac{\partial F}{\partial \xi}\left(\frac{\partial F}{\partial \sigma_{ij}}\cdot\frac{\partial F}{\partial \sigma_{ij}}\right)^{1/2}} \tag{4.12}$$

Now, recall that D can be expressed as [Eq. (3.19a), Chapter 3]

$$D = D_u(1 - e^{-A\xi_D^Z}) \tag{4.13a}$$

Therefore,

$$\frac{dD}{d\xi_D} = D_u A Z \xi_D^{Z-1} e^{-A\xi_D^Z} \tag{4.13b}$$

Substitution of Eqs. (4.11) and (4.13b) in Eq. (4.6) leads to

$$dD = \frac{(D_u A Z \xi_D^{Z-1} e^{-A\xi_D^Z}) \frac{\partial F}{\partial \sigma_{mn}} \cdot C^e_{mnpq} \left(\frac{\partial F}{\partial \sigma_{ij}} \cdot \frac{\partial F}{\partial \sigma_{ij}} - \frac{1}{3} \frac{\partial F}{\partial \sigma_{ii}} \cdot \frac{\partial F}{\partial \sigma_{ii}} \right)^{1/2}}{\frac{\partial F}{\partial \sigma_{mn}} C^e_{mnpq} \frac{\partial F}{\partial \sigma_{pq}} - \frac{\partial F}{\partial \xi} \left(\frac{\partial F}{\partial \sigma_{mn}} \cdot \frac{\partial F}{\partial \sigma_{mn}} \right)^{1/2}} \cdot d\varepsilon^i_{pq} \quad (4.14a)$$

or

$$dD = R_{pq} d\varepsilon^i_{pq} \quad (4.14b)$$

Now, the DSC equations (4.1) become

$$d\sigma^a_{ij} = [(1 - D)C^{i(ep)}_{ijk\ell} + D(1 + \alpha)C^c_{ijk\ell}] d\varepsilon^i_{k\ell} + R_{k\ell} d\varepsilon^i_{k\ell}(\sigma^c_{ij} - \sigma^i_{ij}) \quad (4.15a)$$

or

$$d\sigma^a_{ij} = [(1 - D)C^{i(ep)}_{ijk\ell} + D(1 + \alpha)C^c_{ijk\ell} + R_{k\ell}(\sigma^c_{ij} - \sigma^i_{ij})] d\varepsilon^i_{k\ell} \quad (4.15b)$$

or

$$d\sigma^a_{ij} = C^{DSC}_{ijk\ell} d\varepsilon^i_{k\ell} \quad (4.15c)$$

Here, $C^{DSC}_{ijk\ell}$ is the constitutive tensor given by

$$C^{DSC}_{ijk\ell} = [(1 - D)C^{i(ep)}_{ijk\ell} + D(1 + \alpha)C^c_{ijk\ell} + (\sigma^c_{ij} - \sigma^i_{ij})R_{k\ell}] \quad (4.16a)$$

or in matrix notation:

$$\underset{\sim}{C}^{DSC} = [(1 - D)\underset{\sim}{C}^{i(ep)} + D(1 + \alpha)\underset{\sim}{C}^c + (\underset{\sim}{\sigma}^c - \underset{\sim}{\sigma}^i)\underset{\sim}{R}^T] \quad (4.16b)$$

4.1 Relative Intact Response

The RI response given by $C^i_{ijk\ell}$ in Eq. (4.2) can be defined by using elastic, elastoplastic, elastoviscoplastic, etc., characterizations. For example, consider the δ_0-version of the HISS plasticity model (Chapter 7) in which the yield function, F, is given by

$$F = \overline{J}_{2D} - (\gamma \overline{J}_1^2 - \alpha \overline{J}_1^n)(1 - \beta S_r)^{-0.5} = 0 \quad (4.17)$$

DSC Equations and Specializations

where γ, n, and β are material parameters, α is the hardening or growth function, $\bar{J}_{2D} = J_{2D}/p_a^2$, $\bar{J}_1 = (J_1 + 3R)/p_a$, R is the bonding stress, $S_r = (\sqrt{27}/2) \cdot J_{3D} \cdot J_{2D}^{-3/2}$. Then the constitutive equation for the RI part with the plasticity model is given by

$$d\sigma_{ij}^i = C_{ijk\ell}^{i(ep)} \cdot d\varepsilon_{k\ell}^i \tag{4.18a}$$

or

$$d\underset{\sim}{\sigma}^i = \underset{\sim}{C}^{i(ep)} \cdot d\underset{\sim}{\varepsilon}^i \tag{4.18b}$$

where

$$C_{ijk\ell}^{i(ep)} = C_{ijk\ell}^e - \frac{C_{ijmn}^e \cdot \dfrac{\partial F}{\partial \sigma_{mn}} \cdot \dfrac{\partial F}{\partial \sigma_{uv}} \cdot C_{uvk\ell}^e}{\dfrac{\partial F}{\partial \sigma_{ij}} C_{ijk\ell}^e \dfrac{\partial F}{\partial \sigma_{k\ell}} - \dfrac{\partial F}{\partial \xi}\left(\dfrac{\partial F}{\partial \sigma_{k\ell}} \cdot \dfrac{\partial F}{\partial \sigma_{k\ell}}\right)^{1/2}} \tag{4.18c}$$

or in matrix notation:

$$\underset{\sim}{C}^{i(ep)} = \underset{\sim}{C}^e - \frac{\underset{\sim}{C}^e \left(\dfrac{\partial F}{\partial \underset{\sim}{\sigma}}\right)\left(\dfrac{\partial F}{\partial \underset{\sim}{\sigma}}\right)^T \cdot \underset{\sim}{C}^e}{\dfrac{\partial F}{\partial \underset{\sim}{\sigma}} \cdot \underset{\sim}{C}^e \cdot \dfrac{\partial F}{\partial \underset{\sim}{\sigma}} - \dfrac{\partial F}{\partial \xi}\left[\left(\dfrac{\partial F}{\partial \underset{\sim}{\sigma}}\right)^T \dfrac{\partial F}{\partial \underset{\sim}{\sigma}}\right]^{1/2}} \tag{4.18d}$$

4.1.1 Fully Adjusted Response

If it is assumed that the FA parts carry only the hydrostatic stress and no shear stress (other characterizations have been discussed in Chapter 3) and that the FA can be characterized by using the elastoplastic RI response, the constitutive matrix $\bar{\underset{\sim}{C}}^c$ can be expressed as

$$\bar{\underset{\sim}{C}}^c = \bar{\underset{\sim}{C}}^{i(ep)} = \begin{bmatrix} \bar{C}_1 & \bar{C}_2 & \bar{C}_3 & 0 & 0 & 0 \\ \bar{C}_1 & \bar{C}_2 & \bar{C}_3 & 0 & 0 & 0 \\ \bar{C}_1 & \bar{C}_2 & \bar{C}_3 & 0 & 0 & 0 \\ 0 & 0 & 0 & 0 & 0 & 0 \\ 0 & 0 & 0 & 0 & 0 & 0 \\ 0 & 0 & 0 & 0 & 0 & 0 \end{bmatrix} \tag{4.19a}$$

where

$$\bar{C}_1 = \frac{C_{11} + C_{21} + C_{31}}{3}$$

$$\bar{C}_2 = \frac{C_{12} + C_{22} + C_{32}}{3}$$

$$\bar{C}_3 = \frac{C_{13} + C_{23} + C_{33}}{3}$$

and C_{ij} are the components of the elastoplastic matrix, $\underset{\sim}{C}^{i(ep)}$. If the FA response is assumed to be linear elastic, $\bar{C}_1 = \bar{C}_2 = \bar{C}_3 = K$, which is the bulk modulus.

Hence, the DSC equations are given in matrix notation as

$$d\underset{\sim}{\sigma}^a = [(1 - D)\underset{\sim}{C}^{i(ep)} + D(1 + \alpha)\underset{\sim}{\bar{C}}^{i(ep)} + (\underset{\sim}{\sigma}^c - \underset{\sim}{\sigma}^i)^T \underset{\sim}{R}]d\underset{\sim}{\varepsilon}^i \quad (4.19b)$$

$$= \underset{\sim}{C}^{DSC} d\underset{\sim}{\varepsilon}^i \quad (4.19c)$$

Note that in Eq. (4.19b), the observed or average stress increment is expressed in terms of the strain increment in the RI part. As discussed subsequently in the Section entitled Implementation of the DSC Model in Chapter 13, one of the computational strategies is to implement Eq. (4.19b) and solve the finite-element equations iteratively for a given (or computed) RI strain increment. In other words, under displacement (strain) controlled loading, iterations are performed by holding the applied strains constant. For force (stress) controlled loading, iterations are performed by holding the strains from the computed strain increments. Alternative computational strategies are given in Chapter 13.

4.2 Specializations of DSC Equations

The general form of the DSC, given in Eq. (4.1), can provide a hierarchical basis for specialized models. A number of such specializations are given here.

4.2.1 Linear Elastic

If the material does not involve any disturbance ($D = 0$) and there are no plastic or irreversible deformations, Eq. (4.2) would yield the simple incremental nonlinear elastic (continuum) model:

$$d\sigma^i_{ij} = C^{i(e)}_{ijk\ell} d\varepsilon^i_{k\ell} \quad (4.20)$$

DSC Equations and Specializations

where i denotes the RI state, which in this case involves a nonlinear elastic characterization, and $C_{ijk\ell}^{i(e)}$ is the traditional tangent elastic constitutive tensor corresponding to generalized Hooke's law (see Chapter 5). If the material is linearly elastic, $C_{ijk\ell}^{i(e)}$ will involve constant moduli.

4.2.2 Elastoplastic

If there is no disturbance but the material experiences plastic or irreversible deformations, Eq. (4.2) will specialize to

$$d\sigma_{ij}^i = C_{ijk\ell}^{i(ep)} \cdot d\varepsilon_{k\ell}^i \tag{4.21}$$

where $C_{ijk\ell}^{i(ep)}$ is the elastoplastic constitutive tensor, Eq. (4.18a). It will depend on the yield criteria (F) and flow rule used; details are given in Chapters 6 and 7.

4.2.3 Elastoviscoplastic

With $D = 0$, Eq. (4.2) can specialize to a viscoelastic or elastoviscoplastic characterization for which time integration is required. Details of the development of the DSC with this characterization for the RI response are given in Chapter 8.

4.2.4 Thermal Effects

The effect of temperature on the material response can be included, in a simplified manner, by expressing various elastic, plastic, viscous, and disturbance parameters as functions of temperature. A simple expression that can express the temperature dependence of most of these parameters is given by (1)

$$p = p_r \left(\frac{T}{T_r}\right)^c \tag{4.22}$$

where p is the parameter, p_r is value of the parameter at the reference temperature, T_r (say, 300K), and c is a parameter. The values of p_r and c are found from laboratory stress–strain-strength tests at different temperatures. Details including the effect due to the coefficient of thermal expansion are given in subsequent chapters.

4.2.5 Disturbance Models

As discussed previously, the RI response can be characterized as elastic, elastoplastic, viscoplastic, etc., with temperature dependence. If the disturbance caused in the material's microstructure is included, Eq. (4.2) will represent the general model. Various simplified assumptions lead to specializations of the general model, some of which are discussed below.

4.2.6 Classical Continuum Damage Model

In the continuum damage model (2), it is assumed that the damaged parts can carry no stress at all, and they act as *voids*. In other words, the observed response derives essentially from the undamaged parts, whose stress–strain-strength behavior is degraded because of the existence of the damaged parts. For instance, the damage parameter, ω, is defined as

$$\omega = \frac{V^d}{V} \tag{4.23}$$

where V^d is the volume of the damaged part and V is the total volume of the material element. Then, ω represents the special case of D ($\equiv \omega$), and Eq. (4.2) specializes to

$$d\sigma_{ij}^a = (1 - D)C_{ijk\ell}^i d\varepsilon_{k\ell}^i - dD\sigma_{ij}^i \tag{4.24}$$

as the terms related to the damaged part vanish.

During loading and deformation, microcracks can develop and coalesce and lead to macrocracks. Zones of such distributed cracks in a material may not lose all the strength or act like voids. In other words, they can possess finite but modified (reduced) levels of strength. Such zones deform and interact with the undamaged zones. As a consequence, the observed response of the material is influenced by the coupling or interaction between the undamaged and damaged parts. The classical damage model, Eq. (4.24), does not include this interaction. The lack of the interaction, which can play an important role in the material's response, may not be realistic. Furthermore, it renders the continuum damage model to be a *local* model, which does not allow for *nonlocal* effects due to the influence of the response of neighboring points (regions) on the stress and strain at a point in the material. As a result, the classical damage model does not include the *internal characteristic dimension*, which can lead to pathological or spurious mesh dependence when it is implemented in a computational (finite-element) procedure (3, 4). Further details are given in Chapter 12.

Damage Model with Microcrack Interaction. In order to allow for the nonlocal effects and interaction between the damaged and undamaged parts, various investigators (e.g., 3, 5, 6) have modified the continuum damage model. We shall review a few of these works in Chapter 12. Here we comment that the DSC model can allow implicitly for the factors such as microcrack interaction and characteristic dimension (4); see also Chapter 12.

DSC Equations and Specializations

4.2.7 DSC Model Without Relative Motions

As stated earlier, the general DSC model can include different stresses and strains in the RI and FA parts. Now, the incremental strain equations can be written as

$$d\varepsilon_{ij}^a = dE_{ij}^a + \frac{1}{3}d\varepsilon_{ii}^a \delta_{ij}$$

$$= \bar{f}_1(dE_{ij}^i, dE_{ij}^c) + \frac{1}{3}\bar{f}_2(d\varepsilon_{ii}^i, d\varepsilon_{ii}^c)\delta_{ij} \quad (4.25)$$

where $\bar{f}_1$ and $\bar{f}_2$ show the functional dependence of total strains on the strains in the RI and FA parts. If it is assumed that the observed, RI, and FA strains are equal, that is, there is no relative motion between the RI and FA parts, we have

$$d\varepsilon_{ij}^a = d\varepsilon_{ij}^i = d\varepsilon_{ij}^c \quad (4.26)$$

Then the behavior of the material can be characterized on the basis of only the stress equations, Eq. (4.2). The assumption of compatible strains, Eq. (4.26), can introduce an error; however, the stresses in the two parts can still be different [Eq. (4.2)].

FA Part—Constrained Liquid (Chapter 3). If it is assumed that the FA part can carry no shear stress but can carry hydrostatic stress Eq. (4.2) is simplified further. This assumption implies that the FA parts are constrained by the surrounding RI parts and act like a *constrained liquid*. Then Eq. (4.2) specializes to

$$d\sigma_{ij}^a = (1-D)d\sigma_{ij}^i + D\frac{dJ_1^c}{3}\delta_{ij} + dD\left(\frac{1}{3}J_1^c\delta_{ij} - S_{ij}^i - \frac{1}{3}J_1^i\delta_{ij}\right) \quad (4.27)$$

An additional assumption that the hydrostatic stresses in the RI and FA parts are equal (i.e., $J_1^i = J_1^c$) may sometimes be appropriate. Then Eq. (4.27) reduces to (7, 8)

$$d\sigma_{ij}^a = (1-D)d\sigma_{ij}^i + D\left(\frac{dJ_1^i}{3}\right)\delta_{ij} - dDS_{ij}^i \quad (4.28)$$

In the following, we shall discuss the critical-state characterization of the FA behavior. However, appropriate laboratory tests are not often available for all materials (other than geologic and concrete) to characterize and find parameters for the critical-state behavior. Hence, for practical applications, Eq. (4.28) may provide satisfactory characterization for some materials like metals, alloys, and ceramics. In this case, it is relatively easy to characterize the FA response corresponding to the second term in Eq. (4.28), which requires the (tangent) bulk modulus (K) for the material in the FA state, if it were characterized as elastic, and relevant plasticity parameters if it were characterized as elastoplastic.

4.2.8 Critical-State Characterization for FA Response

As discussed in Chapter 3, an initially loose or dense (granular) material with a given initial mean pressure $p_0 = J_{10}/3$ approaches the critical state at which the material continues to experience shear strains under the constant shear stress reached up to that state. Thus, the material at the critical state continues to carry the specific shear stress for the given initial pressure and can be considered to act like a *constrained liquid–solid*. This state can provide a realistic simulation for the behavior of material at the FA state. It may be noted, however, that the use of the critical-state to represent the FA state in the DSC involves different considerations than those in the classical critical-state concept in soil mechanics (9). In the latter, the main attention has been toward the characterization of the final critical (collapse or failure) behavior of soils (clays). In the DSC, on the other hand, the FA material considered to be at the critical state can exist and grow, at distributed locations, from the very beginning of the loading. Only in the limiting case does the *entire* material element approach the critical state.

The behavior of material at the critical state is characterized based on the following two equations (see also Chapter 3):

$$\sqrt{J_{2D}^c} = \overline{m} J_1^c \tag{4.29a}$$

$$e^c = e_0^c - \lambda \ell n(J_1^c/3p_a) \tag{4.29b}$$

where $\sqrt{J_{2D}^c}$ and J_1^c are the invariants at the critical state (here, $J_1/3$ is nondimensionalized with respect to p_a), $\overline{m}$ is the slope of the J_1^c vs. $\sqrt{J_{2D}^c}$ line [Fig 3.3(d), Chapter 3], e_0^c is the initial value of the critical void ratio, e^c, corresponding to $J_1^c = 3p_a$, p_a is the atmospheric pressure constant, and λ is the slope of the e vs. ℓn ($J_1^c/3p_a$) line (Fig. 3.3(e)). $C_{ijk\ell}^c$ in Eq. (4.2) can now be obtained by using Eq. (4.29), as described below.

4.2.9 DSC Equations with Critical State

Consider the stress and incremental stress relations in Eq. (4.1). The observed stress is now expressed as the sum of the shear and isotropic components as

$$S_{ij}^a + \frac{1}{3}J_1^a \delta_{ij} = (1-D)\left(S_{ij}^i + \frac{1}{3}J_1^i \delta_{ij}\right) + D\left(S_{ij}^c + \frac{1}{3}J_1^c \delta_{ij}\right) \tag{4.30a}$$

If we assume that $J_1^a = J_1^i = J_1^c = J_1$, Eq. (4.30a) reduces to

$$S_{ij}^a = (1-D)S_{ij}^i + DS_{ij}^c \qquad (4.30b)$$

Now, as a simplification, assume the following relation between the shear stresses:

$$S_{ij}^a = F_1(D)S_{ij}^i = F_2(D)S_{ij}^c \qquad (4.31a)$$

which with the relation $1/2 S_{ij}S_{ij} = J_{2D}$ leads to

$$\sqrt{J_{2D}^i} = \frac{F_2}{F_1} \cdot \sqrt{J_{2D}^c} \qquad (4.31b)$$

Now, from Eq. (4.29a), we derive

$$S_{ij}^c = \frac{\overline{m} J_1^c}{\sqrt{J_{2D}^i}} S_{ij}^i = \frac{\overline{m} J_1^i}{\sqrt{J_{2D}^i}} S_{ij}^i = \overline{m}\eta S_{ij}^i \qquad (4.32)$$

where $\eta = J_1^i / \sqrt{J_{2D}^i}$.

Therefore,

$$\sigma_{ij}^c = (\overline{m}\eta)S_{ij}^i + \frac{1}{3}J_1^i \delta_{ij} \qquad (4.33)$$

Now

$$\sigma_{ij}^a = S_{ij}^a + \frac{1}{3}J_1^a \delta_{ij} \qquad (4.34a)$$

Hence, from Eqs. (4.30b) and (4.32), we have

$$\sigma_{ij}^a = (1-D)S_{ij}^i + DS_{ij}^c + \frac{1}{3}J_1^i \delta_{ij} \qquad (4.34b)$$

$$= (1-D)S_{ij}^i + D\overline{m}\eta S_{ij}^i + \frac{1}{3}J_1^i \delta_{ij} \qquad (4.34c)$$

Differentiation of Eq. (4.34c) leads to

$$d\sigma_{ij}^a = (1-D)dS_{ij}^i - dDS_{ij}^i + D\overline{m}\eta dS_{ij}^i$$
$$+ D\overline{m}S_{ij}^i d\eta + \overline{m}\eta S_{ij}^i dD + \frac{1}{3}dJ_1^i \delta_{ij} \qquad (4.35a)$$

or

$$d\sigma_{ij}^a = [D\bar{m}\eta + (1-D)]dS_{ij}^i + [D\bar{m}\eta + (1-D)]\frac{dJ_1^i}{3}\delta_{ij}$$
$$- [D\bar{m}\eta + (1-D)]\frac{dJ_1^i}{3}\delta_{ij} + D\bar{m}S_{ij}^i d\eta$$
$$+ dD(\bar{m}\eta S_{ij}^i - S_{ij}^i) + \frac{1}{3}dJ_1^i\delta_{ij} \quad (4.35b)$$

Finally,

$$d\sigma_{ij}^a = [D\bar{m}\eta + (1-D)]d\sigma_{ij}^i + (D - D\bar{m}\eta)\frac{dJ_1^i}{3}\delta_{ij}$$
$$+ [dD(\bar{m}\eta - 1) + D\bar{m}d\eta]S_{ij}^i \quad (4.35c)$$

in which

$$d\sigma_{ij}^i = C_{ijk\ell}^i d\varepsilon_{k\ell}^i \quad (4.36a)$$

and

$$\frac{dJ_1^i}{3} = K_t d\varepsilon_v^i \quad (4.36b)$$

where K_t is the (tangent) bulk modulus. Equation (4.35c) expresses the observed stress increment in terms of the RI stresses, and stress and strain increments, the critical-state parameter ($\bar{m}$), and the stress ratio, η. It involves the assumption that the mean pressures in the observed, RI, and FA parts are equal, which can be appropriate for many situations. Note that if $\bar{m} = 0$, that is, the FA part cannot carry any shear stress, Eq. (4.35c) will reduce to Eq. (4.28).

Furthermore, as will be discussed later (Chapter 13), Eq. (4.35) leads to the computation of observed stress increments, $d\sigma_{ij}^a$, and the RI strain increments, $d\varepsilon_{ij}^i$. Then we need to evaluate the observed strain increment, $d\varepsilon_{ij}^a$ corresponding to $d\sigma_{ij}^a$. Often it is possible to make the additional assumption that the total (or deviatoric or hydrostatic) strains are equal in the three parts, that is, $d\varepsilon_{ij}^a = d\varepsilon_{ij}^i = d\varepsilon_{ij}^c$. This assumption can be considered to be similar to compatibility of strains (no diffusion) in the components of a porous multiphase material (10), implying that there are no relative motions between the RI and FA parts. With this assumption of compatible strains, the DSC formulation would require use of only the stress equations (4.35), and its implementation in the computer procedures would be simplified.

DSC Equations and Specializations

If the strains and mean pressures are different in the three parts, iterative procedures are needed (Chapter 13). Then it is necessary to develop strain equations, as discussed below.

4.3 Derivation of Strain Equations

In general, the three strains ε_{ij}^a, ε_{ij}^i, and ε_{ij}^c, and their increments can be different, leading to relative motion between the RI and FA parts (Fig. 4.1). The observed incremental strains, $d\varepsilon_{ij}^a$, can be expressed as

$$d\varepsilon_{ij}^a = dE_{ij}^a + \frac{1}{3}d\varepsilon_v^a \delta_{ij} \quad (4.37)$$

As indicated before, various assumptions are possible, the simplest being that $d\varepsilon_{ij}^a = d\varepsilon_{ij}^i = d\varepsilon_{ij}^c$, which would need consideration of only Eq. (4.35). In general, however, if the three strains are different, we can express them as

$$d\varepsilon_{ij}^i = f_1(D) d\varepsilon_{ij}^a \quad (4.38a)$$

$$d\varepsilon_{ij}^c = f_2(D) d\varepsilon_{ij}^i \quad (4.38b)$$

where f_1 and f_2 are functions of D. This situation would require an iterative procedure with the assumption that at the start of the incremental loading, $f_1(D) = f_2(D) = 1.0$.

If it can be assumed that the deviatoric strains are equal, i.e., $dE_{ij}^a = dE_{ij}^i = dE_{ij}^c$, then the volumetric strains will be different in the three parts, which can be expressed by using the critical-state concept, as described below.

4.3.1 Strain Equations Using Critical State*

Let V be the total volume of the material element for solid skeleton. Then,

$$V = V^i + V^c \quad (4.39)$$

where V^i is the volume of solids in the RI state and V^c is the volume of solids in the FA state. As the material is porous, it consists of a solid particle skeleton, and voids or pores. The pores can be fully or partially saturated with a fluid and gas (air). For the present, we consider dry material; the saturated

---

* Taken from Refs. (8, 11, 12).

condition will be treated later in Chapter 9. As a result, we deal with effective stresses.

We decompose the volume of solids, V_s, and the volume of voids, V_v, as follows:

$$V_s = V_s^i + V_s^c \tag{4.40}$$

$$V_v = V_v^i + V_v^c \tag{4.40b}$$

where the superscripts i and c denote volumes in the RI and FA states, respectively. In the case of porous materials, the void ratio (e) is defined to denote the ratio of the volume of voids (V_v) to that of solids (V_s) (13). Hence,

$$e^a = \frac{V_v}{V_s} = \frac{V_v^i + V_v^c}{V_s} = \frac{V_v^i \cdot V_s^i}{V_s \cdot V_s^i} + \frac{V_v^c \cdot V_s^c}{V_s \cdot V_s^c} \tag{4.41a}$$

or

$$e^a = \frac{V_v^i}{V_s^i}\left(\frac{V_s - V_s^c}{V_s}\right) + \frac{V_v^c \cdot V_s^c}{V_s^c \cdot V_s} \tag{4.41b}$$

Now let the disturbance based on the volume of solids at critical state (V_s^c) to the total volume of solids (V_s) be given by

$$D_e = \frac{V_s^c}{V_s} \tag{4.42a}$$

hence,

$$V_s^i/V_s = 1 - D_e \tag{4.42b}$$

Therefore,

$$e^a = (1 - D_e)e^i + D_e e^c \tag{4.42c}$$

where $e^i = V_v^i / V_s^i$ is the void ratio of the RI part and $e^c = V_v^c/V_s^c$ is the void ratio of the FA parts.

Differentiation of e^a in Eq. (4.42c) gives (here, the subscript e on D is dropped for convenience)

$$de^a = (1 - D)de^i + Dde^c + dD(e^c - e^i) \tag{4.43}$$

DSC Equations and Specializations

The relation between incremental volume ($d\varepsilon_v$) and the void ratio (de) is given by (9, 13, 14)

$$d\varepsilon_v = -\frac{de}{1+e_0} \qquad (4.44)$$

where e_0 is the initial void ratio. Then Eq. (4.43) becomes

$$d\varepsilon_v^a = (1-D)d\varepsilon_v^i + Dd\varepsilon_v^c + dD\left(\frac{e^i - e^c}{1+e_0}\right) \qquad (4.45)$$

The observed strain, $d\varepsilon_{ij}^a$, can be written as

$$\begin{aligned} d\varepsilon_{ij}^a &= dE_{ij}^a + \frac{1}{3}d\varepsilon_v^a \delta_{ij} \\ &= dE_{ij}^a + \frac{1}{3}\delta_{ij}\left[(1-D)d\varepsilon_v^i + Dd\varepsilon_v^c + dD\left(\frac{e^i - e^c}{1+e_0}\right)\right] \end{aligned} \qquad (4.46)$$

Now, from Eq. (4.29b), the void ratio e^c can be expressed as

$$e^c = e_0^c - \lambda \ln(J_1^c/3p_a) \qquad (4.29b)$$

Therefore,

$$de^c = -\lambda \frac{dJ_1^c}{J_1^c} \qquad (4.47a)$$

and

$$d\varepsilon_v^a = -\frac{\lambda}{1+e_0}\frac{dJ_1^c}{J_1^c} \qquad (4.47b)$$

Then Eq. (4.46) leads to

$$\begin{aligned} d\varepsilon_{ij}^a &= dE_{ij}^a + \frac{1}{3}\delta_{ij}\left[(1-D)d\varepsilon_v^i + D\left(-\frac{\lambda}{1+e_0}\right)\frac{dJ_1^c}{J_1^c}\right] \\ &\quad + dD\left[\frac{e^i - e_0^c + \lambda \ln(J_1^c/3p_a)}{1+e_0}\right] \end{aligned} \qquad (4.48)$$

Hence, with the FA state simulated by using the critical-state equations, the solution should lead to (approximate) simultaneous satisfaction of the stress, Eq. (4.35), and strain, Eq. (4.48), during the iterative procedures described in the following and in Chapter 13.

The foregoing simulation for the FA material by using the critical-state equations, Eq. (4.29), is used for geologic (soils and rocks) materials and concrete, because volumetric (void ratio) measurements are often available together with the stress–strain response. For materials such as metals and alloys, volumetric measurements are usually not available. Hence, the use of the critical-state equations may not be feasible. In such cases, the alternative simulation for the FA state as a constrained liquid that can carry mean stress but no shear stress ($\bar{m} = 0.0$) can be used [Eq. (4.28)].

4.4 General Formulation of DSC Equations

The assumptions of compatible strains and mean pressures in the three states lead to simplified equations only in terms of observed stress increment. However, in general, the strains and mean pressure can be different, and it becomes necessary to obtain convergent solutions based on both the stress and strain equations, Eqs. (4.2) and (4.48). In the following, we discuss such a general procedure.

The functions f_1 and f_2 in Eq. (4.38) may be assumed to be scalars β_2 and β_1, which can vary during the incremental iterative procedure (15):

$$d\varepsilon^c_{ij} = \beta_1 d\varepsilon^i_{ij} \quad \text{or} \quad d\underline{\varepsilon}^c = \beta_1 d\underline{\varepsilon}^i \tag{4.49a}$$

$$d\varepsilon^i_{ij} = \beta_2 d\varepsilon^a_{ij} \quad \text{or} \quad d\underline{\varepsilon}^i = \beta_2 d\underline{\varepsilon}^a \tag{4.49b}$$

Now Eq. (4.2) can be expressed in matrix notation as

$$d\underline{\sigma}^a = (1 - D)\underline{C}^i d\underline{\varepsilon}^i + D\underline{C}^c d\underline{\varepsilon}^c + \underline{R}^T(\underline{\sigma}^c - \underline{\sigma}^i)d\underline{\varepsilon}^i \tag{4.50a}$$

where $dD = \underline{R}^T d\underline{\varepsilon}^i$ [Eq. (4.14b)]. Substitution of Eq. (4.49) into Eq. (4.50a) leads to

$$d\underline{\sigma}^a = [(1 - D)\beta_2 \underline{C}^i + \beta_1\beta_2 D\underline{C}^c + \beta_2 \underline{R}^T(\underline{\sigma}^c - \underline{\sigma}^i)]d\underline{\varepsilon}^a$$

$$= \underline{L}_n d\underline{\varepsilon}^a_n \tag{4.50b}$$

where

$$\underline{L}_n = (1 - D)\beta_2 \underline{C}^i + \beta_1\beta_2 D\underline{C}^c + \beta_2 \underline{R}^T(\underline{\sigma}^c - \underline{\sigma}^i)$$

DSC Equations and Specializations

The values of β_1 and β_2 are usually not known, except at the beginning of the loading, when they can be assumed to be unity, implying that at the start, the observed, RI, and FA strains are equal. The values of β_1 and β_2 can be calculated (during iterations) as follows, by assuming that the mean pressure in the RI and FA (critical) parts are the same, i.e.,

$$dJ_1^c = dJ_1^i \qquad (4.51a)$$

Now, from Eq. (4.47b),

$$dJ_1^c = \frac{J_1^c}{\lambda}(1 + e_0)d\varepsilon_v^c \qquad (4.51b)$$

and

$$dJ_1^i = C_{iik\ell}^{i(ep)} d\varepsilon_{k\ell}^i \qquad (4.51c)$$

where $C_{iik\ell}^{i(ep)}$ is relevant to the volumetric part of $C_{ijk\ell}^{i(ep)}$.
Therefore,

$$-\frac{J_1^c}{\lambda}(1 + e_0)d\varepsilon_v^c = \frac{-J_1^c}{\lambda}(1 + e_0)\beta_1 d\varepsilon_v^i = C_{iik\ell}^{i(ep)} d\varepsilon_{k\ell}^i \qquad (4.52a)$$

and hence,

$$\beta_1 = -\frac{C_{iik\ell}^{i(ep)} d\varepsilon_{k\ell}^i}{\frac{J_1^c}{\lambda}(1 + e_0)d\varepsilon_v^i} \qquad (4.52b)$$

Now, the observed strain increment can be written as

$$d\varepsilon^a = (1 - \theta)d\varepsilon^i + \theta d\varepsilon^c = (1 - \theta)\beta_2 d\varepsilon^a + \theta\beta_1\beta_2 d\varepsilon^a \qquad (4.53a)$$

and

$$\|d\varepsilon^a\| = (1 - \theta)\beta_2 \|d\varepsilon^a\| + \theta\beta_1\beta_2 \|d\varepsilon^a\| \qquad (4.53b)$$

where $\|\ \|$ denotes the norm, e.g., the sum of the strain increments. Therefore,

$$\beta_2 = \frac{1}{(1 - \theta) + \theta\beta_1} \qquad (4.54)$$

Here θ, the interpolation factor ($0 \leq \theta \leq 1$), can be assumed to be the disturbance, D ($0 \leq D \leq 1$), which varies during the incremental-iterative analysis.

4.5 Alternative Formulations of DSC

Consider a material element composed of two materials, 1 and 2 [Eq. (2.8), Chapter 2]. Elements of each material can involve RI and FA states during its deformation. The incremental equation for materials 1 and 2 can be expressed as

$$d\sigma^a_{1ij} = (1 - D_1)C^{1i}_{ijk\ell}d\varepsilon^{1i}_{k\ell} + D_1 C^{1c}_{ijk\ell}d\varepsilon^{1c}_{k\ell} + dD_1(\sigma^{1c}_{ij} - \sigma^{1i}_{ij}) \quad (4.55a)$$

and

$$d\sigma^a_{2ij} = (1 - D_2)C^{2i}_{ijk\ell}d\varepsilon^{2i}_{k\ell} + DC^{2c}_{ijk\ell}d\varepsilon^{2c}_{k\ell} + dD_2(\sigma^{2c}_{ij} - \sigma^{2c}_{ij}) \quad (4.55b)$$

Then the observed incremental response of the composite element can be written as

$$d\sigma^a_{ij} = (1 - \bar{D})d\sigma^a_{1ij} + \bar{D}d\sigma^a_{2ij} + d\bar{D}(\sigma^{2a}_{ij} - \sigma^{1a}_{ij}) \quad (4.56)$$

For a material element composed of solids and fluid, the incremental form of Eq. (2.14b) is derived as

$$d\sigma^a_{ij} = (1 - D)dp^f \delta_{ij} + Dd\sigma^s_{ij} + dD(\sigma^s_{ij} - p^f \delta_{ij}) \quad (4.57a)$$

For bonded materials, e.g., Eq. (2.30a), the incremental equations are given by

$$d\sigma^a_{ij} = (1 - \bar{D})C^i_{ijk\ell}d\varepsilon^i_{k\ell} + \bar{D}C^b_{ijk\ell}d\varepsilon^b_{k\ell} + d\bar{D}(\sigma^b_{ij} - \sigma^i_{ij}) \quad (4.57b)$$

4.6 Examples

Example 4.1

Consider the uniaxial stress–strain behavior, shown in Fig. 4.2. Assume the RI behavior to be (a) linear elastic and (b) nonlinear elastic simulated by a hyperbola. Assume the FA stress state as $\sigma^c = 17$ N/mm^2. Also assume that the disturbance, D [Eq. (4.13a)], is expressed only in terms of the axial total strain (ε) or plastic strain (ε^p).

DSC Equations and Specializations

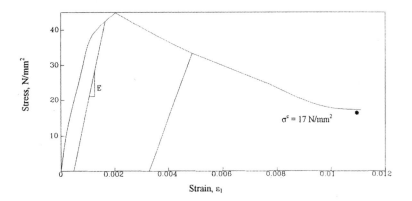

FIGURE 4.2
Uniaxial stress–strain curve.

Evaluate the disturbance D from

$$D = \frac{\sigma^i - \sigma^a}{\sigma^i - \sigma^c} \tag{1a}$$

for the RI response as linear and nonlinear elastic.

Derive the incremental Eq. (4.15) for the one-dimensional behavior by ignoring other strains, in terms of $dD \approx D^m - D^{m-1}$, where m is any point on the stress–strain curve. Also determine dD by using the following equation:

$$D = D_u(1 - e^{-A\varepsilon^z}) \tag{1b}$$

Here, $D_u \approx 1.0$ can be assumed.

Example 4.2

Consider the elastoplastic response shown in Fig. 4.3.

Derive Eq. (4.15) for the one-dimensional case with the assumptions similar to those in Example 4.1. Here, D can be given by

$$D = 1 - e^{-A(\varepsilon^p)^Z} \tag{2a}$$

where ε^p is the irreversible strain. The term dD can be obtained by using Eq. (4.14b). The yield function, F, is given by

$$F = \sigma - \sigma_y = 0 \tag{2b}$$

Hint: Obtain A and Z by plotting $\ln(\varepsilon^p)$ vs. $\ln[-\ln(1 - D)]$.

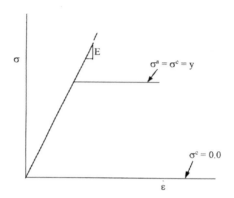

FIGURE 4.3
Elastoplastic response.

Example 4.3

Derive Eq. (4.15) for the one-dimensional case in Example 4.2 by using the von Mises criterion:

$$F = \sqrt{J_{2D}} - k = 0 \tag{3a}$$

Example 4.4

Consider the lateral strains, $\varepsilon_2 = \varepsilon_3$, for the uniaxial stress behavior. Let ξ be given by

$$\xi = \int (d\varepsilon_1^2 + 2d\varepsilon_2^2)^{1/2} \tag{4a}$$

and

$$D = 1 - e^{-A\xi^Z} \tag{4b}$$

Derive Eq. (4.15) by using the von Mises criterion.

References

1. Desai, C.S., Chia, J., Kundu, T., and Prince, J., "Thermomechanical Response of Materials and Interfaces in Electronic Packaging, Part I—Unified Constitutive Model and Calibration; Part II—Unified Constitutive Model, Validation and Design," *J. of Electronic Packaging*, ASME, 119, 4, 1997, 294–300; 301–309.
2. Kachanov, L.M., *Introduction to Continuum Damage Mechanics*, Martinus Nijhoft Publishers, Dordrecht, The Netherlands, 1986.

3. Bazant, Z.P. and Cedolin, L., *Stability of Structures*, Oxford Univ. Press, New York, 1991.
4. Desai, C.S., Basaran, C., and Zhang, W., "Numerical Algorithms and Mesh Dependence in the Disturbed State Concept," *Int. J. Num. Meth. Eng.*, 40, 1997, 3059–3083.
5. Bazant, Z.P., "Nonlocal Damage Theory Based on Micromechanics of Crack Interactions," *J. of Eng. Mech.*, ASCE, 120, 1994, 593–617.
6. Mühlhaus, H.B., *Continuum Models for Materials with Microstructure*, John Wiley & Sons, UK, 1995.
7. Frantziskonis, G. and Desai, C.S., "Constitutive Model with Strain Softening," *Int. J. Solids & Struct.*, 23, 6, 1987, 751–767.
8. Desai, C.S., "Constitutive Modelling Using the Disturbed State as Microstructure Self-Adjustment Approach," Chap. 8 in *Continuum Models for Materials with Microstructure*, John Wiley & Sons, Chichester, UK, 1995.
9. Schofield, A.N. and Wroth, C.P., *Critical State Soil Mechanics*, McGraw-Hill, London, 1968.
10. Bowen, R.M., "Theory of Mixtures" in *Continuum Physics*, Eringen, A.C. (Editor), Academic Press, New York, 1976.
11. Armaleh, S.H. and Desai, C.S., "Modelling and Testing of a Cohesionless Soil Using the Disturbed State Concept," *Int. J. Mech. Behavior of Materials*, 5, 3, 1994.
12. Katti, D.R. and Desai, C.S., "Modelling and Testing of Cohesive Soils Using the Disturbed State Concept," *J. of Eng. Mech., ASCE,* 121, 5, 1995, 648–658.
13. Terzaghi, K., Peck, R.B., and Mesri, G., *Soil Mechanics in Engineering Practice*, John Wiley & Sons, New York, 1996.
14. Desai, C.S. and Siriwardane, H.J., *Constitutive Laws for Engineering Materials*, Prentice-Hall, Englewood Cliffs, NJ, 1984.
15. Shao, C. and Desai, C.S., "Implementation of DSC Model for Dynamic Analysis of Soil-Structure Interaction Problems," *Report to NSF*, Dept. of Civil Engng. and Engng. Mechanics., University of Arizona, Tucson, AZ, USA, 1998.

5
Theory of Elasticity in DSC

CONTENTS
5.1 Linear Elasticity..115
    5.1.1 Nonlinear or Piecewise Linear Behavior......................117
5.2 Variable Parameter Models..118
    5.2.1 Functional Forms...118
    5.2.2 Hyperelastic Models..120
    5.2.3 First-Order Model..121
    5.2.4 Second-Order Cauchy Elastic Model............................124
5.3 Relatively Intact Behavior..124
5.4 Fully Adjusted Behavior...125
5.5 Disturbance Function..126
5.6 Material Parameters..127
    5.6.1 Thermal Effects: Thermoelasticity..................................127
5.7 Examples...129
    5.7.1 Correlation with Crack Density.....................................140

As derived in Chapter 2, the DSC equations are given by

$$d\sigma_{ij}^a = (1 - D)C_{ijk\ell}^i d\varepsilon_{k\ell}^i + DC_{ijk\ell}^c d\varepsilon_{k\ell}^c + dD(\sigma_{ij}^c - \sigma_{ij}^i) \quad (5.1)$$

In this chapter, we present the theory of elasticity as the model for the behavior of the material in the RI state, represented by the constitutive tensor, $C_{ijk\ell}^i$.

5.1 Linear Elasticity

If the behavior of the RI material is characterized as linear elastic with the assumption of isotropy, the constitutive equations will be given by (1–3)

$$\sigma_{ij}^i = C_{ijk\ell}^{i(e)} \varepsilon_{k\ell}^i \quad (5.2a)$$

or

$$\underset{\sim}{\sigma}^i = \underset{\sim}{C}^{i(e)} \underset{\sim}{\varepsilon}^i \tag{5.2b}$$

where σ_{ij}^i is the stress tensor for the RI material

$$\sigma_{ij}^i = \begin{pmatrix} \sigma_{11} & \sigma_{12} & \sigma_{13} \\ \sigma_{12} & \sigma_{22} & \sigma_{23} \\ \sigma_{13} & \sigma_{23} & \sigma_{33} \end{pmatrix}^i \tag{5.3}$$

and ε_{ij}^i is the strain tensor

$$\varepsilon_{ij}^i = \begin{pmatrix} \varepsilon_{11} & \varepsilon_{12} & \varepsilon_{13} \\ \varepsilon_{12} & \varepsilon_{22} & \varepsilon_{23} \\ \varepsilon_{13} & \varepsilon_{23} & \varepsilon_{33} \end{pmatrix}^i \tag{5.4}$$

where $\sigma_{11}, \sigma_{22}, \sigma_{33}$ and $\varepsilon_{11}, \varepsilon_{22}, \varepsilon_{33}$ are the normal, and $\sigma_{12}, \sigma_{23}, \sigma_{13}$ and $\varepsilon_{12}, \varepsilon_{23}, \varepsilon_{13}$ the shear components of stress and strain, respectively. $\underset{\sim}{\sigma}^i$ and $\underset{\sim}{\varepsilon}^i$ are the vectors of stress and strain components, given by

$$\underset{\sim}{\sigma}^i = \begin{Bmatrix} \sigma_{11} \\ \sigma_{22} \\ \sigma_{33} \\ \sigma_{12} \\ \sigma_{23} \\ \sigma_{13} \end{Bmatrix}; \quad \underset{\sim}{\varepsilon}^i = \begin{Bmatrix} \varepsilon_{11} \\ \varepsilon_{22} \\ \varepsilon_{33} \\ \varepsilon_{12} \\ \varepsilon_{23} \\ \varepsilon_{13} \end{Bmatrix} \tag{5.5}$$

The fourth-order constitutive tensor, $C_{ijk\ell}^{i(e)}$, for linear elastic and isotropic materials can be expressed in matrix notation using the generalized Hooke's law as (4)

$$\begin{Bmatrix} \sigma_{11} \\ \sigma_{22} \\ \sigma_{33} \\ \sigma_{12} \\ \sigma_{23} \\ \sigma_{13} \end{Bmatrix}^i = \frac{E}{(1+v)(1-2v)} \begin{bmatrix} 1-v & v & v & 0 & 0 & 0 \\ & 1-v & v & 0 & 0 & 0 \\ & & 1-v & 0 & 0 & 0 \\ & & & 1-2v & 0 & 0 \\ & \text{symmetrical} & & & 1-2v & 0 \\ & & & & & 1-2v \end{bmatrix} \begin{Bmatrix} \varepsilon_{11} \\ \varepsilon_{22} \\ \varepsilon_{33} \\ \varepsilon_{12} \\ \varepsilon_{23} \\ \varepsilon_{13} \end{Bmatrix}^i$$

$$\tag{5.6}$$

Theory of Elasticity in DSC

Here, $2\varepsilon_{12} = \gamma_{xy}$, $2\varepsilon_{23} = \gamma_{yz}$, and $2\varepsilon_{13} = \gamma_{xz}$, where γ_{xy}, γ_{yz}, and γ_{xz} are engineering shear strains, E is the elastic modulus, and v is the Poisson's ratio.

In terms of the shear modulus, G, and the bulk modulus, K, Eq. (5.6) becomes

$$\begin{Bmatrix} \sigma_{11} \\ \sigma_{22} \\ \sigma_{33} \\ \sigma_{12} \\ \sigma_{23} \\ \sigma_{13} \end{Bmatrix}^i = \begin{bmatrix} K+\frac{4G}{3} & K-\frac{2G}{3} & K-\frac{2G}{3} & 0 & 0 & 0 \\ & K+\frac{4G}{3} & K-\frac{2G}{3} & 0 & 0 & 0 \\ & & K+\frac{4G}{3} & 0 & 0 & 0 \\ & & & 2G & 0 & 0 \\ & & & & 2G & 0 \\ & & & & & 2G \end{bmatrix} \begin{Bmatrix} \varepsilon_{11} \\ \varepsilon_{22} \\ \varepsilon_{33} \\ \varepsilon_{12} \\ \varepsilon_{23} \\ \varepsilon_{13} \end{Bmatrix}^i \quad (5.7)$$

The relations between the elastic constants are given by

$$G = \frac{E}{2(1+v)} \quad (5.8a)$$

$$K = \frac{E}{3(1-2v)} \quad (5.8b)$$

5.1.1 Nonlinear or Piecewise Linear Behavior

Note that the material parameters E and v, K and G are evaluated as tangent quantities when Hooke's law, Eq. (5.2), is written in the incremental form

$$d\underline{\sigma}^i = \underline{C}_t^{i(e)} d\underline{\varepsilon}^i \quad (5.9)$$

where the subscript t denotes tangent quantity. In this case, the nonlinear response (see Fig. 5.1) is approximated as piecewise linear by dividing the nonlinear behavior in a number of linear increments. The material parameters are often assumed to be constant within each increment. The monotonic response of some nonlinear materials is often simulated by using the piecewise linear approximation in which the parameters or moduli (E_t, v_t, G_t, K_t) vary from increment to increment. Such models are often referred to as quasi-linear or variable parameter or variable moduli (4).

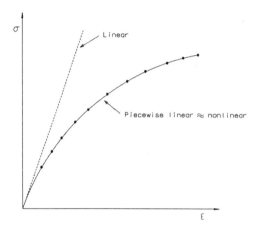

FIGURE 5.1
Piecewise linear approximation to nonlinear behavior.

5.2 Variable Parameter Models

Here, the nonlinear behavior expressed as a stress–strain response can be simulated using mathematical functions. Then, the tangent moduli are evaluated as derivatives of the function at (selected) points in an increment. Some of the commonly used functions are stated below.

5.2.1 Functional Forms

Observed nonlinear stress–strain data can be simulated by using mathematical functions such as polynomials, hyperbola, spline, and exponential (4). For example, schematics of such stress–strain curves for different initial mean pressure (σ_0) and density (ρ_0) are shown in Fig. 5.2; here the stress–strain responses are expressed in terms of stress measure, $\bar{\sigma}$, which can be axial stress (σ_1), stress difference ($\sigma_1 - \sigma_3$), octahedral shear stress, τ_{oct}, or second invariant ($\sqrt{J_{2D}}$) of the deviatoric stress tensor, S_{ij}, given by

$$J_{2D} = \frac{1}{6}[(\sigma_1 - \sigma_2)^2 + (\sigma_2 - \sigma_3)^2 + (\sigma_1 - \sigma_3)^2] \qquad (5.10a)$$

where $S_{ij} = \sigma_{ij} - (1/3)\sigma_{ii}\delta_{ij}$, $\sigma_{ii} = J_1$ = first invariant of σ_{ij}, δ_{ij} = Kronecker delta, σ_i ($i = 1, 2, 3$) = principal stresses, and

$$\tau_{oct}^2 = \frac{2}{3}J_{2D} \qquad (5.10b)$$

Theory of Elasticity in DSC

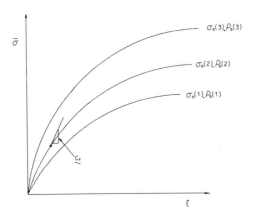

FIGURE 5.2
Functional representation of nonliner stress–strain response.

The strain measure, $\bar{\varepsilon}$ in Fig. 5.2, can be the axial strain, ε_1, strain difference $\varepsilon_1 - \varepsilon_3$, octahedral shear strain, γ_{oct}, or the second invariant (I_{2D}) of the deviatoric strain tensor, E_{ij}:

$$I_{2D} = \frac{1}{6}[(\varepsilon_1 - \varepsilon_2)^2 + (\varepsilon_2 - \varepsilon_3)^2 + (\varepsilon_1 - \varepsilon_3)^2] \quad (5.10c)$$

$$E_{ij} = \varepsilon_{ij} - \frac{1}{3}\varepsilon_{ii}\delta_{ij} \quad (5.11a)$$

$$\varepsilon_{ii} = \varepsilon_v = I_1 \quad (5.11b)$$

where I_1 is the first invariant of the total strain tensor, ε_{ij}, ε_i ($i = 1, 2, 3$) are the principal strains, and

$$\gamma_{oct}^2 = \frac{2}{3}I_{2D} \quad (5.11c)$$

As is indicated in Fig. 5.2, the stress–strain behavior depends on factors such as (initial) mean pressure, p_0 or σ_0, and density (ρ_0). Hence, for a given value of σ_0, the functional form of the stress–strain relation can be expressed as

$$\bar{\sigma} = \bar{\sigma}(\bar{\varepsilon}) \quad (5.12a)$$

and for the behavior dependent on σ_0

$$\bar{\sigma} = \bar{\sigma}(\bar{\varepsilon}, \sigma_0) \quad (5.12b)$$

or

$$\bar{\sigma} = \bar{\sigma}(\bar{\varepsilon}, p_0 \quad \text{or} \quad J_{10}/3)$$

Then, the tangent moduli such as E_t (Young's elastic), v_t (Poisson's ratio), G_t (shear modulus), and K_t (bulk modulus) can be obtained as appropriate derivatives of the function in Eq. (5.12). For example:

$$E_t = \left. \frac{d(\sigma_1 - \sigma_3)}{d\varepsilon_1} \right|_{\sigma_0 = \text{constant}} \quad (5.13a)$$

$$v_t = \left. -\frac{d\varepsilon_3}{d\varepsilon_1} \right|_{\sigma_0 = \text{constant}} \quad (5.13b)$$

$$K_t = \frac{d(J_1/3)}{d\varepsilon_v} \quad (5.13c)$$

$$G_t = \frac{d(\tau_{\text{oct}})}{d(\gamma_{\text{oct}})} \quad (5.13d)$$

The above quasilinear elastic models are based on the idea that the nonlinear response can be simulated incrementally based on the variable values of the tangent material moduli as functions of stress and/or strain. The nonlinear behavior can be simulated more formally by using higher-order or hyperelastic and hypoelastic models; the above quasilinear models can be shown to be special (lowest) cases of the higher-order models (4).

5.2.2 Hyperelastic Models

Cauchy and Green elastic are two of the main models often used to characterize the nonlinear elastic behavior of materials. Details of these models are given in various publications (1, 4–6). Here we give a brief description of the Cauchy elastic models, in which the relation between stress and strain is expressed as

$$\sigma_{ij} = \alpha_0 \delta_{ij} + \alpha_1 \varepsilon_{ij} + \alpha_2 \varepsilon_{im} \varepsilon_{mj} + \alpha_3 \varepsilon_{im} \varepsilon_{mn} \varepsilon_{nj} + \cdots \quad (5.14a)$$

where $\alpha_0, \alpha_1, \ldots, \alpha_n$ represent response functions or parameters. By using the Cayley–Hamilton theorem (7), we can write Eq. (5.14a) as

$$\sigma_{ij} = \phi_0 \delta_{ij} + \phi_1 \varepsilon_{ij} + \phi_2 \varepsilon_{im} \varepsilon_{mj} \quad (5.14b)$$

where ϕ_0, ϕ_1, and ϕ_2 are the response functions expressed in terms of the invariants I_1, I_2, and I_3 of the strain tensor. We can derive various special forms of Eq. (5.14) with different orders, some of which are described below.

5.2.3 First-Order Model

For the first-order model, only the first two terms in Eq. (5.14b) are relevant, and $\phi_2 = 0$. Then ϕ_1 is a constant (say, $= \beta_2$), and ϕ_0 can be a linear function of the first strain invariant, say ($= \beta_0 \delta_{ij} + \beta_1 I_1 \delta_{ij}$), and Eq. (5.14) becomes

$$\sigma_{ij} = \beta_0 \delta_{ij} + \beta_1 I_1 \delta_{ij} + \beta_2 \varepsilon_{ij} \qquad (5.15a)$$

In this equation, $\beta_0 \delta_{ij}$ is the initial (isotropic) stress when the strain $= 0$. Hence, if we start from a stressfree condition, Eq. (5.15a) reduces to

$$\sigma_{ij} = \beta_1 I_1 \delta_{ij} + \beta_2 \varepsilon_{ij} \qquad (5.15b)$$

where β_1 and β_2 are elastic material constants. It can be shown that the first-order Cauchy elastic model, Eq. (5.15b), is the same as linear elastic Hooke's law for isotropic materials. This can be done as follows:

Let us first consider the state of uniform volumetric deformation. Then the strain tensor, ε_{ij}, is given by

$$\varepsilon_{ij} = \left(\frac{I_1}{3}\right)\delta_{ij} = \begin{pmatrix} I_1/3 & 0 & 0 \\ 0 & I_1/3 & 0 \\ 0 & 0 & I_1/3 \end{pmatrix} \qquad (5.16a)$$

Substitution of Eq. (5.16a) in Eq. (5.15) gives

$$\sigma_{ij} = \left(\beta_1 I_1 + \beta_2 \frac{I_1}{3}\right)\delta_{ij} \qquad (5.16b)$$

For isotropic or hydrostatic stress condition, the mean pressure, p, is given by

$$p = \frac{\sigma_{ij}}{3} = \frac{J_1}{3} = \frac{\sigma_{11} + \sigma_{22} + \sigma_{33}}{3} \qquad (5.17a)$$

Therefore, Eq. (5.16b) becomes

$$\frac{J_1}{3} = \left(\beta_1 + \frac{\beta_2}{3}\right)I_1 = K I_1 \qquad (5.17b)$$

where K is the bulk modulus; see Fig. 5.3(a).

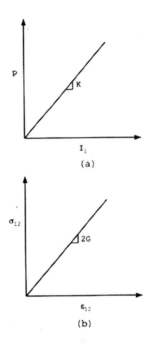

FIGURE 5.3
Bulk and shear moduli.

Now we consider the case of pure shear strain condition, which implies that for an isotropic material, there is no volumetric strain and there exist only nonzero shear strain components, $\varepsilon_{12} = \varepsilon_{21}$. Then Eq. (5.15b) becomes

$$\sigma_{ij} = \beta_2 \varepsilon_{ij} \qquad (5.18a)$$

or

$$\sigma_{ij} = \begin{pmatrix} 0 & \beta_2 \varepsilon_{12} & 0 \\ \beta_2 \varepsilon_{12} & 0 & 0 \\ 0 & 0 & 0 \end{pmatrix} \qquad (5.18b)$$

Here, $\varepsilon_{12} = 1/2\, \gamma_{12}$, where γ_{12} is the engineering shear strain. Then, from Eq. (5.18), we have

$$\sigma_{12} = \beta_2 \left(\frac{\gamma_{12}}{2} \right) \qquad (5.19a)$$

Theory of Elasticity in DSC

Therefore,

$$\sigma_{12} = \frac{\beta_2}{2}\gamma_{12} = G\gamma_{12} \qquad (5.19b)$$

where $G = \beta_2/2$ is the shear modulus; see Fig. 5.3(b). Now, substitution of $\beta_2 = 2G$ in Eq. (5.17b) leads to

$$\frac{J_1}{3} = \left(\beta_1 + \frac{2G}{3}\right)I_1 \qquad (5.20)$$

and

$$K = \beta_1 + \frac{2G}{3} \qquad (5.21a)$$

$$\beta_1 = K - \frac{2G}{3} \qquad (5.21b)$$

With the above values of β_1 and β_2, Eq. (5.15b) is written as

$$\sigma_{ij} = \left(K - \frac{2G}{3}\right)I_1\delta_{ij} + 2G\varepsilon_{ij}$$

$$= KI_1\delta_{ij} + 2G\left(\varepsilon_{ij} - \frac{I_1}{3}\delta_{ij}\right) \qquad (5.22)$$

Note that the strain tensor, ε_{ij}, can be decomposed as

$$\varepsilon_{ij} = E_{ij} + \frac{I_1}{3}\delta_{ij} \qquad (5.23)$$

where E_{ij} is the deviatoric strain tensor. Then, Eq. (5.22) can be expressed as

$$\sigma_{ij} = KI_1\delta_{ij} + 2GE_{ij}$$

$$= \left(\frac{J_1}{3}\delta_{ij} + S_{ij}\right) \qquad (5.24)$$

where S_{ij} is the deviatoric stress tensor and $J_1/3$ is the mean pressure. Note that Eq. (5.24) represents the generalized Hooke's law for linear elastic behavior, from Eq. (5.2).

5.2.4 Second-Order Cauchy Elastic Model

We express ϕ_0, ϕ_1, and ϕ_2 in Eq. (5.14b) in terms of the strain invariants such that the resulting expression provides a second-order expression in strains. Then we can write

$$\phi_0 = \beta_1 I_1 + \beta_2 I_1^2 + \beta_3 I_2 \quad (5.25a)$$

$$\phi_1 = \beta_4 + \beta_5 I_1 \quad (5.25b)$$

$$\phi_2 = \beta_6 \quad (5.25c)$$

where β_i ($i = 1, 2, \ldots, 6$) are material parameters or constants. Substitution of Eq. (5.25) in Eq. (5.14b) results in

$$\sigma_{ij} = (\beta_1 I_1 + \beta_2 I_1^2 + \beta_3 I_2)\delta_{ij} + (\beta_4 + \beta_5 I_1)\varepsilon_{ij} + \beta_6 \varepsilon_{im}\varepsilon_{mj} \quad (5.26)$$

It may be noted that the second-order model, Eq. (5.26), contains the first-order (linear) model as a special case. Hence, the linear part of Eq. (5.26) is

$$\sigma_{ij} = \beta_1 I_1 \delta_{ij} + \beta_4 \varepsilon_{ij} \quad (5.27)$$

This equation is the same as Eq. (5.24) in which $\beta_1 = K - 2G/3$ and $\beta_4 = 2G$. Then Eq. (5.26) is given by

$$\sigma_{ij} = \left[\left(K - \frac{2G}{3}\right)I_1 + \beta_2 I_1^2 + \beta_3 I_2\right]\delta_{ij} + (2G + \beta_5 I_1)\varepsilon_{ij} + \beta_6 \varepsilon_{im}\varepsilon_{mj} \quad (5.28)$$

which includes six material parameters K, G, β_2, β_3, β_5, and β_6. These parameters need to be determined from laboratory tests. Details of other higher-order models including the Green elastic are given in various publications (4–6).

5.3 Relative Intact Behavior

For some materials, it is possible and can be appropriate to simulate the RI response by using the linear elastic models described above, or quasilinear or piecewise linear approximation, Eq. (5.9), or the higher-order or hyperelastic model, e.g., Eq. (5.28), Fig. 5.4. For the linear elastic model, the constitutive

Theory of Elasticity in DSC

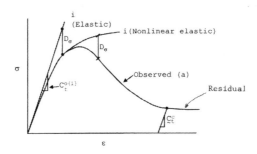

FIGURE 5.4
Elastic model as RI characterization.

tensor $C^i_{ijk\ell}$ for the RI material, Eq. (5.2), can be written as

$$C^{i(e)}_{ijk\ell} = C^e_{ijk\ell}(K, G, \text{ or } E, \nu) \tag{5.29}$$

For the quasilinear and (incremental) hyperelastic model, it can be written as

$$C^{i(e)}_{ijk\ell} = C^e_{ijk\ell}(K_t, G_t; E_t, \nu_t) \tag{5.30}$$

where the subscript t denotes tangent modulus, which is evaluated as the derivative or slope (at a point) based on the function used to simulate the behavior.

If there is no disturbance due to microcracking, damage, softening, or stiffening, D in Eq. (5.1) will be zero, and the observed (denoted by a) and RI responses will be the same. However, if a linear or nonlinear elastic material experiences disturbance, the observed behavior, Fig. 5.4, will deviate from the behavior without the disturbance, i.e., the RI response.

5.4 Fully Adjusted Behavior

If the material experiences microcracks and disturbance (damage), and if the cracked zones are assumed to behave like voids, as in the classical damage models (8), then such FA or damaged zones will have zero strength (Chapter 3). Hence, they can carry no stress, and the corresponding constitutive matrix in Eq. (5.1) will be

$$\underset{\sim}{C}^c = 0 \tag{5.31a}$$

It may be realistic to assign a finite (small) elastic stiffness to the disturbed, cracked, or damaged regions. Then,

$$\underset{\sim}{C}^c = \underset{\sim}{C}^c(K^c_t, G^c_t; E^c_t, \nu^c_t) \tag{5.31b}$$

where the superscript c denotes the FA state, K_t^c, G_t^c, etc. are (tangent) elastic parameters with modified (reduced) values corresponding to the zone in stress–strain response relevant to the FA state; see Fig. 5.4.

If it is assumed that the FA material can carry hydrostatic stress and no shear stress, assuming that the bulk behavior of the FA material is the same as that of the RI material, the constitutive matrix can be expressed as [see Eq. (4.19a), Chapter 4]

$$\bar{\underline{C}}^c = \bar{\underline{C}}^i \tag{5.32a}$$

Now from Eq. (5.7), we can obtain

$$(\sigma_{11} + \sigma_{22} + \sigma_{33})^i = 3K^i(\varepsilon_{11} + \varepsilon_{22} + \varepsilon_{33})^i \tag{5.32b}$$

or

$$K^i = \frac{J_1^i}{3\varepsilon_v^i}$$

Thus, the behavior of the FA material may be defined based on the bulk modulus, K^i, of the RI material. The bulk modulus can also be defined as K^c, determined from the moduli (E^c or G^c) in the residual region of the stress–strain response (Fig. 5.4).

5.5 Disturbance Function

Based on available stress–strain data (Fig. 5.4), the disturbance function, D, can be expressed (approximately) as (Chapter 3)

$$D_\sigma = \frac{\sigma^i - \sigma^a}{\sigma^i - \sigma^c} \tag{5.33}$$

Then, Eq. (5.1) can be used to obtain one-, two-, and three-dimensional idealizations. For example, for the uniaxial case ($\varepsilon_{22} = \varepsilon_{33} = 0.0$) and isotropic material, Eq. (5.1) reduces to

$$d\sigma_{11}^a = (1 - D)E^i d\varepsilon_{11}^i + DE^c d\varepsilon_{11}^c + dD(\sigma_{11}^c - \sigma_{11}^i) \tag{5.34}$$

and for the plane-stress case

$$d\underline{\sigma}^a = (1 - D)\underline{C}^i d\underline{\varepsilon}^i + D\underline{C}^c d\underline{\varepsilon}^c + dD(\underline{\sigma}^c - \underline{\sigma}^i) \tag{5.35}$$

Theory of Elasticity in DSC

where

$$d\underset{\sim}{\sigma}^T = [d\sigma_{11} \ d\sigma_{22} \ d\sigma_{12}]$$

$$\underset{\sim}{\sigma}^{iT} = [\sigma_{11} \ \sigma_{22} \ \sigma_{12}]^i$$

$$\underset{\sim}{\sigma}^{cT} = [\sigma_{11} \ \sigma_{22} \ \sigma_{12}]^c$$

$$\underset{\sim}{C} = \frac{E_t}{1 - v_t^2} \begin{bmatrix} 1 & v_t & 0 \\ & 1 & 0 \\ \text{symm} & & 1 - v_t \end{bmatrix}$$

5.6 Material Parameters

For linear elastic behavior, the material parameters are E, v, G, and K. In general, for nonlinear behavior, these parameters are obtained as (average) slopes of the unloading response. The elastic Young's modulus, E, can be obtained from uniaxial stress (σ_1 vs. ε_1) or triaxial tests (σ_1 or $\sigma_1 - \sigma_3$) vs. ε_1 ($\sigma_3 = $ constant) plots; see Fig. 5.5(a). Poisson's ratio can be obtained from the lateral strain (ε_3) or volumetric strains vs. ε_1 plots, shown in Fig. 5.5(b). The shear modulus, G, is obtained from shear tests as the unloading slope of shear stress (τ or $\sqrt{J_{2D}}$) vs. shear strain (γ or $\sqrt{I_{2D}}$) plots; see Fig. 5.5(c). The bulk modulus is obtained from volumetric or hydrostatic tests as the unloading slope of mean pressure (p or $J_1/3$) vs. volumetric strain (ε_v) plots; see Fig. 5.5(d).

5.6.1 Thermal Effects: Thermoelasticity

The temperature effects on the elastic response can be introduced by expressing the parameters, say, E and v or K and G, as functions of temperature and by modifying the elasticity equations to allow for the effect of the thermal expansion. Then, Eq. (5.2) can be written as

$$d\sigma_{ij}^i = C_{ijk\ell}^{i(e)}(T)[d\varepsilon_{k\ell} - \alpha_T(T)dT\delta_{k\ell}] \tag{5.36}$$

where T is the temperature, α_T is the coefficient of thermal expansion, dT is the change in temperature, δ_{ij} is the Kronecker delta, and $C_{ijk\ell}^{i(e)}(T)$ is the temperature-dependent constitutive tensor.

The material constants are determined from laboratory stress–strain tests at different temperatures and are expressed in terms of the temperature. A number of functional forms are possible and used in the literature (9–11). The following form has often been used and can simulate the temperature-dependent response of many materials (12):

$$p(T) = p_r \left(\frac{T}{T_r}\right)^c \tag{5.37}$$

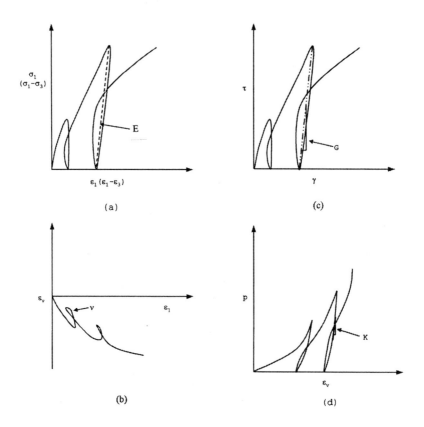

FIGURE 5.5
Elastic constants from laboratory test data.

where T_r is the reference temperature (e.g., 300 K), p_r is the value of the parameter at temperature T_r, and c is a parameter found from laboratory tests. For example, based on available laboratory tests at different temperatures for 60/40 (Sn/Pb) solders (13–15), plots of variations of E, ν, and α_T were obtained; see Fig. 5.6 (16). Then the thermal dependence of these parameters is expressed as

$$E(T) = 23.45 \left(\frac{T}{300}\right)^{-0.292} \tag{5.38a}$$

$$\nu(T) = 0.40 \left(\frac{T}{300}\right)^{0.14} \tag{5.38b}$$

$$\alpha_T(T) = 3 \times 10^{-6} \left(\frac{T}{300}\right)^{0.24} \tag{5.38c}$$

where $E_{300} = 23.45$ GPa, $\nu_{300} = 0.40$, and $\alpha_{T(300)} = 3 \times 10^{-6}/\text{K}$.

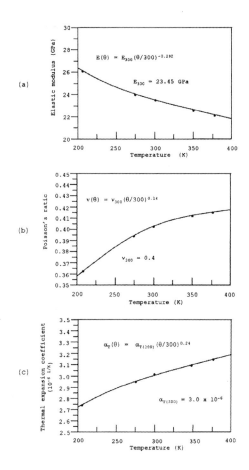

FIGURE 5.6
Plots of relationship between elastic constants and temperature for 60% Sn–40% Pb solders, (a) elastic modulus, (b) Poisson's ratio, (c) thermal expansion coefficient (16).

5.7 Examples

Example 5.1
Consider the one-dimensional specialization of the DSC equations (Eq. 5.1) as

$$d\sigma_1^a = (1 - D)E^i d\varepsilon_1^i + D d\sigma_1^c + dD(\sigma_1^c + \sigma_1^i) \tag{1a}$$

where σ_1 and ε_1 are uniaxial stress and strain, respectively, and E^i is the elastic modulus for the RI response. Here, for simplification, the effects of the other strains are ignored.

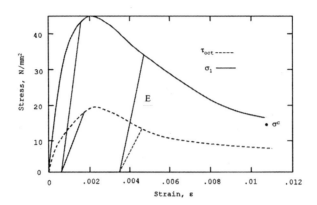

FIGURE 5.7
Uniaxial stress–strain curves.

Figure 5.7 shows octahedral shear (τ_{oct}) or axial stress (σ_1)–strain (ε_1) behavior of a material under uniaxial stress loading. The octahedral shear stress is given by

$$\tau_{oct}^2 = \frac{1}{9}[(\sigma_1 - \sigma_2)^2 + (\sigma_2 - \sigma_3)^2 + (\sigma_1 - \sigma_3)^2] \tag{1b}$$

Therefore,

$$\sigma_1 = \frac{3\tau_{oct}}{\sqrt{2}} \tag{1c}$$

Assume that the RI behavior is simulated as linear elastic with the elastic modulus, E^i, computed as the average of the unloading slopes of σ_1 vs. ε_1 plot; see Fig. 5.7; it is found to be = 43,000 N/mm². The FA response is given by the residual asymptotic value of τ_{oct}^c = 8 N/mm²; hence, σ_1^c = 17 N/mm².

The disturbance, D, can be evaluated from

$$D = \frac{\sigma_1^i - \sigma_1^a}{\sigma_1^i - \sigma_1^c} = \frac{\sigma_1^i - \sigma_1^a}{\sigma_1^i - 17} \tag{1d}$$

where the σ_1^i are different values on the linear elastic and observed curves, respectively, and σ_1^c = 17 N/mm². The irreversible strains, ε_1^p, at different points are computed from

$$\varepsilon_1^p = \varepsilon_1 - \varepsilon_1^e \tag{1e}$$

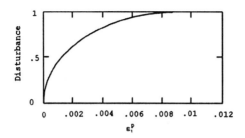

FIGURE 5.8
Disturbance vs. irreversible strains: linear RI simulation.

where

$$\varepsilon_1^e = \frac{\sigma_1}{43{,}000}$$

A plot of D vs. ε_1^p using Eqs. (1d) and (1e) is shown in Fig. 5.8. It can also be appropriate to use total strain, ε_1, with the plot of D vs. ε_1. Here we assumed that D is expressed only in terms of the axial strains; that is, the other two strains (ε_2 and ε_3) are not considered.

Now, the incremental Eq. (1a) can be integrated starting with the zero stress and strain at the origin, by writing it as

$$d\sigma_{1(k)}^a = (1 - D_k) E^i d\varepsilon_{1(k)}^i + D_k d\sigma_{1(k)}^c + dD(\sigma_{1(k-1)}^c - \sigma_{1(k-1)}^i) \quad (1f)$$

where k denotes increment and $d\sigma_1^c = 0$ because σ_1^c is a constant value. Also, we have assumed that the total stresses, σ_1^c and σ_1^i, are evaluated at the previous increment.

Let the first strain increment $d\varepsilon_{1(1)}^i = 0.0005$. The RI stress increment is given by

$$d\sigma_{1(1)}^i = E^i \times 0.0005 = 43{,}000 \times 0.0005$$
$$= 21.5 \text{ N/mm}^2$$

The corresponding observed stress increment from Fig. 5.7 is found to be about 21.0 N/mm². Therefore,

$$D_1 = \frac{21.5 - 21.0}{21.5 - 17.0} = \frac{0.50}{4.5} = 0.111$$

As $D_0 = 0$ at the origin, $dD = 0.111 - 0.000 = 0.111$.
Therefore,

$$d\sigma_{1(1)}^a = (1 - 0.111)21.5 + 0.111(17 - 0)$$
$$= 0.889 \times 21.5 + 5.661$$
$$= 19.11 + 1.887 = 21.0 \text{ N/mm}^2$$

Now, consider the second strain increment, $d\varepsilon^i_{1(2)} = 0.0005$. The total strain $\varepsilon^i_{1(2)} = 0.0005 + 0.0005 = 0.001$.

Therefore,

$$\sigma^i_{1(2)} = 43{,}000 \times 0.001$$
$$= 43.00 \text{ N/mm}^2$$

and

$$\sigma^a_{1(2)} \approx 37.00 \text{ N/mm}^2$$

Hence,

$$D_2 = \frac{43 - 37}{43 - 17} = \frac{6}{26} = 0.23$$

Therefore, $dD_2 = 0.230 - 0.111 = 0.119$, and

$$d\sigma^a_{1(2)} = (1 - 0.230)(43.0 - 21.5) + 0.119(17 - 21)$$
$$= 16.55 - 0.476 = 16.09$$

Therefore,

$$\sigma^a_{1(2)} = 21.0 + 16.08 = 37.08 \text{ N/mm}^2$$

Consider the third increment as $d\varepsilon^i_{1(3)} = 0.002$. Therefore, the total strain = $0.001 + 0.002 = 0.003$.

Therefore,

$$\sigma^i_{1(3)} = 43{,}000 \times 0.003$$
$$= 129.00 \text{ N/mm}^2$$
$$\sigma^a_{1(3)} \approx 39.00 \text{ N/mm}^2$$
$$D_3 = \frac{129 - 39}{129 - 17} = \frac{90}{112} = 0.804$$
$$dD_3 = 0.804 - 0.23 = 0.574$$
$$d\sigma^a_{1(3)} = (1 - 0.804)(129.00 - 37.00) + 0.574(17.00 - 43.00)$$
$$= 0.196 \times 92.00 - 14.92$$
$$= 18.03 - 14.92 = 3.11$$

Theory of Elasticity in DSC

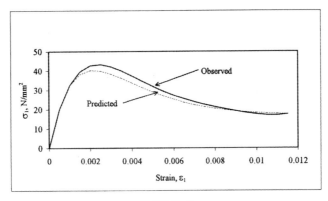

(a) $dD = D_k - D_{k-1}$

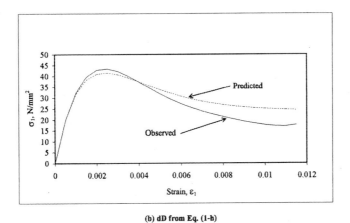

(b) dD from Eq. (1-h)

FIGURE 5.9
Comparison of observed and predicted behavior.

Therefore, $\sigma_{1(3)}^a = 37.08 + 3.11 = 40.19$, which indicates the drop in the observed stress after the peak stress.

The predicted stress–strain curve from the foregoing procedure is shown in Fig. 5.9(a) and compares very well with the observed curve shown in Fig. 5.7.

Note 1. It is possible that in the early range of the stress–strain behavior, D will be negative. In that case, D can be assumed to be zero, as in that range no significant disturbance has occurred, and the RI and observed responses are essentially the same.

Note 2. In the foregoing calculations, we have used the observed stress from the given stress–strain curve so as to calculate the disturbance, D. This is only for illustration purposes. In general, D can be calculated from the graph of D vs. (plastic) strain (Fig. 5.8). Also, D can be expressed as

$$D = D_u(1 - e^{-A\varepsilon^{p(z)}}) \tag{1g}$$

where D_u corresponds to the ultimate or residual value. For instance, D_u can be calculated from the residual value of about 17.0 N/mm² (Fig. 5.7). The irreversible strain trajectory (ξ) is given by the total axial plastic strain, ε_1^p. The parameters A and Z are then found by using the procedure given in Chapter 3. Then, for a given total strain increment, the corresponding plastic strain increment and total plastic strain, ε_1^p, can be found. Substitution of D_u, A, Z, and ε_1^p in Eq. (1g) gives the value of D. The increment or rate dD can also be found by using Eq. (4.13), Chapter 4, as

$$dD_k = AZe^{-A\varepsilon_k^{(p)z}} \cdot \varepsilon_k^{p(Z-1)}(\varepsilon_k^p - \varepsilon_{k-1}^p) \tag{1h}$$

Figure 5.9(b) shows comparison of predicted and observed behavior obtained by using dD_k in Eq. (1h). The value of $D_u = 1.0$ was assumed, and $A = 651.80$ and $Z = 1.03$ were found by using the procedure in Chapter 3.

Note 3. The calculations can be improved by using small strain increments and an iterative procedure. In the iterative procedure, the value of the total stress, σ_1^i, at the end of the previous increment is used for the first iteration during the increment. For subsequent iterations, the value of σ_1^i at the previous iteration can be used. The iterations can be continued until convergence, that is, the two successive iterative values of σ_1^a, show a small difference $= \bar{\varepsilon}$.

Example 5.2
Consider that the RI response for the problem in Example 5.1 is characterized as nonlinear elastic and that the stress–strain curve is simulated by using the hyperbola as

$$\sigma_1^i = \frac{\varepsilon_1^i}{a + b\varepsilon_1^i} \tag{2a}$$

where a and b are parameters related to the initial slope (E_i) and the asymptotic stress, σ_1^f (Fig. 5.10).

In order to evaluate the RI stress, σ_1^i, we need to evaluate a and b, which are given by

$$a = \frac{1}{E_i} = \frac{1}{43{,}000} = 2.33 \times 10^{-5} \tag{2b}$$

$$b = \frac{1}{\sigma_1^f} = \frac{1}{75} = 0.0133 \tag{2c}$$

Theory of Elasticity in DSC   135

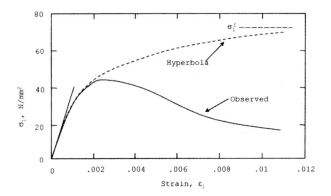

FIGURE 5.10
Stress–strain curve and hyperbolic simulation.

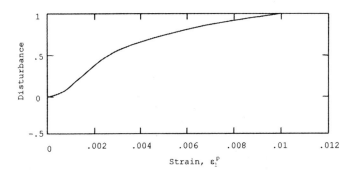

FIGURE 5.11
Disturbance vs. irreversible strains: nonlinear RI simulation.

The value of σ_1^f denotes the asymptotic stress (Fig. 5.10). A plot of the RI stress–strain curve according to Eq. (2a) is shown in Fig. 5.10, together with the observed stress–strain response. The disturbance, D, is now found using Eq. (1d); the plot of D vs. irreversible strain, ε_1^p, is shown in Fig. 5.11.

As in Example 5.1, Eq. (1f) is now integrated to predict stress increments, $d\sigma_1^a$, and total stress, $\sigma_1^a = \sum_{i=1}^{N}(d\sigma_1^a)$, where N is the number of strain increments. A plot of the predicted stress–strain curve is shown in Fig. 5.12, which is in very good agreement with the observed behavior (Fig. 5.10).

Example 5.3

Figure 5.13 shows a linear one-dimensional elastic stress–strain ($\sigma - \varepsilon$) response for a composite element made of two materials, denoted as 1 and 2. The observed response (*a*) of the composite can be expressed by using the DSC (Chapter 2) in terms of the responses of the two materials. The elastic moduli for material 1, observed, and material 2 are 100, 30, and 10, respectively.

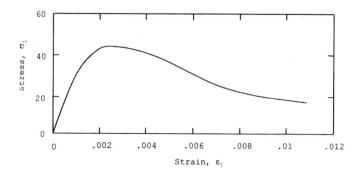

FIGURE 5.12
Predicted stress–strain curves: nonlinear RI simulation.

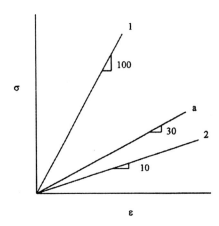

FIGURE 5.13
Composite material: two linear elastic components.

It is seen that the disturbance, D, is constant $= 0.777$; note that for composite materials, disturbance has a different meaning compared to that for the same material element (Chapter 2). For example, at $\varepsilon = 0.10$, $\sigma^1 = 10$, $\sigma^a = 3$, and $\sigma^2 = 1.0$; therefore, $D = (10 - 3)/(10 - 1.0) = 0.777$. Then the observed response, σ^a, can be found from

$$\sigma^a = (1 - D)\sigma^1 + D\sigma^2 \tag{3a}$$

For example, at $\varepsilon = 0.05$:

$$\sigma^a = (1 - 0.777)5 + 0.777 \times 0.5$$
$$= 1.50$$

Example 5.4

The observed nonlinear elastic response (a), Fig. 5.14, is expressed by using hyperbolic relation, Eq. (2a). The responses of the components of the composite material (1 and 2) are also expressed by using the hyperbolic relation. The

Theory of Elasticity in DSC

parameters a and b, from Eq. (2a), are obtained from the initial slopes and asymptotic stresses as given in the following table.

Parameter	Material 1	Observed (a)	Material 2
E	43,000 N/mm²	31,500	20,000
σ_f	75 N/mm²	55	35
a	$1/E^i = 2.33 \times 10^{-5}$	3.18×10^{-5}	5×10^{-5}
b	$1/\sigma_f = 0.0133$	0.0182	0.0286

Computation of stresses for different strains are shown below

DSC Nonlinear Elastic as RI and FA Responses

ε	σ^1	σ^a	σ^2	D
0	0	0	0	0
0.0005	16.69449	12.2399	7.77605	0.499481
0.001	27.3224	20.02002	12.72265	0.500172
0.002	40.08016	29.34703	18.65672	0.500999
0.003	47.46835	34.74233	22.09131	0.501478
0.004	52.28758	38.25921	24.3309	0.50179
0.005	55.67929	40.7332	25.90674	0.502009
0.006	58.19593	42.56829	27.07581	0.502172
0.008	61.6808	45.10854	28.6944	0.502397
0.01	63.97953	46.78363	29.7619	0.502545

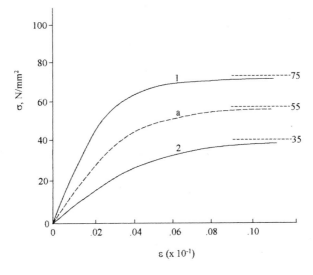

FIGURE 5.14
Composite material: two nonlinear components.

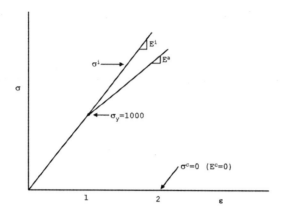

FIGURE 5.15
DSC bilinear elastic response.

The value of D is essentially constant, at 0.50. For example, the observed stress for $\varepsilon = 0.005$ is given by

$$\begin{aligned} \sigma^a &= (1 - 0.5)55.68 + 0.5 \times 25.9 \\ &= (27.84 + 12.95) \\ &= 40.79 \text{ N/mm}^2 \end{aligned}$$

Comment. As discussed in Chapter 2, the formulation of the DSC for an element made of more than one material is referred to as multicomponent DSC, which is also described further in Chapter 8.

Example 5.5

Figure 5.15 shows the observed bilinear response of a material; the yield stress, $\sigma_y = 1000$ units. Assume that the RI response is simulated as linear elastic with $E^i = 1000$ units, and the FA response as $\sigma^c = 0$, i.e., $E^c = 0$. The modulus, E^a, after σ_y is 800 units. Evaluate the total and incremental stress for a typical strain, $\varepsilon = 2$ units. The disturbance, D, is given by

$$D = 1 - \frac{\sigma^a}{\sigma^i} \tag{5a}$$

Consider two points, $\varepsilon = 1$ and $\varepsilon = 2$. Let us evaluate σ^a and $d\sigma^a$ at $\varepsilon = 2$. The values of D at $\varepsilon = 1$ and $\varepsilon = 2$ are

$$D_1 = 1 - \frac{\sigma^i}{\sigma^i} = 0; \quad D_2 = 1 - \frac{1800}{2000} = 0.1; \quad \text{hence, } dD = 0.1.$$

$$\begin{aligned} \sigma_2^a &= (1 - 0.1)2000 + 0.1 \times 0 \\ &= 1800 \end{aligned}$$

Theory of Elasticity in DSC

Alternatively:

$$d\sigma^a = (1 - 0.1)(2000 - 1000) + 0.1 \times 0 + 0.1 \times (0 - 1000)$$
$$= (900 - 100) = 800$$

hence,

$$\sigma_2^a = \sigma_1^a + d\sigma^a = 1000 + 800 = 1800$$

Example 5.6: Ceramic Composite

A ceramic composite was fabricated by using a special thermal liquefaction process (17–19) in which a mixture of finely ground basaltic rock and metal (stainless steel) fibers (about 15% by weight) was heated in a furnace under cycles of heating and cooling around the temperature of about 1100°C. Flat specimens of 3.0 × 6.0 × 0.5 inch (7.6 × 15.2 × 1.3 cm) were tested under tension and compression loading, unloading, and reloading cycles; two tests were performed for each of the plain and fiber-reinforced ceramic. In addition to the stress–strain behavior, ultrasonic P-wave velocity and attenuation measurements were made at different locations on the specimen during the loading cycles (18, 19).

The DSC model was used by assuming the RI behavior to be linear elastic or elastoplastic (δ_0-model); the latter is considered in Chapter 7. The FA material was defined using the assumption that it can carry only the hydrostatic stress. Table 5.1 shows material parameters for the DSC model.

TABLE 5.1

Typical Parameters for Typical Fiber-Reinforced Ceramic Composite Under Uniaxial Tension and Compression Tests (Disturbance parameters below are relevant to linear elastic RI simulation.)

Parameter	Tension	Compression
E	29 GPa	29 GPa
v	0.24	0.24
γ	0.000741	0.1565
β	0.75	0.76
n	6.00	5.88
a_1	1.00×10^{-14}	1.3×10^{-14}
η_1	0.972	1.29
R	167 MPa	25 MPa
h	0.003	0.044
w	4.53	4.53
s	2.03	2.03
D_u	1.0	1.00

The disturbance function was defined by using the following expression [Eq. (3.16a, Chapter 3]:

$$D = D_u \left[1 - \left\{ 1 + \left(\frac{\xi_D}{h} \right)^w \right\}^{-s} \right] \qquad (6a)$$

where h, w, and s are material parameters and ξ_D is the trajectory of deviatoric plastic strains.

Figures 5.16(a) and (b) show typical comparisons between predictions by the DSC model and test data for the tensile and compressive behavior of fiber-reinforced specimens 1 and 2, respectively. Here, parameters from tests on specimen 2 were used to predict the tensile behavior of test specimen 1; and those from test 3 were used to predict the compressive behavior of test 2. Thus, these are considered to be independent validations. It can be seen that the prediction from the linear elastic simulation for the RI behavior provides satisfactory correlations with the test data.

5.7.1 Correlation with Crack Density

Hudson (20) proposed the following equation for crack density in a deforming (elastic) material. It was assumed that the mean shape of randomly distributed cracks is circular and the wavelengths of elastic waves are large compared with the size of cracks and with their separation distances, so that the mean over the statistical ensemble can be used to predict properties of a single sample.

$$C_d = \frac{\frac{15(1 - \nu_0)}{2(1 - 2\nu_0)} \left(1 - \frac{V^{*2}}{V_0^2} \right)}{\left(2U_{11} + \frac{3 - 2\nu_0 + 7\nu_0^2}{(1 - 2\nu_0)^2} \right) \cdot U_{33}} \qquad (6b)$$

$$= \Omega \left(1 - \frac{V^{*2}}{V_0^2} \right) \qquad (6c)$$

where $U_{11} = 16(1 - \nu_0)/3(2 - \nu_0)$, $U_{33} = 8(1 - \nu_0)/3$, ν_0 and V_0 are the Poisson's ratio and P-wave velocity in the RI material, and V^* is the observed root mean square P-wave velocity. Here, $C_d = na^3$, where n is the open cracks per unit volume, and a is the mean dimension (radius) of cracks (20).

Theory of Elasticity in DSC

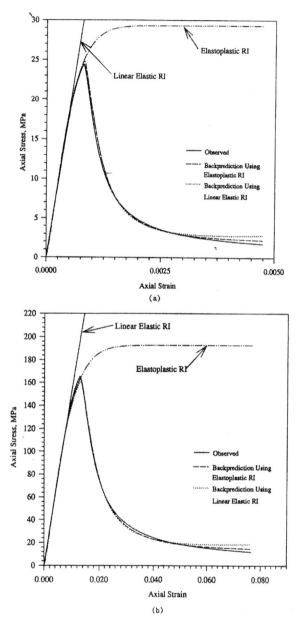

FIGURE 5.16
Comparisons between DSC predictions and test data for ceramic composites (18, 19): (a) tensile response of specimen 1; (b) compressive response of specimen 2. (With permission from Elsevier Science.)

Figure 5.17 shows variations of C_d and disturbance, D_v, [computed by using the ultrasonic P-wave velocities, [Eq. (3.9), Chapter 3] with $\sqrt{J_{2D}}$ for the tensile test on the fiber-reinforced ceramic.] Both C_d and D_v increase during the virgin (monotonic) loading. During unloading, they decrease because as the

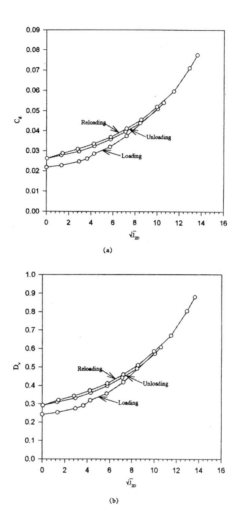

FIGURE 5.17
Crack density and disturbance during uniaxial loading, unloading, and reloading for ceramic composites (18).

tensile stress is removed, the cracks close. Then, during reloading and subsequent virgin loading, both increase. Thus, disturbance represents a measure of crack density, and both show consistent trends for the tension test involving loading, unloading, and reloading.

Example 5.7: Cemented Sand

As discussed in Chapter 2, the DSC can be used to characterize the behavior of bonded materials such as rocks, concrete, and artificially cemented sand. Here, we consider cemented Leighton–Buzzard sand with bonding material (quick-setting cement) 5% by weight and 14% by weight of water (18, 19, 21). Cubic specimens (100 × 100 × 100 mm) sizes were tested using a multiaxial

Theory of Elasticity in DSC

TABLE 5.2
Parameters for Cemented Sand

Parameter	Value
E	120 MPa
v	0.353
m	0.23
λ	0.11
e_0^c	0.27
h	0.007
w (average)	1.30
s (average)	0.60

test device, under different stress paths (Fig. 3.16, Chapter 3) and mean pressures. Measurements were obtained both for the stress–strain and ultrasonic (pressure and attenuation) properties.

The RI behavior was simulated as linearly elastic. The elastic moduli and the disturbance parameters are shown in Table 5.2. The FA response was simulated by using the critical-state model from Chapter 4.

Figures 5.18 and 5.19 show comparisons between the predictions and observed data for CTC (compression) and TE (extension) tests with $\sigma_0 = 30$ and 45 kPa, respectively; descriptions of stress paths are given in Fig. 3.16 and Chapter 7. Both the stress–strain and volume change behavior are predicted well by the DSC model.

Figure 5.20 shows the variations of crack density and disturbance with $\sqrt{J_{2D}}$ for the CTC 30 test. During the virgin loading both increase. During unloading, as the load is removed, the microcracks experience opening, and both C_d and D show an increase. During reloading, both decrease due to the coalescence of the microcracks. Then, after the end of reloading, during the virgin loading, both increase again. These are considered to be consistent trends for the compression test.

The foregoing Examples 5.6 and 5.7 show that even with the assumption of linear elastic behavior for the RI response, the DSC model provides satisfactory predictions of the stress–strain and volumetric responses for the ceramic composite and cemented sand. However, if the response of a (granular) material involves such factors as coupling of volumetric response, it would be more appropriate to adopt a plasticity model (e.g., δ_0-model) for the RI behavior, because such a model allows for the volumetric coupling. Alternatively, it may be possible to consider such coupling by including the volume change response in the disturbance function while using the linear elastic model for the RI response.

Example 5.8

Consider a composite made of three materials, as in Fig. 5.21. The elastic moduli, E_1 and E_2, are equal to 43,000 and 31,500 N/mm² for materials 1 and 2, respectively. The observed behavior and that of material 3 are simulated as

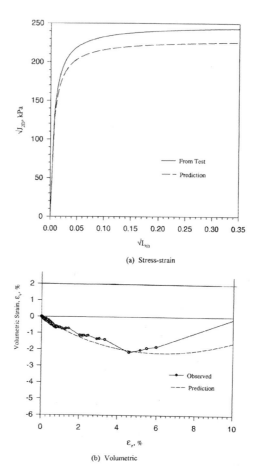

FIGURE 5.18
Comparisons between DSC predictions and multiaxial compression (CTC: $\sigma_0 = 30$ kPa) test for cemented sand. (From Refs. 18 and 19, with permission from Elsevier Science.)

nonlinear elastic using the hyperbolic relation in Eq. (2a); the parameters a and b are 5×10^{-5} and 0.0286, and 1×10^{-4} and 0.057, respectively. Two disturbances are defined with respect to the responses of materials 1 and 3 (D_{13}) and materials 2 and 3 (D_{23}) as

$$D_{13} = \frac{\sigma^1 - \sigma^a}{\sigma^1 - \sigma^3} \tag{8a}$$

$$D_{23} = \frac{\sigma^2 - \sigma^a}{\sigma^2 - \sigma^3} \tag{8b}$$

Theory of Elasticity in DSC

The values of the strains, stresses and disturbances are given below:

ε	σ^1	σ^2	σ^a	σ^3	D_{13}	D_{23}
0	0	0	0	0		
0.0005	21.5	15.75	7.77605	3.891051	0.779374	0.672399
0.001	43	31.5	12.72265	6.369427	0.82656	0.747192
0.002	86	63	18.65672	9.345794	0.878533	0.826464
0.003	129	94.5	22.09131	11.07011	0.906544	0.867899
0.004	172	126	24.3309	12.19512	0.924059	0.893363
0.005	215	157.5	25.90674	12.98701	0.936045	0.910598
0.006	258	189	27.07581	13.57466	0.944764	0.923038
0.008	344	252	28.6944	14.38849	0.956598	0.939793
0.01	430	315	29.7619	14.92537	0.964256	0.950557

The observed response can be found as the average response from the two disturbance conditions as

$$\sigma^a = [(1 - D_{13})\sigma^1 + D_{13}\sigma^3 + (1 - D_{23})\sigma^2 + D_{23}\sigma^3]/2 \qquad (8c)$$

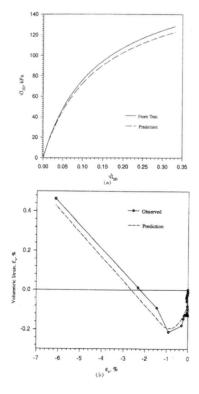

FIGURE 5.19
Comparisons between DSC predictions and multiaxial extension (TE: $\sigma_0 = 45$ kPa) test for cemented sand. (From Refs. 18 and 19, with permission from Elsevier Science.)

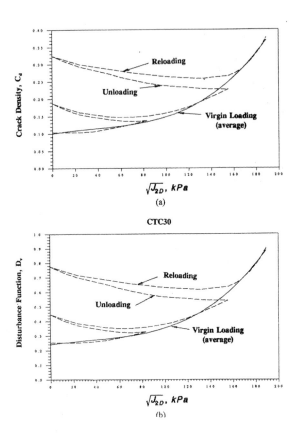

FIGURE 5.20
Crack density and disturbance during multiaxial loading, unloading, and reloading for cemented sand, CTC: $\sigma_0 = 30$ kPa (18).

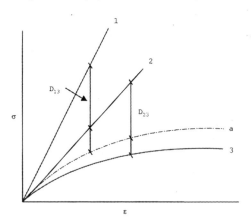

FIGURE 5.21
Composite material with three components: two linear elastic and one nonlinear elastic.

For example, for $\varepsilon = 0.005$, σ^a is computed as

$$\sigma^a = [(1 - 0.936)215 + 0.936 \times 12.98 + (1 - 0.91)157.5 + 0.91 \times 12.98]/2$$
$$= 51.90/2 = 25.95 \text{ N/mm}^2$$

References

1. Fung, Y.C., *Foundations of Solid Mechanics*, Prentice-Hall, Inc., Englewood Cliffs, New Jersey, 1965.
2. Malvern, L.E., *Continuum Mechanics*, MacMillan Publishing Co., New York, 1966.
3. Timoshenko, S.P. and Goodier, J.N., *Theory of Elasticity*, 3rd Edition, McGraw-Hill Book Co., New York, 1970.
4. Desai, C.S. and Siriwardane, H.J., *Constitutive Laws for Engineering Materials*, Prentice-Hall, Englewood Cliffs, NJ, 1984.
5. Eringen, A.C., *Nonlinear Theory of Continuous Media*, McGraw-Hill Book Co., New York, 1962.
6. Truesdell, C., *Continuum Mechanics—The Mechanical Foundations of Elasticity and Fluid Dynamics*, Vol. 1, Gordon and Breach, Science Publishers, New York, 1967.
7. Franklin, J.N., *Matrix Theory*, Prentice-Hall, Englewood Cliffs, NJ, 1968.
8. Kachanov, L.M., *Introduction to Continuum Damage Mechanics*, Martinus Nijhoft Publishers, Dordrecht, The Netherlands, 1986.
9. Seraphim, D., Lasky, R., and Li, C., *Principles of Electronic Packaging*, McGraw-Hill, New York, 1989.
10. Knecht, S. and Fox, L., "Integrated Matrix Creep: Application to Accelerated Testing and Lifetime Prediction," in *Solder Joint Reliability*, Lau, J.H. (Editor), Van Nostrand Reinhold, New York, 1995.
11. Sarihan, V., "Temperature Dependent Viscoplastic Simulation of Controlled Collapse Under Thermal Cycling," *J. of Electronic Packaging*, ASME, 115, 1993, 16–21.
12. Desai, C.S., Chia, J., Kundu, T., and Prince, J.L., "Thermomechanical Response of Materials and Interfaces in Electronic Packaging: Part I – Unified Constitutive Model and Calibration; Part II – Unified Constitutive Models, Validation and Design," *J. of Elect. Packaging*, ASME, 119, 4, 1997, 294–300; 301–309.
13. Riemer, D.E., "Prediction of Temperature Cycling Life for SMT Solder Joints on TCE-Mismatched Substrates," *Proc. Electronic Component Conf.*, IEEE, 1990, 418–423.
14. Skipor, A., Harren, S., and Botsis, J., "Constitutive Characterization of 63/37 Sn/Pb Eutectic Solder Using the Bodner-Partom Unified Creep-Plasticity Model," *Adv. in Elect. Packaging, Proc. Joint ASME/JSME Conf. On Electronic Packaging*, Chen, W.T. and Abe, H. (Editors), Vol. 2, 1992, 661–672.
15. Pan, T.Y., "Thermal Cycling-Induced Plastic Deformation in Solder Joints—Part I: Accumulated Deformation in Surface Mount Joints," *J. Electronic Packaging*, ASME, 113, 1991, 8–15.

16. Chia, J. and Desai, C.S., "Constitutive Modelling of Thermomechanical Response of Materials in Semiconductor Devices Using the Disturbed State Concept," *Report to NSF,* Dept. of Civil Eng. and Eng. Mechs., University of Arizona, Tucson, AZ, USA, 1994.
17. Desai, C.S. and Girdner, K., "Structural Materials from Lunar Simulants Through Thermal Liquefaction," *Proc. Eng. Construction and Operations in Space II, ASCE,* Denver, Colorado, 1992, 528–536.
18. Toth, J. and Desai, C.S., "Development of Lunar Ceramic Composites, Testing and Constitutive Modelling Including Cemented Sand," *Report,* Dept. of Civil Engng. and Engng. Mechanics, University of Arizona, Tucson, Arizona, USA, 1994.
19. Desai, C.S. and Toth, J., "Disturbed State Constitutive Modeling Based on Stress-Strain and Nondestructive Behavior," *Int. J. Solids & Struct.,* 33, 11, 1996, 1619–1650.
20. Hudson, A., "Wave Speed and Attenuation of Elastic Waves in Material Containing Cracks," *Geophys. J.,* Royal Astron. Soc., 64, 1981, 113–150.
21. Desai, C.S., Jagannath, S.V., and Kundu, T., "Mechanical and Ultrasonic Response of Soil," *J. of Eng. Mech., ASCE,* 121, 6, 1995, 744–752.

6

Theory of Plasticity in DSC

CONTENTS
6.1　Introduction 149
　　6.1.1　Mechanisms 151
　　6.1.2　Theoretical Development 152
　　6.1.3　Yield Criteria 152
　　6.1.4　Mohr–Coulomb Yield Criterion 154
　　6.1.5　Continuous Yielding or Hardening Models 156
　　6.1.6　Critical-State Concept 156
　　6.1.7　Yield Surface 160
　　6.1.8　Parameters in the CS Model 161
　　6.1.9　Cap Model 161
　　6.1.10 Advantages and Limitations of the CS and Cap Models 163
6.2　Incremental Equations 164
　　6.2.1　Parameters and Determination from Laboratory Tests 165
　　6.2.2　Cyclic Loading 168
　　6.2.3　Thermoplasticity 168
6.3　Examples 168

6.1 Introduction

The development and application of the theory of plasticity have occurred over the last many years (1–13). The intention here is to describe briefly the basic aspects of the theory and its use in the DSC.

The theory of elasticity (Chapter 5) is applicable if the material is elastic; that is, upon removal of load, it returns to its original configuration along the same path; see Fig. 6.1(a). However, except for limited ranges of loading, most materials do not return to their original configuration; that is, they follow different paths during unloading. As a result, at the end of unloading, the material retains a part of the deformation or strain, which is referred to as irreversible, inelastic, or plastic strain, ε^p. Hence, the total strain, ε, at different points is

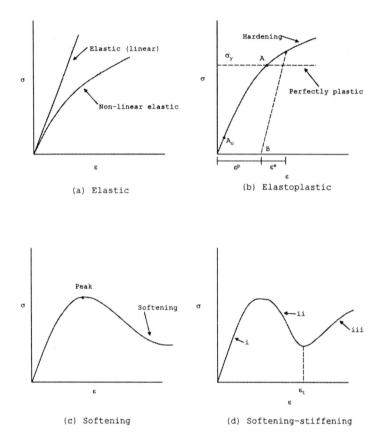

FIGURE 6.1
Schematics of responses.

assumed to be composed of the plastic, ε^p, and the elastic or recoverable parts, ε^e; Fig. 6.1(b).

If the material (element) is elastic up to point A and then yields plastically, as in Fig. 6.1(b), it is referred to as *elastoplastic* material. If, after point A, it experiences continuing deformation under the constant yield stress (σ_y), it is called *elastic perfectly plastic* material. After the yield point A, if the material is unloaded, it will not return to its original configuration and will experience plastic strain, ε^p.

The "stiffness" of some materials may experience gradual decrease during yielding; however, the stress under continuing loading increases after the yield point. In other words, every point during the loading is a "new" yield point, and the next yield stress is greater than the previous yield stress. Such behavior is called *elastic-plastic hardening* response.

Many materials, e.g., geologic, concrete, and polymers, may exhibit very little or no elastic region, and the point A_0 is very near the origin [Fig. 6.1(b)]. That is, they exhibit irreversible strains almost from the very beginning of

Theory of Plasticity in DSC

loading, and harden or yield continuously. Such a response is termed continuous hardening or yielding.

If, after reaching the peak stress, the stress experiences a decrease compared to the peak stress, the response is called *strain softening*; see Fig. 6.1(c). Some materials may first exhibit a drop in stress after peak, i.e., strain-softening response, and then after a certain level of threshold strain (ε_t) may harden or stiffen. Such a behavior is called *strain-softening-stiffening* behavior [Fig. 6.1(d)].

6.1.1 Mechanisms

The manifestation of irreversible or plastic strains can be attributed to the internal changes at the atomic and microlevels. At the atomic level, the bonds can stretch, causing a recoverable elastic response. Then relative slip can occur, causing an irrecoverable response (5). In the elastic range, the material particles, see Fig. 6.2(a), maintain bonding or contact such that upon removal

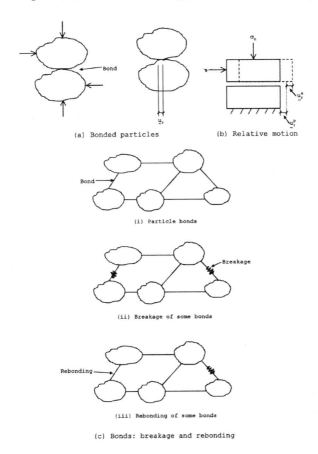

FIGURE 6.2
Mechanisms at the particle level.

of the load, the material springs back to its original state. After yield, which can be considered as a threshold state, the contacts may experience relative motions (u_r) [Fig. 6.2(b)] such as translation along the contact. The measured macrolevel motions or displacements represent accumulation of motions that occur at the atomic and microlevels. A part of the accumulated motion (displacement) is not recovered when particles are unloaded. This is similar to the behavior of two blocks [Fig. 6.2(b)] in which, upon unloading, a part of the displacement (u_r^p) is not recovered. The motion (displacement) that is not recovered is referred to as irreversible or plastic.

The cohesive part of the material strength is mainly present in the case of the elastic-perfectly-plastic materials. The continuing strains occur under constant yield stress, which represents the cohesive strength due to interparticle attraction and chemical bonds. Both the cohesive and frictional properties are often present in many materials; hence, after the yield for an elastic-plastic hardening material [Fig. 6.1(b)], the yield strength can increase with further loading.

In the case of strain-softening response [Fig. 6.1(c)], some of the interparticle bonds may experience breakage, resulting in the reduction in the stress the material can carry. However, for some materials and loading, after a certain reduction in stress and self-adjustment of the microstructure, the contacts may reestablish and rebonding may occur with increasing strains; such a behavior can be called stiffening or healing; see Fig. 6.1(d). During softening, bond breakage can occur, and during stiffening rebonding can occur; see Fig. 6.2(c). Such behavior is considered in Chapter 10.

6.1.2 Theoretical Development

From the above discussion, it is evident that in order to define the behavior of an elastic-plastic material, it is required to define (a) the state at which yielding occurs and (b) the behavior after yield so as to evaluate the plastic deformations during the post-yield region.

6.1.3 Yield Criteria

A yield criterion defines the limit of the elastic regime and is usually expressed in terms of stresses. In the case of one-dimensional or uniaxial loading, the yield criterion can be expressed as

$$F = F(\sigma_y) \tag{6.1}$$

where F is the yield function and σ_y is the uniaxial tensile or compressive stress at which yielding occurs. In general, the yield criterion is expressed in terms of the six components of stress as

$$F = F(\underset{\sim}{\sigma}) \tag{6.2}$$

Theory of Plasticity in DSC

If we assume that the material is isotropic, i.e., its response is independent of the direction, F can be expressed as

$$F = F(\sigma_1, \sigma_2, \sigma_3) \quad (6.3)$$

where σ_1, σ_2, and σ_3 are the principal stresses. Often, the yield function is expressed in terms of the invariants (J_1, J_2, J_3) of the total stress tensor, σ_{ij},

$$F = F(J_1, J_2, J_3) \quad (6.4a)$$

Sometimes, it is appropriate to use the invariants, J_{2D}, J_{3D}, of the deviatoric stress tensor, S_{ij}, as

$$F = F(J_1, J_{2D}, J_{3D}) \quad (6.4b)$$

where

$$J_1 = \sigma_{ii}; \quad J_2 = \frac{1}{2}\sigma_{ij}\sigma_{ji} = \frac{1}{2}\text{tr}(\underline{\sigma})^2; \quad J_3 = \frac{1}{3}\sigma_{ik}\sigma_{km}\sigma_{mi} = \frac{1}{3}\text{tr}(\underline{\sigma})^3;$$

$$J_{2D} = \frac{1}{2}S_{ij}S_{ij} = J_2 - \frac{J_1^2}{6} = \frac{1}{6}[(\sigma_1 - \sigma_2)^2 + (\sigma_2 - \sigma_3)^2 + (\sigma_1 - \sigma_3)^2]$$

and

$$J_{3D} = J_3 - \frac{2}{3}J_1 J_2 + \frac{2}{27}J_1^3 = \frac{1}{3}S_{ij}S_{jm}S_{mi} = \frac{1}{3}\text{tr}(\underline{S})^3$$

If the yield behavior of the material, like that of some metals and saturated cohesive soils under undrained conditions, is affected only by the shear stress, then F is expressed as

$$F = F(J_{2D}) \quad (6.5a)$$

which leads to the well-known von Mises criterion (2, 6, 12)

$$F = J_{2D} - k^2 = 0 \quad (6.5b)$$

where k is the material parameter determined from laboratory tests. Hence,

$$\frac{1}{6}[(\sigma_1 - \sigma_2)^2 + (\sigma_2 - \sigma_3)^2 + (\sigma_1 - \sigma_3)^2] = k^2 \quad (6.5c)$$

For uniaxial tension loading, if the yield stress is σ_y, then Eq. 6.5(c), gives

$$\frac{1}{3}\sigma_1^2 = \frac{1}{3}\sigma_y^2 = k^2 \quad (6.5d)$$

Therefore,

$$k = \frac{\sigma_y}{\sqrt{3}} \qquad (6.5e)$$

In the case of pure shear stress at yield, s, we can derive the following relation:

$$s = \frac{\sigma_y}{\sqrt{3}} \qquad (6.5f)$$

The von Mises criterion (F) plots as a cylinder, with constant radius, Fig. 6.3(a), in the principal stress space, indicating that the behavior is not affected by the mean pressure, $p = J_1/3$. The Drucker–Prager (D–P) yield criterion allows for the effect of mean pressure and is given by

$$F = \sqrt{J_{2D}} - \alpha J_1 - k = 0 \qquad (6.6a)$$

where α and k are material parameters. The D–P criterion plots as a right cone in the stress space, indicating that the yield response is a function of the mean pressure, p. For conventional triaxial compression (CTC) loading ($\sigma_1 > \sigma_2 = \sigma_3$; Fig. 3.16), and for plane strain idealization, the values of α and k are related to the angle of friction, ϕ, and cohesion, c, as (12, 14–16)

CTC

$$\alpha = \frac{2\sin\phi}{\sqrt{3}(3 - \sin\phi)} \qquad (6.6b)$$

$$k = \frac{6c\cos\phi}{\sqrt{3}(3 - \sin\phi)} \qquad (6.6c)$$

Plane strain

$$\alpha = \frac{\tan\phi}{(9 + 12\tan^2\phi)^{1/2}} \qquad (6.6d)$$

$$k = \frac{3c}{(9 + 12\tan^2\phi)^{1/2}} \qquad (6.6e)$$

6.1.4 Mohr–Coulomb Yield Criterion

The Mohr–Coulomb (M–C) yield criterion allows for the effect of friction and plots as an irregular hexagon in the stress space; see Fig. 6.3(b). One of the properties of this criterion is that it includes different strengths under different

Theory of Plasticity in DSC

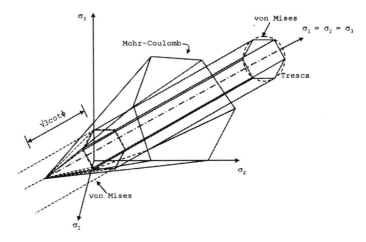

(a) Plots of F for different models in σ_1-σ_2-σ_3 space

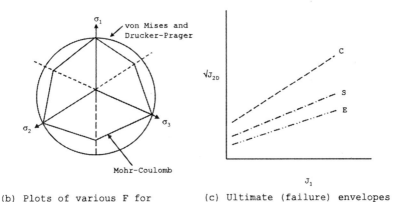

(b) Plots of various F for different models on Π-plane

(c) Ultimate (failure) envelopes under different stress paths

FIGURE 6.3
Representation of plasticity models and effect of stress path.

paths of stress or loading, e.g., compression (C), extension (E), and simple shear (S); see Fig. 6.3(c). As a result, it is often considered to be more appropriate for frictional and geologic materials. The yield function according to the M–C criterion is given by

$$F = J_1 \sin\phi + \sqrt{J_{2D}}\cos\theta - \frac{\sqrt{J_{2D}}}{3}\sin\phi\sin\theta - c\cos\phi = 0 \quad (6.7a)$$

where θ is the Lode angle:

$$\theta = \frac{1}{3}\sin^{-1}\left(-\frac{3\sqrt{3}}{2}\frac{J_{3D}}{J_{2D}^{3/2}}\right) \qquad (6.7b)$$

and

$$-\frac{\pi}{6} \le \theta \le \frac{\pi}{6} \qquad (6.7c)$$

Hence, it includes the effect of the third invariant, J_{3D}, of S_{ij}, which is given after Eq. (6.4b).

6.1.5 Continuous Yielding or Hardening Models

In the classical plasticity models described above, the process of yielding depends on the state of stress; however, no consideration is given to the changing physical state of the material due to internal microstructural transformations. During deformation, the material state changes continuously; for example, the density, specific volume, or void ratio changes under both hydrostatic and shear stresses. This can be particularly true in the case of materials like geologic, concrete, and some ceramics, which exhibit a coupled response due to shear and hydrostatic loadings. As a consequence, the material may exhibit yielding (almost) from the very beginning, e.g., point A_0 in Fig. 6.1(b). That is, every point on the stress–strain response is a yield point, and the yield stress increases continuously. Continuous yielding may be attributed to internal microstructural changes due to relative particle motions such as compaction, sliding, and rotation. The irreversible or plastic part of these motions plays a significant role in continuous yielding, which can be expressed in terms of internal variables such as volumetric and deviatoric plastic strain trajectories, and plastic work or dissipated energy.

In the context of geologic materials, mention may be made of Hvorslev's (17) pioneering work on the behavior of remolded clays, Casagrande's (18) idea of critical density or void ratio, the critical-state concept by Roscoe et al. (19, 20), and the cap models by DiMaggio and Sandler (21). Here, we shall briefly describe the continuous yielding models based on the critical-state and cap concepts.

6.1.6 Critical-State Concept

We first describe the basic idea in the critical-state (CS) concept. Irrespective of the initial density, a (granular) material, under shear stress, passes through various deformation states and finally *approaches the critical* state at which its volume or density or void ratio does not change. In the critical state, the

Theory of Plasticity in DSC

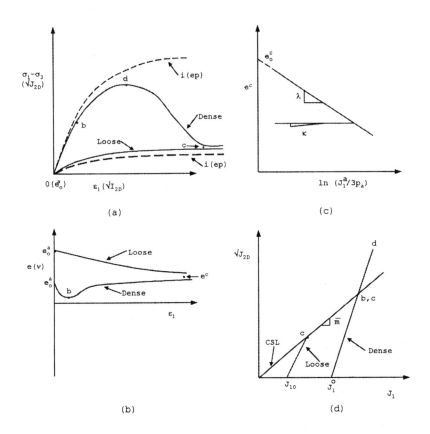

FIGURE 6.4
Behavior of loose and dense granular materials and critical state.

material with a given initial mean pressure continues to deform under the constant shear stress reached up to that state without further change in its volume (see Chapter 3).

Figure 6.4 shows symbolic stress–strain and volumetric responses under compressive loading of initially loose and dense materials with given initial mean pressure, $p_0 = \sigma_0 = J_{10}/3$. In the case of the loose material, the volume continuously decreases and then approaches the critical state (CS), denoted by c. The dense material may first compact (decrease in volume) and after an instantaneous state of constant volume, which represents a *threshold transition*, the material dilates, i.e., its volume increases. Finally, it approaches the same critical state of invariant volume as that for the loose material under the given p_0.

Laboratory testing of normally consolidated cohesive soils under undrained and drained conditions leads to similar critical-state conditions. This is depicted in Fig. 6.5(a) and (b) (12). Here it can be seen that shearing under different initial mean pressures [$p = (\sigma_1 + \sigma_3)/2$] leads to invariant shear stresses ($q^c = \sigma_1 - \sigma_3$) and void ratios ($e^c$) at the critical state. The locus of such shear

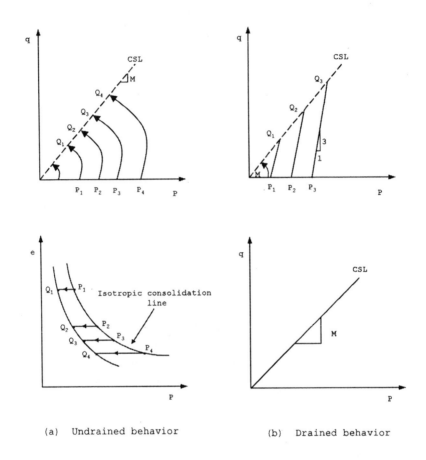

FIGURE 6.5
Undrained and drained behavior of granular materials.

stresses at the critical state is often found to be a straight line on the q–p plot, shown in Fig. 6.5, which is called the *critical-state line* (CSL). The locus of the void ratios at the critical state on the e–ℓnp plot, in Fig. 6.4(c), is often curved but is frequently approximated as the line with a slope of λ.

Now, if the two responses q–p and e–p are combined together, in Fig. 6.6(a), they represent curved surfaces in the q–p–e plot. Thus, the behavior of the material is now not only dependent on the state of stress (q, p), but also on the physical state represented by void ratio (e) or density (ρ). As shown in Fig. 6.6(b), the projection of the curved surface [Fig. 6.6(a)] on the q–p plot represents yield surfaces that grow continuously and intersect the CSL such that the tangent to the yield surface is horizontal; this implies that the change in the volumetric (plastic) strain at the intersection vanishes. Similarly, if the material is assumed to be isotropic, it will experience only volumetric strains under hydrostatic loading and hence, the yield surface would intersect the p-axis at a right angle. In Fig. 6.6(b), ε_s^p and ε_v^p denote

Theory of Plasticity in DSC

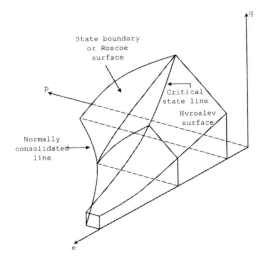

(a) Critical state in q-p-e space

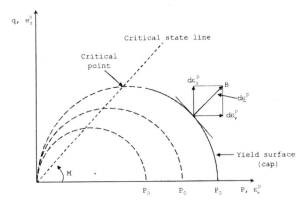

(b) Projection of critical state and yield surfaces on q-p space with shear and volumetric strains

FIGURE 6.6
Critical-state representation (12, 19, 20).

the plastic shear and volumetric strains, respectively, and d denotes increment. In the earlier development of the critical-state concept (19, 20), the yield surfaces were considered elliptical.

The foregoing now permits definition of the material behavior under continuous yielding and at the critical state. It may be noted that in the earlier work (19, 20), the main emphasis was on the definition of the critical state, whereas the formulation, use, and implementation of the continuous yielding aspect became evident later, particularly in the context of computational procedures such as the finite-element method.

There have been a number of subsequent works involving the formalization and use of the critical-state concept for describing behavior of materials such as overconsolidated clays (22) and rocks (23). However, because a main objective here is to discuss the hierarchical single-surface (HISS) approach, which includes the critical-state model as the special case, we present only brief descriptions of some aspects in the earlier critical-state models.

6.1.7 Yield Surface

The equation of the yield surface, F_y [Fig. 6.6(b)], according to the modified *Cam clay* model, is derived as (19, 20)

$$M^2 p^2 - M^2 p_0 p + q^2 = 0 \tag{6.8}$$

where M is the slope of the CSL given by

$$q = Mp \tag{6.9}$$

and p_0 is the varying value of mean pressure at the intersection of the yield surface with the p-axis, Fig. 6.6(b), during shearing, and denotes the yielding or hardening as function of the volumetric plastic strain ($d\varepsilon_v^p$) or void ratio (de_v^p):

$$d\varepsilon_v^p = \frac{de^p}{1 + e_0} = \frac{(\lambda - \kappa)}{1 + e_0} \frac{dp_0}{p_0} \tag{6.10a}$$

Also, at the critical state:

$$e^c = e_0^c - \lambda \ln(J_1^c / 3 p_a) \tag{6.10b}$$

or

$$J_1^c = 3 p_a e^{\frac{e_0^c - e^c}{\lambda}}$$

Here, λ and κ are the slopes of the consolidation (isotropic) loading and unloading responses, respectively [Fig. 6.4(c)], e_0^c is the value of e^c corresponding to $J_1 = 3 p_a$, p_a is the atmospheric pressure constant, and e^c is the void ratio at the critical state.

For the conventional triaxial stress condition ($\sigma_1 > \sigma_2 = \sigma_3$), $q = \sigma_1 - \sigma_3 = \sqrt{3}\sqrt{J_{2D}}$ and $p = (\sigma_1 + 2\sigma_3)/3 = J_1/3$. Substitution in Eq. (6.8) gives the yield surface in terms of the stress invariants (12) as

$$M^2 J_1^2 - M^2 J_1 J_{10} + 27 J_{2D} = 0 \tag{6.11}$$

where J_{10} is the value of J_1 at the intersection of the yield surface with the J_1-axis.

Theory of Plasticity in DSC

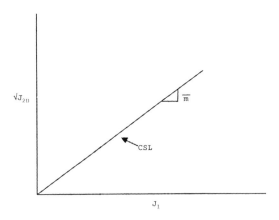

FIGURE 6.7
Critical-state line on $\sqrt{J_{2D}} - J_1$ space.

For the general three-dimensional case:

$$J_{2D} = \frac{1}{6}[(\sigma_1 - \sigma_3)^2 + (\sigma_2 - \sigma_3)^2 + (\sigma_1 - \sigma_3)^2] \quad (6.12a)$$

$$p = (\sigma_1 + \sigma_2 + \sigma_3)/3 \quad (6.12b)$$

hence, the yield surface is expressed as

$$\bar{m}^2 J_1^2 - \bar{m}^2 J_1 J_{10} + J_{2D} = 0 \quad (6.13)$$

where $\bar{m}$ is the slope of the critical-state line in $\sqrt{J_{2D}} - J_1$ space, and $M = 3\sqrt{3}\bar{m}$; see Fig. 6.7.

6.1.8 Parameters in the CS Model

If the CS model is used in the context of elastoplastic behavior with isotropic hardening (the yield surface grows symmetrically with respect to the origin in the stress space), the following six parameters are required for the basic CS model:

Elastic: E and ν or K and G
Plastic: M, λ, κ (for unloading), e_0 (initial void ratio)

Details of their determination from laboratory tests are given subsequently.

6.1.9 Cap Model

The basic idea in the cap model is similar to that in the critical-state model. It was proposed by DiMaggio and Sandler (21) to characterize the behavior of sands. In this model, Fig. 6.8, two functions are defined, one to characterize continuous

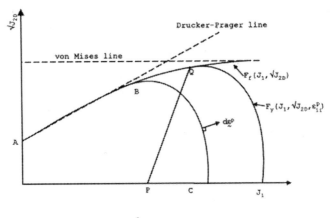

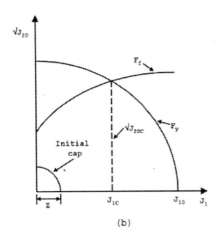

FIGURE 6.8
Representation of cap model in $\sqrt{J_{2D}} - J_1$ stress space.

yielding, F_y, and the other for the failure behavior, F_f, which is composed of an initial portion of the Drucker–Prager surface and the ultimate part of the von Mises criterion, joined smoothly. This implies that in the early stage of loading when the mean pressure is low, the material behaves as cohesive-frictional, and at higher mean pressure, it exhibits essentially cohesive response.

The expression for the yield caps, F_y, assumed to be elliptical, is given by

$$F_y = \left(\frac{J_{10} - J_{1c}}{\sqrt{J_{2DC}}}\right)^2 J_{2D} + (J_1 - J_{1c})^2 = 0 \qquad (6.14)$$

where J_{10} represents the intersection of the yield surface with the J_1-axis [similar to p_0 in the critical state (Cam clay) model], J_{1C} is the value of J_1 at the center

Theory of Plasticity in DSC

of the ellipse, and $\sqrt{J_{2DC}}$ is the value of $\sqrt{J_{2D}}$ when $J_1 = J_{1c}$. The term J_{10} represents hardening or yielding and is expressed in terms of volumetric plastic strain as

$$J_{10} = -\frac{1}{\bar{D}}\ln\left(1 - \frac{\varepsilon_v^p}{W}\right) + Z \tag{6.15}$$

where $\bar{D}$, W, and Z are material parameters; the latter denotes the size of the cap due to the initial (stress) conditions.

The expression for the failure surface is given by

$$F_f = \sqrt{J_{2D}} + \bar{\gamma}e^{-\bar{\beta}J_1} - \bar{\alpha} = 0 \tag{6.16}$$

where $\bar{\alpha}$, $\bar{\beta}$, and $\bar{\gamma}$ are the material parameters.

The yield caps intersect the failure surface, as shown in Fig. 6.8. As the loading is applied, along a given stress path (P to Q), the material passes from one yield surface to the next, and finally, at failure, the surface, F_f, is reached. As at failure there is no volume change, F_y intersects F_f such that the tangent to F_y is horizontal. Also, as the material is assumed to be isotropic, F_y intersects the J_1-axis at a right angle. In computational analysis, it becomes necessary to design special schemes to handle the intersection point (B) while computing plastic strain increments.

The number of parameters in the elastoplastic cap model is *nine*, which includes two elastic (e.g., E and v or K and G), $\bar{\alpha}$, $\bar{\beta}$, and $\bar{\gamma}$ for the failure surface, and D, W, Z and $R = (J_{10} - J_{1C})/\sqrt{J_{2DC}}$ for the yield cap (12, 21).

6.1.10 Advantages and Limitations of the CS and Cap Models

Both the critical-state (Cam-clay) and the cap models can provide improved characterization of materials that exhibit continuous yielding. As a result, a number of studies have used and implemented these models, often with modifications, in computer procedures. However, they suffer from the following limitations in handling a number of important attributes of the behavior of materials.

1. In both, the yielding is assumed to depend only on the volumetric plastic strains. It is found that for many materials, yielding can also depend on the deviatoric plastic strains. Hence, it is felt that both should be included in the definition of yielding.
2. The yield surface in both, Eq. (6.11) and Eq. (6.14), plot as circular in the principal stress space ($\sigma_1-\sigma_2$, σ_3); this implies that the strength of the material is independent of stress path. On the other hand, many (geologic and other) materials exhibit different strengths under different paths, e.g., compression, extension, and simple shear; Fig. 6.3(c).

3. As can be seen from Figs. 6.6 and 6.8, the plastic strain increment, $d\underset{\sim}{\varepsilon}^p$, is directed outward to the yield surfaces. If the compressive volume change is considered to be positive in the positive J_1-axis, this implies that up to the intersection of the yield surface with the CSL or the failure surface [Figs. 6.6(b) and 6.8(a)], the models would predict compressive volume change. Then at the intersection, there will be no volume change. In the case of the CS model, the yield surface is used in the calculations, then after the critical state is reached, the volume change will be zero. In the cap model, if F_f is used, volume change can occur after the failure and can involve dilation or volume increase as the plastic strain increment will now be directed in the negative J_1-direction. In many materials, it is found that the dilation may start before the peak or the failure is reached, point b in Fig. 6.4. Thus, these models cannot allow for the occurrence of dilation before the peak or failure.

4. Although the issue of strain-softening response has been discussed in the context of the CS models and subsequent modifications, there has been no consistent approach available for strain softening, that allows for the discontinuous nature of the material due to microcracking.

5. Many (frictional) materials exhibit nonassociative response, that is, the plastic potential function, Q, is different from the yield function, F. The earlier CS and cap models have been based on associated plasticity, i.e., $Q \equiv F$.

Comments: To Hierarchical Single-Surface Models. The foregoing models are presented for completeness. The hierarchical single-surface (HISS) plasticity models (Chapter 7) provide a unified approach, which includes most of the foregoing and other plasticity models as special cases. It eliminates or reduces the above deficiencies and usually requires fewer parameters, compared to those in the previously available models of comparable capacity. The HISS-δ_0 model (Chapter 7) is mainly used to characterize the RI response in the DSC model.

6.2 Incremental Equations

The consistency condition and the normality rule are given by (1–6)

$$dF = 0 \qquad (6.17a)$$

and

$$d\underset{\sim}{\varepsilon}^p = \lambda \frac{\partial Q}{\partial \underset{\sim}{\sigma}} \qquad (6.17b)$$

Theory of Plasticity in DSC

where $d\underset{\sim}{\varepsilon}^p$ is the vector of plastic strain increments, Q is the plastic potential function ($Q \equiv F$ for associative plasticity), and λ is the scalar proportionality parameter. Use of Eq. (6.17) leads to the incremental equations as (for details of derivations, see Chapter 7)

$$d\underset{\sim}{\sigma} = \underset{\sim}{C}^{ep} d\underset{\sim}{\varepsilon} \tag{6.18a}$$

$$\underset{\sim}{C}^{ep} = \underset{\sim}{C}^e - \underset{\sim}{C}^p \tag{6.18b}$$

where $\underset{\sim}{C}^e$ is the elasticity matrix, and $\underset{\sim}{C}^p$ is the matrix representing the contribution of the plastic response.

6.2.1 Parameters and Determination from Laboratory Tests

Elasticity: For linear elastic and isotropic materials, $\underset{\sim}{C}^e$, Eq. (6.18) is expressed in terms of elastic moduli such as E, v or G and K. Procedures for their determination are given in Chapter 5.

Plasticity: For the von Mises criterion, Eq. (6.5), the yield stress, σ_y, in tension or compression or cohesive strength, c, are needed. Figure 6.9(a) shows a uniaxial stress–strain curve from which the yield stress, σ_y, can be obtained at the point where plastic action initiates. It represents the point after which the material will experience plastic strains upon unloading.

For nonfrictional materials, the cohesive strength, c, can be found from the value of shear stress (τ) at peak or failure [Fig. 6.9(b)], which shows the envelope of shear stress, τ, vs. normal stress, σ. If the material possesses mainly cohesive strength, the Mohr envelope [Fig. 6.9(b)] will be horizontal with c as the intercept along the shear stress (τ) axis. For the Mohr–Coulomb criterion, Eq. (6.7a), the values of c and the angle of friction (ϕ) are found from Mohr diagrams; see Fig. 6.9(b).

For the Drucker–Prager criterion, Eq. (6.6), the values of α and k can be found from the values of cohesion, c, and friction angle, ϕ. Alternatively, α and k can be found from plots of $\sqrt{J_{2D}} - J_1$; see Fig. 6.9(c).

The parameters for the critical-state model (e.g., $\lambda, \overline{m}, e_0$) can be found by using the procedure described in Chapter 3.

The parameters for the cap model are $\overline{\gamma}, \overline{\beta},$ and $\overline{\alpha}$ in Eq. (6.16) and $\overline{D}, W$, and Z and $R(J_{10})$ in Eqs. (6.14) and (6.15). The laboratory stress–strain test results are used to find peak (failure) stress, and plotted as in Fig. 6.10(a). The intercept of the von Mises envelope at larger values of J_1, along $\sqrt{J_{2D}}$, gives the values of $\overline{\alpha}$, and that for the D–P envelope (at $J_1 = 0$) gives the value of $\overline{\alpha} - \overline{\gamma}$. For finding the value of $\overline{\beta}$, we can select a point(s) on the transition curve; see Fig. 6.10(a). Equation (6.16) can be now written as

$$e^{-\beta J_1} = \frac{\overline{\alpha} - \sqrt{J_{2D}}}{\overline{\gamma}} \tag{6.19a}$$

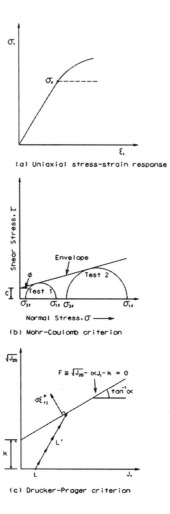

FIGURE 6.9
Parameters for different yield criteria.

Therefore,

$$\beta = -\frac{1}{J_1}\ln\left(\frac{\bar{\alpha} - \sqrt{J_{2D}}}{\bar{\gamma}}\right) \qquad (6.19b)$$

For no initial stress (or yielding) $Z = 0$. The value of $\bar{D}$ and W can be found from hydrostatic test data [Fig. 6.10(b)]. With $Z = 0$, Eq. (6.15) becomes

$$J_{10} = -\frac{1}{\bar{D}}\ln\left(1 - \frac{\varepsilon_v^p}{W}\right) \qquad (6.20a)$$

Theory of Plasticity in DSC

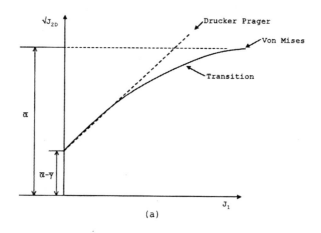

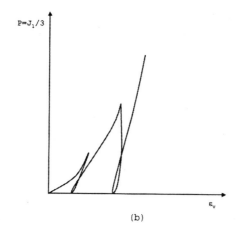

FIGURE 6.10
Parameters for cap model.

where $J_{10} = 3p$, p = mean pressure. Therefore,

$$3p\bar{D} = -\ln\left(1 - \frac{\varepsilon_v^p}{W}\right) \qquad (6.20b)$$

and

$$\varepsilon_v^p = W(1 - e^{-3p\bar{D}}) \qquad (6.20c)$$

$$= \varepsilon_v - \varepsilon_v^e \qquad (6.20d)$$

Here, ε_v is the total volumetric strains, and ε_v^e is the elastic volumetric strain given by

$$\varepsilon_v^e = \frac{p}{K} \qquad (6.20e)$$

where K is the bulk modulus. Hence,

$$\varepsilon_v = W(1 - e^{-3p\bar{D}}) + \frac{p}{K} \qquad (6.20f)$$

Now, we can select pairs of points on the loading part of the p vs. ε_v curves, shown in Fig. 6.10(b). Then, substitution of the values of p and ε_v in Eq. (6.20f) leads to two equations that are solved for W and $\bar{D}$. A number of pairs of points can be used for finding average values of W and D.

The value of $R = (J_{10} - J_{1c})/\sqrt{J_{2DC}}$, Eq. (6.14), is required to define the size and shape of the yield caps, which can be represented as contours of constant volumetric plastic strains. If the yield caps are elliptical, R is given by the ratio of the major to minor axes of the ellipse. For granular materials, the value of R is often found in the range of 1.67 to 2.0.

6.2.2 Cyclic Loading

In this chapter, we have discussed the characterization only for monotonic loading, often referred to as *virgin* loading, in which the applied stress increases continuously. In Chapter 7, we shall discuss DSC models for cyclic loading.

6.2.3 Thermoplasticity

Behavior of materials can be influenced significantly by temperature and its variations, often together with mechanical loading. It is possible to incorporate temperature effects by appropriate modification of the yield function, F. Details are given in Chapter 7, with respect to the HISS models; the models presented here can be considered special cases of the HISS models.

6.3 Examples

Example 6.1

Consider a one-dimensional stress–strain response, Fig. 6.11, under uniaxial stress (σ), and the following two yield functions:

$$F = \sigma - \sigma_y = 0 \qquad (1a)$$

$$F = J_{2D} - k^2 = 0, \quad k = \frac{\sigma_y}{\sqrt{3}} \qquad (1b)$$

Theory of Plasticity in DSC

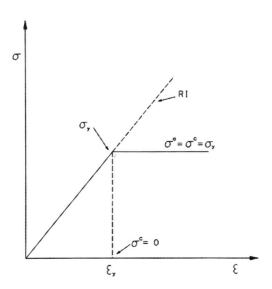

FIGURE 6.11
Uniaxial stress–strain curve.

where σ_y is the yield stress. Derive specialized incremental plasticity equations based on Eq. (6.18):

$$d\underset{\sim}{\sigma} = \underset{\sim}{C}^{ep} d\underset{\sim}{\varepsilon} = (\underset{\sim}{C}^e - \underset{\sim}{C}^p) d\underset{\sim}{\varepsilon}$$

$$= \left[\underset{\sim}{C}^e - \frac{\underset{\sim}{C}^e \cdot \dfrac{\partial F}{\partial \underset{\sim}{\sigma}} \left(\dfrac{\partial F}{\partial \underset{\sim}{\sigma}} \right)^T \underset{\sim}{C}^e}{\left(\dfrac{\partial F}{\partial \underset{\sim}{\sigma}} \right)^T \cdot \underset{\sim}{C}^e \dfrac{\partial F}{\partial \underset{\sim}{\sigma}}} \right] d\underset{\sim}{\varepsilon} \qquad (1c)$$

Details are given in Chapter 7, Eq. (7.44).

For the one-dimensional case, $\underset{\sim}{\sigma} = \sigma$, $\underset{\sim}{\varepsilon} = \varepsilon$, and $\underset{\sim}{C}^e \equiv E$; here we have ignored the effect of other strains. Therefore, from the yield function in Eq. (1a), we have

$$\frac{\partial F}{\partial \sigma} = \frac{\partial}{\partial \sigma}(\sigma - \sigma_y) = 1 \qquad (1d)$$

Hence,

$$d\sigma = \left[E - \frac{E \cdot 1 \cdot 1 \cdot E}{1 \cdot E \cdot 1} \right] d\varepsilon$$

$$= 0.d\varepsilon$$

which indicates that there is no change in stress during yielding for the perfectly plastic model.

For the yield function in Eq. (1b):

$$J_{2D} = \frac{1}{6}(2\sigma^2) = \frac{\sigma^2}{3}$$

$$F = \frac{\sigma^2}{3} - k^2 \qquad (1e)$$

$$\frac{\partial F}{\partial \sigma} = \frac{2}{3}\sigma$$

Therefore,

$$d\sigma = \left(E - \frac{E\frac{2}{3}\sigma \cdot \frac{2}{3}\sigma E}{\frac{2}{3}\sigma E \frac{2}{3}\sigma}\right)d\varepsilon$$

$$= (E - E)d\varepsilon = 0.d\varepsilon$$

Again indicating that for the von Mises criterion with the perfectly plastic model, there is no change in stress during yielding.

Example 6.2

Assume that the RI behavior is simulated as linear elastic with modulus E, Fig. 6.11. Consider that the observed response during yielding is given by

$$\sigma^a = \sigma^c = \sigma_y \qquad (2a)$$

that is, the FA response is assumed to possess strength equal to the yield stress, σ_y.

The disturbance, D, can be evaluated [Eq. (3.7a), (Chapter 3)] as

$$D = \frac{\sigma^i - \sigma^a}{\sigma^i - \sigma_y} \qquad (2b)$$

In the linear elastic range, before σ_y,

$$D = \frac{\sigma^i - \sigma^i}{\sigma^i - \sigma_y} = 0 \qquad (2c)$$

Theory of Plasticity in DSC

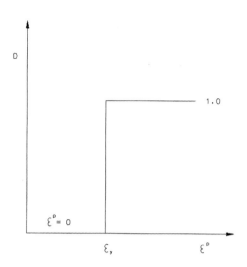

FIGURE 6.12
Disturbance vs. plastic strain: $\sigma^c = \sigma_y$.

and after yield

$$D = \frac{\sigma^i - \sigma^a}{\sigma^i - \sigma_y} = 1 \tag{2d}$$

Hence, D vs. plastic strain, ε^p, is as shown in Fig. 6.12.

The details of the evaluation of the observed stress-strain response are as follows:

Before σ_y:

$$\sigma^a = (1 - D)\sigma^i + D\sigma^c$$
$$= (1 - 0)\sigma^i + 0 \cdot \sigma^c = \sigma^i$$
$$d\sigma^a = (1 - D)d\sigma^i + Dd\sigma^c + dD(\sigma^i - \sigma^c)$$
$$= (1 - 0)d\sigma^i + 0 + 0 = d\sigma^i$$

because $D = 0$ and $dD = 0$.

After σ_y:

$$\sigma^a = (1 - 1)\sigma^i + 1 \cdot \sigma^c = \sigma_y$$
$$d\sigma^a = (1 - 1)d\sigma^i + 1 \cdot d\sigma_y + 0$$
$$= d\sigma_y = 0$$

because $dD = 0$ and $d\sigma_y = 0$ as σ_y is constant. This indicates that there is no change in stress during yielding. Thus, the DSC model provides, as a specialization, the classical perfectly plastic model.

Example 6.3

Consider Example 6.2 and assume that the FA state does not carry any stress; therefore, $\sigma^c = 0$, and $\sigma^i = \sigma_y$ during yielding. Then, the disturbance function, D, is given by

$$D = \frac{\sigma^i - \sigma^a}{\sigma^i} \tag{3a}$$

Before σ_y:

$$D = \frac{\sigma^i - \sigma^a}{\sigma^i} = 0$$

After σ_y:

$$D = \frac{\sigma^i - \sigma^a}{\sigma^i} = 1 - \frac{\sigma_y}{\sigma^i}$$

The plot of D vs. ε^p is given in Fig. 6.13.

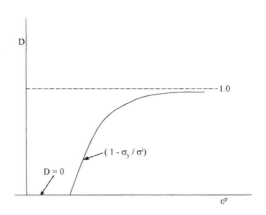

FIGURE 6.13
Disturbance vs. plastic strain: $\sigma^c = 0.0$.

Theory of Plasticity in DSC

Now the observed stress–strain behavior can be predicted as follows:

Before σ_y:

$$\sigma^a = (1 - D)\sigma^i + D\sigma^c$$
$$= (1 - 0)\sigma^i + 0 \cdot \sigma^c = \sigma^i$$
$$d\sigma^a = (1 - 0)d\sigma^i + 0 \cdot d\sigma^c + 0 \cdot (\sigma^c - \sigma^i)$$
$$= d\sigma^i$$

After σ_y:

$$\sigma^a = \left(1 - 1 + \frac{\sigma_y}{\sigma^i}\right)\sigma^i + \left(1 - \frac{\sigma_y}{\sigma^i}\right) \cdot 0$$

$$= \frac{\sigma_y}{\sigma^i} \cdot \sigma^i = \sigma_y$$

$$d\sigma^a = \left(1 - 1 + \frac{\sigma_y}{\sigma^i}\right)d\sigma^i + \left(1 - \frac{\sigma_y}{\sigma^i}\right)0$$
$$+ dD(0 - \sigma^i)$$

$$= \frac{\sigma_y}{\sigma^i} \cdot d\sigma^i + dD(-\sigma^i)$$

Now,

$$D = 1 - \sigma_y(\sigma^i)^{-1}$$
$$dD = 0 - \sigma_y(-1)(\sigma^i)^{-2} \cdot d\sigma^i$$
$$= \frac{\sigma_y}{(\sigma^i)^2} \cdot d\sigma^i$$

Therefore,

$$d\sigma^a = \frac{\sigma_y}{\sigma^i}d\sigma^i + \frac{\sigma_y}{(\sigma^i)^2} \cdot d\sigma^i(-\sigma^i)$$

$$= \frac{\sigma_y}{\sigma^i}d\sigma^i - \frac{\sigma_y}{\sigma^i} \cdot d\sigma^i$$

$$= 0$$

Again indicating that there is no change in stress during yielding and with the assumption that $\sigma^c = 0$, the DSC leads to the classical perfectly plastic model.

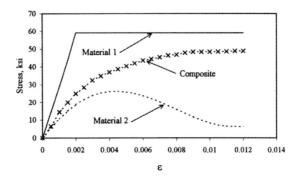

FIGURE 6.14
Behavior of metal, concrete, and composite: 1 psi = 6.89 kPa.

Example 6.4

Consider a composite made of two materials, a metal and concrete (ceramic). Figure 6.14 shows the (measured) behavior of the metal (material 1) as elastoplastic, concrete (material 2) as strain softening, and the composite as plastic hardening. The behavior of materials 1 and 2 is assumed to be simulated by using the DSC model. In the case of the metal, the RI behavior is simulated as linear elastic, and the FA stress, $\sigma^c = 59.00$ ksi, is assumed; see Fig. 6.15(a). For concrete, the RI response is considered as nonlinear (hardening) with $\sigma^c = 6.0$ ksi, Fig. 6.15(b).

The behavior of the metal and that of concrete were first computed by using the following equations at step k (with $d\sigma^c = 0$):

$$d\sigma_k^a = (1 - D_k)(\sigma_k^i - \sigma_{k-1}^i) + dD_k(\sigma_k^c - \sigma_{k-1}^c) \tag{4a}$$

$$D_k = \frac{\sigma_k^i - \sigma_k^a}{\sigma_k^i - \sigma^c} \tag{4b}$$

$$dD_k = D_k - D_{k-1} \tag{4c}$$

$$\sigma_k^a = \sigma_{k-1}^a + d\sigma_k^a \tag{4d}$$

Now, the observed stresses at various points (k) for the metal and concrete were used to evaluate the disturbance for the composite as

$$D_k = \frac{\sigma_k^{1a} - \sigma_k^a}{\sigma_k^{1a} - \sigma_k^{2a}} \tag{4e}$$

where σ_k^a is the observed composite stress at point k at a given strain level. Then the predicted composite behavior was obtained by integrating the

Theory of Plasticity in DSC

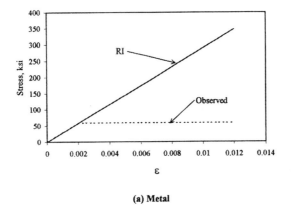

(a) Metal

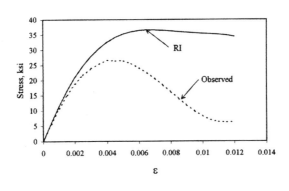

(b) Concrete

FIGURE 6.15
DSC responses for metal and concrete: 1 psi = 6.89 kPa.

following equation:

$$d\sigma_k^a = (1 - D_k)d\sigma_k^{1a} + D_k d\sigma_k^{2a} + dD_k(\sigma_{k-1}^{2a} - \sigma_{k-1}^{1a}) \tag{4f}$$

where $dD_k = D_k - D_{k-1}$. Figure 6.16 shows that the predicted behavior of the composite compares very well with the observed behavior.

Some of the following practice examples may require calculations using iterative procedures. It may be useful to prepare computer routines for such iterative procedures. As necessary, appropriate values of parameters such as E can be adopted.

Example 6.5

Assume that the RI behavior is simulated as bilinear elastic (see Fig. 5.15, Chapter 5). By assuming that (a) $\sigma^a = \sigma^c = \sigma_y$ and (b) $\sigma^c = 0$, $\sigma^a = \sigma_y$, predict the observed (elastic perfectly plastic) response.

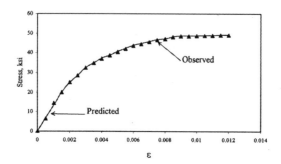

FIGURE 6.16
Comparison of predicted and observed behavior of composite: 1 psi = 6.89 kPa.

Example 6.6
Assume that the RI behavior is simulated by using the hyperbolic relation, Eq. (2a), Example 5.2. Then predict the elastic perfectly plastic stress–strain behavior assuming that (a) $\sigma^a = \sigma^c = \sigma_y$, and (b) $\sigma^c = 0$, $\sigma^a = \sigma_y$.

Example 6.7
Predict the elastic plastic behavior assuming the RI behavior to be elastic (E), and the Drucker–Prager yield criterion, Eq. (6.6a).

References

1. Prager, W., "Recent Developments in Mathematical Theory of Plasticity," *J. of Appl. Phys.*, 2013, 1949, 235–241.
2. Hill, R., *The Mathematical Theory of Plasticity*, Oxford Univ. Press, Oxford, 1950.
3. Drucker, D.C., "A More Fundamental Approach to Plastic Stress-Strain Relations," *Proc. First U.S. Nat. Congr. Appl. Mech.*, 1951, 487–491.
4. Drucker, D.C., "A Definition of Stable Inelastic Material," *ASME Trans.*, 81, 1959, 101–106.
5. Callister, W.D., *Material Science and Engineering—An Introduction*, John Wiley & Sons, New York, 1985.
6. Bridgeman, P.W., *Studies in Large Plastic Flow and Fracture with Special Emphasis on the Effects of Hydrostatic Pressure*, McGraw-Hill Book Co., New York, 1952.
7. Fung, Y.C., *Foundations of Solid Mechanics*, Prentice-Hall, Inc., Englewood Cliffs, NJ, 1965.
8. Malvern, L.E., *Introduction to Mechanics of Continuous Medium*, Prentice-Hall, Inc., Englewood Cliffs, NJ, 1969.
9. Kachanov, L.M., *Foundations of Theory of Plasticity*, North-Holland Publishing Co., Amsterdam, 1971.
10. Mendelson, A., *Plasticity: Theory and Applications*, MacMillan Publ. Co., New York, 1968.

11. Owen, D.R.J. and Hinton, E., *Finite Elements in Plasticity: Theory and Practice*, Pineridge Press, Swansea, U.K., 1980.
12. Desai, C.S. and Siriwardane, H.J., *Constitutive Laws for Engineering Materials*, Prentice-Hall, Inc., Englewood Cliffs, NJ, 1984.
13. Chen, W.F. and Han, D.J., *Plasticity for Structural Engineers*, Springer-Verlag, New York, 1988.
14. Reyes, S.F. and Deere, D.U., "Elastic-plastic Analysis of Underground Openings by the Finite Element Method," *Proc. 1st Int. Cong. Rock Mech.*, Vol. II, Lisbon, 1966, 477–486.
15. Christian, J.T. and Desai, C.S., "Constitutive Laws for Geologic Media," Chapter 12 in *Numerical Methods in Geotechnical Engineering*, C.S. Desai and J.T. Christian (Editors), McGraw-Hill Book Co., New York, 1977.
16. Desai, C.S. and Abel, J.F., *Introduction to the Finite Element Method*, Van Nostrand Reinhold Co., New York, 1972.
17. Hvorslev, M.J., "Physical Properties of Remolded Cohesive Soils," *Translation 69-5*, U.S. Army Corps of Engineers Waterways Experiment Stn., Vicksburg, MS, June 1969.
18. Casagrande, A., "Characteristics of Cohesionless Soils Affecting the Stability of Slopes and Earth Fills," *J. Boston Soc. Civil Eng.*, 1936, 257–276.
19. Roscoe, A.N., Schofield, A., and Wroth, P.C., "On Yielding of Soils," *Geotechnique*, 8, 1958, 22–53.
20. Schofield, A.N. and Wroth, P.C., *Critical State Soil Mechanics*, McGraw-Hill Book Co., London, 1968.
21. DiMaggo, F.L. and Sandler, I., "Material Model for Granular Soils," *J. Eng. Mech.*, ASCE, 97, 3, 1971, 935–950.
22. Banerjee, P.K. and Yousif, N.B., "A Plasticity Model for Mechanical Behaviour of Anisotropically Consolidated Clay," *Int. J. Numer. Analyt. Meth. Geomech.*, 10, 5, 1986, 521–541.
23. Brown, E.T. and Yu, H.S., "A Model for the Ductile Yield of Porous Rock," *Int. J. Numer. Analyt. Meth. Geomech.*, 12, 6, 1988, 679–688.

7
Hierarchical Single-Surface Plasticity Models in DSC

CONTENTS

- 7.1 Introduction 180
  - 7.1.1 Basic HISS Model 180
- 7.2 Specializations of the HISS Model 183
  - 7.2.1 Classical Plasticity Models 183
- 7.3 HISS Versions 186
- 7.4 Material Parameters 186
  - 7.4.1 Curved Ultimate Envelope 190
  - 7.4.2 Relation Between Ultimate Parameters, Cohesion, and Angle of Friction 191
  - 7.4.3 Bonding Stress, R 193
  - 7.4.4 Phase or State Change Parameter 194
  - 7.4.5 Hardening or Growth Parameters 197
  - 7.4.6 Nonassociative δ_1-Model 199
  - 7.4.7 Thermal Effects on Parameters 201
  - 7.4.8 Rate Effects 202
- 7.5 Repetitive Loading 204
  - 7.5.1 Dynamic Loading 205
- 7.6 The Derivation of Elastoplastic Equations 205
  - 7.6.1 Details of Derivatives 209
- 7.7 Incremental Iterative Analysis 210
  - 7.7.1 Possible Stress States 214
  - 7.7.2 Elastic-Plastic Loading 214
  - 7.7.3 Correction Procedures 215
  - 7.7.4 Subincrementation Procedure 216
- 7.8 Drift Correction Procedure 217
- 7.9 Thermoplasticity 220
- 7.10 Examples 222
- 7.11 Stress Path 227
  - 7.11.1 Stress Path Analysis of the HISS Model 228
- 7.12 Examples of Validation 231
  - 7.12.1 Examples of Geologic Materials 236
  - 7.12.2 Repeated Loading and Permanent Deformations ... 251

7.1 Introduction

The hierarchical single-surface (HISS) plasticity models provide a general formulation for the elastoplastic characterization of the material behavior. These models, which can allow for isotropic and anisotropic hardening, and associated and nonassociated plasticity characterizations, can be used to represent material response based on the continuum plasticity theory (1–6). In the case of the DSC, they can be used to represent the RI response; in many cases, the basic and simplest version, HISS-δ_0, that allows for isotropic hardening and associated response has been used. It may be mentioned that use of plasticity theory is one of the possible ways to characterize the RI behavior. However, the DSC can be formulated by simulating the (RI) response as nonlinear with irreversible deformations, without invoking the theory of plasticity.

We will present comprehensive details of the HISS approach including its theoretical background, specializations, nature of parameters and their determination from laboratory tests, details of parameters for typical materials, derivation of elastoplastic equations, incremental iterative analysis, and validations of the models for a number of materials: geologic, concrete, metals, alloys, silicon, etc.

It is useful to emphasize one of the important advantages offered by the hierarchical nature of the HISS model. Most other available models provide for a specific characteristic of the material behavior. On the other hand, the HISS (and DSC) approach provides for hierarchical adoption of models of increasing sophistication, say, linear elastic to nonassociated elastoplastic to elastoplastic with softening (disturbance). Thus, the user can select the most appropriate model for a given material in specific engineering application. In this context, it is shown that the classical such as von Mises, Mohr–Coulomb, and Drucker–Prager, continuous yielding such as critical-state and capped (see Chapter 6), and other models such as by Matsuoka and Nakai (7), Lade and co-workers (8, 9), and Vermeer (10) can be derived as special cases of the HISS model. It is also indicated that the anisotropic and kinematic versions of the HISS model can provide alternative characterizations to models proposed by Mroz and co-workers (11–13), Krieg (14), Prevost (15), and Dafalias and co-workers (16, 17).

7.1.1 Basic HISS Model

Consider the associated plasticity model in the hierarchical single-surface (HISS) approach. The yield function, F, is given by

$$F = \bar{J}_{2D} - (-\alpha \bar{J}_1^n + \gamma \bar{J}_1^2)(1 - \beta S_r)^m = 0$$
$$= \bar{J}_{2D} - F_b F_s = 0 \qquad (7.1)$$

Here, $\bar{J}_{2D} = J_{2D}/p_a^2$, where p_a is the atmospheric pressure constant; $\bar{J}_1 = (J_1 + 3R)/p_a$, where $R =$ the bonding stress (e.g., tensile strength under compressive loading and compressive strength under tensile loading); $n =$ phase change parameter where the volume change transits from compaction to dilation or vanishes; $m = -0.5$ is often used; $\gamma =$ parameter related to the (ultimate) yield surface, β is related to the shape of F in the σ_1–σ_2–σ_3 space, $S_r =$ stress ratio $= (\sqrt{27}/2) \cdot (J_{3D}/J_{2D}^{3/2})$, and α is the growth or hardening parameter, which can be expressed in terms of internal variables such as plastic strain trajectory or accumulated plastic strains, and plastic work or dissipated energy. Based on the development of the hardening function for a wide range of "solid" materials, interfaces and joints, it was found that use of the plastic strain trajectory provides a more consistent formulation than that by plastic work. Moreover, it is relatively easier to compute the plastic strain trajectory from available test data. Hence, the plastic strain trajectory is commonly used in this text.

A simple form of α is given by

$$\alpha = \frac{a_1}{\xi^{n_1}} \tag{7.2}$$

where $\xi = \int (d\varepsilon_{ij}^p d\varepsilon_{ij}^p)^{1/2}$ is the trajectory of plastic strains, ε_{ij}^p, d denotes increment, ξ is composed of deviatoric (ξ_D) and volumetric (ξ_v) plastic strain trajectories:

$$\xi = \xi_D + \xi_v = \int (dE_{ij}^p \cdot dE_{ij}^p)^{1/2} + \frac{1}{\sqrt{3}} |\varepsilon_{ii}^p| \tag{7.3}$$

Here, E_{ij}^p is the deviatoric plastic strains tensor $= \varepsilon_{ij}^p - (1/3)\varepsilon_{ii}\delta_{ij}$, and $\varepsilon_{ii}^p = \varepsilon_v^p =$ volumetric plastic strain. Figure 7.1(a), (b), and (c) show schematics of $F = 0$ in $\sqrt{J_{2D}} - J_1$, octahedral plane, and σ_1–σ_2–σ_3 stress spaces. The expanding yield surfaces are continuous and approach the final or ultimate yield surface when $\alpha = 0$. The points on the yield surfaces with horizontal tangents denote zero volume change or critical-state line (Chapter 6).

In the case of nonassociated plasticity, the plastic potential function, Q, is expressed as (1, 3, 5, 18–20)

$$Q = F + h(J_i, \alpha) \tag{7.4}$$

where $h =$ correction function, which is introduced, for simplicity, through the hardening function as

$$\alpha_Q = \alpha + \kappa(\alpha_0 - \alpha)(1 - r_v) \tag{7.5}$$

where $\alpha_0 =$ value of α at the end of initial (hydrostatic) loading, and κ is the nonassociative parameter, and $r_v = \xi_v/\xi$.

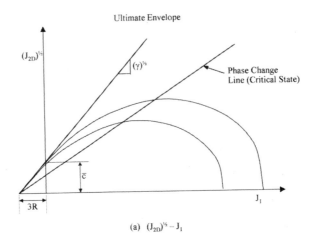

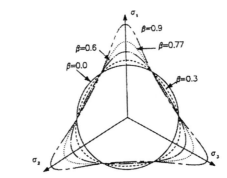

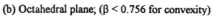

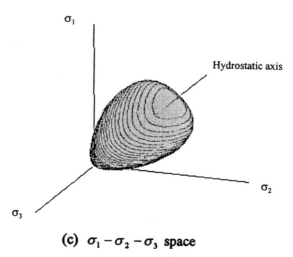

FIGURE 7.1
Plots of F in different stress spaces (1–3).

7.2 Specializations of the HISS Model

Before presenting derivations of the incremental constitutive equations, we discuss various specializations of the HISS model.

7.2.1 Classical Plasticity Models

The parameter β in Eq. (7.1) defines the shape of the yield surface, F, in the principal stress space, Fig. 7.1(b). In the context of the theory of plasticity in which it is required that F be convex, the value of β is limited by $\beta \leq 0.76$ (see Example 7.1). If it is assumed that ($\beta = 0$) and $n = 2$, Eq. (7.1) becomes (here, the overbar is deleted for convenience)

$$F = J_{2D} + \alpha(J_1 + 3R)^2 - \gamma(J_1 + 3R)^2 = 0 \tag{7.6}$$

where $3R$ is the intercept along the J_1-axis denoting tensile or compressive bonding stress; see Fig. 7.1(a).

Equation (7.6) can represent various classical plasticity models such as von Mises, Tresca, Drucker–Prager, and Mohr–Coulomb if there is no continuous yielding involved before the failure (ultimate) yield envelope is reached. Hence, as $\alpha = 0$ at the ultimate condition, we have

$$F = J_{2D} - \gamma J_1^2 - 6\gamma J_1 R - 9\gamma R^2 = 0 \tag{7.7a}$$

Now, $3R$ in Fig. 7.1(a) can be expressed approximately as

$$3R = \frac{\bar{c}}{\sqrt{\gamma}} \tag{7.7b}$$

where $\bar{c}$ is the intercept along the $\sqrt{J_{2D}}$-axis at $J_1 = 0$ and is related to the cohesive (or tensile) strength of the material. Substitution of $3R$ in Eq. (7.7a) leads to

$$J_{2D} - \gamma J_1^2 - 2\sqrt{\gamma}\bar{c}J_1 - \bar{c}^2 = 0 \tag{7.7c}$$

If the effect of J_1^2 is ignored, Eq. (7.7c) reduces to

$$J_{2D} - (2\sqrt{\gamma} \cdot \bar{c})J_1 - \bar{c}^2 = 0 \tag{7.7d}$$

which has the same form as the Drucker–Prager yield criterion, Eq. (6.6a) of Chapter 6. Now, if the effect of J_1 is ignored, Eq. (7.7d) becomes

$$J_{2D} - \bar{c}^2 = 0 \tag{7.7e}$$

which has the same form as the von Mises criterion, Eq. (6.5b) of Chapter 6.

If we assume that the material (soil) is normally consolidated or cohesionless and $3R = 0$, Eq. (7.6) becomes

$$J_{2D} + \alpha J_1^2 - \gamma J_1^2 = 0 \tag{7.8}$$

which has the form similar to the modified Cam Clay model in the critical state soil mechanics, Eq. (6.11) of Chapter 6. In the Cam Clay model, p_0 denotes the hardening parameter, which is dependent on the volumetric plastic strain or void ratio. In the HISS model, the hardening function, α, can depend on the total plastic strain trajectory ξ, Eq. (7.2), which includes both the volumetric and deviatoric plastic strains, thereby providing a generalization of the continuous yielding or hardening process.

In the cap model, failure and continuous yield surfaces are defined (see Chapter 6). In the HISS model, at the ultimate, $\alpha = 0$, and we have Eq. (7.7d), which is analogous to the failure surface, F_f [Eq. (6.16), Chapter 6] in the cap model. Also, Eq. (7.6) is analogous to the continuous yield surface, F_y, in Eq. (6.14) of Chapter 6.

Matsuoka and Nakai (7) proposed the following (yield) criterion (for cohesionless materials) as expanding open surfaces (Fig. 7.2):

$$F = J_1 J_2 - \bar{\lambda}_1 J_3 = 0 \tag{7.9a}$$

and Lade and co-workers (8, 9), proposed the following similar criterion (Fig. 7.2):

$$F = J_1^3 - \bar{\lambda}_2 J_3 = 0 \tag{7.9b}$$

Both above criteria can be expressed as special cases of the F [Eq. (7.1)] in the HISS model. Subsequently, Lade and co-workers (9) modified their criterion to include closed and continuous yield surfaces, similar to the closed continuous yield surfaces in the HISS model (21, 22).

The model proposed by Vermeer (10) introduces the effect of both shear and plastic strains on the hardening response. The total strain increment (or rate), $d\varepsilon$, is expressed as

$$d\underset{\sim}{\varepsilon} = d\underset{\sim}{\varepsilon}^e + d\underset{\sim}{\varepsilon}^{ps} + d\underset{\sim}{\varepsilon}^{pc} \tag{7.9c}$$

where $d\varepsilon^{ps}$ and $d\varepsilon^{pc}$ are the incremental plastic strains due to shear and isotropic compression, respectively. It uses the criterion by Matsuoka and Nakai (7), Eq. (7.9a), to define the yield function relevant to plastic shear strains as

$$F_p^s = J_1 J_2 - \bar{\alpha} J_3 = 0 \tag{7.9d}$$

where the hardening function, $\bar{\alpha}$, is dependent on the plastic shear strain γ^p, mobilized angle of friction (ϕ_m), peak friction angle (at failure) ϕ_p, and shear

Hierarchical Single-Surface Plasticity Models in DSC

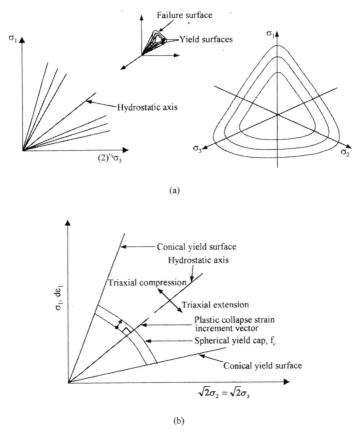

FIGURE 7.2
Plots of open yield and failure surfaces in Lade et al. (8). ©1975, with permission from Elsevier Science.

modulus G_0 at isotropic stress ($J_{10}/3$). A plastic potential function is defined to introduce the effect of volume dilitancy. A specific yield function is defined for volumetric response as

$$F_p^c = \varepsilon_0^c \left(\frac{J_1}{J_{10}}\right)^{\bar{\beta}} - \varepsilon^{pc} \qquad (7.9e)$$

where ε_0^c and $\bar{\beta}$ are material constants, ε^{pc} is another hardening function for volumetric response. The model can involve two (or more) constants to define elastic response, and five for the plastic response for cohesionless materials. For cohesive materials, the number of constants will be higher. Although this model involves effects of both the volumetric and deviatoric plastic strains on the hardening or yielding behavior, it is felt that their separation may not be necessary. The HISS models unify them and also allow the flexibility of using one or both terms in Eq. (7.3).

Table 7.1(a) to (c) shows outlines of various models and their chronological development and comparison of HISS models with other models. It can be seen from the descriptions herein that the HISS models provide improved characterization with an equal or fewer number of parameters for comparable capabilities of other available models.

7.3 HISS Versions

The basic δ_0 version of the HISS approach is based on associative plasticity. Various hierarchical models for the inclusion of additional factors have been developed: nonassociative δ_1-model (19, 20), kinematic and anisotropic hardening δ_2-models for sands and clays (26–28), and viscoplastic δ_{vp} (29–31). It is found that the basic δ_0-model is usually sufficient to characterize the RI response of many materials. With the disturbance, D, in the DSC, and the RI response modelled using the δ_0-model, the nonassociative and anisotropic behavior are accounted for, to some extent. This is because the disturbance, D, causes the deviation from normality of the plastic strains with respect to F (32). Hence, the major attention here is given to the δ_0-model.

7.4 Material Parameters

The elasticity parameters E, ν, G, and K are determined from laboratory stress–strain and volumetric behavior, as described in Chapter 5. Their values will depend on the measured response under a given stress path; see Fig. 7.3(c). For example, E and ν can be found from different stress paths as follows:

Test	E	ν
CTC RTE	$\dfrac{3}{\sqrt{2}} S_1$	$\dfrac{2\|S_1\|}{\|S_2\| + \|S_3\|}$
CTE RTC	$\dfrac{3\sqrt{2}\|S_1\|(\|S_2\| + \|S_3\|)}{(4\|S_1\| + \|S_2\| + \|S_3\|)}$	$\dfrac{\|S_2\| + \|S_3\|}{4\|S_1\|}$
TC TE	$\dfrac{\sqrt{2}}{3}(1 + \nu)(\|S_1\| + \|S_2\| + \|S_3\|)$	—
SS	$\dfrac{\sqrt{3}}{2\sqrt{2}}(1 + \nu)(\|S_1\| + \|S_3\|)$	—

[1] S_i = (average) slope of the unloading/reloading curve, τ_{oct} vs. $\varepsilon_i (i = 1, 2, 3)$ plot.
[2] CTC ($\sigma_1 > \sigma_2 = \sigma_3$), and so on.

TABLE 7.1(a)
Review of Various Models (From Ref. 21, with permission from Elsevier Science)

Model	Yield function, F Potential function, Q	Number of Constants and Comments
1. Classical plasticity (see Chapter 6)	$F = (J_1, J_{2D}, c, \phi) = 0$ $Q \equiv F$	Constants: 3 or 4 One failure surface defines plastic behavior.
2. Critical state (23; see also Chapter 6)	$F_y = F_y[J_1(\text{or } p),$ $J_{2D}(\text{or } q), \lambda,$ $\kappa, M, e_0] = 0$ $F_c = F_f[J_1(\text{or } p),$ $J_{2D}(\text{or } q), M] = 0$ $Q \equiv F$	Constants: 6 F_y (below F_c) defines continuous yielding, and F_c defines critical or failure state. Later modifications consider nonassociative behavior.
3. Cap models (see Chapter 6)	$F_y = F_y[J_1, J_{2D}, R, D,$ $W, Z] = 0$ $F_c = F_f[J_1, J_{2D}, \bar{\alpha},$ $\bar{\gamma}, \bar{\beta}] = 0$ $Q \equiv F$	Constant: 9 F_y (below F_f) defines continuous yielding, and F_f defines failure surface.
4. Matsuoka and Nakai (7)	$F = \dfrac{J_1 J_2}{J_3} - \text{constant} = 0$	F is open failure surface.
5. Lade (8)	$F = \dfrac{J_1^3}{J_3} - \text{constant} = 0$ $F_c = $ spherical cap $Q = J_1^3 - \kappa J_3$	Constants: 13 F gives open (failure) surface.
6. Desai (HISS) (1–6)	$F = \left(\dfrac{J_{2D}}{p_a}\right)^2 - F_b F_s = 0$	Constants: 7 to 8 for δ_0 8 to 9 for δ_1 Single surface F for continuous yielding and ultimate, including failure, peak, and critical state.
7. Schreyer (24)	$F = \gamma \dfrac{J_1}{3} - L - \sigma_3 = 0$	Constants: 10 Single surface F defines continuous yielding approaching the failure or stationary state.
8. Kim and Lade (9)	$F = f_1 \cdot f_2 = 0$ $f_1 = \left(\psi_1 \dfrac{J_1^3}{J_3} - \dfrac{J_1^2}{J_2}\right)$ $f_2 = \left(\dfrac{J_1}{p_a}\right)^h e^q$ $Q = q_1 \cdot q_2$ $q_1 = \left(\psi_1 \dfrac{J_1^3}{J_3} - \dfrac{J_1^2}{J_2} + \psi_2\right)$ $q_2 = \left(\dfrac{J_1}{p_a}\right)^\mu$	Constants: 11 Single surface F defines continuous yielding. Failure is defined by a separate function.

TABLE 7.1(b)

Comparison of δ_0 and Cap (Critical-State) and Lade Models

Model	δ_0	Cap Model
Number of functions for the yield surface	1	2
Number of constants	5 + 2 (linear elastic constants)	7 + 2 (linear elastic) +2 (nonlinear elastic)
Capabilities		
Hardening	Total plastic strains	Volumetric plastic strains
Shear dilation	Before peak stress	At the peak stress only
Different strength in different stress paths	Yes	No
Modifications for added features	Systematic hierarchical additions	No systematic hierarchical method available

Comparison with Lade's Model
Lade and co-workers' model with nonassociative plasticity model involved about 13 constants. Recently they modified their model by incorporating single continuous surfaces as in the HISS models; with the modification, the number of constants is 11 (9).

TABLE 7.1(c)

Comparison of Models for Cyclic Behavior of Clays

Model	HISS-δ_2 (26)	Bounding Surface Model (16, 17)	Mroz, Prevost's Model (12, 13, 15, 25)
Number of functions for the yield surface	1	3	Multiple moving surfaces with similar shape
Number of constants	9	19	5 + 10 × m where m is the number of yield surfaces. For the example presented, the number of constants is 95.
Induced anisotropy	Yes	No	Yes
Prediction of behavior of NC soil	Yes	Yes	Yes
Predicting shear-induced volumetric changes in drained tests or shear-induced pore pressures in undrained test for OC clay	Yes	Yes	Yes

The parameters in the HISS δ_0-model are γ and β associated with the ultimate yield envelope [Figs. 7-1(a) and (b)], n associated with the transition (or phase change) from contractive to dilative or zero volume change, and α the hardening or growth function, which is expressed as a function of the plastic strain trajectory, ξ, Eq. (7.3), which is composed of the deviatoric plastic strain trajectory, ξ_D, and volumetric plastic strain trajectory ξ_V, given by

$$\xi_D = \int (dE_{ij}^p \, dE_{ij}^p)^{1/2} \qquad (7.10a)$$

and

$$\xi_v = \frac{1}{\sqrt{3}}|\varepsilon_v^p| \qquad (7.10b)$$

The hardening function can be expressed in different forms, e.g.,

$$\alpha = \frac{a_1}{\xi^{n_1}} \qquad (7.11a)$$

$$\alpha = \frac{h_1}{[\xi_v + h_3 \xi_D^{h_1}]^{h_2}} \qquad (7.11b)$$

$$\alpha = \bar{h}_1 \exp\left[-\bar{h}_2 \xi \left(1 - \frac{\xi_D}{\bar{h}_3 + \bar{h}_4 \xi_D}\right)\right] \qquad (7.11c)$$

where a_1, η_1, h_i ($i = 1,\ldots,4$) and $\bar{h}_i$ ($i = 1,\ldots,4$) are material parameters.

γ and β. The ultimate yield envelope, which can often be curved, but may be approximated as an average straight line, represents the locus of stress states asymptotic to the observed stress–strain response, Fig. 7.3(a); Fig. 7.3(b) shows such points plotted on $\sqrt{J_{2D}} - J_1$ stress space. Note that the ultimate yield envelope can be different for different stress paths, e.g., compression, extensions, and simple shear, Fig. 7.3(c); details of stress paths are given later. One of the attributes of the ultimate yield envelope in the HISS model is that it defines the asymptotic stress state. As a result, the traditional failure or peak stress states are included as special cases; they are below the ultimate or may coincide with it. There exists a yield surface below (or at the ultimate) corresponding to the failure or peak stress.

In the case of nonfrictional materials like metals and alloys (solders), the effect of mean pressure and stress path may not be significant. In that case, the ultimate envelope will be essentially horizontal, and the intercept along the $\sqrt{J_{2D}}$-axis at $J_1 = 0$ will represent the cohesive or tensile strength of the material. If the behavior is affected by the mean pressure to a small extent, it can be appropriate to assign a small value to the slope (γ) of the ultimate yield envelope.

As indicated in Fig. 7.1(a), the yield surfaces grow with continuous hardening, and finally they approach the ultimate yield or envelope. According to the definition of α, Eq. (7.11), at the ultimate α = 0. Therefore, Eq. (7.1) gives F_u at ultimate as (here the overbar is deleted for convenience)

$$F_u = J_{2D} - \gamma J_1^2 (1 - \beta S_r)^{-0.5} = 0 \qquad (7.12a)$$

Depending on the variation of β, the plot of the ultimate yield surface will be as shown in Fig. 7.1(b), triangular with rounded corners. When β = 0, the yield surface will be circular like in the Drucker–Prager criterion.

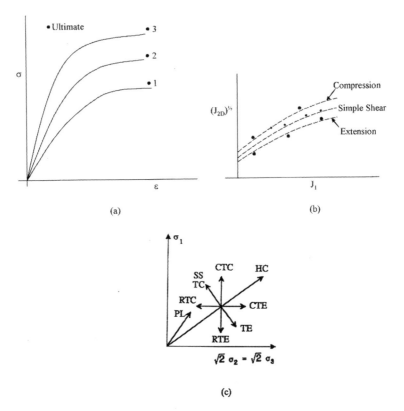

FIGURE 7.3
Parameters γ and β in HISS model.

The values of γ and β are found by substituting values of ($\sqrt{J_{2D}}$, J_1) for various points on the ultimate envelopes, Fig. 7.3(b). If the material behavior is dependent on the stress path, it is appropriate to include points from tests under different stress paths tests such as compression, extension, and simple shear. Then Eq. (7.12a) yields a number of simultaneous equations in γ and β, which are solved by using a least-square or an optimization procedure; these are described in Appendix II. It is found that the value of $\beta \leq 0.76$ to ensure that the yield surface is convex when the theory of plasticity is employed; Fig. 7.1(b) shows that for $\beta = 0.77$, the yield surface is concave.

7.4.1 Curved Ultimate Envelope

For some materials, the ultimate envelope will be curved or nonlinear. To account for a curved envelope, Eq. (7.12a) can be expressed as

$$F_u = J_{2D} - \gamma J_1^q (1 - \beta S_r)^{-0.5} = 0 \qquad (7.12b)$$

where q is a material parameter that can be found by using a least-square or optimization procedure.

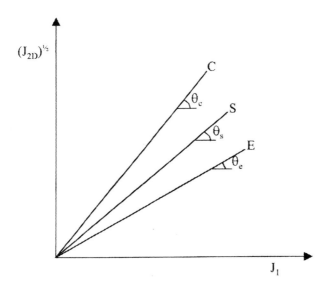

(a) Ultimate Envelopes in $(J_{2D})^{1/2}$ - J_1 Space

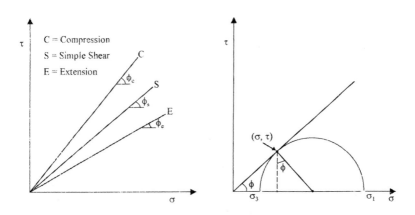

(b) Ultimate Envelopes in Mohr – Coulomb (σ, τ) Space

FIGURE 7.4
Ultimate envelopes in different stress spaces.

7.4.2 Relation Between Ultimate Parameters, Cohesion, and Angle of Friction

If the failure envelope in the Mohr–Coulomb criterion is assumed to be the ultimate envelope, parameters γ and β can be related to the traditional angle of friction, ϕ, from the shear stress, τ, vs. normal stress, σ, plots; see Fig 7.4.

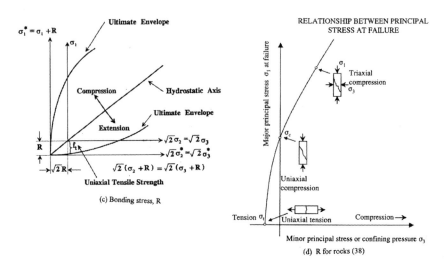

FIGURE 7.4
(continued)

The relations are given below (see Example 7.1 for derivations):

$$p_1 = \tan\theta_C = \left[\sqrt{\gamma}(1-\beta)^{-\frac{1}{4}}\right]_C = \frac{2}{\sqrt{3}}\left(\frac{\sin\phi_C}{3-\sin\phi_C}\right) \quad (7.13a)$$

$$p_2 = \tan\theta_E = \left[\sqrt{\gamma}(1-\beta)^{-\frac{1}{4}}\right]_E = \frac{2}{\sqrt{3}}\left(\frac{\sin\phi_E}{3+\sin\phi_E}\right) \quad (7.13b)$$

as $S_r = 1$ for compression (CTC) and $S_r = -1$ for extension (RTE) paths. Therefore,

$$\sqrt{\gamma} = \frac{\tan\theta_C}{(1-\beta)^{0.5m}} = \frac{\tan\theta_E}{(1-\beta)^{0.5m}} \quad (7.13c)$$

and

$$\beta = \frac{p_1^4 - p_2^4}{p_1^4 + p_2^4} \quad (7.13d)$$

Note that $\sqrt{\gamma}$ refers to the simple shear (SS) stress path, Fig. 7.3(b) and (c), when $S_r = 0$. In the foregoing equations ϕ_C, ϕ_S, and ϕ_E are the slopes of the compression, simple shear, and extension envelopes in the τ–σ stress space, and θ_C, θ_S, and θ_E are the corresponding slopes in the $\sqrt{J_{2D}}$–J_1 stress space, Fig. 7.4(a).

If β is constant, the ultimate envelope is a straight line. For some materials, the ultimate envelope is curved, and then β can be expressed as a function of mean pressure ($J_1/3$) as (33–36)

$$\beta = \beta(J_1) \tag{7.14}$$

or

$$\beta = \beta_0 e^{-\beta_1 J_1} \tag{7.15a}$$

or

$$F_s = \left[\exp\left(\frac{\beta_1}{\beta_0}J_1\right) - \beta S_r\right]^m \tag{7.15b}$$

where β_0 and β_1 are parameters, and $\bar{\beta}_0$ is the reference value. Thus, in this case, it is necessary to consider additional parameters in Eq. (7.14), which can be found by using least-square fit procedures on plots of the ultimate envelope as a function of J_1 (33, 36).

7.4.3 Bonding Stress, R

The bonding stress, R, is introduced by transforming the stress tensor as

$$\sigma^*_{ij} = \sigma_{ij} + R\delta_{ij} \tag{7.16}$$

where δ_{ij} is the Kronecker delta. In Fig. 7.1(a), $3R$ denotes the distance from the intersection of the ultimate yield surface with the J_1-axis to the origin. In the case of geologic materials, where compressive behavior may be of the primary interest, R is related to the tensile strength of the material. In the case of metals, where tensile behavior may be of the main interest, R is related to the compressive strength.

The values of R can be obtained by assuming the ultimate envelope to be a straight line, from

$$R = \frac{\bar{c}}{3\sqrt{\gamma}} \tag{7.17}$$

where $\sqrt{\gamma}$ is the slope of the ultimate envelope and $\bar{c}$ is proportional to the cohesive strength of the material.

The uniaxial tensile strength, f_t (say for a concrete), can be used to evaluate R. As indicated in Fig. 7.4(c), the value of R is slightly greater than f_t (33, 34, 37):

$$1.003 f_t \leq R \leq 1.014 f_t \qquad (7.18)$$

For rocks, f_t can be evaluated from the following empirical expression (38, 35, 36), Fig. 7.4(d):

$$f_t = \frac{1}{2} f_c (m' - \sqrt{m'^2 + 4s}) \qquad (7.19)$$

where f_c is the uniaxial compressive strength, $s = 1.0$ for intact rock, and m' is found from compression test results, which varies from about 0.001 for highly disturbed rock masses to about 25 for hard, intact rock (38).

In the case of metals and metallic materials like solders, tensile behavior may be relevant and the bonding stress, R, will represent (high) compressive strength, as the ultimate yield envelope is essentially horizontal in the $\sqrt{J_{2D}} - J_1$ stress space; i.e., $\sqrt{\gamma}$ is small. For uniaxial tensile behavior, axial stress σ_1 is applied, for which $J_1 = \sigma_1 + 0 + 0 = \sigma_1$ and $\sqrt{J_{2D}} = (1/\sqrt{3})\sigma_1$. At the ultimate yield condition [Fig. 7.1(a)]:

$$\sqrt{J_{2D_u}} = \frac{1}{\sqrt{3}} \sigma_{1u} \qquad (7.20a)$$

$$3R = \frac{\sqrt{J_{2D_u}}}{\sqrt{\gamma}} - \sigma_{1u} \qquad (7.20b)$$

from which the value of γ can be estimated. For example, we can assign a small value of about 95% of $\sqrt{J_{2D_u}}$ at $J_1 = 0$ and compute the slope ($\sqrt{\gamma}$), Fig. 7.1(a) (39, 40).

7.4.4 Phase or State Change Parameter

The phase change (PC) parameter, n, is related to the state of stress at which the material passes through the state of zero volume change. In the case of metallic materials that can be characterized as elastic perfectly plastic, the onset of the yield stress can represent the state of zero volume change, Fig. 7.5(a).

In the case of plastic hardening materials, such a state may occur during the hardening process, Fig. 7.5(b). For initially loose granular materials, it is approached at higher deformations, Fig. 7.5(b). For initially dense materials, the state often is reached somewhere before the peak stress [point b, Fig. 7.5(b)], at which the material experiences transition from compactive to

Hierarchical Single-Surface Plasticity Models in DSC

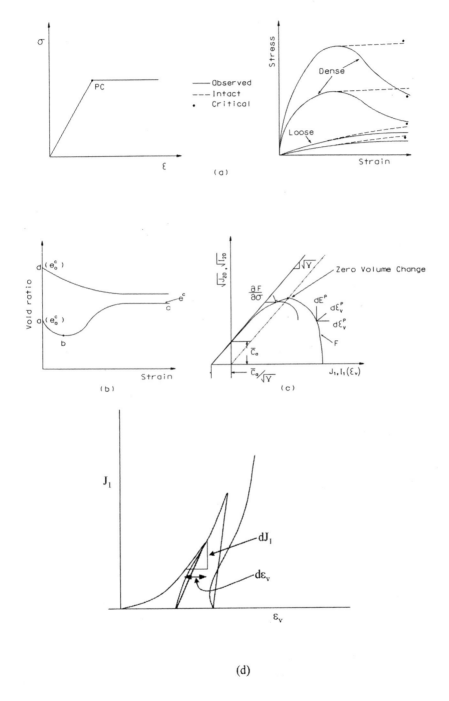

FIGURE 7.5
Stress-strain response and phase change or volume transition.

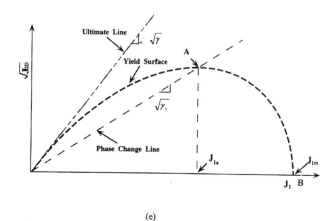

(e)

FIGURE 7.5
(continued)

dilative (volume increase) behavior. In the ultimate region, the state of zero volume change is approached in the residual stress (critical-state) condition. The zero plastic volume change would occur when the volumetric plastic strain increment vanishes, Fig. 7.5(c), that is, when $\partial F/(\partial J_1) = 0$. Based on Eq. (7.1), this leads to the expression for n as (1–6)

$$n = \left. \frac{2}{1 - \left(\frac{J_{2D}}{J_1^2}\right) \cdot \frac{1}{F_s \gamma}} \right|_{d\varepsilon_v = 0} \tag{7.21}$$

where the stresses are relevant to the point (from test data) where the volume change is zero; Fig. 7.5(b) and (c). Details are given in Example 7.1.

The value of n can also be found from hydrostatic tests by using the following equation (4, 28, 41):

$$J_1^{n-1} d\varepsilon_v = \sqrt{3}(n-2)\gamma \cdot dJ_1 \tag{7.22}$$

where dJ_1 and $d\varepsilon_v$ are the increments in the J_1 vs. ε_v response, Fig. 7.5(d).

The value of n should be greater than 2.0 for a convex yield surface and may depend on such factors as initial density; however, as a simplification, an average value of n can be used. For dense granular materials like sands, n may be around 3.0, while for loose granular materials and other materials such as rock and concrete, it would be higher, often of the order of 6 to 10.

For metallic materials and cohesive (saturated) soils, a value somewhat higher than 2, of the order of 2.05 to 2.40, is often found to be appropriate; such a value would imply a (nearly) elliptical yield surface. The following procedure may be used to evaluate n for cohesive materials. For these materials, "failure" may occur long before the load along a stress path reaches the ultimate envelope, Fig. 7.1(a). In other words, failure may occur when the phase change or the critical-state line, Fig. 7.5(e), is approached. Then the value of n is found from (28, 41)

$$\frac{J_{1a}}{J_{1m}} = \left(\frac{2}{n}\right)^{\frac{1}{n-2}} \tag{7.23a}$$

where J_{1m} is the maximum value of J_1 for a yield surface, and J_{1a} is at the intersection (A) of the phase change line and the yield surface. The value of J_{1m} is found from the effective consolidation pressure (p') as

$$J_{1m} = 3p' \tag{7.23b}$$

The value of n can also be found from the slopes of the phase change line, γ_t, and the ultimate line (γ) as

$$\frac{\gamma_t}{\gamma} = \left(\frac{n-2}{n}\right)^{1/2} \tag{7.24}$$

Details are given in Example 7.2.

7.4.5 Hardening or Growth Parameters

These parameters, e.g., a_1 and η_1 for α in (Eq. 7.11a), are found from laboratory stress–strain tests plotted, for example, in terms of principal stresses, σ_i ($i = 1, 2, 3$), vs. principal strains ε_i ($i = 1, 2, 3$); see Fig. 7.6(a). This implies that the test specimen is subjected to the principal stresses with no (applied) shear stresses. If the results are available from pure shear tests, then the plots in terms of shear stress τ and shear stress γ can be used; see Fig. 7.6(b). The main quantity needed here is the plastic strain trajectory given by

$$\xi = \int (d\varepsilon_1^p \cdot d\varepsilon_1^p + d\varepsilon_2^p \cdot d\varepsilon_2^p + d\varepsilon_3^p \cdot d\varepsilon_3^p)^{1/2} \tag{7.25a}$$

or

$$\xi = \xi_D = \int (d\gamma^p \cdot d\gamma^p)^{1/2} \tag{7.25b}$$

The increments of plastic strains $d\varepsilon_1^p$, $d\varepsilon_2^p$, $d\varepsilon_3^p$ or $d\gamma^p$ are found from stress–strain curves, Fig. 7.6(a) or (b), based on the knowledge of the unloading modulus (E_u) or (G_u). Then, the increment of the trajectory $d\xi$ is computed for each stress increment and the total value ξ is found as

$$\xi = \xi_0 + \sum_{i=1}^{j} d\xi_i \tag{7.26}$$

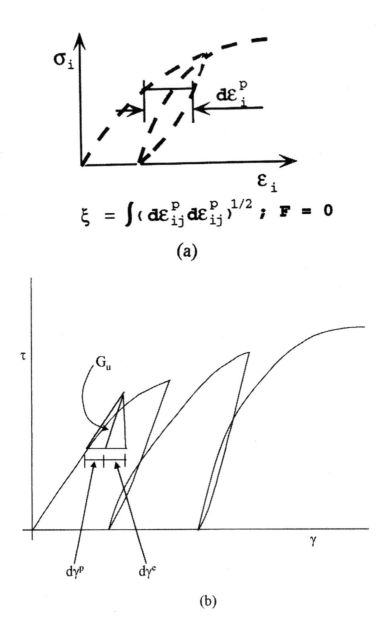

FIGURE 7.6
Hardening or growth parameters, a_1 and η_1.

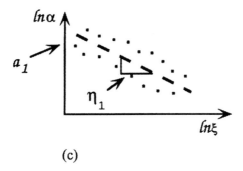

FIGURE 7.6
(continued)

where i is the increment and ξ_0 is the initial value of ξ before the (shear) test. The latter denotes the plastic strains due to initial conditions such as initial stresses or strains (due to hydrostatic loading), microcracking, and defects. If the material has no such prior history, $\xi_0 = 0$.

Now Eq. (7.11a) is expressed as

$$\ln \alpha + \eta_1 \ln \xi = \ln a_1 \tag{7.27}$$

The value of the hardening function, α, for the state of stress corresponding to the total stress after the stress increment, i, is found using Eq. (7.1), i.e., $F = 0$. Thus, values of $\ln \alpha$ and $\ln \xi$ are found at various points along the stress-strain curves and are plotted as shown in Fig. 7.6(c). If the points are not widely scattered, an average straight line is drawn through the points. The slope of the line gives η_1, which denotes the rate of hardening. The intercept along $\ln \alpha$, when $\ln(\xi) = 0$, i.e., $\xi = 1$, gives the value of a_1. For some materials, the values of a_1 and η_1 may depend on factors such as initial density, stress path, and temperature. In that case, additional constants are needed to define a_1 and η_1. The values of parameters in other forms of α, Eq. (7.11), can be obtained similarly from laboratory test data (4–6, 19, 41).

7.4.6 Nonassociative δ_1-Model

The parameter κ in Eq. (7.5) allows for the correction (deviation from normality) with respect to the δ_0-model; see Fig. 7.7(a). If it is assumed that the correction is affected significantly by the volume change behavior, this parameter is obtained from the volumetric response, Fig. 7.7(b). In general, κ can vary with strain; however, as a simplification, it is assumed to be constant, related to the slope, S_u, of the volumetric response near the ultimate region, Fig. 7.7(b).

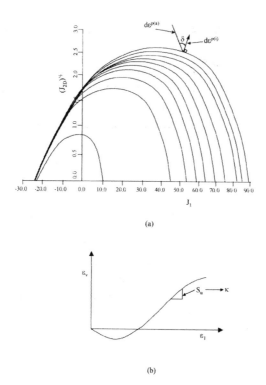

FIGURE 7.7
Nonassociative response with DSC and parameter, κ.

Then κ is found from (4–6, 19, 20)

$$\kappa = \frac{1}{(\alpha_0 - \alpha)(1 - r_v)}\left(\frac{Y}{Z} - \alpha\right) \tag{7.28}$$

where Y and Z are related to the state of stress in the ultimate region and S_u.

The foregoing (eight) parameters (E, ν; γ, β; n; a_1, η_1, and R) are required to define the associative (isotropic hardening) δ_0-version in the HISS model. The δ_0-version can be used to characterize the behavior of material in the RI state in the DSC.

The introduction of the disturbance function, D, in the DSC, allows for the deviation from normality of the observed strain increment, $d\varepsilon^{p(a)}$. In other words, the plastic strain increment $d\varepsilon^{p(i)}$ from the δ_0-model is normal to F, while $d\varepsilon^{p(a)}$ is not normal to F and shows deviation, δ, from normality; Fig. 7.7(a) (32). Such deviation can allow for the response as affected by frictional characteristics, similar to that given by nonassociative plasticity models (19). Also, such deviation can represent a measure of induced anisotropy (26). As a result, when the DSC model is used, it may not be necessary to employ the classical nonassociative and anisotropic hardening models.

Hierarchical Single-Surface Plasticity Models in DSC

TABLE 7.2
Parameters in HISS Models

Model	Constants for δ_0-Model	Additional Constants Beyond δ_0-Model	Total
δ_0-Associative*	7 to 8	—	7 to 8
δ_0-Nonassociative	7 to 8	1	8 to 9
δ_{0+vp}-Viscoplastic	7 to 8	2	9 to 11
δ_{0+D}-Disturbance	7 to 8	3	10 to 12
δ_2-Anisotropic (sands) cyclic loading	7	4	11 to 12
δ_0^* (clays) cyclic loading	8	2	10
Fluid pressure, p	7 to 8	2	9 to 10
Temperature (T)	7 to 8	m	7 to 8 + m (depends on how many parameters are functions of T)

*For granular materials, $R = 0$; hence constants = 7.

As the δ_0-model is mainly used (sometimes δ_1 is used) to represent the RI response, details of the other versions, e.g., the anisotropic hardening (δ_2), are not given here. They are available in various publications (3–6, 26–28).

Computer Program. A computer program has been developed to evaluate the foregoing parameters based on given sets of laboratory test data. It is described in Appendix II. Table 7.2 shows a summary of parameters for various HISS models including hierarchical addition of constants to allow for factors for increasing capabilities.

7.4.7 Thermal Effects on Parameters

As discussed in Chapters 5 and 6, the thermal effects can be included in the plasticity (HISS) models by expressing the parameters in terms of temperature. A simple expression is often used (39, 40):

$$p = p_r \left(\frac{T}{T_r}\right)^c \tag{7.29}$$

where p is any parameter, p_r is its value at response temperature T_r (say, 300K), T is temperature, and c is a parameter. The values of p_r and c are found from laboratory stress–strain–volume change data at different temperatures. The procedure for finding the elastic parameters as a function of temperature is given in Chapter 5 [Eq. (5.37)].

The procedures for finding the parameters γ, β, a_1, η_1, n, and R in the δ_0-model from data at a given temperature are described earlier in this chapter. Their variation vs. temperature leads to functions such as that in Eq. (7.29). Example of such variations for solder (60Sn/40Pb) are given below (39, 40);

they are derived from uniaxial tension tests under two strain rates, $\dot{\varepsilon} = 0.0002$ and $0.002/\text{sec}$ (42); see Fig. 7.8.

Strain Rate $\dot{\varepsilon} = 0.0002$:

$$\gamma(T) = \gamma_{300}\left(\frac{T}{300}\right)^{-0.072} ; \quad \gamma_{300} = 0.00082$$

$$\alpha(T) = \alpha_{300}\left(\frac{T}{300}\right)^{-2.578} ; \quad \alpha_{300} = \frac{a_1(300)}{\xi^{n_1(300)}} \quad (7.30\text{a})$$

$$a_1(300) = 0.024 \times 10^{-4}; \quad \eta_1(\text{average}) = 0.394$$

$$R(T) = R_{300}\left(\frac{T}{300}\right)^{-2.95} ; \quad R_{300} = 217.5 \text{ mPa}$$

$$E(T) = E_{300}\left(\frac{T}{300}\right)^{-0.292} ; \quad E_{300} = 23.45 \text{ GPa}$$

$$\nu(T) = \nu_{300}\left(\frac{T}{300}\right)^{0.14} ; \quad \nu_{300} = 0.40 \quad (7.30\text{b})$$

$$\alpha_1(T) = \alpha_{1(300)}\left(\frac{T}{300}\right)^{0.24} ; \quad \alpha_{T(300)} = 3 \times 10^{-6}/\text{K}$$

Strain Rate $\dot{\varepsilon} = 0.002$:

$$\gamma(T) = \gamma_{300}\left(\frac{T}{300}\right)^{-0.034} ; \quad \gamma_{300} = 0.00082$$

$$\alpha(T) = \alpha_{300}\left(\frac{T}{300}\right)^{-5.5} ; \quad \alpha_{300} = \frac{a_1(300)}{\xi^{n_1(300)}} \quad (7.31)$$

$$a_1(300) = 2.0 \times 10^{-6}; \quad \eta_1(\text{average}) = 0.615$$

$$R(T) = R_{300}\left(\frac{T}{300}\right)^{-1.91} ; \quad R_{300} = 240.67 \text{ MPa}$$

The value of $\beta = 0$ implies essentially a circular yield surface in the $\sigma_1 - \sigma_2 - \sigma_3$ space, and $n = 2.1$ is assumed for both strain rates. Variations for elastic parameters for the same solder are presented in Chapter 5.

7.4.8 Rate Effects

The responses of some materials are affected by the rate of loading. In the context of the laboratory behavior, the rate effect is often defined through the strain rate, $\dot{\varepsilon}$. The rate effect can be included in the HISS (and DSC) models

Hierarchical Single-Surface Plasticity Models in DSC

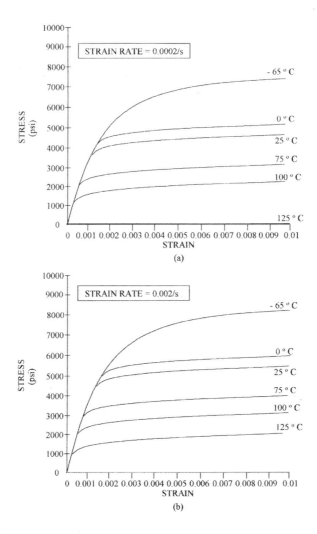

FIGURE 7.8
Uniaxial tension stress–strain curves of 60% Sn–40% PB solders: (a) strain rate = 0.0002/sec, (b) strain rate = 0.002/sec (42). ©1990, IEEE. With permission.

by expressing the parameters as functions of the rate. For example, for both the thermal and rate effects, any parameter p can be expressed as

$$p = p(T, \dot{\varepsilon}) \qquad (7.32)$$

The parameters for appropriate functions in Eq. (7.32) can be determined from laboratory tests under different T and $\dot{\varepsilon}$. Strain rate effect is discussed later in Chapter 8.

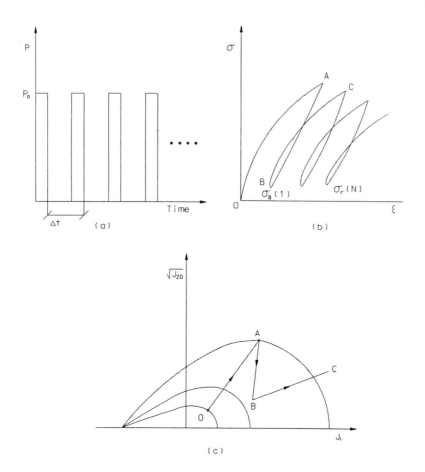

FIGURE 7.9
Repetitive load and stress conditions.

7.5 Repetitive Loading

Under repetitive loading, Fig. 7.9(a), the applied load amplitude is increased (to P_0) and decreased to zero, and then increased to P_0 after a lapse of time, Δt. If we ignore, for the time being, the effect of time, we can consider that the load is applied as above repeatedly at different cycles, N, and the behavior of the material may be assumed to be independent of time.

Under the application of the external load, P_0, a material element would undergo loading–unloading–reloading cycles as depicted in Fig. 7.9(b). Here, upon removal of the load, P_0, the state of stress in a material element, may not return to zero. In other words, at the end of unloading in a given cycle, the material may retain a residual or initial stress, $\sigma_r(N)$, Fig. 7.9(b).

Figure 7.9(c) depicts the situation with respect to yield surfaces in the $J_1-\sqrt{J_{2D}}$ space. Loading from 0 to A starts from the initial state (σ_0, α_0), and at the end of the loading, the state will be (σ_A, α_A). Upon unloading, $(A-B)$, during which plastic-strain change may occur, the state will be (σ_B, α_B). Then reloading to P_0 $(B-C)$ will start from the residual or initial state (σ_B, α_B); in general, the residual state will be $[\sigma_r(N), \alpha_r(N)]$, where N is the cycle number. Thus, the application of full load P_0 will start from the residual condition, which can be considered as the initial state for the current cycle, and will result in the modified matrix, $\underline{C}^{ep}$, corresponding to the residual state. Hence, repetitive loading will lead to continuously increasing plastic strains as the cycles increase. However, the rate of increase in plastic strains may decrease with cycles.

The foregoing approach can be used to simulate behavior of materials (metals, geologic, asphalt, etc.) under repetitive loading. When implemented in a computer (finite-element) method, the procedure would allow calculation of displacements, stresses, strain, and pore water pressures under repetitive loading (see Chapter 13). Examples of materials such as asphalt and soils, with simplified procedures for computing plastic or permanent strains (deformations), are given later.

7.5.1 Dynamic Loading

Implementation of the HISS and DSC models for dynamic (earthquake) analysis is given in Chapter 13. Here, the loading and the stress–strain behavior can involve both positive and negative values of loads and stresses, which is often referred to as two-way loading; see Fig. 7.10.

7.6 The Derivation of Elastoplastic Equations

The total (incremental) strain, $d\underline{\varepsilon}^i$, is decomposed and expressed in matrix notation as

$$d\underline{\varepsilon}^i = d\underline{\varepsilon}^e + d\underline{\varepsilon}^p \qquad (7.33)$$

where $d\underline{\varepsilon}^e$ = elastic strain vector, and $d\underline{\varepsilon}^p$ = plastic strain vector.

The incremental elastic strain, $d\underline{\varepsilon}^e$, is related to the increment stresses, $d\underline{\sigma}$, as

$$d\underline{\sigma} = \underline{C}^e \, d\underline{\varepsilon}^e \qquad (7.34)$$

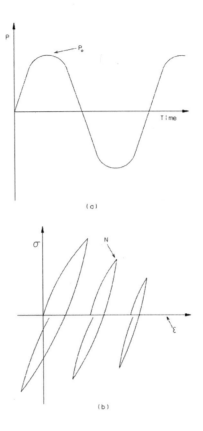

FIGURE 7.10
Dynamic loading.

Here, the elastic constitutive matrix, $\underline{C}^e$, for isotropic material is expressed in terms of the elastic modulus, E, and Poisson's ratio, ν, or the shear modulus, G, and the bulk modulus, K. Often, these parameters are assumed to be constant for all increments during the incremental analysis. However, they can be considered as variable and expressed as tangent quantities:

$$E_t = E_t[(\sigma_1 - \sigma_3), p] \quad \text{and} \quad \nu_t = \nu_t[(\sigma_1 - \sigma_3), p] \quad (7.35)$$

or

$$G_t = G_t(\sqrt{J_{2D}}, J_1) \quad \text{and} \quad K_t = K_t(J_1, \sqrt{J_{2D}}) \quad (7.36)$$

where the subscript t denotes tangent quantity, which can be found as the derivative (slope) of the function used. The latter can be such mathematical functions as hyperbola, parabola, splines, and exponential (43).

Now, from the theory of plasticity, we can use the normality rule (44, 45)

$$d\underset{\sim}{\varepsilon}^p = \lambda \frac{\partial Q}{\partial \underset{\sim}{\sigma}} \tag{7.37}$$

and the consistency condition as

$$dF = 0 \tag{7.38}$$

where $d\underset{\sim}{\varepsilon}^p$ = vector of incremental plastic strains, and λ is a scalar (positive) proportionality parameter.

Equation (7.38) gives

$$\left(\frac{\partial F}{\partial \underset{\sim}{\sigma}}\right)^T \cdot d\underset{\sim}{\sigma} + \frac{\partial F}{\partial \xi} \cdot d\xi = 0 \tag{7.39}$$

The term $d\xi$ is given by

$$\begin{aligned}
d\xi &= \left[(d\underset{\sim}{\varepsilon}^p)^T \cdot d\underset{\sim}{\varepsilon}^p\right]^{1/2} \\
&= \left[\lambda\left(\frac{\partial Q}{\partial \underset{\sim}{\sigma}}\right)^T \cdot \lambda\left(\frac{\partial Q}{\partial \underset{\sim}{\sigma}}\right)\right]^{1/2} \\
&= \lambda\left[\left(\frac{\partial Q}{\partial \underset{\sim}{\sigma}}\right)^T \cdot \frac{\partial Q}{\partial \underset{\sim}{\sigma}}\right]^{1/2} \\
&= \lambda \gamma_F
\end{aligned} \tag{7.40}$$

Therefore, Eq. (7.39) leads to

$$\left(\frac{\partial F}{\partial \underset{\sim}{\sigma}}\right)^T \cdot d\underset{\sim}{\sigma} + \frac{\partial F}{\partial \xi} \cdot \lambda \cdot \gamma_F = 0 \tag{7.41}$$

Substitution for $d\underset{\sim}{\sigma}$ from Eq. (7.34) gives

$$\left(\frac{\partial F}{\partial \underset{\sim}{\sigma}}\right)^T \underset{\sim}{C}^e \cdot d\underset{\sim}{\varepsilon}^e + \frac{\partial F}{\partial \xi} \cdot \lambda \cdot \gamma_F = 0 \tag{7.42a}$$

and substitution for $d\underset{\sim}{\varepsilon}^e$ from Eq. (7.33) leads to

$$\left(\frac{\partial F}{\partial \underset{\sim}{\sigma}}\right)^T \underset{\sim}{C}^e \cdot (d\underset{\sim}{\varepsilon} - d\underset{\sim}{\varepsilon}^p) + \frac{\partial F}{\partial \xi} \cdot \lambda \cdot \gamma_F = 0 \tag{7.42b}$$

which, with $d\varepsilon^p$ from Eq. (7.37), gives

$$\left(\frac{\partial F}{\partial \sigma}\right)^T \mathcal{C}^e d\varepsilon - \left(\frac{\partial F}{\partial \sigma}\right)^T \mathcal{C}^e \cdot \lambda \frac{\partial Q}{\partial \sigma} + \frac{\partial F}{\partial \xi} \cdot \lambda \cdot \gamma_F = 0 \qquad (7.42c)$$

Hence,

$$\left(\frac{\partial F}{\partial \sigma}\right)^T \cdot \mathcal{C}^e \cdot d\varepsilon - \lambda \left[\left(\frac{\partial F}{\partial \sigma}\right)^T \mathcal{C}^e \cdot \frac{\partial Q}{\partial \sigma} - \frac{\partial F}{\partial \xi} \cdot \gamma_F\right] = 0 \qquad (7.42d)$$

which leads to the expression for λ as

$$\lambda = \frac{\left(\frac{\partial F}{\partial \sigma}\right)^T \mathcal{C}^e \cdot d\varepsilon}{\left(\frac{\partial F}{\partial \sigma}\right)^T \mathcal{C}^e \cdot \frac{\partial Q}{\partial \sigma} - \frac{\partial F}{\partial \xi} \cdot \gamma_F} \qquad (7.43)$$

Now, use of Eq. (7.37) with λ and substitution in Eq. (7.34) gives

$$d\sigma = \mathcal{C}^e(d\varepsilon - d\varepsilon^p) \qquad (7.44a)$$

$$= \left[\mathcal{C}^e - \frac{\mathcal{C}^e \frac{\partial Q}{\partial \sigma}\left(\frac{\partial F}{\partial \sigma}\right)^T \mathcal{C}^e}{\left(\frac{\partial F}{\partial \sigma}\right)^T \mathcal{C}^e \cdot \frac{\partial Q}{\partial \sigma} - \frac{\partial F}{\partial \xi} \cdot \gamma_F}\right] d\varepsilon \qquad (7.44b)$$

or

$$d\sigma = (\mathcal{C}^e - \mathcal{C}^p) d\varepsilon \qquad (7.44c)$$

$$d\sigma = \mathcal{C}^{ep} \cdot d\varepsilon \qquad (7.44d)$$

Here, $\mathcal{C}^{ep}$ is called the *elasto-plastic constitutive matrix*.

Note that for associative plasticity, $Q \equiv F$ will be substituted in Eq. (7.44), and that Eq. (7.44) can be used for any F and Q, including those from Chapter 6 and Eqs. (7.1) and (7.4).

7.6.1 Details of Derivatives

We now give details of the derivatives involved in Eq. (7.44). The quantities $\bar{J}_1$ and $\bar{J}_{2D}$ will be used as J_1 and J_{2D} (without overbars) with the understanding that they are nondimensionalized with respect to p_a; hence, although not shown in the following derivations, p_a will occur in the denominator.

From the chain rule of differentiation, we have from Eq. (7.1)

$$\frac{\partial F(J_1, J_{2D}, J_{3D})}{\partial \sigma_{ij}} = \frac{\partial F}{\partial J_1} \cdot \frac{\partial J_1}{\partial \sigma_{ij}} + \frac{\partial F}{\partial J_{2D}} \cdot \frac{\partial J_{2D}}{\partial \sigma_{ij}} + \frac{\partial F}{\partial J_{3D}} \cdot \frac{\partial J_{3D}}{\partial \sigma_{ij}} \quad (7.45a)$$

where

$$\frac{\partial F}{\partial J_1} = (n\alpha J_1^{n-1} - 2\gamma J_1)F_s \quad (7.46a)$$

and

$$\frac{\partial J_1}{\partial \sigma_{ij}} = \delta_{ij} \quad \text{or} \quad \left(\frac{\partial J_1}{\partial \underline{\sigma}}\right)^T = \begin{bmatrix} 1 & 1 & 1 & 0 & 0 & 0 \end{bmatrix} \quad (7.47b)$$

$$\frac{\partial F}{\partial J_{2D}} = 1 - F_b[m(1 - \beta S_r)^{m-1}]\left(\beta \cdot \frac{3}{2} \frac{\sqrt{27}}{2} \cdot J_{3D} \cdot J_{2D}^{-5/2}\right) \quad (7.47a)$$

Here

$$\frac{\partial J_{2D}}{\partial \sigma_{ij}} = S_{ij} \quad (7.47b)$$

where S_{ij} is the deviatoric stress tensor:

$$S_{ij} = \sigma_{ij} - \frac{J_1}{3} \cdot \delta_{ij} \quad (7.47c)$$

$$\frac{\partial F}{\partial J_{3D}} = m\beta \frac{\sqrt{27}}{2} F_b (1 - \beta S_r)^{m-1} \cdot J_{2D}^{-3/2} \quad (7.48a)$$

and

$$\frac{\partial J_{3D}}{\partial \sigma_{ij}} = T_{ij} - \frac{2}{3} J_{2D} \cdot \delta_{ij} \quad (7.48b)$$

where $T_{ij} = S_{ik} \cdot S_{kj}$.

The above expressions can be written in the matrix notation, e.g., Eq. (7.47b), as

$$\left(\frac{\partial J_{2D}}{\partial \sigma}\right)^T = \underset{\sim}{S}^T \qquad (7.47c)$$

where $\underset{\sim}{S} = [S_1\ S_2\ S_3\ S_4\ S_5\ S_6] = [S_{11}\ S_{22}\ S_{33}\ 2S_{12}\ 2S_{23}\ 2S_{31}]$ denote the components of the deviatoric stresses, and Eq. (7.48b)

$$\left(\frac{\partial J_{3D}}{\partial \sigma}\right)^T = \left\{\begin{array}{c} T_1 - \frac{2}{3}J_{2D} \\ T_2 - \frac{2}{3}J_{2D} \\ T_3 - \frac{2}{3}J_{2D} \\ T_4 \\ T_5 \\ T_6 \end{array}\right\} \qquad (7.48c)$$

where $\underset{\sim}{T}^T = [T_1\ T_2\ T_3\ T_4\ T_5\ T_6] = [T_{11}\ T_{22}\ T_{33}\ 2T_{12}\ 2T_{23}\ 2T_{31}]$.

The expressions for $\partial Q/\partial J_1$, $\partial Q/\partial J_{2D}$, and $\partial Q/\partial J_{3D}$ can be calculated similarly. The main difference will be in the expression of α, i.e.,

For the δ_0-msodel:

$$F_{b0} = -\alpha J_1^n + \gamma J_1^2 \qquad (7.49a)$$

For the δ_1-model:

$$F_{b1} = -\alpha_Q J_1^n + \gamma J_1^2 \qquad (7.49b)$$

where α_Q is given in Eq. (7.5).

7.7 Incremental Iterative Analysis

The verification of a constitutive model constitutes an important phase toward reliable use of the model for practical application. Such verification is usually performed in two steps: (1) prediction of laboratory tests for which the parameters are determined and of *independent* tests not used in finding the parameters; and (2) implementation of the model in solution (computer finite-element) procedures and then prediction of the behavior of practical (field) boundary-value problems and/or those simulated in the laboratory. In this chapter, we present details of the implementation of the HISS model in computational procedures, and validations with respect to laboratory behavior of a number of materials. Validations with respect to (field) boundary-value problems are given in Chapter 13.

In dealing with nonlinear material behavior, e.g., elastoplastic characterization, it is usually necessary to perform incremental analysis, very often accompanied by iterations so as to ensure convergence to equilibrium states. The finite-element incremental equations in the displacement-based approach are derived as (46)

$$\underline{k}\, d\underline{q}_i = d\underline{Q}_i \tag{7.50}$$

where $\underline{k}$ = tangent stiffness matrix, q = vector of nodal displacements, Q = vector of nodal-applied loads, d denotes increment, and i denotes incremental stage of loading. Here, the total load Q is divided into (small) increments = dQ_i; see Fig. 7.11(a); the number of increments, N, is chosen by the analyst, depending on the properties of the problems on hand and experience. The stiffness matrix $\underline{k}$ is given by

$$\underline{k}_{i-1} = \int \underline{B}^T \underline{C}^{ep}_{i-1} \underline{B}\, dV \tag{7.51}$$

where V is the volume (of finite element) and $\underline{B}$ is the strain-displacement transformation matrix in (46, 47)

$$d\underline{\varepsilon}_i = \underline{B}\, dq_i \tag{7.52}$$

ε = vector of strain components. Note that in Eq. (7.51) the elastoplastic constitutive matrix is evaluated at the start of the current increment (i) or at the end of the previous increment ($i - 1$); Fig. 7.11(b).

The *assembled* equations are solved for incremental displacements at all node points, which provides the incremental element strains from Eq. (7.52). Then the incremental stress, $d\underline{\sigma}$ is evaluated from

$$d\underline{\sigma}_i = \underline{C}^{ep}_{i-1}\, d\underline{\varepsilon}_i \tag{7.53a}$$

and the total stress after the increment is given as

$$\underline{\sigma}_i = \underline{\sigma}_{i-1} + d\underline{\sigma}_i \tag{7.53b}$$

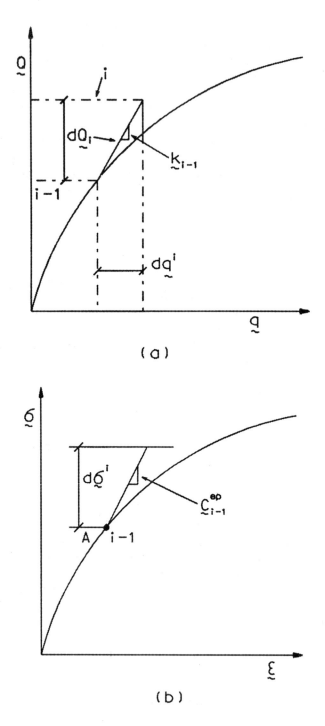

FIGURE 7.11
Incremental loading and drift correction.

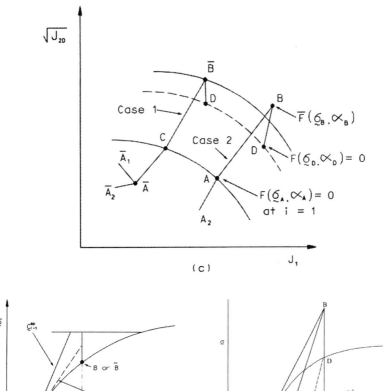
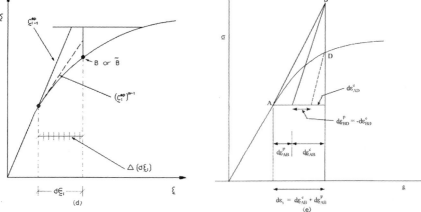

FIGURE 7.11
(continued)

Now, if (by chance) the stress, σ_i, and the corresponding hardening function, α_i, after the stress increment, $d\sigma_i$, satisfy the yield function, F, at point B [Fig. 7.11(c)], then

$$F(\sigma_B, \alpha_B) = 0 \tag{7.54}$$

Here, $\sigma_B = \sigma_i$ and α_B = hardening function at B (see below). However, such an occurrence is rare, and usually Eq. (7.54) will not be satisfied. The "correct" yield surface F, shown by the dashed curve in Fig. 7.11(c), will, therefore, be different from $\bar{F}$; the difference between them is often referred to as the *drift*. Hence, it becomes necessary to correct for the drift so as to obtain $F \approx 0$. This process is called *drift correction*.

7.7.1 Possible Stress States

The above case, i.e., when point A is exactly on the previous yield surface at $(i-1)$, may not occur for all Gauss points (elements). In other words, at $(i-1)$ some points may be in the elastic range, as indicated by $\bar{A}$ in Fig. 7.11(c). In the case of incremental analysis, it becomes necessary to consider both possibilities.

In order to check whether or not the current surface $F = 0$ is satisfied, we need to calculate the hardening function, α, Eq. (7.11). This is achieved by calculating the plastic strain increment as

$$d\underset{\sim}{\varepsilon}_i^p = d\underset{\sim}{\varepsilon}_i - d\underset{\sim}{\varepsilon}_i^e \qquad (7.55)$$

in which the elastic strains vector, $d\underset{\sim}{\varepsilon}_i^e$, is obtained from

$$d\underset{\sim}{\varepsilon}_i^e = (\underset{\sim}{C}^e)^{-1} d\underset{\sim}{\sigma}_i \qquad (7.56)$$

Once $d\underset{\sim}{\varepsilon}_i^p$ is found, the total plastic strain trajectory [Eq. (7.26)] is found as

$$\xi_i = \xi_{i-1} + [(d\underset{\sim}{\varepsilon}_i^p)^T d\underset{\sim}{\varepsilon}_i^p]^{1/2} \qquad (7.57)$$

Then, α_i is found from Eq. (7.11).

Now, $\underset{\sim}{\sigma}_i$ and α_i are substituted in Eq. (7.1) as

$$F(\underset{\sim}{\sigma}_i, \alpha_i) \gtreqless 0 \qquad (7.58)$$

If $F \approx 0$, then the new stress point lies on the corresponding yield surface, and we proceed to the next increment.

Elastic State. If $F < 0$, the new stress point is in the elastic (unloading) region; and if $F > 0$, (drift) correction is needed. For elastic unloading [A to A_2 or $\bar{A}$ to $\bar{A}_2$, Fig. 7.11(c)], assuming that during unloading, no plastic strains occur (i.e., α_A remains the same), the total stress is found as

$$\underset{\sim}{\sigma}_i = \underset{\sim}{\sigma}_{i-1} + \underset{\sim}{C}^e \cdot d\underset{\sim}{\varepsilon}_i \qquad (7.59)$$

7.7.2 Elastic-Plastic Loading

If $F > 0$, i.e., there is elastoplastic loading, two cases need consideration.

Case 1. The material is in the elastic range at the start of the current increment and yields during this increment, $\bar{A}$ to $\bar{B}$; Fig. 7.11(c). Since the behavior is elastic from $\bar{A}$ to C, it is necessary to calculate the stress state at C so as to apply the drift correction scheme for C to $\bar{B}$. Here, a Newton–Raphson-type procedure can be used to find, iteratively, from iterations ($j = 1, 2, \ldots, m$) the fraction of the stress increment that provides the location of point C. For this, we write the yield surface, F, as

$$F[(\sigma_{i-1}^{j-1} + \delta^j d\sigma_i, \alpha_{i-1}^{j-1})] = F(\sigma_{i-1}^{j-1}, \alpha_{i-1}^{j-1}) + \delta^j \frac{\partial F}{\partial \sigma} \cdot d\sigma_i = 0 \quad (7.60)$$

Therefore,

$$\delta^j = -\frac{F(\sigma_{i-1}^{j-1}, \alpha_{i-1}^{j-1})}{\frac{\partial F}{\partial \sigma} \cdot d\sigma_i} \quad (7.61)$$

which represents the portion of the total stress increment that represents the correction to locate the stress state at point C. Then the stress at the jth iteration is given by

$$\sigma_{i-1}^j = \sigma_{i-1}^{j-1} + \delta^j d\sigma_i \quad (7.62)$$

Here, $\sigma_{i-1}^0 = \sigma_{i-1}$. The iterations j are continued until convergence is reached. For instance, the convergence is considered to have been reached when

$$\left| F(\sigma_{i-1}^j, \alpha_{i-1}) \right| \leq \bar{\varepsilon} \quad (7.63)$$

where $\bar{\varepsilon}$ is the tolerance, which can be equal to 10^{-4} (or less).

Now, the actual strain increment, $d\varepsilon^j$ from C to $\bar{B}$, is computed as the strain increment $d\underset{\sim}{\varepsilon}$ minus the elastic strain increment from $\bar{A}$ to C, i.e.,

$$d\underset{\sim}{\varepsilon} = d\underset{\sim}{\varepsilon}_i - \sum_j \delta^j (\underset{\sim}{C}^e)^{-1} d\sigma_i \quad (7.64)$$

With the above information, we have the stress at point $C = \sigma_{i-1}^j = \sigma_{i-1}$ and the stress at point $\bar{B}$ (or B) $= \sigma_i$; and the strain increment $d\varepsilon^j$ from C to $\bar{B}$. Now we can apply the correction so as to locate $F \approx 0$ at point D, Fig. 7.11(c).

Case 2. Here, the increment starts at point A when $F_A = 0$, which is a specialization of case 1.

7.7.3 Correction Procedures

A number of correction procedures have been proposed in the context of the incremental-iterative analysis for the integration of elastoplastic equations, e.g., subincrementation (2, 28, 48), drift correction (49, 28), and substepping

(50). The subincrementation procedure may lead to convergence if the subincrements (of strain) are adopted to be very small. It can be appropriate for integrating the incremental constitutive equations, Eq. (7.44), for backpredicting stress–strain responses. However, when the finite-element analysis, with many elements, is to be performed, the subincrementation procedure can be time-consuming and may not often provide convergence. In that case, the drift correction based on predictor-correction procedures, often in combination with the subincrementation procedure, can provide improved and robust schemes. In the substepping scheme, the stress increments are computed on the basis of known strain increments by adjusting the size of each substep automatically (50). The subincrementation and drift correction procedures are used commonly with the HISS-plasticity models. Hence, their details are given below.

7.7.4 Subincrementation Procedure

It is first required to set initial values of various quantities as

$$\sigma_i^0 = \sigma_{i-1}; \quad C_i^{ep(0)} = C_{i-1}^{ep}; \quad \xi_i^0 = \xi_{i-1} \tag{7.65}$$

Here the strain increment corresponding to C to $\bar{B}$ (or A to B) is divided into a number of small strain increments $\Delta(d\varepsilon_i) = d\varepsilon_i^j/n$, where n is the number of subincrements, Fig. 7.11(c).

For each subincrement, say the mth, the total stress is computed as

$$\sigma_i^m = \sigma_i^{m-1} + C_i^{ep(m-1)} \cdot \Delta(d\varepsilon_i) \tag{7.66}$$

and the corresponding hardening parameter, α_i^j, is found by computing the trajectory, ξ_i^j, as

$$\xi_i^m = \xi_i^{m-1} + \Delta(d\xi) \tag{7.67}$$

where

$$d\xi = \lambda\left[\left(\frac{\partial Q}{\partial \sigma}\right)^T \cdot \frac{\partial Q}{\partial \sigma}\right]^{1/2}$$

and λ is computed using Eq. (7.43).

Now check if $F(\sigma_i^m, \alpha_i^m) \leq$ tolerance. If the subincrement is small (and if the tolerance is satisfied), we consider the subincremental results satisfactory and proceed to the next subincrement. If the subincrement is large and the above criterion is not satisfied, it is possible to perform the drift correction

Hierarchical Single-Surface Plasticity Models in DSC

(described below) for the current subincrement so as to improve the accuracy, before going to the next strain subincrement $(m + 1)$.

For the next subincrement, compute the constitutive matrix, $(C_i^{ep})^m$, using Eq. (7.44). Then apply the next subincrement, $\Delta(d\underset{\sim}{\varepsilon}_i)$, and go to Eq. (7.66). When *all* subincrements are applied, Fig. 7.11(c),

$$\underset{\sim}{\sigma}_i = \underset{\sim}{\sigma}_i^n; \quad \underset{\sim}{C}_i^{ep} = (\underset{\sim}{C}_i^{ep})^n; \quad \xi_i = \xi_i^n; \quad \text{and} \quad \alpha_i = \alpha_i^n \qquad (7.68)$$

In the case of the integration of the incremental equations for back predictions of stress–strain responses, it may be necessary to perform the above procedure only once before proceeding to the next increment. However, for finite-element analysis, it needs to be performed for all Gauss points in the elements, and the quantities in Eq. (7.68) are stored for all such points before proceeding to the next (load) increment.

7.8 Drift Correction Procedure

A drift correction procedure can be used by itself or together with the subincremental strain scheme described earlier. It is believed that the drift correction procedures are needed in finite-element analysis with most nonlinear (plasticity) models.

Assume that for both cases 1 and 2, the yield surface at the start (at C or A) is satisfied [Fig. 7.11(c)], i.e.,

$$F(\underset{\sim}{\sigma}_A, \alpha_A) = 0 \qquad (7.69)$$

Now, the current increment, i, causes the stress increment, $d\underset{\sim}{\sigma}^B$, which takes us to point B [Fig. 7.11(c)], at which $F(\underset{\sim}{\sigma}_B, \alpha_B) = 0$ is not satisfied. The yield surface condition is satisfied at the (desired) point D, where $F(\underset{\sim}{\sigma}_D, \alpha_D) \approx 0$. Hence, the drift correction procedure would entail modifications in $\underset{\sim}{\sigma}_B$ and α_B so as to lead to $\underset{\sim}{\sigma}_D, \alpha_D$.

The quantities at point B are found as follows:

$$\underset{\sim}{\sigma}_B = \underset{\sim}{\sigma}_A + (d\underset{\sim}{\sigma}_i)_{A-B} \qquad (7.70a)$$

where

$$(d\underset{\sim}{\sigma}_i)_{A-B} = \underset{\sim}{C}^e (d\underset{\sim}{\varepsilon}^e)_{A-B} \qquad (7.70b)$$

$$\xi_B = \xi_A + d\xi_{A-B} \qquad (7.71a)$$

$$d\xi_{A-B} = \left[(d\underset{\sim}{\varepsilon}^p)^T \cdot d\underset{\sim}{\varepsilon}^p\right]^{1/2}$$

$$= f_1(d\underset{\sim}{\varepsilon}^p) \quad (7.71\text{b})$$

$$\alpha_B = \frac{a_1}{\xi_B^{n_1}} = f_2(\xi) \quad (7.71\text{c})$$

Here, $d\underset{\sim}{\varepsilon}^p = d\underset{\sim}{\varepsilon} - d\underset{\sim}{\varepsilon}^e$ and $d\underset{\sim}{\varepsilon}$ is found using Eq. (7.52).

Now let the errors or modification required in the elastic and plastic incremental strains be denoted as $(d\underset{\sim}{\varepsilon}^e)_{BD}$ and $(d\underset{\sim}{\varepsilon}^p)_{BD}$. Then [Fig. 7.11(d)], respectively.

$$d\underset{\sim}{\varepsilon}_i = d\underset{\sim}{\varepsilon}^e_{AB} + d\underset{\sim}{\varepsilon}^p_{AB} + d\underset{\sim}{\varepsilon}^e_{BD} + d\underset{\sim}{\varepsilon}^p_{BD} \quad (7.72)$$

Considering that the strain increment $d\underset{\sim}{\varepsilon}_i = d\underset{\sim}{\varepsilon}^e_{AB} + d\underset{\sim}{\varepsilon}^p_{AB}$ is the same at point D [Fig. 7.11(d)], we have

$$d\underset{\sim}{\varepsilon}^e_{BD} + d\underset{\sim}{\varepsilon}^p_{BD} = 0 \quad (7.73\text{a})$$

Therefore,

$$d\underset{\sim}{\varepsilon}^e_{B-D} = -d\underset{\sim}{\varepsilon}^p_{B-D} \quad (7.73\text{b})$$

In other words, the plastic strains from $B - D$ are accompanied by an equal decrease in the elastic strains.

Then, at the converged state, D, we have

$$\underset{\sim}{\sigma}_D = \underset{\sim}{\sigma}_B + d\underset{\sim}{\sigma}_{BD} \quad (7.74\text{a})$$

where

$$d\underset{\sim}{\sigma}_{BD} = \underset{\sim}{C}^e \cdot d\underset{\sim}{\varepsilon}^e_{BD} \quad (7.74\text{b})$$

and the hardening parameters are

$$\xi_D = \xi_B + d\xi_{BD} \quad (7.75\text{a})$$

where

$$d\xi_{B-D} = f_1(d\underset{\sim}{\varepsilon}^p)_{BD} \quad \text{and} \quad \alpha_D = f_2(\xi)_D \quad (7.75\text{b})$$

where the functions f_1 and f_2 will depend on the hardening function used, e.g., Eq. (7.11).

Substitution of Eqs. (7.74a), (7.75a), and (7.75c) in the equation for F gives

$$F[(\sigma_B + d\sigma_{BD}), f_2(\xi_B + d\xi_{BD})] = 0 \qquad (7.76)$$

which, when expanded into the Taylor series, gives

$$F\left[\sigma_B, f_2(\xi)_B\right] + \left[\left(\frac{\partial F}{\partial \sigma}\right)^T \cdot d\sigma_{B-D} + \left(\frac{\partial F}{\partial f_2}\right) df_2\right]_B + \cdots = 0 \qquad (7.77)$$

Now, from the normality rule,

$$d\varepsilon^p_{BD} = \lambda_{BD}\left(\frac{\partial Q}{\partial \sigma}\right)_B \qquad (7.78)$$

Here, $Q \equiv F$ for associated plasticity.
Substitution of Eq. (7.78) into (7.75b) gives

$$d\xi_{BD} = \lambda_{BD} \cdot f_1\left(\frac{\partial Q}{\partial \sigma}\right) \qquad (7.79)$$

Substitution of Eqs. (7.73b) and (7.78) into Eq. (7.74b) gives

$$d\sigma_{BD} = -\lambda_{BD}\, \underline{C}^e \frac{\partial Q}{\partial \sigma} \qquad (7.80)$$

Now, substitution of Eqs. (7.79) and (7.80) into Eq. (7.77) gives

$$F\left[\sigma_B, f_2(\xi)_B\right] + \left\{-\left(\frac{\partial F}{\partial \sigma}\right)^T_B \underline{C}^e \frac{\partial Q}{\partial \sigma} + \left(\frac{\partial F}{\partial \alpha}\right)\left(\frac{\partial f_2}{\partial \xi}\right)_B \cdot f_1 \frac{\partial Q}{\partial \sigma}\right\}_B \lambda_{B-D} + \cdots = 0 \qquad (7.81)$$

Ignoring higher-order terms, we find the expression for λ_{BD} as

$$\lambda_{BD} = \frac{-F[\sigma_B, f_2(\xi)_B]}{\left\{-\left(\frac{\partial F}{\partial \sigma}\right)^T_B \underline{C}^e \frac{\partial Q}{\partial \sigma} + \left(\frac{\partial F}{\partial \alpha} \cdot \frac{\partial f_2}{\partial \xi}\right)_B \cdot f_1 \frac{\partial Q}{\partial \sigma}\right\}} \qquad (7.82)$$

Hence, the quantities at D can be found as

$$\sigma_D = \sigma_B - \lambda_{BD} \cdot \underline{C}^e \frac{\partial Q}{\partial \sigma} \qquad (7.83a)$$

$$\xi_D = \xi_B + \lambda_{BD} \cdot f_1\left(\frac{\partial Q}{\partial \sigma}\right) \qquad (7.83b)$$

and

$$\alpha_D = f_2(\xi)_D \qquad (7.83c)$$

Hence,

$$F(\sigma_D, \alpha_D) \approx 0 \qquad (7.84)$$

should be satisfied, for a given tolerance (10^{-3} to 10^{-6}). Then, Eq. (7.84) will represent the converged yield surface for increment i.

As the higher-order terms in Eq. (7.81) are ignored, it may be appropriate and sometimes required to perform iterations using the result in Eq. (7.82). Here, the solutions obtained in Eq. (7.83) can be substituted in Eq. (7.82), and the next set of the quantities are calculated. The procedure is continued until Eq. (7.84) is satisfied.

7.9 Thermoplasticity

The temperature effects in the plasticity model can be introduced through the modification of various terms. For example, the yield function in the HISS (δ_0) model can be expressed as (39, 40)

$$F(T) = \bar{J}_{2D} - [-\alpha(T)\bar{J}_1^{n(T)} + \gamma(T)\bar{J}_1^2\,][1 - \beta(T)S_r]^{-0.5} = 0 \qquad (7.85)$$

where T is the temperature, $\bar{J}_1 = J_1 + 3R$, $R(T)$ is the bonding stress; here, the invariant quantities denote nondimensionalized values with respect to p_a. The temperature-dependent hardening function, α, can be written as

$$\alpha(T) = \frac{a_1}{\xi^{\eta_1}} \qquad (7.86)$$

where a_1 and η_1 are hardening parameters that may be expressed as functions of T, and ξ is the thermoplastic strain trajectory. Depending on the effect of temperature, a_1 and/or η_1 may be essentially constant with temperature.

The normality rule can now be used (51, 52):

$$d\underset{\sim}{\varepsilon}^p(T) = \lambda \frac{\partial Q(\underset{\sim}{\sigma}, \xi, T)}{\partial \underset{\sim}{\sigma}} \tag{7.87}$$

The total incremental strain vector $d\underset{\sim}{\varepsilon}^t(T)$ can be decomposed as

$$d\underset{\sim}{\varepsilon}^t(T) = d\underset{\sim}{\varepsilon}^e(T) + d\underset{\sim}{\varepsilon}^p(T) + d\underset{\sim}{\varepsilon}(T) \tag{7.88}$$

where $d\underset{\sim}{\varepsilon}(T)$ is the strain vector due to the temperature change, dT. Hence,

$$d\underset{\sim}{\varepsilon}^e(T) = d\underset{\sim}{\varepsilon}^t - \lambda \frac{\partial Q}{\partial \underset{\sim}{\sigma}} - d\underset{\sim}{\varepsilon}(T) \tag{7.89}$$

and

$$\begin{aligned} d\underset{\sim}{\sigma} &= \underset{\sim}{C}^e(T) d\underset{\sim}{\varepsilon}^e \\ &= \underset{\sim}{C}^e(T) \left[d\underset{\sim}{\varepsilon}^t - \lambda \frac{\partial Q}{\partial \underset{\sim}{\sigma}} - \alpha_T \underset{\sim}{I}_0 \, dT \right] \end{aligned} \tag{7.90}$$

where $\underset{\sim}{I}_0 = [1 \ 1 \ 0]$ and $[1 \ 1 \ 1 \ 0 \ 0 \ 0]$ for two-dimensional and three-dimensional cases, respectively.

Now the consistency condition gives

$$dF(\underset{\sim}{\sigma}, \xi, T) = 0 \tag{7.91}$$

Therefore,

$$dF = \left(\frac{\partial F}{\partial \underset{\sim}{\sigma}}\right)^T d\underset{\sim}{\sigma} + \frac{\partial F}{\partial \xi} d\xi + \frac{\partial F}{\partial T} dT \tag{7.92}$$

The use of Eqs. (7.89), (7.90), and (7.91) leads to the expression of λ as

$$\lambda = \frac{\left(\frac{\partial F}{\partial \underset{\sim}{\sigma}}\right)^T \underset{\sim}{C}^e(T) d\underset{\sim}{\varepsilon} + \frac{\partial F}{\partial T} dT - \alpha_T \frac{\partial F}{\partial \underset{\sim}{\sigma}} \underset{\sim}{C}^e(T) \underset{\sim}{I}_0 \, dT}{\frac{\partial F}{\partial \underset{\sim}{\sigma}} \underset{\sim}{C}^e(T) \frac{\partial Q}{\partial \underset{\sim}{\sigma}} - \frac{\partial F}{\partial \xi} \left[\left(\frac{\partial Q}{\partial \underset{\sim}{\sigma}}\right)^T \cdot \frac{\partial Q}{\partial \underset{\sim}{\sigma}} \right]^{1/2}} \tag{7.93}$$

Substitution of λ in Eq. (7.90) gives

$$d\underline{\sigma} = \underline{C}^e(T)\left[\underline{I} - \frac{\frac{\partial Q}{\partial \underline{\sigma}} \cdot \left(\frac{\partial F}{\partial \underline{\sigma}}\right)^T \underline{C}^e(T)}{\left(\frac{\partial F}{\partial \underline{\sigma}}\right)^T \underline{C}^e(T)\frac{\partial Q}{\partial \underline{\sigma}} - \frac{\partial F}{\partial \xi}\left[\left(\frac{\partial Q}{\partial \underline{\sigma}}\right)^T \cdot \frac{\partial Q}{\partial \underline{\sigma}}\right]^{1/2}}\right]d\underline{\varepsilon}$$

(7.94a)

$$-\underline{C}^e(T)\left[\alpha_T \underline{I}_0 - \frac{\alpha_T \frac{\partial F}{\partial \underline{\sigma}}\underline{C}^e(T) \cdot \underline{I}_0 \frac{\partial Q}{\partial \underline{\sigma}} - \frac{\partial F}{\partial T}\frac{\partial Q}{\partial \underline{\sigma}}}{\left(\frac{\partial F}{\partial \underline{\sigma}}\right)^T \underline{C}^e(T)\frac{\partial Q}{\partial \underline{\sigma}} - \frac{\partial F}{\partial \xi}\left[\left(\frac{\partial Q}{\partial \underline{\sigma}}\right)^T \frac{\partial Q}{\partial \underline{\sigma}}\right]^{1/2}}\right]dT$$

$$= \underline{C}^{ep}(T)d\underline{\varepsilon} + d\underline{\sigma}(T) \qquad (7.94b)$$

where $\underline{I}^T$ is the unit matrix.

Material Parameters. The variation of the elasticity and plasticity parameters and α_T is shown in Eq. (7.29). Their determination has been discussed earlier in Section 7.4.7, "Thermal Effects on Parameters."

7.10 Examples

We now present examples of derivations and properties of some of the expressions in the HISS models, analysis of the effect of stress paths, and validation of the models with respect to the laboratory test behavior of a number of materials.

Example 7.1

Derive the conditions on parameters n, γ, and β in the yield function, F of Eq. (7.1).

Parameter n:

The state when the volume transits from compaction to dilation, or change in volume vanishes, i.e., $d\varepsilon_v = 0$, is related to the highest point on the yield surface (Fig. 7.1), where $\partial F/\partial \bar{J}_1 = 0$:

$$\frac{\partial F}{\partial \bar{J}_1} = (\alpha n \bar{J}_1^{n-1} - 2\gamma \bar{J}_1)F_s - 0 \qquad (1a)$$

$\bar{J}_1$, F_s, and α refer to the state of stress at the transition. Equation (1a) leads to (as $F_s \neq 0$)

$$(\bar{J}_1)_t = \left(\frac{2\gamma}{\alpha_t \cdot n}\right)^{\frac{1}{n-2}} \qquad (1b)$$

where the subscript t denotes transition. Now Eq. (7.1) at the transition becomes

$$(\bar{J}_{2D})_t = \left[-\alpha_t (\bar{J}_1)_t^n + \gamma(\bar{J}_1)^2 \right] F_{st} \tag{1c}$$

Substitution of Eq. (1b) in Eq. (1c) leads to

$$\left(\frac{\bar{J}_{2D}}{\bar{J}_1^2} \right)_t = \gamma_t = \left[\gamma - \frac{2\gamma}{n} \right] F_{st} \tag{1d}$$

Because F_s and γ_t are positive, the term in the brackets should be nonnegative. Therefore, $n \geq 2$.

Alternatively, for $\beta = 0$, Eq. (7.1) gives

$$\sqrt{\bar{J}_{2D}} = (-\alpha \bar{J}_1^n + \gamma \bar{J}_1^2)^{-0.5} \tag{1e}$$

The curvature, Ψ, of the projection surface (curve) in Eq. (1e) can be expressed as

$$\Psi = \frac{\frac{\partial^2 \sqrt{\bar{J}_{2D}}}{\partial \bar{J}_1^2}}{\left[1 + \left(\frac{\partial \sqrt{\bar{J}_{2D}}}{\partial \bar{J}_1} \right)^2 \right]^{3/2}} \tag{1f}$$

For the yield curve to be convex, $\Psi \geq 0$. Solution of Eq. (1f) numerically leads to the condition that $n > 2$ (53).

Parameter γ:
At the ultimate condition, $\alpha = 0$. Therefore, from Eq. (7.1) we have

$$\frac{\bar{J}_{2D}}{\bar{J}_1^2} = \gamma(1 - \beta S_r)^{-0.5} = \gamma F_s \tag{1g}$$

As $\bar{J}_{2D} > 0$ and $\bar{J}_1^2 > 0$, and as $\beta < 1$ (see below), $(1 - \beta S_r)^{-0.5} > 0$. Hence, $\gamma > 0$.

Now for hydrostatic or spherical state of stress, $J_{2D} = 0$ and Eq. (7.1) gives

$$\frac{\gamma}{\alpha} = (\bar{J}_1)^{n-2} \tag{1h}$$

when $\alpha = 0$, as $\gamma > 0$, $\bar{J}_1 = \infty$. That is, the ultimate envelopes intersect with the J_1-axis at infinity. For $\alpha > 0$, the yield surface (Fig. 7.1) will intersect the J_1-axis at finite values. For $\alpha \approx \infty$, the state of stress will be at $\bar{J}_1 = 0$.

Parameter β:
The slope (θ) of an ultimate envelope [Fig. 7.4(a)] is given by

$$\tan^2 \theta = \frac{\bar{J}_{2D}}{\bar{J}_1^2} = \gamma F_s = \gamma(1 - \beta S_r)^{-0.5} \tag{1i}$$

For $F_s = 1$ when $S_r = 0$, i.e., for the simple shear (S) stress path, $\tan^2 \theta = \gamma$. For other stress paths, i.e., $S_r = \pm 1$ ($S_r = 1$ for compression and $= -1$ for extension paths):

$$\tan^2 \theta_C = \gamma(1 - \beta)^{-0.5} \tag{1j}$$

or

$$\tan^2 \theta_E = \gamma(1 + \beta)^{-0.5} \tag{1k}$$

If $\beta = 0$, $\tan^2 \theta = \gamma$. If $\beta = 1$, Eq. (1j) gives $\tan^2 \theta = 0$, which is inadmissible. If $\beta > 1$, $(1 - \beta)$ in Eq. (1j) would give a negative value, which is not admissible. Hence, $0 \leq \beta < 1$.

Alternatively, from Eq. (7.1) we have

$$\bar{J}_{2D} = C_1 \left(1 - \frac{\sqrt{27}}{2} \cdot \frac{J_{2D}}{J_{2D}^{3/2}} \cdot \beta \right)^{-0.5} \tag{1l}$$

where $C_1 = \alpha(\bar{J}_1)^n + \gamma(\bar{J}_1^2) = $ constant. On the π-plane in the principal stress space (Fig. 7.12), we have (53)

$$\begin{aligned} r^2 &= 2J_{2D} \\ y_i &= S_i \\ \theta_i &= \frac{y_i}{r} \end{aligned} \tag{1m}$$

where S_i ($i = 1, 2, 3$) are the normal components of the deviatoric stress tensor, S_{ij}; x, and y are the Cartesian coordinates; and r is the radius of the polar coordinate system. Then Eq. (1m) leads to

$$r = \sqrt{2C_1}(1 - \beta \sin 3\theta)^{-0.25} \tag{1n}$$

Hierarchical Single-Surface Plasticity Models in DSC

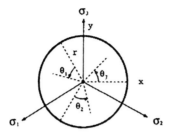

FIGURE 7.12
Plot on π-phase (53).

For the yield surface to be convex, β should be such that the curvature, $\psi > 0$,

$$\psi = \frac{2r'^2 - 2rr'' + r^2}{(r'^2 + r^2)^{3/2}} \tag{1o}$$

The numerical solution of Eq. (1o) shows that

$$0 \leq \beta < 0.755923.$$

Example 7.2 Derive Eq. (7.24)

At point B [Fig. 7.5(e)] $J_{2D} = 0$. Assuming $S_r = 0$, we have, from Eq. (7.1),

$$\alpha \frac{J_{1m}^n}{p_a^n} = \frac{\gamma J_{1m}^2}{p_a^2} \tag{2a}$$

Therefore,

$$\frac{J_{1m}}{p_a} = \left(\frac{\gamma}{\alpha}\right)^{\frac{1}{n-2}} \tag{2b}$$

Now, the use of Eq. (1b) in Eq. (2b) gives

$$\frac{J_{1a}}{J_{1m}} = \left(\frac{2}{n}\right)^{\frac{1}{n-2}} \tag{2c}$$

where J_{1a} is related to the phase change point A [Fig. 7.5(e)].
At the ultimate condition, $\alpha = 0$; therefore, Eq. (7.1) leads to

$$\gamma = \frac{\sqrt{J_{2D}}}{J_1} = \sqrt{\gamma F_s} \tag{2d}$$

where γ is the slope of the ultimate line [Fig. 7.5(e)]. Equation (7.1) can be written as

$$\frac{J_{2D}}{J_1^2} - \left(-\alpha \frac{J_1^{n-2}}{p_a^{n-2}} + \gamma\right) F_s = 0 \tag{2e}$$

Now, substitution of Eq. (1b) at the transition state in Eq. (2e) gives

$$\gamma_t = \frac{\sqrt{J_{2D}}}{J_1} = \left[\left(1 - \frac{2}{n}\right)\gamma \cdot F_s\right]^{1/2} \tag{2f}$$

where γ_t is the slope of the phase change line [Fig. 7.5(e)]. Therefore,

$$\frac{\gamma_t}{\gamma} = \left(\frac{n-2}{n}\right)^{1/2} \tag{2g}$$

Example 7.3

Derive the relations between the slopes (compression and extension) in the J_1–$\sqrt{J_{2D}}$ and τ–σ (Mohr–Coulomb) stress spaces.

It is useful to derive the relations (Eq. 7.13) between the slope of the ultimate surface and ($\theta = \sqrt{\gamma F_s}$) [Fig. 7.4(a)], the traditional slope (ϕ) of the failure envelope in the Mohr–Coulomb criterion [Fig. 7.4(b)]. Then, γ can be computed from the angle of friction, ϕ.

First consider compression path. From Fig. 7.4(b), we have

$$\frac{\sigma_1 - \sigma_3}{\sigma_1 + \sigma_3} = \sin\phi \tag{3a}$$

At the ultimate condition, $\alpha = 0$; therefore,

$$\frac{\sqrt{J_{2D}}}{J_1} = \sqrt{\gamma(1 - \beta S_r)^{-0.5}} \tag{3b}$$

Now, by substituting $\sigma_2 = \sigma_3$, we have

$$J_{2D} = \frac{1}{6} \cdot 2(\sigma_1 - \sigma_3)^2 \quad \text{and} \quad J_1 = \sigma_1 + 2\sigma_3$$

Therefore,

$$\frac{\sqrt{J_{2D}}}{J_1} = \frac{1}{\sqrt{3}} \frac{\sigma_1 - \sigma_3}{\sigma_1 + 2\sigma_3} = \frac{1}{\sqrt{3}} \frac{\sigma_1 - \sigma_3}{(\sigma_1 + \sigma_3) + \sigma_3}$$
$$= \frac{1}{\sqrt{3}} \frac{(\sigma_1 - \sigma_3)/(\sigma_1 + \sigma_3)}{1 + \sigma_3/(\sigma_1 + \sigma_3)} \tag{3c}$$

Then Eqs. (3a), (3c), and (3d) give

$$\tan \theta_C = \frac{\sqrt{J_{2D}}}{J_1} = \frac{2}{\sqrt{3}} \frac{\sin \phi_c}{3 - \sin \phi_c} \tag{3d}$$

where $\theta_c = \sqrt{\gamma(1 - \beta S_r)^{-0.5}}$ is the slope of the ultimate envelope [Fig. 7.4(a)]. Similarly, for the extension path,

$$\tan \phi_E = \sqrt{\gamma(1 - \beta S_r)^{-0.5}} = \frac{2}{\sqrt{3}} \frac{\sin \phi_E}{3 + \sin \phi_E} \tag{3e}$$

7.11 Stress Path

The behavior of most materials is influenced by the path of loading or stress path. The stress paths are usually depicted on stress spaces such as $J_1 - \sqrt{J_{2D}}$, $\sigma_1 - \sqrt{2}\sigma_2 (=\sqrt{2}\sigma_3)$, $\sigma_1 - \sigma_2$, and $\sigma_1 - \sigma_2 - \sigma_3$. In the case of materials such as metals and alloys, tensile loading is often more significant. Here, tension is considered to be positive. Figure 7.13(a) shows uniaxial tension and compression test ($\sigma_1 \neq 0$, $\sigma_2 = \sigma_3 = 0$) and a schematic of stress–strain responses. The stress path followed during the uniaxial tension (UT_0) and uniaxial compression (UC_0) tests are depicted in three stress spaces [Fig. 7.13(a)]. In the $J_1 - \sqrt{J_{2D}}$ space, the stress path has the slope of

$$1/\sqrt{3}\left(\sqrt{J_{2D}}/J_1 = \frac{1}{\sqrt{3}} \frac{\sigma_1}{\sigma_1}\right)$$

In the $\sigma_1 - \sigma_2$ space, the stress paths are vertical. In the $\sigma_1 - \sigma_2 - \sigma_3$ space, the stress paths follow the σ_1-axis.

Figure 7.13(b) shows straight-line stress paths used commonly for frictional (geologic) materials. Often, other stress paths such as proportional loading (PL) and circular (CSP) are followed. The latter can capture the combined effect of a number of stress paths in a single test. Table 7.3 shows an

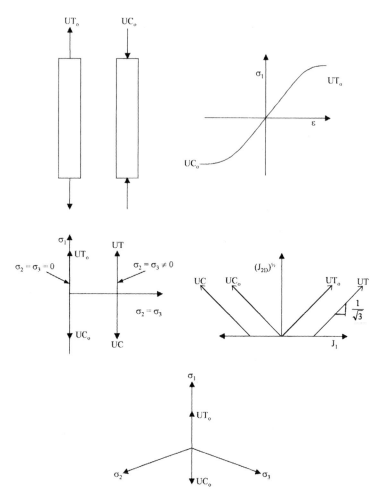

(a) Uniaxial tension or compression tests and stress paths: Tension is assumed positive in this figure

FIGURE 7.13
Stress paths in testing.

explanation of various stress paths and their loading conditions in terms of principal stresses, σ_1, σ_2, and σ_3.

7.11.1 Stress Path Analysis of the HISS Model

Material points in any engineering structure may experience changes in the stress paths during loading, unloading, and reloading. For instance, during increasing load, a point may follow the compression stress (CTC) path [Fig. 7.13(b)]; however, upon unloading, it may follow the opposite path

Hierarchical Single-Surface Plasticity Models in DSC   229

TABLE 7.3

Explanations of Stress Paths

Symbol	Explanation	Stress Increments During Loading and Unloading	
		Initial	Increment
HC	Hydrostatic compression	$\sigma_1 = \sigma_2 = \sigma_3 = 0$	$d\sigma_1 = d\sigma_2 = d\sigma_3 = d\sigma_0$
CTC	Conventional triaxial compression	$\sigma_1 = \sigma_2 = \sigma_3 = \sigma_0$	$d\sigma_1 > 0, d\sigma_2 = d\sigma_3 = 0$
RTE	Reduced triaxial extension	$\sigma_1 = \sigma_2 = \sigma_3 = \sigma_0$	$d\sigma_1 < 0, d\sigma_2 = d\sigma_3 = 0$
CTE	Conventional triaxial extension	$\sigma_1 = \sigma_2 = \sigma_3 = \sigma_0$	$d\sigma_1 = 0, d\sigma_2 = d\sigma_3 > 0$
RTC	Reduced triaxial compression	$\sigma_1 = \sigma_2 = \sigma_3 = \sigma_0$	$d\sigma_1 = 0, d\sigma_2 = d\sigma_3 < 0$
TC	Triaxial compression	$\sigma_1 = \sigma_2 = \sigma_3 = \sigma_0$	$d\sigma_1 > 0, d\sigma_2 = d\sigma_3 = -\dfrac{d\sigma_1}{2}$
TE	Triaxial extension	$\sigma_1 = \sigma_2 = \sigma_3 = \sigma_0$	$d\sigma_1 < 0, d\sigma_2 = d\sigma_3 = \dfrac{-d\sigma_1}{2}$
SS	Simple shear	$\sigma_1 = \sigma_2 = \sigma_3 = \sigma_0$	$d\sigma_1 = -d\sigma_3, d\sigma_2 = 0$
PL	Proportional loading	$\sigma_1/\sigma_2 = \sigma_3/\sigma_0 =$ constant	
A	Arbitrary (e.g., circular)	Design any path in the stress space.	

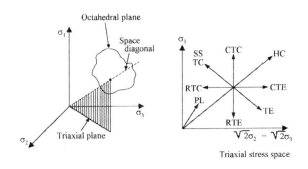

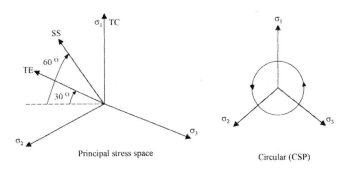

(b) General stress paths: Multiaxial testing: Compression is assumed positive in this figure

FIGURE 7.13
(continued)

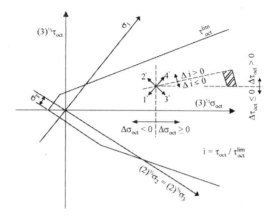

FIGURE 7.14
Stress path groups and switch functions (From Ref. 55. Reprinted with permission from A.A. Balkema).

(RTE). In fact, points in different regions in the same structure under tension or compression loading may follow different stress paths. Hence, a constitutive model should be able to handle the changing or switching stress paths. Gudehus (54) proposed a procedure for analyzing the capability of constitutive models for the stress-path dependence. It is based on the idea of application of unit strain rate (increment) along different stress paths and then evaluating the corresponding induced stresses along different paths in the stress space. Dolezalova and co-workers (55, 56) have presented the application of this procedure for the evaluation of a number of constitutive models such as nonlinear elastic (hyperbolic and variable moduli, Chapter 5), associative plasticity (Cam clay, Chapter 6), and nonassociative plasticity by Lade et al. (8, 9), and the HISS-δ_1 model. Here, we present a brief review in the context of the HISS-δ_1 model.

Figure 7.14 depicts the stress-path groups and switch functions considered for the analysis (55, 56). Here, the typical stress-path groups are identified as 11, 12, 41, 42, and the switch conditions are defined on the basis of $\Delta\sigma_{oct}$, $\Delta\tau_{oct}$, and i, where σ_{oct} ($\sigma_1 + \sigma_2 + \sigma_3 = J_1$) is the octahedral stress, τ_{oct} is the octahedral shear stress, i is the ratio $\tau_{oct}/\tau_{oct}^{lim}$, and τ_{oct}^{lim} is the limiting octahedral stress. Figure 7.15(a) and (b) show applied strain increments in different directions and the corresponding stress increments, respectively. The latter are computed based on the incremental constitutive equations for the model analyzed. For the triaxial loading condition, the unit strain increment (rate 1, $\dot{\varepsilon}$ = $\sqrt{\dot{\varepsilon}_1^2 + 2\dot{\varepsilon}_3^2}$ and the incremental stress response $r_\sigma = \sqrt{\dot{\sigma}_1^2 + 2\dot{\sigma}_3^2}$, respectively.

Figure 7.16 shows the unit response envelopes for the nonassociative HISS-δ_1 model. Similar envelopes for other models have been reported by Dolezalova et al. (55). Based on these analyses, it was reported that the nonlinear elastic (hyperbolic) and the associative Cam clay models show weak response and do not account properly for the stress-path group 12. Also, as the yielding is controlled only by the volumetric strains, the Cam clay model, if used for

Hierarchical Single-Surface Plasticity Models in DSC

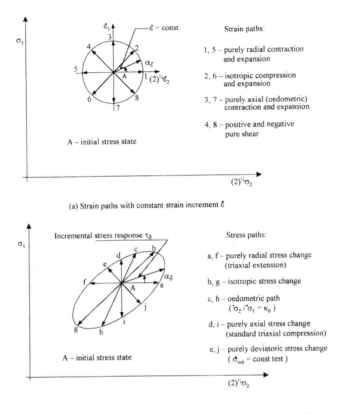

FIGURE 7.15
Strain increment and corresponding stress response (56).

frictional materials, leads to overestimation of computed displacements during finite-element analysis. The nonassociative models do provide satisfactory prediction of stress responses under various stress paths. In particular, it is reported (55) that the HISS-δ_1 model provides realistic predictions, partly because it allows for the effect of both volumetric and deviatoric (plastic) strains in the characterization of the yielding behavior.

7.12 Examples of Validation

We now present a number of examples of the application of the DSC and its specialized versions, e.g., the HISS-δ_0/δ_1 plasticity models. When disturbance is considered, the HISS-plasticity (δ_0) model is used commonly to characterize the behavior of the RI material. The FA response is characterized as

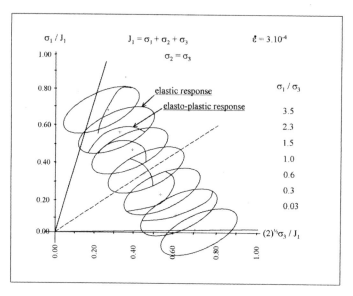

FIGURE 7.16
Unit response envelopes for Leighton Buzzard sand with HISS-δ_1 mode. (From Ref. 55. With permission.)

constrained-liquid or constrained liquid-solid (critical state), Chapter 4. Only a few validations are presented, while references are given for other materials.

The observed laboratory behavior of the materials considered is used to find the parameters by employing the procedures described in this chapter and in Chapters 5 and 6. The values of parameters are given for most of the examples.

The incremental equations, Eq. (7.44), are used to predict the responses under different factors such as stress path, type of loading, and initial conditions such as confining pressure and density. The incremental equations are integrated for the initial loading such as hydrostatic. Based on the state of stress at the end of the initial loading, the values of parameters such as the hardening function (α_0) and the disturbance (D_0) are computed. Then the equations are integrated for the subsequent shear loading under a given stress path. The validations are often obtained for the test data used to find the parameters, and for *independent* tests not used in finding the parameters; the latter provides a more rigorous validation.

In the case of frictional and geologic materials, the response is affected by mean pressure. Hence, tests are usually performed under different initial mean pressures or confining stresses. Also, compressive behavior is often significant; hence, tests are usually performed under compression loading, with the assumption that compression is positive. Geologic materials involve initial compressive stresses. Under loading, the compressive stresses may increase or decrease. The latter is often termed as an *extension*. To determine tensile behavior, sometimes direct tension or indirect (e.g., Brazilian split cylinder) tests are performed.

Example 7.4 Initial Conditions

Hardening:
The expression of α can be obtained from Eq. (7.1) as

$$\alpha = \left[\gamma - \frac{J_{2D}}{J_1^2 \cdot (1 - \beta S_r)^m}\right]\left(\frac{J_1}{p_a}\right)^{2-n} \quad (4a)$$

The initial value α_0 can be found by substituting in Eq. (4a), J_1, J_{2D}, and S_r computed from the initial or *in situ* stresses, σ_x, σ_y, τ_{xy}, for two-dimensional and σ_x, σ_y, σ_z, τ_{xy}, τ_{yz}, τ_{zx} for three-dimensional problems. Now from Eq. (7.11a),

$$\xi = (a_1/\alpha_0)1/\eta_1 \quad (4b)$$

Then, substitution of α_0 in Eq. (4b) gives the value of the initial plastic strain trajectory, ξ_0.

If the initial loading can be assumed to be proportional, the (initial) volumetric plastic strain trajectory, ξ_{v0}, can be found as (57)

$$\xi_{v0} = \frac{\xi_0 \frac{\partial Q}{\partial \sigma_{ii}}}{\sqrt{3}\left(\frac{\partial Q}{\partial \underline{\sigma}} \cdot \frac{\partial Q}{\partial \underline{\sigma}}\right)^{1/2}} \quad (4c)$$

Then the initial deviatoric plastic strain trajectory, ξ_{D0}, is evaluated from

$$\xi_{D0} = \sqrt{\xi_0^2 - \xi_{v0}^2} \quad (4d)$$

In the case of hydrostatic initial stress ($\sigma_x = \sigma_y = \sigma_z$; $\tau_{xy} = \tau_{yz} = \tau_{zx} = 0$), for example, in the triaxial test, Eq. (4a) simplifies to

$$\alpha_0 = \gamma\left(\frac{J_1}{p_a}\right)^{2-n}$$
$$\xi_0 = \xi_{v0} = (a_1/\alpha_0)^{1/\eta_1} \quad (4e)$$
$$\xi_{D0} = 0$$

Disturbance. The initial disturbance, D_0, can be found using, e.g., Eq. (3.15) or (3.16). Here it is necessary to have the knowledge of the initial (deviatoric) plastic strain trajectory, which is often difficult to find. However, laboratory tests can be performed, say, on an initially anisotropic material and/or with initial flaws (microcracks), under a small hydrostatic loading. Such a loading will cause shear strains in the specimens, which can be used to

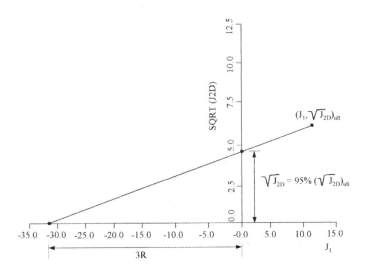

FIGURE 7.17
Determination of γ and $3R$ for Pb/Sn solder (39, 40).

evaluate the plastic strains. Alternatively, nondestructive (ultrasonic P-wave) tests can be performed to measure initial velocities in the principal (three) directions. Then, D_0 can be found based on measured velocities (58).

Example 7.5 Metal Alloys (Solders)

The elastoplastic behavior of metal alloys such as solders (Pb/Sn) can be characterized using the δ_0 model, and the DSC model can be used for their cyclic response involving softening or degradation. Often, behavior of these materials can be assumed not to be affected by the mean pressure, and hence, the value of the ultimate parameter, γ, will be small, and $\beta \approx 0$, implying that the yield surface in the σ_1–σ_2–σ_3 space [Fig. 7.1(b)] is essentially circular. At the same time, the behavior is usually nonlinear and plastic hardening or yielding (Fig. 7.8). Hence, it is often appropriate to use the δ_0-model with the hardening function, α [Eq. (7.11a)].

The value of γ can be estimated by using Eq. (7.20). If test results at different confining stresses are not available, and if the response is not affected significantly by the confining stress, an ad-hoc procedure can be used (39). For the stress–strain response under a given temperature, Fig. 7.8, the ultimate values of J_1 and $\sqrt{J_{2D}}$ are plotted on the J_1–$\sqrt{J_{2D}}$ stress space, Fig. 7.17. A value of $\sqrt{J_{2D}}$ of about 95% of the $(\sqrt{J_{2D}})_{ult}$ is adopted at $J_1 = 0.0$ (Fig. 7.17). The slope of the line joining the two stress states yields the value of $\sqrt{\gamma}$, while its intersection with the J_1-axis gives the value of the bonding (compressive) stress ($3R$). The same value of γ can be used for other temperatures. However, the values of the bonding stress will be different; it will decrease with temperature.

The parameters for the δ_0-plasticity model for 60/40 (Sn/Pb) solders tested under tension loading by Riemer (42) and Skipor et al. (59), with different

TABLE 7.4(a)
Parameters for Solder (Pb40/Sn60) at $\dot{\varepsilon} = 0.002/\text{sec}$: δ_0-Model (39)

Temperature (K)	208	273	348	373
E (GPa)	26.1	24.1	22.45	22.00
ν	0.380	0.395	0.408	0.412
α_T (1/K) × 10^{-6}	2.75	2.93	3.11	3.16
γ	0.00083	0.00082	0.00082	0.00081
β	0.0	0.0	0.0	0.0
n	2.1	2.1	2.1	2.1
a_1 (× 10^{-6})	8.3	2.93	1.25	0.195
η_1	0.431	0.553	0.626	0.849
η_1 (average)		0.615		
σ_y, yield stress (MPa)	37.241	31.724	20.690	15.172
Bonding stress, R (MPa)	395.80	288.20	175.20	122.10

TABLE 7.4(b)
Parameters for Temperature Dependence, Eq. (7.29) (39)

Parameter	p_{300}	c
E	23.45 (GPa)	−0.292
ν	0.40	0.14
α_T	3 × 10^{-6} (1/K)	0.24
γ	0.00082	−0.034
α	0.05 × 10^{-4}	−5.5
R	240.67 (MPa)	−1.91

temperatures and strain rates, were computed by using the foregoing procedures; details are given elsewhere (39, 40). Tables 7.4(a) and 7.5(a) show typical parameters for the solder at $\dot{\varepsilon} = 0.002/\text{sec}$ and $0.0002/\text{sec}$, respectively, and for different temperatures. The parameters for the temperature dependence, Eq. (7.29), are shown in Tables 7.4(b) and 7.5(b), respectively. Details of the viscoplastic model are given in Chapter 8.

The disturbance parameters were found from the cyclic test data (Fig. 7.18) reported in (60). Their values are shown in Table 7.6. A number of other test data for different solders (61–64) were also used to find parameters for different and similar solders.

Figure 7.19 (a)–(d) show typical comparisons between predicted and observed stress–strain data for different solders (64). Figure 7.19(a)–(c) show plastic hardening results for 63/37 (Sn/Pb) solder at $T = 373\text{K}$, $\dot{\varepsilon} = 0.01/\text{sec}$ (59); for 95/3 (Pb/Sn) solder, $T = 100°\text{C}$ (61), and for 60/40 (Sn/Pb) solder, $T = -50°\text{C}$ and $\dot{\varepsilon} = 0.001/\text{sec}$ (63). The cyclic stress–strain response showing degradation was predicted using the DSC with the δ_0-model for the RI behavior, for 95/5 (Pb/Sn) solder at $T = 20°\text{C}$ (62). It can be seen that the HISS-δ_0 and DSC models provide very good predictions for the plastic yielding and degradation responses of Pb/Sn solders.

TABLE 7.5(a)

Material Parameters for 60/40 (Sn/Pb) Solder at Different Temperature for $\dot{\varepsilon} = 0.0002$/sec: δ_0-Model (39)

Temperature (K)	208	273	298	348	373
E (GPa)	26.10	24.10	23.50	22.45	22.00
ν	0.380	0.395	0.400	0.408	0.412
α_T (1/K) $\times 10^{-6}$	2.75	2.93	3.00	3.11	3.16
γ	0.00083	0.00082	0.00082	0.00081	0.00079
β	0.0	0.0	0.0	0.0	0.0
n	2.1	2.1	2.1	2.1	2.1
a_1 ($\times 10^{-4}$)	0.30	0.16	0.024	0.04	0.11
η_1	0.270	0.330	0.570	0.45	0.35
η_1 (average)		3.94			
σ_y, yield stress (MPa)	34.48	27.59	22.41	14.48	8.28
R (MPa)	433.80	284.58	217.47	116.57	73.90

TABLE 7.5(b)

Parameters for Temperature Dependence, Eq. (7.29) (39)

Parameter	p_{300}	c
E	23.45 (GPa)	-0.292
ν	0.40	0.14
α_T	3.0×10^{-6} (1/K)	0.24
γ	0.00082	-0.072
α	0.35×10^{-4}	-2.60
R	217.47 (MPa)	-2.95

7.12.1 Examples of Geologic Materials

Comprehensive laboratory triaxial tests on cylindrical specimens and multiaxial tests on cubical specimens have been performed and were available for a number of geologic materials such as sands, clays, and rocks. These tests are performed at different initial confining pressures and densities, and stress paths [Fig. 7.13(b)]. Only typical materials are considered here.

Parameters for the δ_0- and δ_1-HISS plasticity models were determined using the foregoing procedures. Table 7.7 shows the values of the parameters for three dry sands: Ottawa, Leighton Buzzard, and Munich for the δ_0- and δ_1-models (3, 19, 20, 65, 66). Table 7.8(a) and (b) shows the δ_0- and disturbance parameters for a saturated marine clay (32) and Leighton Buzzard sand (66), respectively, and Table 7.8(c) shows parameters for a saturated sand (67, 68). Table 7.9 shows parameters for three rocks (33, 34).

TABLE 7.6
Disturbance Parameter for Pn40/Sn60 Solders at Different Temperatures, Obtained from Solomon (39)

Temperature	223° K	308° K	398° K	423° K
Plastic strain	0.103/0.307	0.04/0.082	0.022/0.102	0.036/0.097
Z average	0.7329/0.8697	0.5214/0.6031	0.6973/0.5914	0.6612/0.7224
	0.676	0.676	0.676	0.676
A^*	0.056/0.072	0.188/0.1298	0.0496/0.146	0.197/0.169

$^*A = A_{300}\left(\dfrac{T}{300}\right)^{1.55}$; $A_{300} = 0.102$; disturbance function $D = (1 - e^{-A\xi_D^Z})$.

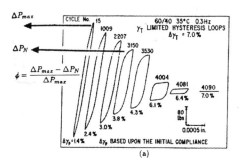

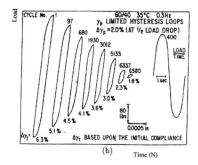

FIGURE 7.18
Cyclic thermomechanical tests for eutectic (Pb/Sn) solder by Solomon (60): (a) total strain controlled; (b) plastic strain controlled.

Example 7.6 Sands δ_0/δ_1 Models

Figures 7.20, 7.21, and 7.22 show comparisons between the test data and predictions for typical stress paths, CTC and TE, and SS obtained by using parameters in Table 7.7, for the Ottawa, Leighton Buzzard, and Munich sands, respectively. Both δ_0 and δ_1 models were used. It is usually found that the δ_0-associative model does not provide satisfactory predictions, particularly for the volumetric response; this is indicated in Figs. 7.20 and 7.21.

TABLE 7.7

Parameters for Dry Sands

Parameter	Sand		
	Ottawa	Leighton Buzzard	Munich
E MPa	262	79	63
(psi)	(38,000)	(11,500)	(9,200)
ν	0.37	0.29	0.21
γ	0.124	0.102	0.105
β	0.494	0.362	0.747
N	3.0	2.5	3.20
$\bar{h}_1^{**}$		0.135	0.1258
a_1^*	2.5×10^{-3}		
$\bar{h}_2^{**}$		450	1355
η_1	0.370		
$\bar{h}_3^{**}$		0.0047	.001
$\bar{h}_4^{**}$		1.02	1.11
R	0.00	0.00	0.00
κ	0.265	0.290	0.35

*Eq. (7.11a)
**Eq. (7.11c)

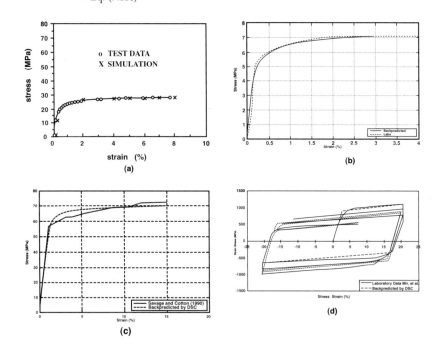

FIGURE 7.19
Typical comparisons between model predictions and test data (39, 64): (a) 63/37 (Sn/Pb), $T = 373$ K, $\dot{\varepsilon} = 0.01$/sec (59); (b) 95/3 (Pb/Sn), $T = 100°$C (61); (c) 60/40 (Sn/Pb), $T = -50°$C, $\dot{\varepsilon} = 0.01$/sec (63); (d) 95/5 (Pb/Sn), $T = 20°$C (62, 64).

TABLE 7.8a
Parameters for Saturated Clay: DSC Model (32)

Parameter	Value
E, MPa (psi)	10 (1500)
ν	0.35
γ	0.047
β	0.00
n	2.80
h_1 *	0.0001
h_2	0.78
$\bar{m}$	0.0694
λ	0.169
e_{oc}	0.903
A	1.73
Z	0.309
D_u	0.75

*In Eq. (7.11b) with $h_3 = h_4 = 0$.

TABLE 7.8b
Parameters for Dry Leighton Buzzard Sand: DSC Model (66)

Parameter	Value	
E, MPa (psi)	118 (17200)	
ν	0.30	
γ	0.065	
β	0.65	
n	2.50	
a_0	0.00342	
a_2	0.01841	in Eqs. (7.b) and (7.c)
η_0	0.3294	
η_2	0.83	
$\bar{m}$	0.20	
λ	0.13	
e_{oc}	0.75	
A_0	3.49	
A_3	−3.34	
A_4	3.28	in Eqs. (7.d), (7.e), and (7.f)
Z_0	0.580	
Z_1	0.098	

Example 7.7 Saturated Clay and Dry Sand: DSC Model

Clay: Figures 7.23 and 7.24 show comparisons between DSC predictions and observed laboratory undrained behavior of one-way cyclic (stress does not change sign) and two-way cyclic (stress changes sign) behavior for a saturated marine clay tested under both cylindrical triaxial and multiaxial

TABLE 7.8(c)

Parameters for Saturated Sand: δ_0-Model (67)

E, MPa	140
(psi)	(20,420)
ν	0.15
γ	0.636
β	0.60
n	3.0
a_1	0.16×10^{-4}
η_1	1.17
R	0.00

TABLE 7.9

Parameters for Rocks (35, 36)

		Rock	
Parameter	Soap Stone	Rock Salt	Sandstone
E, MPa	9,150	21,000	25,500
(psi)	$(1,328 \times 10^3)$	$(3,050 \times 10^3)$	$(3,700 \times 10^3)$
ν	0.0792	0.27	0.11
γ	0.0470	0.0945	0.0774
β_0 \}*	−0.750	β \}** 0.990	0.767
β_1	0.0465	β_1 0.00048	0.0020
n	7.0	3.0	7.20
a_1	0.177×10^{-2}	1.80×10^{-5}	0.467×10^{-2}
η_1	0.747	0.2322	0.345
R, MPa	1.067	1.80	2.90
(psi)	(155)	(260)	(420)
κ	—	0.275	—

*Eq. (7.15a)
**Eq. (7.15b)

conditions (32); the parameters are given in Table 7.8(a). Here, σ'_0 = initial effective mean stress and e_0 = void ratio. The stress–strain behavior in terms of $(\sigma_1 - \sigma_3)$ vs. $\varepsilon_1, \varepsilon_2,$ and ε_3 (Fig. 7.23) is simulated very well by the DSC model, including degradation in the later stages. These figures also show the predicted relatively intact (RI) behavior according to the δ_0-associative plasticity model.

The two-way cyclic behavior is also predicted very well by the DSC model, including the degradation with cycles of loading [Fig. 7.24(a)] and the stress paths followed [Fig. 7.24(b)]. The predicted pore water pressure response also shows satisfactory correlation with the observations [Fig. 7.24(c)]. The pore water pressure, p, was computed using the following equation:

$$p = \frac{J_{1t} - J_1^a}{3} \tag{7a}$$

Hierarchical Single-Surface Plasticity Models in DSC

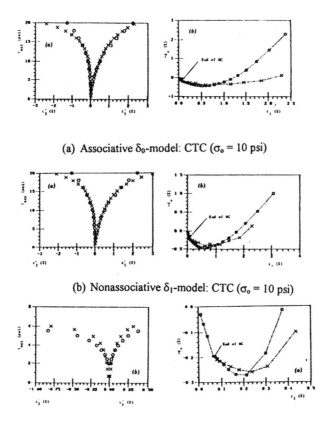

(a) Associative δ_0-model: CTC ($\sigma_o = 10$ psi)

(b) Nonassociative δ_1-model: CTC ($\sigma_o = 10$ psi)

FIGURE 7.20
Comparisons between model predictions and test data for Ottawa sand: $\gamma_0 = 1.73$ g/cc; 1 psi = 6.89 kPa (20).

where J_{1t} is the first invariant of the total stress tensor, and J_1^a is the first invariant of the average (predicted) effective stress.

Dry Sand: The Leighton Buzzard (LB) sand was tested using the cylindrical triaxial device (66). The DSC model, Chapter 4, with the δ_0-model for the RI behavior was used to characterize the behavior of the sand, which exhibited softening response for higher densities (D_r) and lower mean pressures ($\sigma_o = p_0$). The parameters for the sand are shown in Table 7.8(b). Here, the hardening and disturbance parameters were expressed as dependent on relative density, D_r, and/or initial confining stress, σ_0:

$$a_1 = \ln a_0 + D_r \ln a_2 \tag{7b}$$

$$\eta_1 = \eta_0 + \eta_2 D_r \tag{7c}$$

$$A = A_0 + A_2 D_r \tag{7d}$$

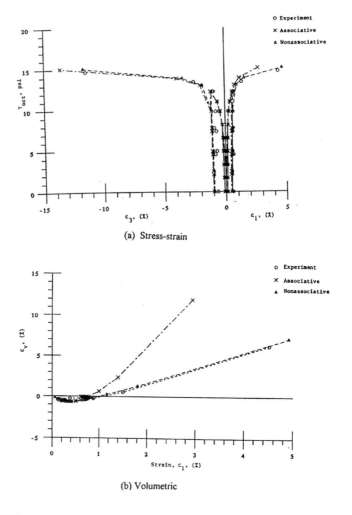

FIGURE 7.21
Comparisons between model predictions and test data for Leighton Buzzard sand: SS ($\sigma_0 = 20$ psi), $\gamma_0 = 1.74$ g/cc, 1 psi = 6.89 kPa (4, 19, 66).

$$A_2 = A_3 + A_4\left(\frac{\sigma_0}{p_a}\right) \qquad (7e)$$

$$Z = Z_0 + Z_1\left(\frac{\sigma_0}{p_a}\right) \qquad (7f)$$

Figure 7.25 shows comparisons between the DSC predictions and test data for $D_r = 95\%$ and $\sigma_0 = 276$ kPa (40 psi) under the CTC stress path. The figure also shows predictions according to the δ_0-model used to simulate the RI behavior. It can be seen in Fig. 7.25(b) that the δ_0-associative model

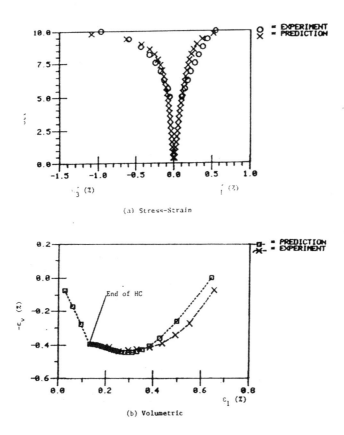

FIGURE 7.22
Comparisons between model predictions and test data for Munich sand: nonassociative δ_1-model: SS (σ_0 = 13 psi), γ_0 = 2.03 g/cc; 1 psi = 6.89 kPa (20).

(RI) predicts a much higher volume change compared to the test data. However, the use of the DSC model provides much improved predictions. Thus, the constraining effect due to friction is included in the DSC model, and it may not be necessary to use a nonassociative model.

Example 7.8 Saturated Sand (67): δ_0-Model

Table 7.8(c) shows the parameters from test on sand C reported by Castro (68). The sand is a natural beach sand from deposits at Huachipato, Chile, and it is uniform fine sand with specific gravity, G_s = 2.87 and uniformity coefficient = 2.3.

Figure 7.26 shows comparisons between prediction (δ_0-plasticity model) and test data for the undrained behavior of sand under CTC path with σ_0 = 98 kPa (14.21 psi). Both the stress–strain and pore water pressure responses are predicted well by the plasticity model.

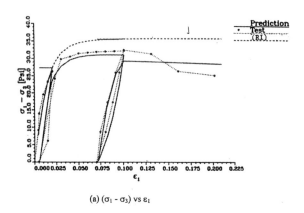

(a) $(\sigma_1 - \sigma_3)$ vs ε_1

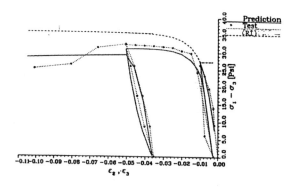

(b) $(\sigma_1 - \sigma_3)$ vs ε_2 and ε_3

FIGURE 7.23
Comparisons between model predictions and one-way cyclic test data for marine clay: CTC ($\sigma'_0 = 110$ psi); $e_0 = 0796$ (32), 1 psi = 6.89 kPa.

Example 7.9 Rocks

Laboratory stress–strain behavior of a soap stone and a rock salt, Table 7.9, were obtained from tests with cubical specimens (33, 34, 36). The test data using cubical specimens for a sandstone found in Japan, reported by Nishida et al. (69), were used.

Figures 7.27 to 7.29 show comparisons between predictions using the δ_0-model for the soap stone, rock salt, and sandstone; both the δ_0- and δ_1-models were used for the rock salt. The results for the soap stone and sandstone are under the SS and TC paths, while those for the rock salt are for the CTC path. It can be seen that the volumetric strains are predicted better for the rock salt by the δ_1-model (Fig. 7.28), while the δ_0-model shows discrepancies in the volumetric response for the other two rocks, Figs. 7.27 and 7.29; use of the δ_1-model would improve the predictions.

Hierarchical Single-Surface Plasticity Models in DSC

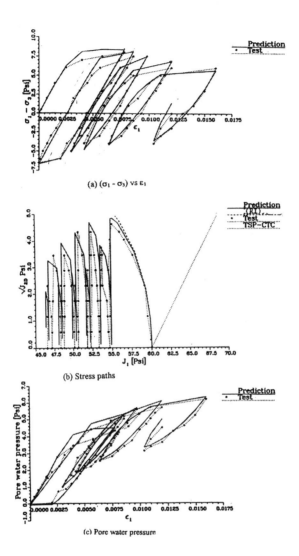

FIGURE 7.24
Comparisons between model predictions and test data for marine clay; two-way cyclic tests; $\sigma_0' = 20$ psi; $e_0 = 0.9123$ (32).

Figure 7.30 shows comparisons between predictions and test data under the CTC path (cylindrical triaxial tests) reported by Waversik and Hanum (70) for a similar rock salt. The parameters in Table 7.9 for the rock salt tested under multiaxial tests (36) were used to predict the behavior of rock salt reported in (70). Thus, this is an *independent* validation and, overall, shows good correlation.

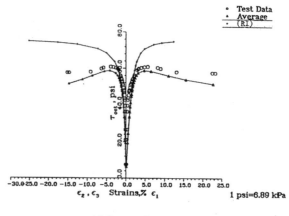

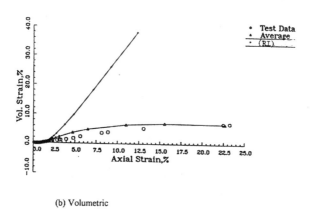

(b) Volumetric

FIGURE 7.25
Comparisons between model (DSC) predictions and test data for Leighton Buzzard sand (CTC, $\sigma_0 = 40$ psi); $D_r = 95\%$ (66).

Example 7.10 Concrete: δ_0-Plasticity Model

Table 7.10 shows material parameters for the δ_0-model for a plain concrete tested under stress-controlled multiaxial loading. Cubical specimens were tested under different initial confining pressures and stress paths [Fig. 7.13(b)] (35). Figure 7.31(a) shows comparisons between predictions and test data (solid circles and triangles) for the ultimate behavior under different stress paths and in different stress spaces. Figure 7.31(b) shows a typical comparison between predictions of stress–strain and volumetric responses and test data for the SS stress path with $\sigma_0 = 31$ MPa (4.5×10^3 psi) and the TE stress path with $\sigma_0 = 28$ MPa (4.0×10^3 psi).

The parameters (Table 7.10) were determined from test data under various straight-line stress paths [Fig. 7.13(b)]. An independent test was performed in

Hierarchical Single-Surface Plasticity Models in DSC

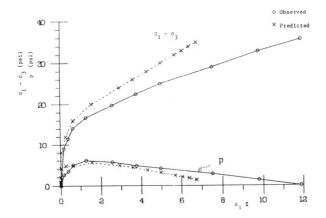

FIGURE 7.26
Comparisons between model predictions and test data for saturated sand: CTC ($\sigma_0 = 14.21$ psi); 1 psi = 6.89 kPa (67).

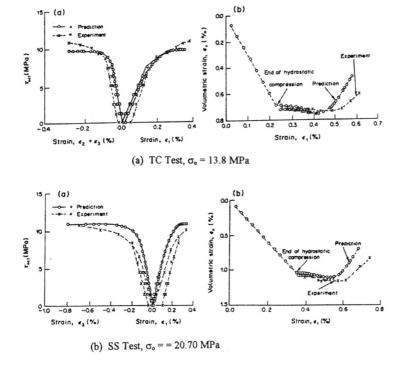

(a) TC Test, $\sigma_o = 13.8$ MPa

(b) SS Test, $\sigma_o = 20.70$ MPa

FIGURE 7.27
Comparisons between model predictions and test data for soap stone. (From Ref. 33. Reprinted with permission from Elsevier Science.)

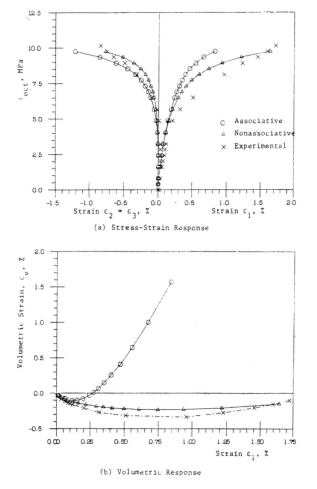

FIGURE 7.28
Comparisons between model predictions and test data for rock salt: CTC ($\sigma_0 = 0.0$ MPa). (From Ref. 36, ©American Geophysical Union. Reprinted with permission.)

which the stress path followed was circular [Fig. 7.13(b)]. This test involved $\tau_{oct} = 7.0$ kPa and $\sigma_0 = 28.0$ kPa. Figure 7.32 shows comparisons between predicted and observed strains under the circular stress path [Fig. 7.13(b)]. This is an independent validation and shows satisfactory predictions by the δ_0-model.

Example 7.11 Concrete: DSC Model

Strain-controlled tests on cubical specimens of plain concrete were reported by Van Mier (71). The parameters for the DSC model were obtained using the tests under uniaxial and proportional loading stress paths (72, 73); they are shown in Table 7.10. In the DSC model, the RI behavior was simulated by

Hierarchical Single-Surface Plasticity Models in DSC 249

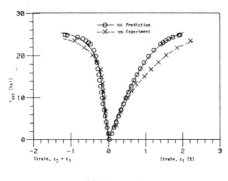

(a) Stress-strain

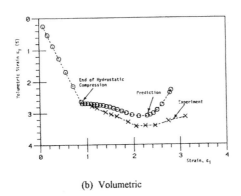

(b) Volumetric

FIGURE 7.29.
Comparisons between model predictions and test data for sandstone: TC ($\sigma_0 = 28.45$ psi) (34); 1 psi = 6.89 kPa (35).

using the δ_0-plasticity model, and for the FA behavior it was assumed that material in that state can carry hydrostatic stress but no shear stress.

Figure 7.33 shows comparisons between predicted and observed test results in terms of τ_{oct} vs. ε_1, τ_{oct} vs. ε_2, and ε_v vs. ε_1 responses for the loading path $\sigma_1/\sigma_2 = \sigma_1/\sigma_3 = 10$ (71).

Example 7.12   Geologic (Unbound) Materials in Pavements: Repeated Loading

Bonaquist (74) and Bonaquist and Witczak (75) conducted a comprehensive series of laboratory triaxial tests on geologic materials (base, subbase, subgrade) in pavement structures. They used a simplified form of the HISS δ_0-model, without the term F_s in Eq. (7.1), to characterize the elastoplastic behavior of the materials. They also developed procedures for reducing the parameters (e.g., by expressing some in terms of others) toward practical application, and considered important factors such as pore water pressure effects, partial saturation, and repeated loading permanent deformations

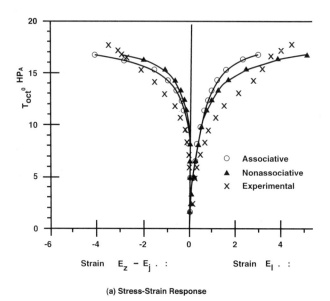

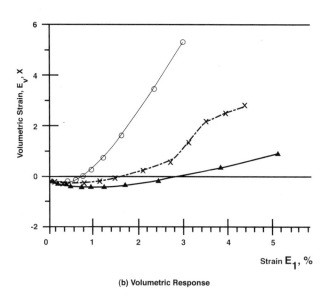

FIGURE 7.30
Comparisons between model predictions and test data for similar rock salt: CTC ($\sigma_0 = 3.45$ MPa). (From Ref. 36. ©American Geophysical Union. Reprinted with permission.)

leading to rutting in pavements. Table 7.11 shows the parameters for the three materials. Determination of parameters involved numerical optimization in which the values of n and η_1 were maintained at their average values of 3.25 and 0.50, respectively. The hardening parameter, a_1, was expressed in terms

TABLE 7.10
Parameters for Plain Concrete (35)

Parameter	Concrete – 1 δ_0-Model (35)		Concrete – 2 DSC Model (72, 73)
E, MPa (psi)	7000 (10^6)		5400 (0.78×10^6)
ν	0.154		0.25
γ	0.1130		0.0678
β_0	0.8437	β	0.755
β_1	.027		
n	7.0		5.237
a_1	9×10^{-3}		4.614×10^{-11}
η_1	0.44		0.826
R, MPa (psi)	2.72 (395)		15.85 (2300)
A	—		668
Z	—		1.50
D_u	—		0.875

of γ as

$$a_1 = 0.0001056 \, e^{4\sqrt{\gamma}} \tag{12a}$$

Thus, knowledge of the ultimate (failure) parameter $\sqrt{\gamma}$ can be used to evaluate the hardening parameter a_1.

7.12.2 Repeated Loading and Permanent Deformations

A special procedure was developed, in the context of the HISS model, to calculate the permanent deformations under repeated loading, based on the bounding surface approach [Mroz et al. (12); Bonaquist and Witczak (75)]. During cyclic loading (loading, unloading, and reloading), the yield surface (F_i) that defines the elastic limit expands with the number of cycles, $i = N$ [Fig. 7.34(a)], which causes cyclic hardening. It is defined based on the accumulated plastic strains (ξ_i) at the end of a given cycle:

$$\xi_i = \xi_0 + \left(1 - \frac{1}{N^{h_c}}\right)(\xi_b - \xi_0) \tag{12b}$$

where ξ_0 is the trajectory corresponding to the initial or *in situ* stress conditions, and ξ_b corresponds to the bounding surface for the applied load (stress) amplitude, $P_{\max}$ [Fig. 7.34(b)], or the maximum applied load (stress) [Fig. 7.34(c)], and h_c is the cyclic hardening parameter. The value of h_c is found from measured test results in terms of accumulated plastic strains vs. the number of cycles (Fig. 7.35). These figures show the measured and computed

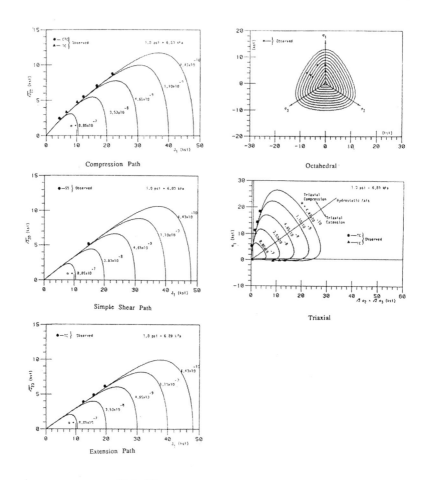

(a) In different stress spaces and paths

FIGURE 7.31
Comparison between model predictions and test data for plain concrete (35).

plastic strains for subgrade and base materials, respectively. Figure 7.36 shows comparisons between predictions and measurements for a multiple load-level test (74). Here, during a load-level, the given stress (J_1, $\sqrt{J_{2D}}$) was cycled for a fixed number of cycles; each level involved different values of J_1 and $\sqrt{J_{2D}}$.

Example 7.13 Rock (Granite) under High Pressure

Triaxial compression tests were performed by Alheid (76) on granite specimens under confining pressures up to 4.5 kb (450 MPa) and temperature up to 350°C. Figure 7.37 shows laboratory test behavior of the granite in terms of stress difference ($\sigma_1 - \sigma_3$) vs. ε_1 and ($\sigma_1 - \sigma_3$) vs. ε_v ($= \varepsilon_1 + \varepsilon_2 + \varepsilon_3$). The behavior exhibits strain softening. However, the δ_0-plasticity model was used

Hierarchical Single-Surface Plasticity Models in DSC

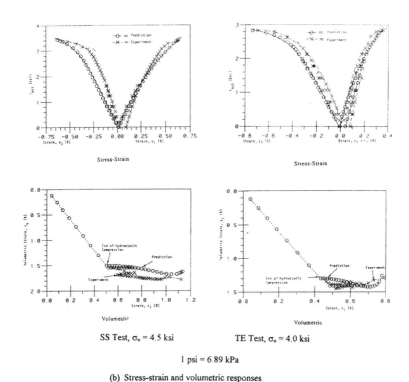

(b) Stress-strain and volumetric responses

FIGURE 7.31
(continued)

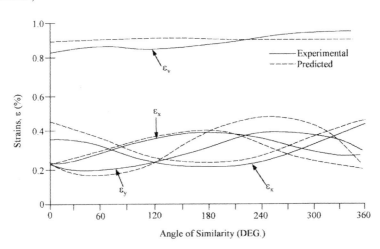

FIGURE 7.32
Comparisons between model predictions and test data for plain concrete: circular stress path, $\sigma_0 = 4$ ksi, $\tau_{oct} = 1$ ksi; 1 psi = 6.89 kPa (35).

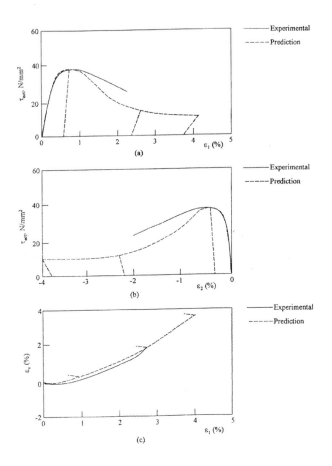

FIGURE 7.33
Comparisons between model predictions and test data for plain concrete: loading path, $\sigma_1/\sigma_2 = \sigma_2/\sigma_3 = 10.0$. (From Ref. 72, with permission from Elsevier Science.)

TABLE 7.11

Parameters for Pavement Materials (74)

	Material		
Parameter	Base: Crushed Aggregate	Subbase: Silty Sand	Subgrade: Silty Sand
γ	0.265	0.0756	0.0328
a_1	8.31×10^{-4}	2.93×10^{-4}	2.45×10^{-4}
η_1	0.50	0.50	0.50
n	3.25	3.25	3.25
R (kPa)	0.00	14.00	31.00

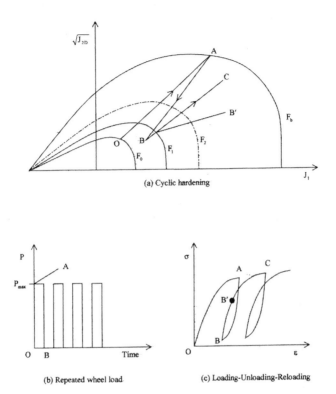

FIGURE 7.34
Cyclic hardening under repeated loading.

to characterize the elastoplastic hardening behavior (77). The parameters for the δ_0-model are given in Table 7.12.

Test data on a similar granite reported by Mogi (78) using a multiaxial device under compression and extension paths were used to find the value of β using Eq. (7.13d), with $\tan \theta_C = 0.418$ and $\tan \theta_E = 0.258$. The value of R was adopted equal to about 10% of the uniaxial compressive strength for the granite of about 180 MPa.

Typical comparisons for the $(\sigma_1 - \sigma_3)$ vs. ε_1 responses for $\sigma_0 = 78$ and 157 MPa are shown in Fig. 7.38(a) and (b), respectively. It can be seen that the δ_0-model predicts the elastoplastic behavior very well. It is reported (76) that the rock experiences microcracking and slip, resulting in degradation and strain softening (Fig. 7.37). Hence, to characterize the complete response, it would be appropriate to use the DSC model.

Example 7.14 Optimum Tests and Sensitivity of Parameters

To determine material parameters, it is desirable to use laboratory tests under as many stress paths and (initial) conditions as possible. Then the constitutive model calibrated on the basis of the tests would include effects of the stress

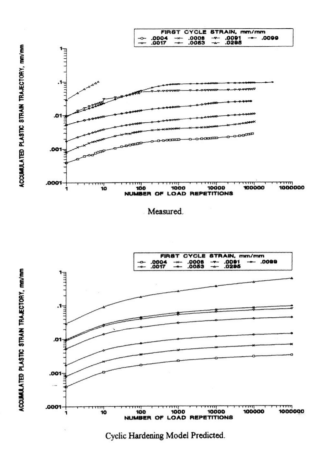

FIGURE 7.35(A)
Comparisons between model (δ_0-cyclic hardening) predictions and test data for accumulated plastic strains: subgrade (74).

paths and conditions that influence the field behavior. However, in practice, it may not be possible to perform many such tests, and it may become necessary to estimate the parameters based on the optimum (or minimum) number of available tests.

In order to determine the optimum number of tests, detailed parametric evaluations were performed. Here, predictions for stress–strain–volume change behavior of a sand (79) tested using the standard and commonly available triaxial device with cylindrical specimens were compared using parameters determined from various sets of tests under different stress paths.

The parameter β was expressed as function of J_1 as in Eq. (7.15b). The nonassociative parameter κ was assumed to be a function of the stress ratio, S_r [Eq. (7.1)] as

$$\kappa = \kappa_1 + \kappa_2 S_r \tag{13a}$$

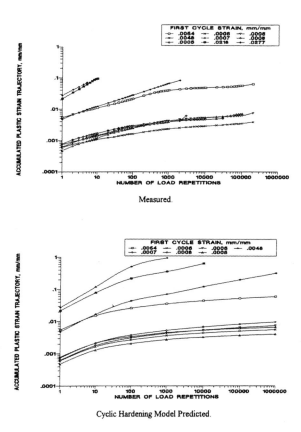

FIGURE 7.35(B)
Comparisons between model (δ_0-cyclic hardening) predictions and test data for accumulated plastic strains: base (74).

where κ_1 and κ_2 are the constants. For the nonassociative model, the hardening function, α_Q, is expressed as in Eq. (7.5).

As noted before, the behavior of a material under different stress paths is different; for instance, Fig. 7.4 shows the ultimate envelopes under compression (C), simple stress (S), and extension (E) stress paths. Thus, the angles of friction ϕ (in τ vs. σ_n plots) corresponding to the ultimate slopes (γ) in $\sqrt{J_{2D}} - J_1$ plots, for the three conditions, will be different. Hence, it is desirable to include stress points from the above stress paths in the evaluation of γ and β. However, if only compression tests are available, it may be assumed that the angle of friction in compression (ϕ_C) and that in extension (ϕ_E) are equal. Then, γ and β can be found based on Eq. (7.13).

With the above considerations, a number of cases were analyzed in which the following factors were considered:

- associative (δ_0-) model with five plasticity constants ($\gamma, \beta, n, a_1, \eta_1$);
- nonassociative (δ_1-) model with six or seven constants, including the preceding five plus one (κ) or two (κ_1 and κ_2) constants;

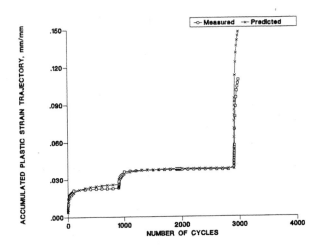

FIGURE 7.36
Comparisons between model (δ_0-cyclic hardening) predictions and test data for accumulated plastic strains: multiple-load permanent deformation test (74).

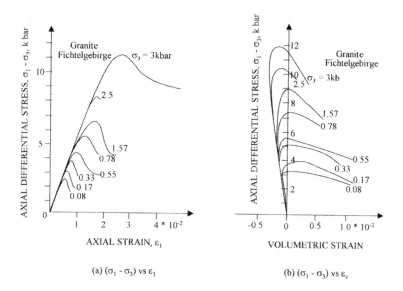

(a) $(\sigma_1 - \sigma_3)$ vs ε_1 (b) $(\sigma_1 - \sigma_3)$ vs ε_v

FIGURE 7.37
Triaxial test data for granite at different initial confining pressures, $\sigma_0 = \sigma_3$ (76, 77).

- straight and curved ultimate envelope with one (β) or two (β and β_1) constants [Eq. (7.15b)];
- constants found as average values from nine tests for ultimate envelopes, five shear tests under different stress paths (CTC, TC, and TE) or two tests (one CTC and one TE) or only one test (CTC) with the assumption of $\phi_C = \phi_E$.

Hierarchical Single-Surface Plasticity Models in DSC

TABLE 7.12
Parameters for Granite (77)(With Permission)

Parameter	Value
E	6.67×10^4 MPa
ν	0.26
γ	0.038
β	0.75
N	8.30
a_1	9.8×10^{-30}
η_1	0.72
R	18 MPa

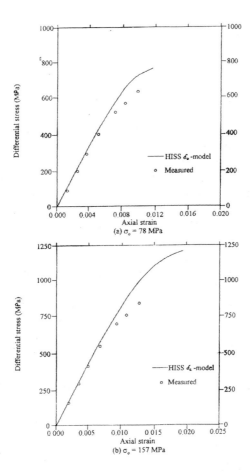

FIGURE 7.38
Comparisons between model (δ_0) predictions and test data for granite (77).

TABLE 7.13
Material Parameters for Optimum Number of Tests (79) (With Permission)

	Curved Envelope	Straight Envelope	Straight Envelope	
Tests used	—Nine tests for ultimate envelope —Five tests for other plasticity parameters	—Two tests: one CTC one TE	—One CTC test —$\phi_C = \phi_E$	
Elasticity	$E = 150$ MPa		$\nu = 0.3$	
Plasticity				
γ	0.0702	0.0702	0.0698	0.0619
β	-0.690	0.70	0.66	0.74
β_1	0.00003	—	—	—
n	3.00	3.00	2.80	2.90
a_1	31.66×10^{-4}	38.5×10^{-4}	36.64×10^{-4}	31.31×10^{-4}
η_1	0.493	0.450	0.505	0.504
κ_1	0.402	0.400	0.42	—
κ_2	0.100	0.101	0.111	—
κ	—	0.400	—	0.271

Table 7.13 shows material constants for various factors considered in the parametric study.

Figure 7.39 shows comparisons between predictions and test data for yield surfaces and the ultimate envelopes for the Badarpur sand (79). Overall, the inclusion of the curved ultimate envelope provides better predictions (see also below). For high mean pressure levels, e.g., in the case of high (earth) dams, the ultimate envelope is curved and it would be appropriate to include it in the model.

Figures 7.40 and 7.41 show comparisons between predictions and laboratory data for the typical factors listed above. It can be seen from both figures that the nonassociative model, curved ultimate envelope, stress-ratio-dependent nonassociative parameter, and constants determined from two tests provide improved predictions. It may be noted, however, that predictions even from one (CTC) test with $\phi_C = \phi_E$ provide satisfactory predictions.

Based on the above, it was concluded that the parameters for the δ_0- and δ_1-model can be evaluated from two tests under CTC and TE paths and can be estimated with one CTC test with the assumption $\phi_C = \phi_E$. Indeed, if a greater number of tests are available, an improved quality of parameters is obtained. A formal procedure for the optimization of parameters in the DSC and HISS models is described in Appendix II.

Example 7.15 Rockfill Material: Particle Size-dependent δ_1-Plasticity Model

Rockfill material involves particles of different sizes. Laboratory drained triaxial tests on a rockfill materials with different particle sizes were reported by

Hierarchical Single-Surface Plasticity Models in DSC

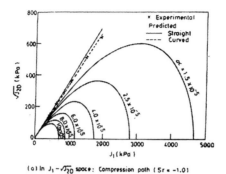

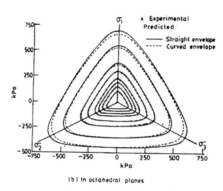

FIGURE 7.39
Comparisons between model predictions and test data for sand: yield surfaces (79)(with permission).

Varadarajan et al. (80). Table 7.14 shows the material parameters as a function of the particle size (average diameter), D_{max}, for materials used in the Ranjit Sagar and Purulia dams. The (initial) elastic modulus was expressed as a nonlinear function of the (initial) mean pressure (σ_0) as (80, 81)

$$E_i = kp_a \left(\frac{\sigma_0}{p_a}\right)^{n'} \tag{15a}$$

where k and n' are parameters.

It can be seen from Table 7.14 that the parameters show consistent increasing or decreasing trends with D_{max}.

Figures 7.42 and 7.43 show typical comparisons between predictions by the δ_1-model and laboratory data. The predicted yield surfaces (Fig. 7.42) compare well with the test data for different values of D_{max}. The stress–strain and volumetric responses are also predicted very well by the δ_1-model (Fig. 7.43). Note that the influence of the particle size can be included in the model by expressing the parameters as functions of D_{max}.

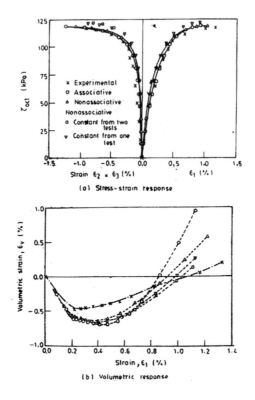

FIGURE 7.40
Comparisons between model predictions and test data: parametric sensitivity analyses, TC (σ_0 = 137.3 kPa) (79)(with permission).

Example 7.16 Silicon Crystal with Dislocation: Disturbance Model

Dillon et al. (82) proposed a three-dimensional generalization of the constitutive model by Haasen (83) for the characterization of thermomechanical behavior of silicon crystals (ribbons) with dislocations. Figure 7.44 shows stress–strain behavior (predicted by their model); it is affected by factors such as (initial) dislocation density (N_0), temperature (T), strain rate ($\dot{\varepsilon}$), and impurities such as oxygen and nitrogen.

The DSC model was used to characterize the behavior of the silicon crystals, and a correlation was established between disturbance and dislocation (see Eq. 3.25a, Chapter 3) (84). In the DSC, the RI behavior was simulated by using a linear elastic model (Chapter 5) in which the elastic moduli were expressed as (82)

$$E = 1.7 \times 10^{11} - 2.771 \times 10^4 (T)^2 p_a \tag{16a}$$

$$G = 6.81 E / 17 p_a \tag{16b}$$

Hierarchical Single-Surface Plasticity Models in DSC

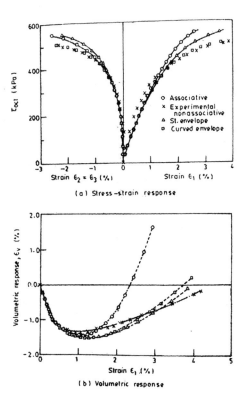

FIGURE 7.41
Comparisons between model predictions and test data: parametric sensitivity analyses, CTC ($\sigma_0 = 274.6$ kPa) (79) (with permission).

It can also be simulated using the δ_0-plasticity model. The disturbance function, Eq. (3.16b), was used and its values from the stress–strain data (Fig. 7.44) were evaluated based on the following equation (Eq. 3.7):

$$D = \frac{\sigma^i - \sigma^a}{\sigma^i - \sigma^c} \tag{16c}$$

The parameters in Eq. (3.16b), h, $\bar{w}$, and s, were found using the procedure described in Chapter 3 (Fig. 3.18). The parameters $\bar{w}$ and s were found to be relatively constant under T, N_0, and $\dot{\varepsilon}$, and their average values were found to be 1.354 and 1.20, respectively. The parameter h was found to be dependent on T and N_0 and was expressed as

$$h(N_0, T) = h_{300}(N_0)\left(\frac{T}{300}\right)^{\bar{n}(N_0)} \tag{16d}$$

TABLE 7.14
Rockfill Materials Parameters (80) (with permission)

Material Constants	Ranjit Sagar Dam Material D_{max} (mm)				Purulia Dam Material D_{max} (mm)			
	10	25	50	80	25	50	80
Elasticity							
k	156.03	193.69	220.34	253.63	167.07	286.02	451.13
n'	0.5768	0.6386	0.6683	0.7146	0.8162	0.6550	0.4068
ν	0.36	0.31	0.30	0.29	0.34	0.33	0.31
Ultimate							
γ	0.0577	0.0630	0.0781	0.0811	0.0619	0.0617	0.0615
β	0.71	0.72	0.73	0.74	0.72	0.72	0.72
Phase change, n	3	3	3	3	3	3	3
Hardening							
a_1	0.158E-3	0.722E-4	0.605E-4	0.985E-5	0.310E-4	0.850E-4	0.130E-3
η_1	0.355	0.453	0.476	0.762	0.600	0.440	0.380
Nonassociative, κ	0.20	0.25	0.22	0.22	0.23	0.20	0.17

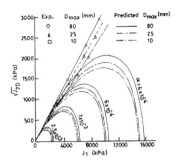

FIGURE 7.42
Comparisons between predictions and test data for yield surface for typical values of D_{max} (80) (with permission).

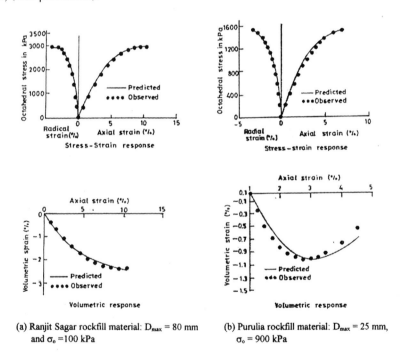

(a) Ranjit Sagar rockfill material: $D_{max} = 80$ mm and $\sigma_o = 100$ kPa

(b) Purulia rockfill material: $D_{max} = 25$ mm, $\sigma_o = 900$ kPa

FIGURE 7.43
Comparisons between predictions and test data for stress–strain and volume change responses (80) (with permission).

where h_{300} is the value of h at $T = 300$ K and $\bar{n}$ is the exponent; both are dependent on N_0 and were expressed as

$$h_{300}(N_0) = 1.195 + 3.675 \times 10^{-9} N_0 \quad (16e)$$

$$\bar{n}(N_0) = -6.735 - 2.577 \times 10^{-9} N_0 \quad (16f)$$

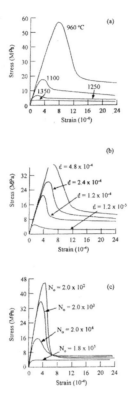

FIGURE 7.44
Stress–strain data for silicon ribbon (82): (a) temperature, (b) strain rate, (c) initial dislocation density.

Hence, the disturbance function was expressed as

$$D = D_u \left[1.0 - \left(1.0 + \frac{\xi_D}{h(N_0, T)}\right)^{1.254}\right]^{-1.2} \tag{16g}$$

The DSC equations [Eq. (3.1)] were integrated to predict the laboratory behavior under different temperatures, strain rates, and dislocation densities (84). Typical results are presented in Figure 7.45(a) and (b), which show typical comparisons between predictions and test data for two conditions: (a) $T = 960°C$, $\dot{\varepsilon} = 1.2 \times 10^{-4}\,\text{s}^{-1}$, and $N_0 = 2 \times 10^4\,\text{cm}^{-2}$ and (b) $T = 1100°C$, $\dot{\varepsilon} = 4.8 \times 10^{-4}\,\text{s}^{-1}$, and $N_0 = 1.8 \times 10^5\,\text{cm}^{-2}$. The correlation is considered to be highly satisfactory.

Application of the DSC model for the softening and stiffening response of silicon with high levels of N_0 and oxygen impurities is presented in Chapter 10.

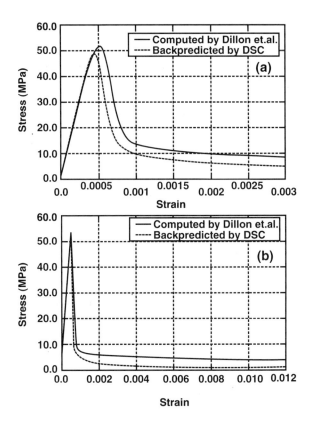

FIGURE 7.45
Comparisons between given and predicted curves for silicon ribbon for independent tests: (a) $T = 960°C$, $\dot{\varepsilon} = 1.2 \times 10^{-4}$ s^{-1}, $N_0 = 2 \times 10^4$ cm^{-2}, (b) $T = 1100°C$, $\dot{\varepsilon} = 4.8 \times 10^{-4}$ s^{-1}, $N_0 = 1.8 \times 10^5$ cm^{-2} (84) (with permission).

References

1. Desai, C.S., "A General Basis for Yield, Failure and Potential Functions in Plasticity," *Int. J. Numer. Analy. Methods Geomech.*, 4, 1980, 361–375.
2. Desai, C.S. and Faruque, M.O., "Constitutive Model for Geologic Materials," *J. Eng. Mech., ASCE*, 110, 9, 1984, 1391–1408.
3. Desai, C.S., Somasundaram, S., and Frantziskonis, G., "A Hierarchical Approach for Constitutive Modelling of Geologic Materials," *Int. J. Numer. Anal. Methods in Geomech.*, 10, 3, 1986, 225–257.
4. Desai, C.S. and Wathugala, G.N., "Hierarchical and Unified Models for Solids and Discontinuities," Notes for Short Course, Tucson, Arizona, USA, 1987.
5. Desai, C.S., "Hierarchical Single Surface and the Disturbed State Constitutive Models with Emphasis on Geotechnical Applications," Chapter 5 in *Geotechnical Engineering*, K.R. Saxena (editor), Oxford & IBH Publ. Co., New Delhi, India, 1994.

6. Desai, C.S., "Constitutive Modelling Using the Disturbed State as Microstructure Self-Adjustment Concept," Chapter 8 in *Continuum Models for Materials with Microstructure*, H.B. Mühlhaus (editor), John Wiley, U.K, 1995.
7. Matsuoka, H. and Nakai, T., "Stress-Deformation and Strength Characteristics of Soil Under Three Different Principal Stresses," *Proc. Jap. Soc. Civil Engrs.*, No. 232, 1974, 59–70.
8. Lade, P.V. and Duncan, J.H., "Elastoplastic Stress-Strain Theory for Cohesionless Soil," *J. Geotech. Eng., ASCE*, 101, 10, 1975, 1037–1053.
9. Lade, P.V. and Kim, M.K., "Single Hardening Constitutive Model for Frictional Materials. III. Comparisons with Experimental Data," *Computers and Geotechnics*, 6, 1988, 31–47.
10. Vermeer, P.A., "A Five-Constant Model Unifying Well-Established Concepts," *Proc. Int. Workshop on Constitutive Relations for Soils*, Grenoble, 1982, 1984, 175–197.
11. Mroz, Z., "On the Description of Anisotropic Work Hardening," *J. Mech. Phys. Solids*, 15, 1967, 163–175.
12. Mroz, Z., Norris, V.A., and Zienkiewicz, O.C., "An Anisotropic Hardening Model for Soils and Its Application to Cyclic Loading," *Int. J. Numer. Analy. Methods Geomech.*, 2, 1978, 203–221.
13. Mroz, Z., Norris, V.A., and Zienkiewicz, O.C., "Application of an Anisotropic Hardening Model in the Analysis of Elasto-plastic Deformation of Soils," *Geotechnique*, 29, 1, 1979, 1–34.
14. Krieg, R.D., "A Practical Two-Surface Plasticity Theory," *J. Appl. Mech.*, 42, 1975, 641–646.
15. Prevost, J.H., "Plasticity Theory of Soil Stress–Strain Behavior," *J. Eng. Mech., ASCE*, 104, 5, 1978, 1177–1194.
16. Dafalias, Y.F. and Popov, E. P., "A Model for Nonlinearly Hardening Materials for Complex Loading," *Acta Mech.*, 21, 3, 1975, 173–192.
17. Dafalias, Y.F., "A Bounding Surface Plasticity Model," *Proc. 7th Can. Cong. Appl. Mech.*, Sherbrooke, Canada, 1979.
18. Baker, R. and Desai, C.S., "Consequences of Deviatoric Normality in Plasticity with Isotropic Hardening," *Int. J. Num. Analyt. Meth. Geomech.*, 6, 3, 1982, 383–390.
19. Desai, C.S. and Hashmi, Q.S.E., "Analysis, Evaluation and Implementation of a Nonassociative Model for Geologic Materials," *Int. J. Plasticity*, 5, 1989, 397–420.
20. Frantziskonis, G., Desai, C.S., and Somasundaram, S., "Constitutive Model for Nonassociative Behavior," *J. of Eng. Mech., ASCE*, 112, 1986, 932–946.
21. Desai, C.S., Letter to the Editor for "Single Surface Yield and Potential Function Plasticity," by Lade, P.V. and Kim. M.K., *Computers and Geotechnics*, 7, 1989, 319–335.
22. Desai, C.S., Discussion to "Single Hardening Model with Application to NC Clay," *J. Geotech. Eng., ASCE*, 118, 2, 1990, 337–340.
23. Roscoe, K.H., Schofield, A.N., and Wroth, C.P., "On Yielding of Soils," *Geotechnique*, 8, 1958, 22–53.
24. Schreyer, H.L., "A Third Invariant Plasticity Theory for Frictional Materials," *J. of Structural Mech.*, June 1983.
25. Sture, S., "Model by Prevost," *Proc., Workshop Session, Symposium on Implementation of Computer Procedures and Stress–Strain Laws in Geotech. Eng.*, Chicago, IL, 1981.

26. Somasundaram, S. and Desai, C.S., "Modelling and Testing for Anisotropic Behavior of Soils," *J. Eng. Mech., ASCE,* 114, 1988, 1473–1496.
27. Desai, C.S. and Galagoda, H.M., "Earthquake Analysis with Generalized Plasticity Model for Saturated Soils," *J. Earthquake Eng. and Struct. Dyn.,* 18, 6, 1989, 903–919.
28. Wathugala, G.W. and Desai, C.S., "Constitutive Model for Cyclic Behavior of Cohesive Soils, I: Theory," *J. Geotech. Eng., ASCE,* 119, 4, 714–729.
29. Desai, C.S. and Zhang, D., "Viscoplastic Model with Generalized Yield Function," *Int. J. Numer. Methods Geomech.,* 11, 1987, 603–620.
30. Desai, C.S., Samtani, N.C., and Vulliet, L., "Constitutive Modeling and Analysis of Creeping Slopes," *J. Geotech. Eng., ASCE,* 121, 1, 1995, 43–56.
31. Samtani, N.C., Desai, C.S., and Vulliet, L., "An Interface Model to Describe Viscoplastic Behavior," *Int. J. Numer. Analyt. Methods Geomech.,* 20, 4, 1996, 231–252.
32. Katti, D.R. and Desai, C.S., "Modelling and Testing of Cohesive Soil Using the Disturbed State Concept," *J. Eng. Mech., ASCE,* 121, 1994, 648–658.
33. Desai, C.S. and Salami, M.R., "A Constitutive Model and Associated Testing for Soft Rock," *Int. J. Rock Mech. & Min. Sc.,* 24, 5, 1987, 299–307.
34. Desai, C.S. and Salami, M.R., "A Constitutive Model for Rocks," *J. Geotech. Eng., ASCE,* 113, 5, 1987, 407–423.
35. Salami, M.R. and Desai, C.S., "Constitutive Modelling of Concrete and Rocks Under Multiaxial Compressive Loading," Report, Dept. of Civil Eng. and Eng. Mechs., University of Arizona, Tucson, AZ, 1986
36. Desai, C.S. and Varadarajan, A., "A Constitutive Model for Quasistatic Behavior of Rock Salt," *J. of Geophys. Research,* 92, B11, 1987, 11, 445–11, 456.
37. Lade, P.V., "Three-Parameter Failure Criterion for Concrete," *Proc. ASCE,* 108, 5, 1982, 850–863.
38. Hoek, E. and Brown, E.T., "Empirical Strength Criterion for Rock Masses," *J. Geotech. Eng., ASCE,* 106, 9, 1980, 1013–1035.
39. Chia, J. and Desai, C.S., "Constitutive Modelling of Thermomechanical Response of Materials in Semiconductor Devices Using the Disturbed State Concept," *Report to NSF,* Dept. of Civil Eng. and Eng. Mechs., University of Arizona, Tucson, AZ, 1994.
40. Desai, C.S., Chia, J., Kundu, T., and Prince, J.L., "Thermomechanical Response of Materials and Interfaces in Electronic Packaging, Parts I & II," *J. of Elect. Packaging, ASME,* 119, 1997, 294–300; 301–309.
41. Wathugala, G.W. and Desai, C.S., "Finite Element Dynamic Analysis of Nonlinear Porous Media with Application to Piles in Saturated Clays," *Report to NSF,* Dept. of Civil Eng. and Eng. Mechs., University of Arizona, Tucson, AZ, 1990.
42. Riemer, D.C., "Prediction of Temperature Cycling Life for SMT Solder Joints on TCE-Mismatched Substrates," *Proc. Elect. Comp., IEEE,* New Jersey, 1990, 418–423.
43. Desai, C.S. and Siriwardane, H.J., *Constitutive Laws for Engineering Materials,* Prentice-Hall, Englewood Cliffs, NJ, 1984.
44. Hill, R., *The Mathematical Theory of Plasticity,* Oxford Univ. Press, London.
45. Drucker, D.C., "Some Implications of Work Hardening and Ideal Plasticity," *Quart. Appl. Math.,* 7, 4, 1950, 411–418.
46. Desai, C.S. and Abel, J.F., *Introduction to the Finite Element Method,* Van Nostrand Reinhold Co., New York, 1972.

47. Desai, C.S. and Kundu, T., *Introductory Finite Element Method*, CRC Press, Boca Raton, FL, 2000, under publication.
48. Faruque, M. and Desai, C.S., "Implementation of a General Constitutive Model for Geologic Materials," *Int. J. Numer. Analyt. Methods Geomech.*, 9, 5, 1985.
49. Potts, D.M. and Gens, A., "A Critical Assessment of Methods of Corrections for Drift from Time-Yield Surface in Elastoplastic Finite Element Analysis," *Int. J. Numer. Analy. Methods Geomech.*, 9, 1985, 149–159.
50. Sloan, S.W., "Substepping Schemes for the Numerical Integration of Elastoplastic Stress–Strain Relations," *Int. J. Num. Meth. in Eng.*, 24, 1987, 893–911.
51. Prager, W., "Non-isothermal Plastic Deformation," Bol. Koninke, Nederl. Acad. Wet, 8(61/3), 1958, 176–182.
52. Ziegler, H., "A Modification of Prager's Hardening Rule," Quart. Appl. Math., 17, 1959, 55–65.
53. Chen, J. and Desai, C.S., "Optimization of the Disturbed State Concept for Constitutive Modelling, and Application in Finite Element Analysis, *Report to NSF*, Dept. of Civil Engng. and Engng. Mechanics, University of Arizona, Tucson, Arizona, 1997.
54. Gudehus, G., "Requirements for Constitutive Relations for Soils," Chapter 4 in *Mechanics of Geomaterials*, Z. Bazant (Editor), John Wiley & Sons, U.K., 1985, 47–63.
55. Dolezalova, M., "Stress Path Analysis of Geotechnical Structures and Selection of Constitutive Models for Geomaterials," *Proc. 9th Int. Conf. on Computer Methods and Advances in Geomech.*, J.X. Yuan (Editor), A.A. Balkema, The Netherlands, 1997, 123–132.
56. Dolezalova, M., Boudik, Z., and Hladik, I., "Numerical Testing and Comparison of Constitutive Models of Geomaterials," *Acta Montana (Prague), Series A: Geodynamics*, 1, 87, 1992, 33–56.
57. Desai, C.S., Sharma, K.G., Wathugala, G.W., and Rigby, D., "Implementation of Hierarchical Single Surface δ_0 and δ_1 Models in Finite Element Procedure," *Int. J. Num. Analyt. Meth. in Geomech.*, 15, 1991, 649–680.
58. Desai, C.S. and Toth, J., "Disturbed State Constitutive Modelling Based on Stress–Strain and Nondestructive Behavior," *Int. J. Solids & Struct.*, 33, 11, 1996, 1619–1650.
59. Skipor, A., Harren, S., and Bostsis, J., "Constitutive Characterization of 63/37 Sn/Pb Eutectic Solder Using the Bodner-Parton Unified Creep-Plasticity Model," *Adv. in Elect. Packaging*, Vol. 2, Proc. of 1992 Joint ASME/JSMT *Conf. on Electronic Packaging*, Chen, W.T. and Abe, H. (Editors), ASME, NY, 661–672.
60. Solomon, H.D., "Low Cycle Fatigue of 60/40 Solder-Plastic Strain Limited vs. Displacement Limited Testing," *Proc. ASM Elect. Packaging: Materials and Processes*, ASM, 1986, 29–49.
61. Cole, M., Canfield, T., Banks, D., Winton, M., Walsh, A., and Goyna, S., "Constant Strain Rate Tensile Properties of Various Lead Based Solder Alloys at 0, 50, and 100°C," *Proc. Materials Development in Microelectronic Packaging Conf.*, Montreal, Canada, 1991.
62. Nir, N., Dudderar, T.D., Wong, C.C., and Storm, A. R., "Fatigue Properties of Microelectronic Solder Joints," *J. of Elect. Packaging, ASME*, 113, 1991.
63. Savage, E. and Getzan, G., "Mechanical Behavior of Gotin/40 Lead Solder at Various Strain Rates and Temperatures," *Proc. Int. Soc. Hybrid Microelectronics (ISHM) Conf.*, Nashville, TN, USA, 1990.

64. Dishongh, T. and Desai, C.S., "Disturbed State Concept for Materials and Interfaces with Applications in Electronic Packaging," *Report to NSF*, Dept. of Civil Eng. and Eng. Mechanics, University of Arizona, Tucson, AZ, 1996.
65. Scheele, F., Desai, C.S., and Muqtadir, A., "Testing and Modelling for 'Munich' Sand," *Soils & Foundations*, Jap. Soc. of Soil Mech. & Found. Eng., 26, 3, 1986.
66. Armaleh, S. H. and Desai, C.S., "Modelling Including Testing of Cohesionless Soils Using Disturbed State Concept," Report, Dept. of Civil Eng. and Eng. Mechanics, University of Arizona, Tucson, AZ, 1990.
67. Galagoda, H.M. and Desai, C.S., "Nonlinear Analysis of Porous Soil Media and Application," *Report to NSF*, Dept. of Civil Engng. and Engng. Mechanics, University of Arizona, Tucson, AZ, 1986.
68. Castro, G., "Liquefaction of Sands," *Ph.D. Dissertation*, Harvard Univ., Cambridge, MA, USA, 1969.
69. Nishida, T., Esaki, T., Aoki, K., and Kimura, T., "On the Stress Loci by Strain Controlled Tests on Sandstone Under Generalized Triaxial Stress," *Proc. 6th Japan Symp. on Rock Mechanics*, Japan, 1984, 43–48.
70. Waversik, W.R. and Hanum, D.W., "Mechanical Behavior of New Mexico Salt in Triaxial Compression up to 200°C." *J. Geophys. Research*, 85(B2), 1980, 891–900.
71. Van Mier, J.G.M., "Strain Softening of Concrete Under Multiaxial Loading Conditions," *Doctoral Dissertation*, Eindover Univ. of Tech., The Netherlands, 1984.
72. Frantziskonis, G. and Desai, C.S., "Constitutive Model with Strain Softening," *Int. J. Solids & Struct.*, 23, 6, 1987.
73. Desai, C.S. and Woo, L., "Damage Model and Implementation in Nonlinear Dynamic Problems," *Computational Mechanics*, 11, 1993, 189–206.
74. Bonaquist, R.F., "Development and Application of a Comprehensive Constitutive Model for Granular Materials in Flexible Pavement Structures," *Doctoral Dissertation*, Dept. of Civil Engng., Univ. of Maryland, College Park, MD, 1996.
75. Bonaquist, R.F. and Witczak, M.W., "A Comprehensive Constitutive Model for Granular Materials in Flexible Pavements," *Proc. Eighth Intl. Conf. on Asphalt Pavements*, Seattle, WA, 1987, 783–802.
76. Alheid, H.J., "Friction Processes on Shear Surfaces in Granite at High Pressure and Temperature," in *High-Pressure Researches in Geoscience*, W. Schreyer (Editor), 1982, 95–102.
77. Freed, A., Private communication, 1996.
78. Mogi, K., "Effect of the Intermediate Principal Stress on Rock Failure," *J. of Geophysical Research*, 72, 20, 1967, 5117–5131.
79. Varadarajan, A. and Desai, C.S., "Material Constants of a Constitutive Model: Determination and Use," *Int. Geotech. Journal*, 23(3), 1993, 291–313.
80. Varadarajan, A., Sharma, K.G., Venkatachalam, K., and Gupta, A.K., "Constitutive Modelling of Rockfill Materials," *Proc. 4th Intl. Conf. on Constitutive Laws of Engineering Materials*, RPI, Troy, NY, July 1999.
81. Janbu, N., "Soil Compressibility as Determined by Odometer and Triaxial Tests," *Proc. Euro. Conf. Soil Mech. Found. Eng.*, Wiesbaden, Germany, Vol. 1, 1963, 19–25.
82. Dillon, D.W., Tsai, C.T., and De Angelis, R.J., "Dislocation Dynamics During Growth of Silicon Ribbon," *J. of Applied Physics*, 60, 5, 1986, 1784–1792.

83. Haasen, P., in *Dislocation Dynamics,* edited by A. R. Rosenfield et al., Battle Inst. of Materials Sc. Colloquia, 1967, McGraw-Hill, New York, 1968.
84. Desai, C.S., Dishongh, T.J., and Deneke, P., "Disturbed State Constitutive Model for Thermomechanical Behavior of Dislocated Silicon with Impurities," *J. of Applied Physics,* 84, 11, 1998, 5977–5984.

8

Creep Behavior: Viscoelastic and Viscoplastic Models in DSC

CONTENTS
- 8.1 Elastoviscoplastic Model 275
  - 8.1.1 Theoretical Details 275
  - 8.1.2 Mechanics of Viscoplastic Solution 277
  - 8.1.3 Viscoplastic Strain Increment 279
  - 8.1.4 Stress Increment 280
  - 8.1.5 Elastoviscoplastic Finite-Element Equations 280
  - 8.1.6 One-dimensional Formulation of Perzyna (evp) Model 282
  - 8.1.7 Selection of Time Step 286
- 8.2 Disturbance Function 286
  - 8.2.1 Finite-Element Equations 288
- 8.3 Rate-Dependent Behavior 289
- 8.4 Parameters for Elastoviscoplastic (evp) Model 290
  - 8.4.1 Determination of Parameters 290
  - 8.4.2 Disturbance Parameters 292
  - 8.4.3 Laboratory Tests and Examples 293
  - 8.4.4 Temperature Dependence 296
- 8.5 Multicomponent DSC and Overlay Model 297
  - 8.5.1 Multicomponent DSC 297
  - 8.5.2 Disturbance Due to Viscoelastic and Viscoplastic Creep 303
- 8.6 Material Parameters in Overlay Model 303
  - 8.6.1 Viscoelastic (ve) Overlay Model 303
  - 8.6.2 Poisson's Ratio 306
  - 8.6.3 Elastoviscoplastic (evp) Overlay Model 307
  - 8.6.4 Viscoelasticviscoplastic (vevp) Overlay Model 307
  - 8.6.5 Parameters for Viscoelastic Model 309
    - 8.6.5.1 Increment 4; $\Delta\sigma_1 = 699$ psi 309
    - 8.6.5.2 Verification 310
    - 8.6.5.3 Increments 5, 6, and 7 311
  - 8.6.6 Physical Meanings of Parameters 312
  - 8.6.7 Disturbance Parameters 312

    8.6.8   Finite-Element Equations 313
    8.6.9   Other Models 313
    8.6.10 Advantages of Overlay Models 313
8.7 Examples 314

In the theory of plasticity (covered in Chapters 6 and 7), we usually restrict attention to quasi-static processes in which time-rate effects are not considered, and the time-dependent viscous (creep or relaxation) effects do not need direct consideration. On the other hand, almost all materials under load experience time-dependent deformations to some extent. These deformations can be elastic or recoverable, and plastic or irrecoverable. Thus, a general model should allow for elastic and plastic, and viscous or creep deformations.

Figure 8.1 shows a schematic of a strain vs. time plot for a material subjected to a constant stress. The instantaneous (elastic) response (0–a) is followed by primary or elastic creep (a–b) during which, if unloading occurs (say at time t_1), elastic recovery (1–2) results, followed by delayed viscoelastic recovery (2–3). If the loading continues after b, secondary creep begins, accompanied by permanent deformations (strains). Unloading at any time (during b–c), say at 4, will involve elastic recovery (4–5), followed by viscoelastic recovery (5–6), and then permanent (viscoplastic) strain. Tertiary creep (c–d), leading to eventual failure, occurs after the secondary creep.

Some materials may experience a significant level of viscoelastic creep, whereas others may experience predominant viscoplastic creep. Indeed, for some materials both may need consideration. Such a general model to account for the creep behavior (excluding tertiary creep) is referred to here as *viscoelasticviscoplastic* (vevp). Before considering the general case, we first present the

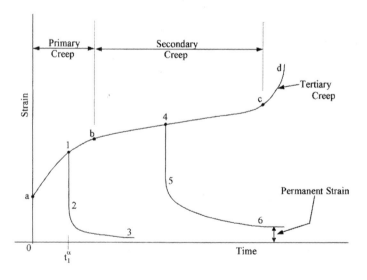

FIGURE 8.1
Schematic of creep behavior.

elastoviscoplastic (evp) model based on the theory proposed by Perzyna (1). It has been used often, in the context of numerical procedures such as the finite-element method, and has provided satisfactory solutions for a number of engineering problems (2–15).

8.1 Elastoviscoplastic Model

Details of the elastoviscoplastic (evp) model according to Perzyna (1), including factors such as identification of material parameters and their determination from laboratory tests, calibration and validation of the models, and implementation in computer procedures, are presented in this section. Use of the models to characterize the RI response in the context of the DSC is also discussed. Similar details for the viscoelastic (ve), elastoviscoplastic (evp), and viscoelastic viscoplastic (vevp) models as special cases of the multicomponent DSC, or overlay concept, are given subsequently. Examples involving the solution of practical problems using computer procedures are given in Chapter 13.

Figure 8.2(a) shows a schematic of the elastoviscoplastic behavior according to the Perzyna (evp) model. A constant stress, σ_0, is applied at time $t = 0$ and held constant up to time $\bar{t}$. At time $\bar{t}$, the stress, σ_0, is removed. The material experiences instantaneous elastic strain, ε^e, at $t = 0$. During the constant application of σ_0, i.e., from time $t = 0$ to $\bar{t}$, viscoplastic strains occur (ε^{vp}). When the stress is removed, the elastic strain is recovered, and the material retains the irreversible viscoplastic strain (ε^{vp}).

The viscoplastic strains in the evp model vary with time from 0 to $\bar{t}$; however, its final value after sufficiently large time is the same as that in the inviscid plasticity corresponding to a given yield function. This is illustrated in Fig. 8.2(b). It shows that for a given stress increment, $d\sigma$, the plastic strains from inviscid plasticity are given by (Chapter 7)

$$d\underline{\sigma} = \underline{C}^{ep} d\underline{\varepsilon} \qquad (8.1a)$$

$$d\underline{\varepsilon}^p = d\underline{\varepsilon} - d\underline{\varepsilon}^e \qquad (8.1b)$$

If the stress increment is held constant under which creep deformation occurs, the viscoplastic strains will grow with time (see ahead) and finally reach the value of $d\underline{\varepsilon}^{vp}$ equal to $d\underline{\varepsilon}^p$.

8.1.1 Theoretical Details

According to Perzyna's theory, the total strain rate, $\dot{\underline{\varepsilon}}$, can be decomposed as

$$\dot{\underline{\varepsilon}} = \dot{\underline{\varepsilon}}^e + \dot{\underline{\varepsilon}}^{vp} \qquad (8.2a)$$

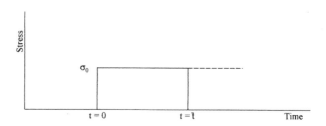
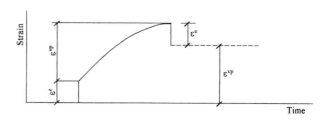
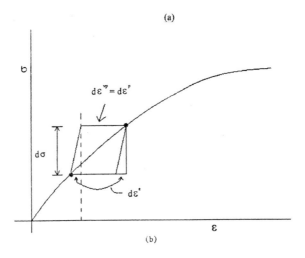

FIGURE 8.2
Schematic of behavior: evp (Perzyna) model.

where

$$\dot{\underline{\sigma}} = \underline{C}^e \dot{\underline{\varepsilon}}^e \tag{8.2b}$$

or

$$\dot{\underline{\varepsilon}}^e = \underline{C}^{(e)-1} \dot{\underline{\sigma}} \tag{8.2c}$$

$$\dot{\underline{\varepsilon}}^{vp} = \Gamma <\phi> \frac{\partial Q}{\partial \underline{\sigma}} \tag{8.2d}$$

where σ is the stress vector, C^e is the elastic constitutive matrix for isotropic material, Q is the plastic potential function (see Chapter 7); $Q \equiv F$ for associated plasticity, Γ is the fluidity parameter, ϕ is the flow function, which is expressed in terms of F, the overdot denotes time rate, and the angle bracket $<>$ has the meaning of a switch-on–switch-off operator as

$$\left\langle \phi\left(\frac{F}{F_0}\right) \right\rangle = \begin{cases} \phi\left(\frac{F}{F_0}\right) & \text{if } \frac{F}{F_0} > 0 \\ 0 & \text{if } \frac{F}{F_0} \leq 0 \end{cases} \quad (8.3a)$$

where F_0 is a reference value of F or any appropriate constant (e.g., yield stress, σ_y, atmospheric pressure constant, p_a) so as to render F/F_0 dimensionless. The flow function ϕ can be expressed in different forms (1–14), e.g.,

$$\phi = \left(\frac{F}{F_0}\right)^N \quad (8.3b)$$

$$\phi = \exp\left(\frac{F}{F_0}\right)^{\bar{N}} - 1.0 \quad (8.3c)$$

where N and $\bar{N}$ are material parameters.

8.1.2 Mechanics of Viscoplastic Solution

The mechanism of the viscoplastic process, according to the foregoing (evp) theory, can be explained by considering a creep test in which a constant stress increment, $\Delta \sigma$, is applied to a material at time $t = 0$. The initial stress or the stress at the end of the previous stress increment is σ_0; see Fig. 8.3(a). Let us consider that yield function F from the δ_0-model (Chapter 7):

$$F = \bar{J}_{2D} - (-\alpha \bar{J}_1^n + \gamma \bar{J}_1^2)(1 - \beta S_r)^{-0.5} \quad (8.4a)$$

in which the hardening function, α, is given by

$$\alpha = \frac{a_1}{\xi_{vp}^{\eta_1}} \quad (8.4b)$$

where ξ_{vp} is the trajectory of viscoplastic strains $\underline{\varepsilon}^{vp}$ given by

$$\xi_{vp} = \int [(d\underline{\varepsilon}^{vp})^T d\underline{\varepsilon}^{vp}]^{1/2} \quad (8.4c)$$

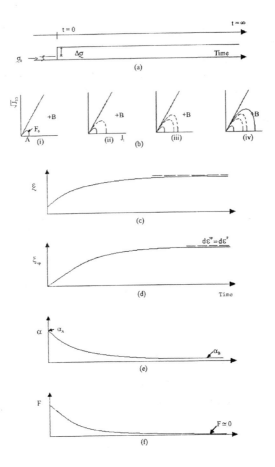

FIGURE 8.3
Mechanics of viscoplastic solution (9, 10, 16).

At time $t = 0$, when the stress increment is applied, let F_0 denote the initial yield surface corresponding to the state of stress, σ_0. This surface is often called the *static* yield surface $F_s (= F_0)$. Under the constant stress, the material experiences viscoplastic deformations, and the yield surface changes from F_s to F_d due to the change in the stress, $\Delta\sigma$, which is called the *overstress*, and the hardening function, α. The modified surface, F_d, may be referred to as the *dynamic* yield surface, and its value is greater than zero ($F_d > 0$) because the stress and hardening, which are changing with time, do not satisfy the condition that $F_d = 0$. The viscoplastic strain is caused by the overstress and grows with time; see Fig. 8.3(d). Its rate of increase, however, decreases continuously and it tends to zero at greater time levels. After sufficiently large time, F_d approaches the equilibrium state at point B [Fig. 8.3(b)], when the yield function $F_B = 0$ is satisfied.

As the time passes, the total strain (ε) (elastic plus viscoplastic) accumulates [Fig. 8.3(c)], leading to the accumulation of viscoplastic strain trajectory

Creep Behavior: Viscoelastic and Viscoplastic Models in DSC

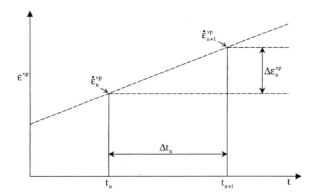

FIGURE 8.4
Time integration in viscoplastic strains.

ξ_{vp}, [Fig. 8.3(d)]. As a consequence, the value of the hardening function, α, changes (decreases) from the initial value of α_A to the final value of α_B [Fig. 8.3(e)], during which the yield surfaces expand from A to B [Fig. 8.3(b)]. At B, the steady condition is reached and the viscoplastic strain rate ceases. As $t \to \infty$, the solution from the viscoplastic model tends toward that from the inviscid plasticity model [Fig. 8.2(b)].

8.1.3 Viscoplastic Strain Increment

The viscoplastic strain grows with time under the stress increment. It can be evaluated with time by performing time integration as described next.

Consider the time increment $\Delta t_n = t_{n+1} - t_n$ in Fig. 8.4. Then the strain increment from time t_n to t_{n+1} can be expressed as

$$\Delta \varepsilon_n^{vp} = \Delta t_n [(1 - \theta) \dot{\varepsilon}_n^{vp} + \theta \dot{\varepsilon}_{n+1}^{vp}] \qquad (8.5a)$$

where $0 \le \theta \le 1$. We obtain different time integration schemes depending on the value of θ; e.g., $\theta = 0$ leads to the simple Euler scheme, $\theta = 0.5$ to the semi-implicit Crank–Nicolson scheme, and $\theta = 1$ to the fully implicit scheme (7–10).

The viscoplastic strain rate at step $n + 1$ in Eq. (8.5a) can be written (Fig. 8.4) using the Taylor series expansion, and by ignoring higher-order terms, as

$$\dot{\varepsilon}_{n+1}^{vp} = \dot{\varepsilon}_n^{vp} + G_n \cdot \Delta \sigma_n \qquad (8.6a)$$

where $\Delta \sigma_n$ is the stress increment from t_n to t_{n+1} and G_n is the gradient matrix given by

$$G_n = \frac{\partial \dot{\varepsilon}_n^{vp}}{\partial \sigma} \qquad (8.6b)$$

Then, based on Eq. (8.2d), F, and ϕ, G_n is derived as

$$G_n = \Gamma\left\{\phi\frac{\partial}{\partial \sigma}\left(\frac{\partial F}{\partial \sigma}\right)^T + \frac{d\phi}{dF}\cdot\frac{\partial F}{\partial \sigma}\left(\frac{\partial F}{\partial \sigma}\right)^T\right\} \quad (8.6c)$$

The substitution of G_n in Eq. (8.5a) leads to

$$\Delta \varepsilon_n^{vp} = \Delta t_n[\dot{\varepsilon}_n^{vp} + \theta \cdot G_n \Delta \sigma_n] \quad (8.5b)$$

8.1.4 Stress Increment

The stress increment is written as

$$\Delta \sigma_n = C_n^e(\Delta \varepsilon_n - \Delta \varepsilon_n^{vp}) \quad (8.7a)$$

where $\Delta \varepsilon_n = B_n \Delta q_n$, B is the strain-displacement transformation matrix, q_n is the displacement vector, and C^e is the elastic constitutive matrix.

Now, substitution of $\Delta \varepsilon_n^{vp}$ from Eq. (8.5b) in Eq. (8.7a) gives

$$\Delta \sigma_n = \bar{C}(B_n \cdot \Delta q_n - \dot{\varepsilon}_n^{vp} \cdot \Delta t_n) \quad (8.7b)$$

$$= \bar{C} \cdot \Delta \varepsilon^e(t) \quad (8.7c)$$

where $\bar{C} = (I + C_n^e \cdot \theta \cdot \Delta t_n \cdot G_n)^{-1} C_n^e$, and $\Delta \varepsilon^e(t)$ is the time-dependent, equivalent "elastic" strain.

8.1.5 Elastoviscoplastic Finite-Element Equations

In the evp approach, we solve the equations of equilibrium below, by performing time integration for strains for a given load increment, ΔQ_n ($n = 1, 2,...$); see Fig. 8.5. The finite-element incremental equations of equilibrium are written as

$$\int_V B_n^T \Delta \sigma_n^i \, dV + \Delta Q_n = 0 \quad (8.8a)$$

where V is the volume of finite element. Now, substitution of Eq. (8.7b) in Eq. (8.8a) gives

$$\int_V B_n^T \bar{C}(B_n \Delta q_n - \dot{\varepsilon}_n^{vp}\Delta t_n) \, dV + \Delta Q_n = 0 \quad (8.8b)$$

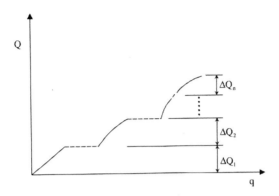

FIGURE 8.5
Incremental loading in evp solution.

or

$$\int_V (B_n^T \bar{C} B_n \, dV) \Delta q_n - \int_V B_n^T \bar{C} \dot{\varepsilon}_n^{vp} \Delta t_n \, dV + \Delta Q_n = 0 \quad (8.8c)$$

or

$$\Delta q_n = k_{t(n)}^{-1} \left[\int_V B_n^T \bar{C} \cdot \dot{\varepsilon}_n^{vp} \Delta t_n \, dV - \Delta Q_n \right]$$

$$= k_{t(n)}^{-1} \Delta \bar{Q} \quad (8.8d)$$

where $k_{t(n)}^{-1}$ is the tangent stiffness matrix given by

$$k_{t(n)} = \int_V B_n^T \bar{C} B_n \, dV \quad (8.9)$$

and $\Delta \bar{Q}$ is the equivalent, residual, or pseudo-incremental load vector, given by

$$\Delta \bar{Q} = \int_V B_n^T \bar{C} \cdot \dot{\varepsilon}_n^{vp} \Delta t_n \, dV - \Delta Q_n \quad (8.10)$$

Once the increment of displacement Δq_n is computed by solving Eq. (8.8d), the increment of stress $\Delta \sigma^n$ can be found from Eq. (8.7b). Then the total quantities at step $n + 1$ are found as

$$q_{n+1} = q_n + \Delta q_n \quad (8.11a)$$

$$\sigma_{n+1} = \sigma_n + \Delta \sigma_n \quad (8.11b)$$

Now the increment of viscoplastic strain, $\Delta \varepsilon_n^{vp}$, is found using Eq. (8.7a):

$$\Delta \varepsilon_n^{vp} = \underline{B}_n \Delta \underline{q}_n - \underline{C}^{(e)-1} \Delta \underline{\sigma}_n \tag{8.12a}$$

and the total quantity is

$$\underline{\varepsilon}_{n+1}^{vp} = \underline{\varepsilon}_n^{vp} + \underline{\varepsilon}_n^{vp} \tag{8.12b}$$

Then Eq. (8.6a) is used to evaluate the viscoplastic strain rate for the time interval Δt_n [Fig. (8.4)]. When the strain rate becomes small, as in Fig. 8.3, the time integration process is stopped. The next load increment, $\Delta \underline{Q}_{n+1}$, is applied and the process is repeated.

8.1.6 One-dimensional Formulation of Perzyna (evp) Model

The rheological model for the uniaxial case is shown in Fig. 8.6 (7, 16). The spring and dashpot provide for (linear) elastic and viscous responses, respectively. The slider provides the plastic yielding response depending on the plasticity model (F) adopted (Chapters 6 and 7). Under a given stress (increment), σ, at time $t = 0$, the instantaneous elastic response, ε^e, is given by

$$\varepsilon^e = \frac{\sigma}{E} \tag{8.13}$$

where E is the elastic modulus of the linear spring. Note that at time $t = 0$, the dashpot and the slider are not operational; however, for $t > 0$, the dashpot and the slider become operational. The slider will move only after the stress in it, σ_s, becomes greater than the yield stress, σ_y. The excess or overstress, σ_d, in the dashpot is given by

$$\sigma_d = \sigma - \sigma_s \tag{8.14}$$

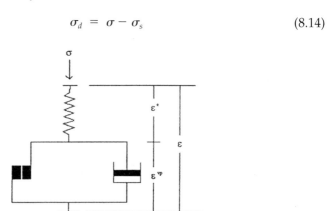

FIGURE 8.6
Elastoviscoplastic (evp) model.

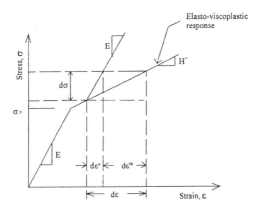

FIGURE 8.7
Elastic linear, strain-hardening, stress–strain behavior: uniaxial case (7, 16) (with permission).

The initiation of viscoplastic strains will take place after the yield stress, σ_y, is induced in the slider. Hence, the stress, $\bar{\sigma}$, for linear strain-hardening response (Fig. 8.7), during which viscoplastic strains occur, is given by

$$\bar{\sigma} = \sigma_y + H' \varepsilon^{vp} \qquad (8.15)$$

where H' is the slope of the linear strain-hardening part of the stress–strain curve. Then the stress in the slider is given by

$$\sigma_s = \begin{cases} \sigma & \text{if } \sigma_s < \bar{\sigma} \\ \bar{\sigma} & \text{if } \sigma_s \geq \bar{\sigma} \end{cases} \qquad (8.16)$$

Now the total strain, ε, is given by

$$\varepsilon = \varepsilon^e + \varepsilon^{vp} \qquad (8.17)$$

where ε^e is the elastic strain, and the stress in the dashpot, σ_d, is given by

$$\sigma_d = \mu \frac{d\varepsilon^{vp}}{dt} \qquad (8.18)$$

where μ is the coefficient of viscosity and t is the time. Before yielding, $\varepsilon^{vp} = 0$, therefore, $\sigma_d = 0$ [Eq. (8.18)], and hence $\sigma_s = \sigma$ [Eq. (8.14)].

Substitution of Eqs. (8.15), (8.16), and (8.18) in Eq. (8.14) gives

$$\mu \frac{d\varepsilon^{vp}}{dt} + H' \varepsilon^{vp} + \sigma_y = \sigma \qquad (8.19)$$

and substitution of ε^{vp}, Eq. (8.17), with Eq. (8.13), in Eq. (8.19) leads to

$$H'E\varepsilon + \mu E\frac{d\varepsilon}{dt} = H'\sigma + E(\sigma + \sigma_y) + \mu\frac{d\sigma}{dt} \quad (8.20a)$$

which represents a first-order ordinary differential equation for the time-dependent relation between stress and strain. Equation (8.20a) can be written in terms of the fluidity parameter, Γ, as

$$\dot{\varepsilon} = \frac{\dot{\sigma}}{E} + \Gamma[\sigma - (\sigma_y + H'\varepsilon^{vp})] \quad (8.20b)$$

where $\Gamma = 1/\mu$.

$$\dot{\varepsilon} = \dot{\varepsilon}^e + \dot{\varepsilon}^{vp} = \frac{\dot{\sigma}}{E} + \dot{\varepsilon}^{vp} \quad (8.21)$$

Comparing Eqs. (8.21) and (8.20b), we have

$$\dot{\varepsilon}^{vp} = \Gamma[\sigma - (\sigma_y + H'\varepsilon^{vp})] \quad (8.22)$$

Here, $\sigma - (\sigma_y + H'\varepsilon^{vp})$ denotes the overstress that causes viscoplastic strains; in other words, in this model, the viscoplastic strain rate is defined uniquely in terms of the overstress.

We can obtain the closed-form solution to Eq. (8.20a) for a constant-applied stress σ. Equation (8.20a) can now be written as

$$\Gamma H'\varepsilon + \frac{d\varepsilon}{dt} = \frac{\Gamma H'}{E}\sigma + \Gamma(\sigma - \sigma_y) \quad (8.23)$$

The solution to Eq. (8.23) is given in (7, 16) as

$$\varepsilon = \frac{\sigma}{E} + \frac{\sigma - \sigma_y}{H'}(1 - e^{-H'\Gamma t}) \quad (8.24)$$

where $H' > 0$. Figure 8.8(a) shows a schematic of the solution in Eq. (8.24). It shows that the instantaneous elastic strain, $\varepsilon^e = \sigma/E$, will be followed by the time-dependent viscoplastic strains, whose limiting value will be $(\sigma - \sigma_y)/H'$ as $t \to \infty$ [Eq. (8.24)], which is the same as the plastic strain from inviscid plasticity.

For an elastic perfectly plastic material ($H' = 0$), the solution is obtained by using l'Hospital's rule such that $H' \to 0$ (7):

$$\varepsilon = \frac{\sigma}{E} + (\sigma - \sigma_y)\Gamma t \quad (8.25)$$

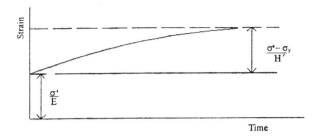

(a) Elastoviscoplastic strain hardening response

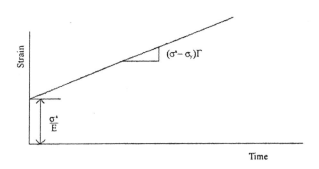

(b) Elastoviscoplastic response: perfectly plastic model

FIGURE 8.8
Responses of evp model: uniaxial case.

Figure 8.8(b) shows a plot of the solution, Eq. (8.25). It can be seen that for the perfectly plastic material, the steady-state condition for ε^{vp} is not reached, and ε^{vp} can continue indefinitely at a constant strain rate.

It may be noted that the model in Eq. (8.24) for a strain-hardening material leads to the equilibrated or steady-state viscoplastic strain, which is the same as the corresponding plastic strain from the inviscid plasticity (Chapter 7) [Fig. 8.2(b)]. Thus, the Perzyna model provides the timewise variation of the irreversible or plastic strain; however, its final magnitude is the same as that from the plasticity model based on the given yield function, F.

Details of the one-dimensional model for stress relaxation are given in Example 8.10.

8.1.7 Selection of Time Step

The robustness and reliability of the solution based on the time integration scheme, Eq. (8.5), depends on the selection of the appropriate size of the time step, Δt. For $\theta \geq 0.5$, it has been shown [Hughes and Taylor (17)] that the scheme in Eq. (8.5) is unconditionally stable. However, stability may not necessarily imply accuracy of the solution, and it thus becomes necessary to limit the size of the time step.

The size of the time step is affected by a number of factors such as material properties and strain rate. An empirical relation between the time step size and strain rate proposed by Cormeau (2), Zienkiewicz and Cormeau (3), and Dinis and Owen (18) for the explicit integration is given by

$$\Delta t \leq \eta \frac{\varepsilon}{\dot{\varepsilon}^{vp}} = \frac{\eta}{\dot{\varepsilon}^{vp}} \left(\frac{\sigma}{E} + \varepsilon^{vp} \right) \tag{8.26}$$

where η is a problem-dependent parameter whose value can be in the range $0.01 \geq \eta \geq 0.15$. It is often economical to vary the time step size during the time integration. One such empirical formula is given by Zienkiewicz and Cormeau (3):

$$\Delta t_{n+1} \leq \eta_0 \Delta t_n \tag{8.27}$$

where $1.2 \geq \eta_0 \geq 2$. Owen and Hinton (7) have given time-size criteria for various yield functions such as Tresca, von Mises, and Mohr–Coulomb for the associated flow rule.

8.2 Disturbance Function

The DSC model can be developed using the constitutive incremental equations for the evp model so as to characterize the RI behavior. Figure 8.9(a) shows a schematic of stress–strain behavior from a creep test in which increments of stress are applied at a slow rate, and the steady viscoplastic strains are measured under the constant values of increments. Such a behavior is referred to as a *static response*. The hardening curve in Fig. 8.9(a) shows the locus of the stresses and steady-state strains, which are the same as that for inviscid plasticity response. The observed (*a*) stress–strain response, Fig. 8.9(a), may exhibit degradation and softening. Then the static hardening response can be treated as RI, and the FA (*c*) response can be identified on the basis of the asymptotic stress states (Chapter 4). The disturbance, D, can now be defined as (Chapter 4)

$$D = \frac{\sigma^i - \sigma^a}{\sigma^i - \sigma^c} \tag{8.28a}$$

Creep Behavior: Viscoelastic and Viscoplastic Models in DSC

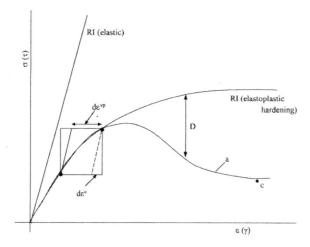

(a) Stress-strain response

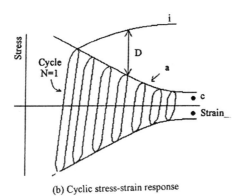

(b) Cyclic stress-strain response

FIGURE 8.9
Disturbance for creep models.

where σ^i denotes a measure of stress such as τ or $\sqrt{J_{2D}}$, σ^a is the observed stress, and σ^c is the FA stress.

For cyclic (repetitive) loading [Fig. 8.9(b)], the disturbance treated as the function of $\xi(N)$, where N is the number of cycles, can be expressed as

$$D(N) = \frac{\sigma^i - \sigma_p}{\sigma^i - \sigma^c} \tag{8.28b}$$

where σ_p is the peak stress for a given cycle N. D in Eq. (8.28b) can be expressed in terms of a viscoplastic (deviatoric) strain trajectory, Eq. (8.4c). In the case of cyclic loading, the RI behavior can be simulated using the inviscid plasticity model, say, δ_0, Chapter 7, by using the initial and monotonic response of the first cycle, extended beyond the peak point; see Fig. 8.9(b).

Now the observed stress increment, $\Delta \underset{\sim}{\sigma}^a$, can be expressed as

$$\Delta \underset{\sim}{\sigma}^a = (1 - D)\Delta \underset{\sim}{\sigma}^i + D\Delta \underset{\sim}{\sigma}^c + dD(\underset{\sim}{\sigma}^c - \underset{\sim}{\sigma}^i) \qquad (8.29)$$

where $\Delta \underset{\sim}{\sigma}^i$ is the intact viscoplastic stress increment as given in Eq. (8.7b). If it is assumed that the strains $\Delta \underset{\sim}{\varepsilon}^e(t)$ [Eq. 8.7(c)] in the RI and FA parts are equal, Eq. (8.29) can be written as

$$\Delta \underset{\sim}{\sigma}^a = (1 - D)\bar{\underset{\sim}{C}}\Delta \underset{\sim}{\varepsilon}^e + D\underset{\sim}{C}^c \Delta \underset{\sim}{\varepsilon}^e + dD(\underset{\sim}{\sigma}^c - \underset{\sim}{\sigma}^i) \qquad (8.30a)$$

$$= [(1 - D)\bar{\underset{\sim}{C}} + D\underset{\sim}{C}^c]\Delta \underset{\sim}{\varepsilon}^e + dD(\underset{\sim}{\sigma}^c - \underset{\sim}{\sigma}^i) \qquad (8.30b)$$

where $\underset{\sim}{C}^c$ is the constitutive matrix for the FA response and will depend on the simulation used (Chapter 4). For example, for the constrained-liquid assumption, $\underset{\sim}{C}^c$ in Eq. (8.30) will specialize to Eq. (4.19) in Chapter 4.

8.2.1 Finite-Element Equations

The equilibrium equations (8.8a) can now be written as

$$\int \underset{\sim}{B}_n^T \Delta \underset{\sim}{\sigma}_n^a \, dV + \Delta \underset{\sim}{Q}_n = 0 \qquad (8.31)$$

Substitution from Eq. (8.30) in Eq. (8.31) and the use of Eq. (8.7b) leads to

$$\int_V \underset{\sim}{B}_n^T [\{(1 - D)\bar{\underset{\sim}{C}} + D\underset{\sim}{C}^c\}\{\underset{\sim}{B}_n \Delta \underset{\sim}{q}_n^i - \dot{\underset{\sim}{\varepsilon}}_n^{vp}\Delta_n^t\} + dD(\underset{\sim}{\sigma}^c - \underset{\sim}{\sigma}^i)] dV + \Delta \underset{\sim}{Q}_n = 0 \qquad (8.32a)$$

or

$$\underset{\sim}{k}_t^{DSC} \cdot \Delta \underset{\sim}{q}_n^i = \underset{\sim}{Q}_1 - \underset{\sim}{Q}_2 - \Delta \underset{\sim}{Q}_n = \Delta \bar{\underset{\sim}{Q}}^{DSC} \qquad (8.32b)$$

where

$$\underset{\sim}{k}_t^{DSC} = \int_V \underset{\sim}{B}_n^T[(1 - D)\bar{\underset{\sim}{C}} + D\underset{\sim}{C}^c]\underset{\sim}{B}_n \, dV$$

$$\underset{\sim}{Q}_1 = \int_V \underset{\sim}{B}_n^T[(1 - D)\bar{\underset{\sim}{C}} + D\underset{\sim}{C}^c]\dot{\underset{\sim}{\varepsilon}}_n^{vp}\Delta t_n \, dV$$

$$\underset{\sim}{Q}_2 = \int_V \underset{\sim}{B}_n^T dD(\underset{\sim}{\sigma}^c - \underset{\sim}{\sigma}^i) \, dV$$

and $\Delta \overline{Q}^{DSC}$ is the equivalent or residual load vector including the effect of disturbance.

8.3 Rate-Dependent Behavior

The behavior of most engineering materials depends on the rate of loading (Fig. 8.10), which is often defined by strain rate, $\dot{\varepsilon}$, or displacement rate, $\dot{\delta}$. The static response refers to the behavior under a slow rate of loading. As the loading rate increases, the material exhibits stiffer response and higher ultimate or failure strengths; Fig. 8.10(a) and (b) show schematics of such behavior at a given initial (continuing) stress or mean pressure. Hence, there exists a unique ultimate envelope corresponding to each strain rate. It is possible that a material may exhibit hardening response under static loading. However, if it is loaded under variable strain rates (under dynamic loading) in which the strain rate decreases after an increasing strain rate, the material may exhibit strain softening behavior (5) [Fig. 8.10(c)].

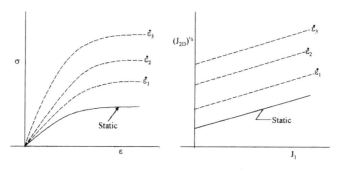

(a) Stress – strain behavior at different rates (b) Ultimate (failure) envelopes for different rates

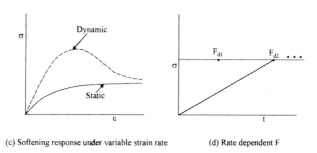

(c) Softening response under variable strain rate (d) Rate dependent F

FIGURE 8.10
Rate-dependent behavior (5). ©1984 John Wiley & Sons Ltd. Reproduced with permission.

The evp model can be used for the strain rate-dependent behavior. In this case, the dynamic yield surface, F_d, will depend on the rate of loading. In other words, if the same stress increment is applied at different rates, the dynamic yield surface, F_{di} ($i = 1, 2, \ldots$) [Fig. 8.10(d)], σ will be different. However, the procedures for the evaluation of the viscoplastic strain and for the convergence of $F_d \to 0$ to the equilibrated yield surface ($F_s = 0$) for the stress increment will be essentially the same as before.

8.4 Parameters for Elastoviscoplastic (evp) Model

The procedures for finding the elasticity and plasticity constants have been described in Chapters 5 to 7. For example, if the δ_0-model in the HISS family is used, the parameters involved are those discussed in Chapter 7. Here we consider the parameters required for the viscous behavior in the evp model.

8.4.1 Determination of Parameters

There are two parameters, Γ [Eq. (8.2d)] and N or $\bar{N}$ [Eq. (8.3)], that need to be determined from appropriate laboratory tests.

Fluidity Parameter, Γ. Squaring both sides of Eq. (8.2d) and multiplying them by 1/2 leads to

$$\frac{1}{2}(\dot{\varepsilon}^{vp})^T \dot{\varepsilon}^{vp} = \frac{1}{2}(\Gamma\langle\phi\rangle)^2 \left(\frac{\partial F}{\partial \sigma}\right)^T \frac{\partial F}{\partial \sigma} \qquad (8.33a)$$

or

$$\dot{I}_2^{vp} = \frac{1}{2}(\Gamma\langle\phi\rangle)^2 \left(\frac{\partial F}{\partial \sigma}\right)^T \frac{\partial F}{\partial \sigma} \qquad (8.33b)$$

where $\dot{I}_2^{vp}$ is the second invariant of the viscoplastic strain rate tensor, $\dot{\varepsilon}^{vp}$. Therefore,

$$\Gamma\langle\phi\rangle = \sqrt{\frac{2\dot{I}_2^{vp}}{\left(\frac{\partial F}{\partial \sigma}\right)^T \frac{\partial F}{\partial \sigma}}} = a \qquad (8.34a)$$

where

$$\frac{\partial F}{\partial \sigma} = \frac{\partial F}{\partial J_1} I + \frac{\partial F}{\partial J_{2D}} S + \frac{\partial F}{\partial J_{3D}} \left(S^T \cdot S - \frac{2}{3} J_{2D} I\right) \qquad (8.34b)$$

Creep Behavior: Viscoelastic and Viscoplastic Models in DSC

$\underline{S}$ is the vector of deviatoric stresses, and the invariant, I_2, is given by

$$I_2 = \frac{1}{2}\varepsilon_{ij}\varepsilon_{ij} = \frac{1}{2}\text{tr}(\varepsilon)^2 \tag{8.34c}$$

in which ε_{ij} is the strain tensor.

The value of a in Eq. (8.34a) can be found for various points during a creep test. Although the applied overstress, σ, is constant in a creep test, the yield function ($F_d > 0$) changes during the creep behavior because the viscoplastic strain rate, Eq. (8.6a), leads to changes in the hardening function, α, Eq. (8.4b), with the viscoplastic strain trajectory, ξ_{vp}, Eq. (8.4c). As the strains are measured during the test, the values of I_2^{vp} can be found for various points.

The flow function, ϕ, as power law is given by Eq. (8.3b) as (2, 3, 7–12)

$$\phi = \left(\frac{F}{F_0}\right)^N \tag{8.35a}$$

Now Eq. (8.34a) is written as

$$\Gamma\left\langle\left(\frac{F}{F_0}\right)^N\right\rangle = a \tag{8.35b}$$

Therefore,

$$\ln\Gamma + N\ln\left(\frac{F}{F_0}\right) = \ln a \tag{8.35c}$$

Variation of a vs. F/F_0 is shown (schematically) in Fig. 8.11(a), whereas in the ln–ln plot it is shown in Fig. 8.11(b). The slope of the (average) straight line gives N, and the intercept when $F/F_0 = 1$ gives Γ.

Other forms of the flow function are possible, e.g., the exponential form is given by (2–4, 7–10, 16) [Eq. (8.3c)]:

$$\phi = \exp\left(\frac{F}{F_0}\right)^{\bar{N}} - 1.0 \tag{8.36a}$$

Then Eq. (8.34a) leads to

$$\Gamma\left[\exp\left(\frac{F}{F_0}\right)^{\bar{N}} - 1.0\right] = a \tag{8.36b}$$

Therefore,

$$\ln\Gamma + \left(\frac{F}{F_0}\right)^{\bar{N}} = \ln a \tag{8.36c}$$

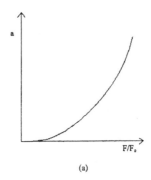

(a)

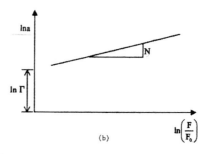

(b)

FIGURE 8.11
Determination of creep parameters.

The values of Γ and $\overline{N}$ are found from a plot of $\ln\Gamma$ vs. F/F_0 evaluated at different points on the creep curve.

8.4.2 Disturbance Parameters

Procedures for the determination of the disturbance parameters, A, Z, and D_u, are described in Chapters 2 and 4. For quasistatic (one-way loading and unloading), results such as in Fig. 8.9(a) can be used with Eq. (8.28a). For cyclic loading [Fig. 8.9(b)], D can be found using Eq. (8.28b). Consider the disturbance given by (Chapter 3)

$$D = D_u(1 - e^{-A\xi_D^Z}) \qquad (8.36c)$$

Here, D_u can be found using Eq. (8.28a) or (8.28b) in which σ^c, is the residual or saturation stress after a large number of cycles. Often, it is appropriate to use $D_u = 1.0$. The values of D at different stress levels are found, and ξ is found from viscoplastic strains at those points. Then plot of $\ln[-\ln(1 - D)]$ vs. $\ln(\xi_{vp})$ provides the values of A and Z (Chapter 3).

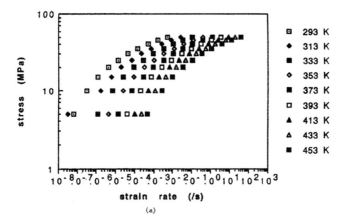

FIGURE 8.12
Plots of measured (a) strain rate vs. stress (19), and (b) $\ln(F/F_0)$ vs. $\ln(\dot{\varepsilon}_1^{vp})$ for determination of Γ and N (11).

8.4.3 Laboratory Tests and Examples

One-dimensional. Usually, results are available from tests with one-dimensional constant stress and/or constant strain loading. For instance, steady-state (uniaxial) creep strain rate ($\dot{\varepsilon}^{vp}$) vs. uniaxial stress (σ_1) results are often available. Figure 8.12(a) shows such temperature-dependent data for a 60/40 (Sn/Pb) solder (11, 19), which can be used to find approximate values of Γ and N. Here, because only one component of strain is available, Eq. (8.2d) reduces to

$$\dot{\varepsilon}_1^{vp} = \Gamma \left(\frac{F}{F_0}\right)^N \tag{8.37a}$$

which leads to

$$\ln \dot{\varepsilon}_1^{vp} = \ln \Gamma + N \ln(F/F_0) \tag{8.37b}$$

The values of F can be found from Eq. (8.4a) as the state of stress and other (plasticity) material parameters are known. Here, $F_0 = p_a$ is used. Typical plots of $\ln(\dot{\varepsilon}_1^{vp})$ vs. $\ln(F/F_0)$ for different temperatures are shown in Fig. 8.12(b) (11); the temperature dependence is discussed later. The slope of an average line gives the value of N and the intercept along $\ln(\dot{\varepsilon}_1^{vp})$ for $F/F_0 = 1$, that is, $\ln(F/F_0) = 0$ gives the value of Γ.

Shear Tests. For geologic and other materials, shear tests are often performed, in which the specimen is first consolidated under the K_0 condition, i.e., lateral strain = 0. After the consolidation, the specimen is subjected to a

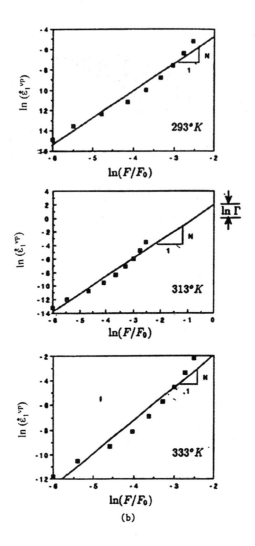

FIGURE 8.12
(Continued).

constant effective normal stress, $\sigma'_n = \sigma'_3$, and a constant shear stress, τ, is applied under undrained conditions; that is, the fluid is not allowed to drain. The shear strains are measured with time under the constant τ. For this undrained case, the vertical strain, ε_1, is zero. Then the viscoplastic strain rate tensor is given by

$$(\dot{\varepsilon}^{vp}) = \begin{pmatrix} 0 & 0 & 0 \\ 0 & 0 & \dot{\gamma}/2 \\ 0 & \dot{\gamma}/2 & 0 \end{pmatrix} \qquad (8.38)$$

where $\dot{\gamma}$ is the rate of measured shear strain. Thus, the value of I_2^{vp} can be evaluated using $\dot{\gamma}/2$ as $\dot{\varepsilon}_{12}^{vp}$ i$_n$ Eq. (8.34c). Now, Eq. (8.34a) is expressed as (9-11):

$$X = \sqrt{\frac{I_2^{vp} p_a^2}{\frac{1}{2} \underset{\sim}{S}^T \cdot \underset{\sim}{S}}} = p_a \sqrt{\frac{I_2^{vp}}{J_{2D}}} \tag{8.39}$$

The value of $\partial F/\partial \underset{\sim}{\sigma}$ in Eq. (8.34b) is found by assuming that the second term is predominant, and the first and third terms can be ignored. Hence,

$$\frac{\partial F}{\partial \underset{\sim}{\sigma}} = \frac{\partial F}{\partial J_{2D}} \underset{\sim}{S} = \frac{\underset{\sim}{S}}{p_a} \tag{8.40a}$$

$$\frac{1}{2}\left(\frac{\partial F}{\partial \underset{\sim}{\sigma}}\right)^T \frac{\partial F}{\partial \underset{\sim}{\sigma}} = \frac{1}{2} \frac{\underset{\sim}{S}^T}{p_a} \cdot \frac{\underset{\sim}{S}}{p_a} = \frac{J_{2D}}{p_a^2} \tag{8.40b}$$

and

$$J_{2D} = \frac{1}{6}(\sigma_3' - \sigma_1')^2 + \tau^2 \tag{8.40c}$$

If it is assumed that after a long time, $\sigma_3' \approx \sigma_1'$, then $J_{2D} = \tau^2$. Now the results are plotted in terms of $\ln(\chi)$ vs. $\ln(F/F_0)$, where $F_0 = p_a$. Figure 8.13 shows a typical plot for a clay (9, 10). The slope of the average line gives N and the intercept along $\ln(\chi)$ for $F/F_0 = 1$ leads to the value of Γ.

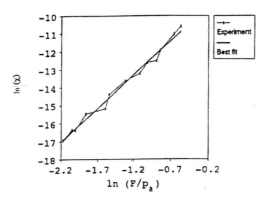

FIGURE 8.13
Determination of creep parameters from shear test for soil (9, 10).

8.4.4 Temperature Dependence

The temperature (T) dependence can be introduced in the viscoplastic model by writing Eq. (8.2d) for associated flow ($Q \equiv F$) as

$$\dot{\varepsilon}^{vp}(T) = \Gamma(T)\langle\phi(F)\rangle\frac{\partial F}{\partial \sigma} \quad (8.41)$$

where $\dot{\varepsilon}_{ij}^{vp}(T)$ is the thermoviscoplastic strain rate tensor, $F(\sigma_{ij}, \alpha, T)$ is the temperature-dependent yield function, and $\Gamma(T)$ is the temperature-dependent fluidity parameter.

As discussed earlier, the rate of increase in the thermoviscoplastic strain is affected by the excess stress above the yield stress as well as by the temperature, which may result in thermal degradation. Assuming that the viscous behavior is exhibited after the passage to the transient thermoviscoplastic state and that it is not significant in the elastic region, the total strain rate, $\dot{\varepsilon}$, can be expressed in terms of the thermoelastic strain rate, $\dot{\varepsilon}^e$, thermoviscoplastic strain rate, $\dot{\varepsilon}_{ij}^{vp}(T)$, and the elastic contribution due to the coefficient of thermal expansion (α_T) effect as

$$\dot{\varepsilon}(T) = \dot{\varepsilon}^e(T) + \dot{\varepsilon}^{vp}(T) + \alpha_T(T)\dot{T}\underline{I} \quad (8.42a)$$

or

$$\dot{\varepsilon}(T) = \dot{\varepsilon}^e(T) + \Gamma(T)\left\langle\phi\left(\frac{F}{F_0}\right)\right\rangle\frac{\partial F}{\partial \sigma} + \alpha_T(T)\dot{T}\underline{I} \quad (8.42b)$$

The constitutive stress–strain equation can now be written as

$$\dot{\sigma} = \underline{C}^e(T)\left[\dot{\varepsilon}(T) - \Gamma(T)\left\langle\phi\left(\frac{F}{F_0}\right)\right\rangle\frac{\partial F}{\partial \sigma} - \alpha_T\dot{T}\underline{I}\right] \quad (8.43)$$

The thermoviscoplastic strain rate at time step $n+1$ is now expressed using Taylor series (ignoring higher-order terms) as

$$\dot{\varepsilon}_{n+1}^{vp}(T) = \dot{\varepsilon}_n^{vp}(T) + \frac{\partial \dot{\varepsilon}_n^{vp}}{\partial \sigma}\cdot\Delta\sigma_n + \frac{\partial \dot{\varepsilon}_n^{vp}}{T}\cdot\Delta T_n\underline{I} \quad (8.44a)$$

$$= \dot{\varepsilon}_n^{vp}(T) + \underline{G}_{n1}\Delta\sigma_n + \underline{G}_{n2}\Delta T_n\underline{I} \quad (8.44b)$$

where $\Delta\sigma_n$ is the stress increments and $\underline{G}_{n1}$ and $\underline{G}_{n2}$ are the gradient matrices

Creep Behavior: Viscoelastic and Viscoplastic Models in DSC   297

at step n, given by

$$G_{n1} = \left(\frac{\partial \dot{\varepsilon}^{vp}(T)}{\partial \underline{\sigma}}\right) = \left(\frac{\partial \Phi}{\partial F} \cdot \frac{\partial F}{\partial \underline{\sigma}}\right)\frac{\partial F}{\partial \underline{\sigma}} + \Phi \frac{\partial^2 F}{\partial \underline{\sigma}^2} \qquad (8.45a)$$

and

$$G_{n2} = \left(\frac{\partial \dot{\varepsilon}^{vp}(T)}{\partial T}\right) = \left(\Gamma(T) \cdot \frac{\partial \phi}{\partial F} \cdot \frac{\partial F}{\partial T}\right)\frac{\partial F}{\partial \underline{\sigma}} + \phi \frac{\partial \Gamma(T)}{\partial T} \cdot \frac{\partial F}{\partial \underline{\sigma}} \qquad (8.45b)$$

where

$$\Phi = \Gamma(T)\phi, \quad \text{and} \quad \Phi = \left(\frac{F}{F_0}\right)^{N(T)}$$

and

$$\frac{\partial \Phi}{\partial F} = \Gamma(T)\frac{N(T)}{F_0}\left(\frac{F}{F_0}\right)^{N-1}$$

Here, the exponent, N, can depend on the temperature, T.

8.5 Multicomponent DSC and Overlay Model

The evp model based on Perzyna's theory simulates the viscoplastic response, particularly the time-dependent plastic response during the secondary creep regime (Fig. 8.1). However, in general, a material may experience both viscoelastic and viscoplastic responses, which include creep deformation during both the primary and secondary creep regimes. Hence, it is appropriate to develop and use a model that can simulate the combined response as closely as possible. In the following, we develop a model, called viscoelastic-viscoplastic (vevp), that can allow consideration of the combined response. Viscoelastic (ve) and elastoviscoplastic (evp) models arise as special hierarchical options of the vevp model.

8.5.1 Multicomponent DSC

In the basic DSC (Chapters 2 and 4), we first considered an element of the *same* material, which is composed of two material parts in the RI and FA reference states. During deformation, the extents of the RI and FA parts change, which is defined through the disturbance, D. In Chapter 2, we also developed the DSC for a material element composed of more than one or two, or a

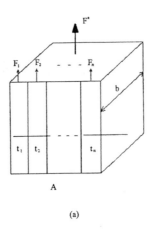

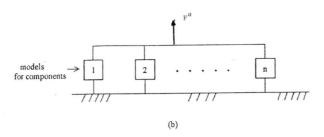

FIGURE 8.14
Multicomponent DSC.

higher number of, different materials. In that case, the behavior of the component materials provide reference state responses. Then the observed stress, σ^a, can be expressed as

$$\sigma^a = \sigma^1 \frac{A_1}{A} + \sigma^2 \frac{A_2}{A} + \cdots + \sigma^n \frac{A_n}{A} \tag{8.46}$$

where $A_1, A_2, \ldots, A_n$ are the areas of the components (Fig. 8.14), with total area $A = \Sigma A_i$ ($i = 1, 2, \ldots, n$), and $\sigma^1, \sigma^2, \ldots, \sigma^n$ are the corresponding stresses. For convenience, we assume that the width of the material element is constant and equals b (Fig. 8.14). Equation (8.46) can be written as

$$\sigma^a = \sigma^1 d_1 + \sigma^2 d_2 + \cdots + \sigma^n d_n \tag{8.47}$$

where $d_i = A_i/A$ is the ratio of the area of component i to the total area A; here, $\Sigma d_i = 1$. Note that Eq. (8.47) is similar to that for the single material element with RI and FA parts in which the disturbance, D, is the ratio of the FA area to the total area, i.e., $D = A^c/A$. In the basic DSC, $D_i(i = 1, 2)$ varies with deformation, while in the multicomponent DSC, $d_i (i = 1, 2, \ldots, n)$ can be constant. Equation (8.47) can also be expressed as

$$\sigma^a = \sigma^1 \frac{t_1 b}{\Sigma t_i b} + \sigma^2 \frac{t_2 b}{\Sigma t_i b} + \cdots + \sigma^n \frac{t_n b}{\Sigma t_i b} \tag{8.48a}$$

or

$$\sigma^a = \sigma^1 t_1 + \sigma^2 t_2 + \cdots + \sigma^n t_n \tag{8.48b}$$

where b = width of the element and $t_1, t_2, \ldots, t_n \, (= t_i/\Sigma t_i)$ are the nondimensionalized thicknesses of the components of the material element (Fig. 8.14); hence, $\Sigma t_i = 1$. Then each component $(i = 1, 2, \ldots, n)$ of the material element can be characterized using different models such as elastic, viscoelastic, and elastoviscoplastic.

The three-dimensional incremental form of Eq. (8.48) can be written as

$$d\underline{\sigma}^a = d\underline{\sigma}^1 t_1 + d\underline{\sigma}^2 t_2 + \cdots + d\underline{\sigma}^n t_n \tag{8.49a}$$

$$= \sum_{j=1}^{n} d\underline{\sigma}^j t_j \tag{8.49b}$$

As indicated before, d_i do not change and hence, t_i also do not change during small deformation. Equation (8.49a) can now be written as

$$d\underline{\sigma}^a = t_1 \underline{C}^1 d\underline{\varepsilon}_1 + t_2 \underline{C}^2 d\underline{\varepsilon}_2 + \cdots + t_n \underline{C}^n d\underline{\varepsilon}_n \tag{8.50}$$

where $\underline{C}^m$ $(m = 1, 2, \ldots, n)$ are the (tangent) constitutive matrices for the components, and $d\varepsilon_m$ $(m = 1, 2, \ldots, n)$ are the corresponding incremental strain vectors, which, in general, can be different. If it is assumed that the strains in all components are equal $(= d\varepsilon)$, Eq. 8.50 becomes

$$d\underline{\sigma}^a = (t_1 \underline{C}^1_1 + t_2 \underline{C}^2_2 + \cdots + t_n \underline{C}) d\underline{\varepsilon} \tag{8.51a}$$

or

$$d\underline{\sigma}^a = \underline{C}_{eq} d\underline{\varepsilon} \tag{8.51b}$$

where $\underline{C}_{eq}$ is the equivalent constitutive matrix for the material element. Figure 8.14(b) shows a symbolic representation of the multicomponent DSC model from Eq. (8.51). Each component unit $(1, 2, \ldots, n)$ can be characterized using elastic, viscoelastic, elastoviscoplastic, or any appropriate model. Thus, it can provide a hierarchical framework from which elastic (e), viscoelastic (ve),

TABLE 8.1
Specializations of Overlay Model

Specialization	Plasticity Model	No. of Overlays	Thickness	Parameters
Elastic (e)	von Mises	1	1.0	E, v, Γ, N, and very high σ_y
Maxwell	von Mises	2	0.5, 0.5	$E_1, v_1, \Gamma_1, N_1, \sigma_{y1} = 0$
				$E_2, v_2, \Gamma_2, N_2, \sigma_{y2} = 0$
Viscoelastic (ve)	von Mises	2	0.5, 0.5	$E_1, w_1, \Gamma_1, N_1, \sigma_{y1} = 0$
				$E_2, v_2, \Gamma_2, N_2, \sigma_{y2} =$ very high
Elastoviscoplastic (evp) (Perzyna type)	Any	1	1.0	E, v, Γ, N, F
Viscoelasticvisco-plastic (vevp)	von Mises	$\left. \begin{array}{c} 1 \\ 1 \end{array} \right\} = 2$	0.5	$E_1, v_1, \Gamma_1, N_1, \sigma_{y1} = 0$
	Any		0.5	$E_2, v_2, \Gamma_2, N_2, \sigma_{y2}$, or F

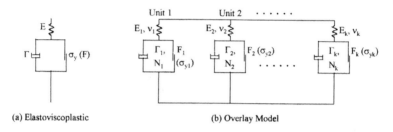

FIGURE 8.15
Overlay model and elastoviscoplastic specialization.

elastoviscoplastic (evp), and viscoelasticviscoplastic (vevp) models can be extracted. Note that the evp model can be the same as that discussed earlier, based on the Perzyna theory. The special form, Eq. (8.51), with compatible strains of the multicomponent DSC model, Eq. (8.50), is the same as the overlay or mechanical sublayer model proposed and used in (7, 20–22).

A rheological representation of the overlay model considered here is shown in Fig. 8.15(b), while that for the elastoviscoplastic (evp-Perzyna) model is shown in Fig. 8.15(a), as a special case. By adopting a suitable number of overlays with different thicknesses such that $\Sigma t_i (i = 1, 2, \ldots, k) = 1$, and by assigning different material properties such as elasticity (E, v), plasticity (perfectly plastic, σ_y; δ_0-plasticity, $\gamma, \beta, a_1, \eta_1, n, 3R$, etc.), and viscous ($\Gamma, N$) to different rheological units, a wide range of special versions can be obtained. Table 8.1 shows a number of possible versions, whose explanations follow.

Elastic (e). For the single unit, Fig. 8.15(a), consider one overlay with thickness $t = 1$. The spring and dashpot are assigned elastic (E, v) and viscous (Γ, N) parameters, respectively, while the yield stress, σ_y, in the slider, as per von Mises, perfectly plasticity model is assigned a very high value, of the order of 10^8 (Table 8.1). Then, as σ_y is very high, the dashpot will not be operational, and only the elastic spring will operate, which will result in an elastic response.

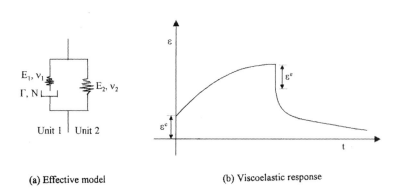

FIGURE 8.16
Viscoelastic model.

Maxwell Model. Here, two units with thickness each equal to 0.50 can be used. Then adoption of the yield stress, $\sigma_{y1} = \sigma_{y2} = 0.0$, leads to two parallel Maxwell models with spring and dashpot in series.

Viscoelastic (ve). Consider the two units from Fig. 8.15(b), with two overlays with $t = 0.5$ each. Units 1 and 2 are assigned the properties as shown in Table 8.1. Because of the very high value of σ_{y2}, the dashpot in unit 2 will not operate, and because $\sigma_{y1} = 0$ in unit 1, only the spring and dashpot will operate; the effective model is shown in Fig. 8.16(a). At time $t = 0$, under a stress of σ_0, instantaneous elastic strain, ε^e, will occur because of the springs in units 1 and 2, given by

$$\varepsilon^e = \frac{\sigma_0}{E_1 t_1 + E_2 t_2} \tag{8.52}$$

For time $t > 0$, viscoelastic deformations will occur due to the operation of the dashpot in unit 1 [Fig. 8.16(b)]. Then, if the stress is removed, the elastic strain will be recovered first instantaneously, and then $\varepsilon \to 0$, as $t \to \infty$, because the elastic spring in unit 2 will force the dashpot to return to the original state. Thus, this specialization will yield viscoelastic response.

Elastoviscoplastic (evp)—Perzyna Model. We consider one unit, Fig. 8.15(a), and one overlay with thickness $t = 1$. The elastic (E, v), viscous (Γ, N), and σ_y (classical plasticity) or F (e.g., δ_0-model) are assigned to the spring, dashpot, and slider, respectively (Table 8.1).

At time $t = 0$, the dashpot will not be operational; hence, the instantaneous elastic strain, $\varepsilon^e = \sigma_0/E$, will occur. Then, for time $t > 0$, the dashpot will be operational, and viscoplastic strains will occur. On unloading, ε^e will be recovered, and the viscoplastic strains (ε^{vp}), [Fig. 8.17(b)], due to the slider (σ_y or F) [Fig. 8.17(a)], will remain. The final viscoplastic strain will be equal to that from inviscid plasticity governed by the yield function, F. In other words, the evp-type model will give the timewise variation of the plastic strains, but the final magnitude of the viscoplastic will be equal to the plastic strain from inviscid plasticity.

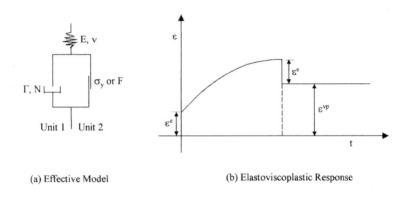

FIGURE 8.17
Elastoviscoplastic model.

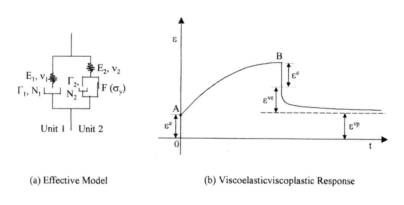

FIGURE 8.18
Viscoelasticviscoplastic model.

Viscoelasticviscoplastic (vevp) Model. Consider the two units in the model, Fig. 8.15(b), and assume two overlays, each with thickness = 0.5. The parameters are assigned as shown in Table 8.1. The resulting model is shown in Fig. 8.18(a).

At time $t = 0$, only the springs will deform as the dashpots are not operational. The instantaneous elastic strain ε^e [0–A, Fig. 8.18(b)], will be given by Eq. (8.52). Then, for time $t > 0$, the dashpot in unit 1 will operate first, as the slider in unit 2 has nonzero yield stress. This will lead to viscoelastic deformations. Now for later time $t > 0$, when the yield condition in slider 2 is reached, the dashpot in unit 2 will also become operational. If continuous yielding (e.g., δ_0-model) is considered for the slider, both the viscoelastic and viscoplastic deformations can occur simultaneously during A–B. Upon unloading, ε^e will be recovered first, then the viscoplastic strain, ε^{ve}, will be recovered, and finally the viscoplastic strain will remain [Fig. 8.18(b)].

Creep Behavior: Viscoelastic and Viscoplastic Models in DSC

8.5.2 Disturbance Due to Viscoelastic and Viscoplastic Creep

Disturbance or damage in many materials such as asphalt concrete can result due to both the viscoelastic and viscoplastic creep. Then the disturbance (D) can be expressed in terms of the total trajectory of creep strains, ξ_t, or energy (work), W, as

$$\xi_t = \int [(d\underline{\varepsilon}^{ve})^T d\underline{\varepsilon}^{ve}]^{1/2} + [(d\underline{\varepsilon}^{vp})^T d\underline{\varepsilon}^{vp}]^{1/2} \tag{8.53a}$$

$$W = \int \underline{\sigma}^T d\underline{\varepsilon}^{ve} + \int \underline{\sigma}^T d\underline{\varepsilon}^{vp} \tag{8.53b}$$

where $\underline{\varepsilon}^{ve}$ and $\underline{\varepsilon}^{vp}$ are the vectors of viscoelastic and viscoplastic strains, respectively, and $\underline{\sigma}$ is the stress vector. Then D can be expressed as

$$D = D_u(1 - e^{-A\xi_t^Z}) \tag{8.54}$$

8.6 Material Parameters in Overlay Model

The parameters in the overlay models, Table 8.1, need to be determined from appropriate creep and/or relaxation test data, e.g., uniaxial, shear, triaxial, and/or multiaxial. The procedure entails suitable adjustment of parameters and optimization schemes such that the predictions from the equations governing the response of overlay models fit the test data. We next consider details of procedures for typical overlay models.

8.6.1 Viscoelastic (ve) Overlay Model

Consider the model in Fig. 8.16(a). It is required to determine four elastic (E_1, ν_1; E_2, ν_2) and two viscous (Γ_1, N_1) parameters. They can be found on the basis of creep test data [Fig. 8.16(b)]. We first derive the equation governing the response of the overlay model.

Consider first the uniaxial (or triaxial) loading response. Let σ_0 be the applied stress, which is held constant, and let t_1 and t_2 be the thicknesses of unit 1 and unit 2, respectively. Then equilibrium of forces gives

$$\sigma_1 t_1 + \sigma_2 t_2 = \sigma_0(t_1 + t_2) \tag{8.55a}$$

Here, $t_1 + t_2 = 1$. Therefore,

$$\frac{d\sigma_2}{dt} = -\frac{t_1 d\sigma_1}{t_2 \, dt} \tag{8.55b}$$

At time $t = 0$, the dashpot is not operational; hence, the strain ε_0 is given by

$$\frac{\sigma_0}{E_1 t_1 + E_2 t_2} \quad \text{and} \quad \sigma_1(0) = \frac{E_1 \sigma_0}{E_1 t_1 + E_2 t_2} \tag{8.56}$$

Now, with respect to unit 1:

$$\varepsilon = \varepsilon^e + \varepsilon^d \tag{8.57a}$$

where ε^e and ε^d are the strains in the spring and dashpot, respectively. Hence,

$$\frac{d\varepsilon}{dt} = \frac{1}{E_1}\frac{d\sigma_1}{dt} + \Gamma \sigma_1 \tag{8.57b}$$

where σ_1 is the stress in unit 1.
With respect to unit 2:

$$\varepsilon = \frac{\sigma_2}{E_2} \tag{8.58a}$$

Hence, using Eq. (8.55b),

$$\frac{d\varepsilon}{dt} = -\frac{1}{E_2 t_2}\frac{t_1 d\sigma_1}{dt} \tag{8.58b}$$

Combining Eqs. (8.57b) and (8.58b), we have

$$\left(\frac{1}{E_1} + \frac{t_1}{t_2}\frac{1}{E_2}\right)\frac{d\sigma_1}{dt} + \Gamma \sigma_1 = 0 \tag{8.59}$$

the solution of which can be obtained as

$$\sigma_1 = \sigma_1(0) \cdot e^{\frac{-t\Gamma a}{b}} \tag{8.60}$$

where $a = E_1 E_2 t_2$, $b = E_1 t_1 + E_2 t_2$. Therefore,

$$\frac{d\sigma_1}{dt} = \sigma_1(0)\frac{-t\Gamma a}{b} e^{\frac{-t\Gamma a}{b}} \tag{8.61a}$$

and from Eq. (8.55b):

$$\frac{d\sigma_2}{dt} = \frac{\Gamma E_1^2 E_2 t_1 \sigma_0}{b^2} e^{\frac{-t\Gamma a}{b}} \qquad (8.61b)$$

Substitution of Eq. (8.61a) in Eq. (8.58b) gives

$$\frac{d\varepsilon}{dt} = \frac{\Gamma E_1^2 t_1 \sigma_0}{b^2} e^{\frac{-t\Gamma a}{b}} \qquad (8.62a)$$

Integration of Eq. (8.62a) leads to the solution for the overlay strain, ε, as

$$\varepsilon = \frac{-\sigma_0 E_1 t_1}{b \; E_2 t_2} e^{\frac{-t\Gamma a}{b}} + \frac{\sigma_0}{E_2 t_2} \qquad (8.62b)$$

As indicated before, the parameters E_1, E_2, and Γ can be found by their suitable adjustment, such that the computed strains ε for given σ_0 fit the observed response. They can be obtained by using the following simplified procedure:
For $t = 0$, Eq. (8.62b) gives

$$\varepsilon_0 = \frac{\sigma_0}{E_1 t_1 + E_2 t_2} \qquad (8.63a)$$

As $t \to \infty$, Eq. (8.62b) gives the ultimate strain, ε_u, as

$$\varepsilon_u = \frac{\sigma_0}{E_2 t_2} \qquad (8.63b)$$

The values of ε_0 and ε_u can be obtained from the measured response [Fig. 8.19(a)]; the latter can be adopted as the asymptotic value. Then the solution of Eq. (8.63) provides the values of E_1 and E_2.

Now the gradient of $d\varepsilon/dt\,(0)$ at $t = 0$ can be measured approximately from Fig. 8.19(a) as

$$\frac{d\varepsilon}{dt}(0) \approx \left(\frac{\varepsilon_{t+1} - \varepsilon_t}{\Delta t}\right)_0 \qquad (8.64a)$$

where ε_{t+1} and ε_t are values of strains near $t = 0$, and Δt is a (small) time step. Substitution of Eq. (8.64a) in Eq. (8.62a) gives

$$\frac{d\varepsilon}{dt}(0) = \frac{\Gamma E_1^2 t_1 \sigma_0}{(E_1 t_1 + E_2 t_2)^2} \qquad (8.64b)$$

Substitution of E_1 and E_2 computed above in Eq. (8.64b) gives the value of Γ.

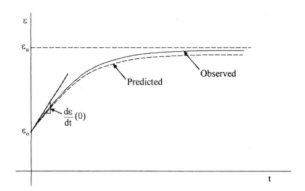

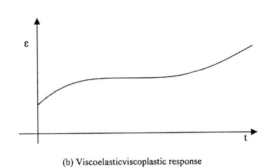

FIGURE 8.19
Determination of parameters for overlay models.

8.6.2 Poisson's Ratio

The values of v_1 and v_2 can be found if multi- (two-) dimensional test data are available. For instance, if the values of lateral strains $\varepsilon_2 = \varepsilon_3$ are available from cylindrical (unconfined) triaxial tests, the volumetric strain is given by

$$\varepsilon_v = \varepsilon_1 = 2\varepsilon_3 \tag{8.65a}$$

Then the bulk modulus, K, which may be assumed to be independent of time, and the same for both units, can be calculated as

$$K = \frac{\sigma_1'}{3\varepsilon_v} \tag{8.65b}$$

where $\sigma_1/3 = p$ is the mean pressure. Then v_1 and v_2 can be found as

$$v_1 = \frac{1}{2}\left(1 - \frac{E_1}{3K}\right) \quad \text{and} \quad v_2 = \frac{1}{2}\left(1 - \frac{E_2}{3K}\right) \tag{8.66a}$$

The parameter N for unit 2 can be assumed to be unity. This implies linear relation between the strain rate, $\dot{\varepsilon}^d$, and the stress, σ_1, in the dashpot:

$$\dot{\varepsilon}^d = \Gamma \sigma_1^N \qquad (8.67)$$

where σ_1 is the stress in unit 1. An example for the determination of the parameters is given subsequently.

8.6.3 Elastoviscoplastic (evp) Overlay Model

The parameters in this model are shown in Fig. 8.17(a). They are found by using the procedures described in Chapters 5 to 7 for elasticity and plasticity parameters (E, v; σ_y for classical models such as the von Mises, and γ, β, n, a_1, η_1, R for the HISS-δ_0 model). Procedures for finding the creep parameters (Γ, N) are given earlier in this chapter.

8.6.4 Viscoelasticviscoplastic (vevp) Overlay Model

The model and parameters are shown in Fig. 8.18(a). Assume that the significant viscoplastic strains will not occur during early times. This will approximately be the situation if the irreversible strains do not occur until the yield stress, σ_y, in the slider is exceeded. In the case of the continuous yield plasticity models such as the HISS-δ_0, irreversible strains can occur from the beginning; however, they can be assumed to be small. In that case, parameters (E_1, v_1, Γ_1, N_1, E_2, v_2) can be adopted to be the same as those for the viscoelastic model.

The parameters Γ_2, N_2, and $\sigma_y(F)$ can be adopted to be the same as in the elastoviscoplastic (evp) model. Indeed, it will be desirable and often necessary to use these parameters and backpredict the test behavior to verify that a satisfactory fit is obtained between the predictions and test data. Brief details of the governing equations for the vevp model, which can be used for the predictions, are given below.

Consider that a stress, σ_0, is applied and then held constant. The deformations in the vevp overlay can be considered to occur in two stages. When the stress, σ_2, in unit 2 is less than the yield stress, σ_y, in the slider (or $F < 0$), the model will give viscoelastic (ve) response. When the stress reaches σ_y at time = t^*, the slider will yield; then both viscoelastic and viscoplastic strains will occur.

For the viscoelastic stage, the solution for strain, ε, will be given by Eq. (8.62b). At $t = \infty$, the stress in unit 1, $\sigma_1 = 0$; hence

$$\sigma_0 = t_2 \sigma_2 \qquad (8.68a)$$

Therefore, the limiting condition when the slider will yield can be expressed as

$$\sigma_0 > t_2 \sigma_y \qquad (8.68b)$$

Now, at $t = t^*$, the following relations will hold:

$$\sigma_y = E_2\varepsilon^* = E_2\left(\frac{-\sigma_0}{b}\cdot\frac{E_1t_1}{E_2t_2}e^{\frac{-t\Gamma a}{b}} + \frac{\sigma_0}{E_2t_2}\right) \quad (8.69a)$$

where ε^* is the strain in the overlay at time $= t^*$. Hence,

$$t^* = \frac{b}{\Gamma E_1 E_2 t_2}\ln\left(\frac{\frac{\sigma_y}{E_2} - \frac{\sigma_0}{E_2 t_2}}{-\frac{\sigma_0}{b}\cdot\frac{E_1 t_1}{E_2 t_2}}\right) \quad (8.69b)$$

Then by using the following equations:

$$\sigma_0 = E_1 t_1 \varepsilon_1^e + E_2 t_2 \varepsilon_2^e \quad (8.70a)$$

$$\varepsilon = \varepsilon_1^e + \varepsilon_1^d \quad (8.70b)$$

$$\varepsilon = \varepsilon_2^e + \varepsilon_2^d \quad (8.70c)$$

the solution for ε for time $\geq t^*$ is obtained as

$$\varepsilon = A(\lambda_1 p_1 e^{\lambda_1 t} + \lambda_2 p_2 e^{\lambda_2 t}) + B(p_1 e^{\lambda_1 t} + p_2 e^{\lambda_2 t}) + C \quad (8.71)$$

where

$$A = -\frac{1}{\Gamma_2 H'}\left(1 + \frac{E_1 t_1}{E_2 t_2}\right)$$

$$B = -\frac{1}{\Gamma_2 H'}\left(\frac{\Gamma_2 E_1 t_1}{t_2} + \frac{\Gamma_2 E_1 t_1 H'}{E_2 t_2} - \frac{\Gamma_1 E_1^2 t_1}{E_2 t_2}\right)$$

$$C = \frac{1}{H'}\left[\sigma_0\left(\frac{1}{t_2} + \frac{H}{E_2 t_2}\right) - \sigma_y\right]$$

$$\lambda_{1,2} = \frac{(1-B) \pm \sqrt{(B-1)^2 + 4A\Gamma_1 E_1}}{2A}$$

$$p_1 = \frac{e^{-\lambda_1 t^*}}{A(\lambda_2 - \lambda_1)}\left[(A\lambda_2 + B)\left(\frac{\sigma_0 - t_2\sigma_y}{E_1 t_1}\right) - \frac{\sigma_y}{E_2} + C\right]$$

$$p_2 = \frac{e^{-\lambda_2 t^*}}{A(\lambda_1 - \lambda_2)}\left[(A\lambda_1 + B)\left(\frac{\sigma_0 - t_2\sigma_y}{E_1 t_1}\right) - \frac{\sigma_y}{E_2} + C\right]$$

Creep Behavior: Viscoelastic and Viscoplastic Models in DSC

TABLE 8.2
Creep Test Data for Indiana Limestone (24)

Increment	Stress Increment $\Delta\sigma_1$, psi	Initial Axial Strain $\varepsilon_0 \times 10^{-6}$	Initial Lateral Strain $\varepsilon_0 \times 10^{-6}$
4	699	125	−33
5	782	150	−39
6	781	147	−41
7	782	142	−42

© John Wiley & Sons Ltd. Reproduced with permission.

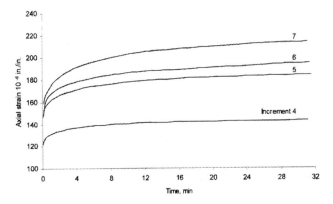

FIGURE 8.20
Creep test data for Indiana limestone (24). © John Wiley & Sons Ltd. Reproduced with permission.

8.6.5 Parameters for Viscoelastic Model

Consider the creep test data for Indiana limestone reported by Hardy et al. (23) and Goodman (24). Details of the applied stress increments ($\Delta\sigma_1$) are given below. It was reported that for increments 1 to 3, there was insignificant time dependence (24); hence, results for only increments 4 to 7 are considered in Table 8.2.

Axial strain vs. time curves for the typical stress increments, 4, 6, and 7, which are used here to evaluate the parameters, are shown in Fig. 8.20.

8.6.5.1 Increment 4; $\Delta\sigma_1 = 699$ psi

The values of strains and the derivative at $t = 0$ are

$$\varepsilon(0) = 125 \text{ in./in.}$$
$$\varepsilon_u = 141 \text{ in./in.}$$
$$\frac{d\varepsilon}{dt}(0) = \frac{127 - 125}{0.25} \times 10^{-6}$$
$$= 8 \times 10^{-6} \text{ in./in./sec}$$

Here the value of strain after $t = 0.25$ min was measured to be about 127×10^{-6} in/in. Now, using Eqs. (8.63a), (8.63b), and (8.64), we can evaluate E_1, E_2, and Γ as follows:

$$\varepsilon(0) = 125 \times 10^{-6} = \frac{699}{0.5(E_1 + E_2)} = \frac{1398}{E_1 + E_2}$$

$$\varepsilon_u = \frac{699}{E_2 \times 0.50} = \frac{1398}{E_2} = 141 \times 10^{-6}$$

Here the applied stress increment is 699 psi and $t_1 = t_2 = 0.50$. Solution of the above equations gives

$$E_2 = \frac{1398}{141} \times 10^6 = 10 \times 10^6 \text{ psi} \quad \text{and} \quad E_1 = 10^6 \text{ psi}$$

Now,

$$\frac{d\varepsilon}{dt}(0) = 8 \times 10^{-6} = \frac{\Gamma \times 1.0 \times 10^6 \times 1.0 \times 10^6 \times 0.5 \times 699}{0.5 \times 0.5 \times 11.0 \times 10^6 \times 11.0 \times 10^6} = \Gamma \times 12$$

$$\therefore \Gamma = 6.67 \times 10^{-7} \text{ 1/psi min}$$

Therefore, the viscosity coefficient, μ, of the dashpot is

$$\mu = \frac{1}{\Gamma} = 1.5 \times 10^6 \text{ psi min}$$

8.6.5.2 Verification

Let us consider two typical time levels, $t = 1.0$ and 4.0 minutes, and use Eq. (8.62b) to compute the corresponding strains by using E_1, E_2, and Γ.

$$a = E_1 E_2 t_2 = 1.0 \times 10^6 \times 10 \times 10^6 \times 0.5 = 5.0 \times 10^6 \times 10^6$$

$$b = 0.5 \times 11 \times 10^6 = 5.5 \times 10^6$$

$$\frac{E_1 t_1}{E_2 t_2} = \frac{1.0 \times 10^6}{10 \times 10^6} \times \frac{0.5}{0.5} = 0.10$$

Therefore,

$$\varepsilon = \frac{-699 \times 0.10}{5.5 \times 10^6} e^{\frac{-t \times 6.67 \times 10^{-7} \times 5.0 \times 10^6 \times 10^6}{5.5 \times 10^6}} + \frac{699}{10 \times 10^6 \times 0.5}$$

$$= (-12.7 \times 10^{-6})e^{-0.61t} + 140 \times 10^{-6}$$

For $t = 1.0$ min,

$$\varepsilon(t = 1) = (-12.7 \times 0.54 + 140) \times 10^{-6} = 130 \times 10^{-6} \text{ in./in.}$$

For $t = 4.0$ min,

$$\varepsilon(t = 4) = (-12.70 \times 0.087 + 140) \times 10^{-6} = 139 \times 10^{-6} \text{ in./in.}$$

These computed values compare well with the corresponding measured values of 131 and 138 × 10^{-6} in./in., respectively (Fig. 8.20).

8.6.5.3 Increments 5, 6, and 7

Increment 5: $\quad \varepsilon_0 = 150 \times 10^{-6}, \; \varepsilon_u = 182 \times 10^{-6}$

$$\frac{d\varepsilon}{dt}(0) = \frac{157 - 150}{0.25} \times 10^{-6} = 28 \times 10^{-6}$$

$$\Delta\sigma_1 = 782 \text{ psi}$$

Increment 6: $\quad \varepsilon_0 = 147 \times 10^{-6}, \; \varepsilon_u = 193 \times 10^{-6}$

$$\frac{d\varepsilon}{d\sigma}(0) = \frac{161 - 147}{0.25} \times 10^{-6} = 56 \times 10^{-6}$$

$$\Delta\sigma_1 = 781 \text{ psi}$$

Increment 7: $\quad \varepsilon_0 = 142 \times 10^{-6}, \; \varepsilon_u = 215 \times 10^{-6}$

$$\frac{d\varepsilon}{dt}(0) = \frac{170 - 142}{0.25} \times 10^{-6} = 112 \times 10^{-6}$$

$$\Delta\sigma_1 = 782 \text{ psi}$$

The values of E_1, E_2, and Γ for increments 5, 6, and 7 were found using the foregoing procedure.

Bulk Modulus. The bulk modulus, K, is found using Eq. (8.65b), where $\Delta\sigma_1/3 = p$ is the mean pressure and $\varepsilon_v = \varepsilon_0 + 2\varepsilon_3$, (Table 3.2). Their values for increments 4 to 7 were found to be 3.9, 3.2, 4.0, and 4.5 × 10^6 psi, with the average value = 3.9 × 10^6 psi. Then the Poisson ratios, v_1 and v_2, are found by using Eq. (8.66). The values of the parameters are listed in Table 8.3.

It can be seen that the values of the parameters vary with the stress increment; E_1 and E_2 increase and decrease, while v_1 and v_2 decrease and increase, respectively. The values of Γ and K do not vary significantly. If, as a simplification, single values of the parameters are desired, it may be necessary to optimize them such that the predictions provide acceptable correlation with the test data (Fig. 8.20).

TABLE 8.3
Parameters for Viscoelastic Overlay Model: Indiana Limestone

Increment	$E_1 \times 10^6$ psi	$E_2 \times 10^6$ psi	v_1	v_2	Γ 1/psi min	K psi	N
4	1.00	10.00	0.46	0.075	6.67		
5	1.83	8.60	0.42	0.132	5.80	3.9×10^6	1.00
6	2.60	8.00	0.39	0.160	5.96		1/psi min
7	3.70	7.30	0.34	0.190	6.32		
5	1.83	8.60	0.42	0.132	5.80		

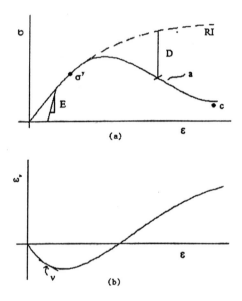

FIGURE 8.21
Schematic of tests for parameters in overlay models.

8.6.6 Physical Meanings of Parameters

In the vevp model, $E_{eq} = E_1 t_1 + E_2 t_2$ and (Γ_1, N_1) denote delayed elasticity moduli, and (Γ_2, N_2, F) denote delayed viscoplastic parameters.

8.6.7 Disturbance Parameters

Figure 8.21(a) and (b) show schematics of stress–strain and volume change behavior; the curves represent the locus of equilibrated strains for various stress increments. As explained earlier, the disturbance can be defined based on the stress–strain response or volumetric response (Fig. 8.21).

8.6.8 Finite-Element Equations

The finite–element equations (8.31) can now be written for the overlay models using Eq. (8.49b) as

$$\int_V B_n^T \left(\sum_{i=1}^n \Delta\sigma_n^j t_j \right) dV + \Delta Q_n = 0 \qquad (8.72)$$

If there is no disturbance (microcracking and degradation), the incremental stress, $\Delta\sigma_n^j$, can be expressed as in Eq. (8.7c). If disturbance is included, $\Delta\sigma_n^j$ can be expressed as in Eq. (8.30). Then the solution procedure involving time integration will be similar to that described earlier in this chapter.

8.6.9 Other Models

Many models have been proposed for the analysis of the creep behavior of engineering materials; these include the classical models (25–28), the endochronic models (29, 30), and the viscoplastic theory based on the overstress (VBO) models (31, 32). In the VBO approach, the observed stress–strain–time response is characterized on the basis of the equilibrium and kinematic (reference) responses as components of the behavior. The difference between the observed and equilibrium responses is considered as the "overstress," which causes viscous or creep deformation similar to that in Perzyna's viscoplasticity model described in this chapter. It is felt that there may be similarities between the VBO and DSC models in that the observed behavior is considered to be composed of the behavior of the deforming materials at certain reference states.

8.6.10 Advantages of Overlay Models

The overlay models offer a number of advantages compared to the classical viscoelastic and viscoplastic formulations based on the closed-form solutions of governing equations (25–28). They are the following:

1. The classical models are often complicated and may not be suitable for implementation in computer procedures and for solving practical engineering problems. On the other hand, the elastoviscoplastic and overlay models are found to be more appropriate for the implementation and practical problems (1, 2, 5, 7, 13).
2. The material parameters (Table 8.1) have physical meanings and can be obtained directly from laboratory stress–strain–time response.
3. Determination of the parameters involves much reduced curve fitting and regression, and their number is less than that in the classical models of comparable capability.
4. The models can be implemented easily in nonlinear computer (finite-element) procedures, as the characteristics of the equations

and matrices involved are similar to those in traditional finite element procedures.

5. The overlay model is consistent with the underlying hierarchical nature of the DSC. As a result, models of increasing sophistication can be adopted by adding parameters corresponding to additional behaviorial features of a material.

6. The multicomponent DSC can lead to more general models in which relative motions can be permitted; that is, the requirement of capability of strains between units in the overlay models (Fig. 8.14) can be relaxed.

8.7 Examples

Example 8.1 Elastoviscoplastic (evp) Models for Solders

Details of the characterization using the elastoplastic (δ_0) model for solders (Pb/Sn) with different compositions used as joining materials in electronic packaging and semiconductors problems are given in Example 7.5, Chapter 7. Here, we consider the viscous or creep behavior by using the evp–Perzyna-type model, which allows essentially for viscoplastic deformations in the secondary creep regime (Fig. 8.1).

The measured stress vs. strain rates under different temperatures for a 60% Sn–40% Pb solder are shown in Fig. 8.12(a), as reported by Pan (19). This data was used to determine the temperature dependent-viscous parameters Γ and N by following the procedure described earlier; see Fig. 8.12(b). The values of the parameters at different temperatures are given in Table 8.2 (11, 12).

The parameter N does not vary significantly with temperature; hence, an average value of 2.67 was used. The temperature dependence for the fluidity parameter, $\Gamma(t)$, is given approximately by

$$\ln \Gamma(T) = \ln \Gamma_{300} \left(\frac{T}{300}\right)^{6.185} \tag{1a}$$

where Γ_{300} is the value of Γ at the reference temperature of 300 K, which was found to be $\ln \Gamma_{300} = 1.8/\text{sec}$.

Predictions. An example of the predicted behavior of the solder showing stress relaxation is shown in Fig. 8.22 (11). A shear stress of 20 MPa was applied and the corresponding shear strain was computed based on the viscoplastic equations (evp) model. Then the strain was held constant and the viscoplastic equations were used to predict the stress relaxation. The elastic and plastic (δ_0-model) parameters are given in Chapter 7, and the viscous

TABLE 8.4
Viscous Parameters for Pb40/Sn60 Solders at Different Temperatures, Obtained from Pan (11, 19)

Temperature	293 K	313 K	333 K	373 K	393 K
Fluidity parameter $\ln(\Gamma)$	0.578	2.058	3.475	4.61	6.96
Parameter N	2.665	2.645	2.667	2.448	2.74
(average)	2.67	2.67	2.67	2.67	2.67

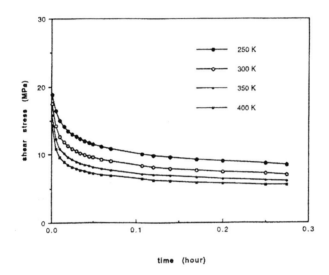

FIGURE 8.22
Simulated stress relaxation for 60% Sn–40% Pb solders at different temperatures (11).

parameters (Table 8.4) were used. The predictions in Fig. 8.22 were obtained for four different temperatures and show consistent trends.

Example 8.2 Elastoviscoplastic (evp) Model for Rock Salt

Appropriate characterization of the thermomechanical behavior of rock salt is important for analysis and design of underground excavations, e.g., storage chambers for nuclear waste. Application of the thermoplastic and thermoviscoplastic models to characterize behavior of rock salt is described in this example (14). Available test data (33–36) for a rock salt were used to determine various temperature-dependent parameters, as follows:

Table 8.5 shows material parameters at different temperatures obtained by using the test data (14, 33–35). Expressions for their temperature dependence are also shown in Table 8.5. The elastic parameters were found from the test data by Burke (33) and Yang (34). The plasticity parameters for the

TABLE 8.5
Material Parameters for Rock Salt at Different Temperatures (14) (with permission)

Temperature parameter	296 K	336 K	350 K	473 K	573 K	673 K
E (GPa)	34.13	32.19	31.59	27.49	25.15	23.35
v	0.279	0.287	0.290	0.310	0.324	0.336
α_T (1/K) ($\times 10^{-5}$)	3.8	4.2	4.3	5.5	6.4	7.2
γ	0.0516	0.0384	0.0349	0.0173	0.0111	0.0076
β	0.690	0.620	0.590	0.450	0.380	0.33
n (average)	—	3.92	—	—	—	—
a_1 ($\times 10^{-9}$)	1.80	0.97	0.95	—	—	—
η_1 (average)	—	—	0.474	—	—	—
Γ ($\times 10^{-3}$/day)	4.95	6.11	6.54	10.77	14.8	19.35
N (average)	—	3.0	—	—	—	—

$$E(T) = 33.92\left(\frac{T}{300}\right)^{-0.462} ; \quad v(T) = 0.28\left(\frac{T}{300}\right)^{-0.224}$$

$$\alpha_T(T) = 3.85 \times 10^{-5}\left(\frac{T}{300}\right)^{0.778}$$

$$\gamma(T) = 0.05\left(\frac{T}{300}\right)^{-2.326} ; \quad \beta(T) = 0.68\left(\frac{T}{300}\right)^{-0.91}$$

$$\alpha(T) = \left(\frac{a_1}{\xi^{n_1}}\right)_{300}\left(\frac{T}{300}\right)^{-0.334} ; \quad \Gamma(T) = 5.0\left(\frac{T}{300}\right)^{1.70}$$

HISS δ_0-model were found from the test data by Kern and Franke (35). The creep parameters were found based on the test data by LeComte (36).

Figure 8.23 shows comparisons between the test results (35) and predictions for stress–strain behavior at three temperatures—300, 336, and 350 K—for initial confining stress = 30 MPa. Here, the temperature-dependent δ_0-plasticity model is used. The temperature-dependent viscoplastic model was used to predict the test behavior (33). Figure 8.24 shows strain vs. time response for creep at two temperatures, 302 and 377 K. It can be seen that the models provide very good predictions of the observed behavior.

Example 8.3 Stress Relaxation: Overlay (ve) Model

Figure 8.25(a) shows a one-element (eight-noded) finite-element mesh, with axisymmetric idealization (Chapter 13). The viscoelastic overlay model with parameters shown in Fig. 8.25(b) is adopted to analyze stress relaxation under a constant displacement of $\delta = 0.1$ units applied at the top; each unit has thickness = 0.50. Since $\sigma_{y1} = \sigma_{y2} = 0$, this model represents two Maxwell units in parallel (Table 8.1). In the time integration, $\Delta t = 0.02$ was used, with integration parameter $\theta = 0.0$. The finite-element computer procedure (37) was used for calculations in this example and the subsequent examples 8.4 to 8.6;

Creep Behavior: Viscoelastic and Viscoplastic Models in DSC

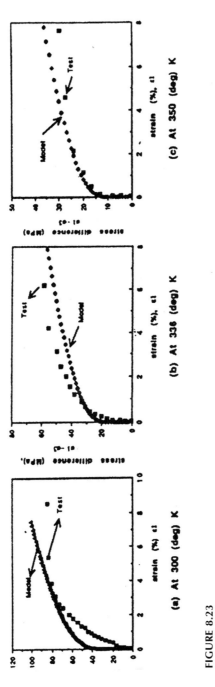

FIGURE 8.23
Comparisons between HISS δ_0-model predictions and test data at different temperatures (14).

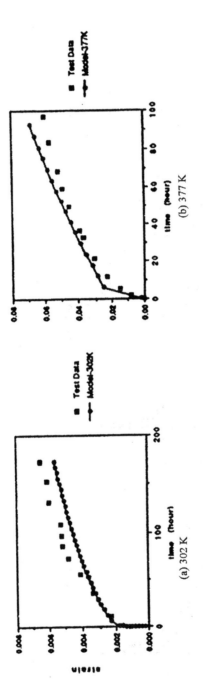

FIGURE 8.24
Comparisons between evp model and test data at different temperatures (14) (with permission).

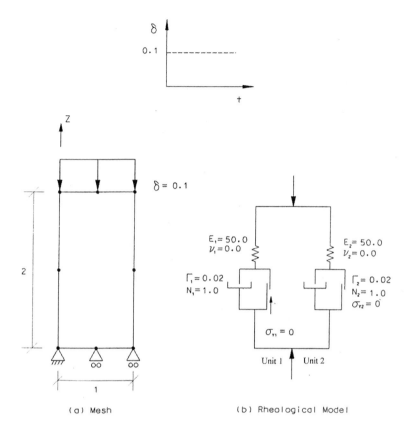

FIGURE 8.25
Stress relaxation: ve (Maxwell) model.

Sharma (38) provided assistance in these calculations. Figure 8.26 shows (axial) stress vs. time computed from the finite-element analysis. The results show the same trends as the viscoelastic solution reported by Yamada and Iwata (39).

Example 8.4 Thermal Creep in Restrained Bar: evp Model

Figure 8.27(a) shows a five-element (eight-noded) mesh for a bar restrained at each end. The bar is subjected to a temperature change $\Delta T = 10$ degrees, which varies linearly up to time $t = 0.1$ and then is held constant [Fig. 8.27(b)]. The parameters for the evp (Perzyna) model (Fig. 8.17) are as follows:

$$E = 50 \text{ units}, v = 0.0$$
$$\Gamma = 0.02, N = 1.0$$
$$\alpha_T = 0.00096$$
$$\sigma_y = 0.0$$

Time step $\Delta t = 0.02$ was used with $\theta = 0$ in the integration.

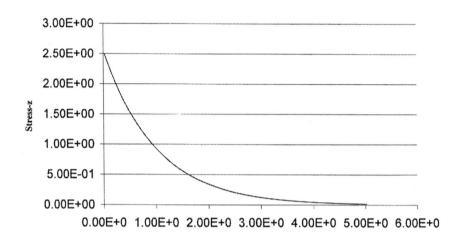

FIGURE 8.26
Stress vs. time response.

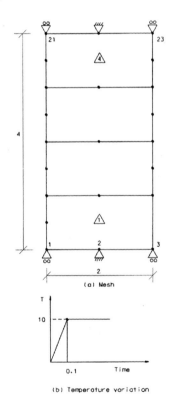

FIGURE 8.27
Mesh and temperature variation.

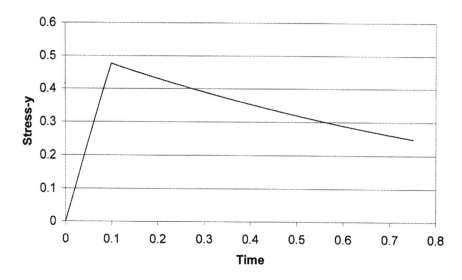

FIGURE 8.28
Stress vs. time behavior for restrained bar.

Figure 8.28 shows the computed axial stress vs. time curve. The stress increases to a value of about 0.48, and thereafter it decreases with time. This result is qualitatively similar to that reported by Yamada and Iwata (39) based on an analytical solution and a finite-element procedure.

Example 8.5 Creep Analysis: ve, evp, and vevp Overlay Models with von Mises Plasticity Criterion

Figure 8.29 shows a five-element mesh. The overlay model is used for ve, evp, and vevp versions (Table 8.1). A stress σ_0 = 800 units is applied at the top, kept constant up to four time units, and is then removed. In the time integration, the time step Δt = 0.10 is used with θ = 0.0. The parameters used are shown in Table 8.6(a). Figure 8.30 shows computed strain vs. time plots for the three models, which show consistent trends.

Example 8.6 Creep Analysis: ve, vp, and vevp Models with HISS Plasticity

The finite-element mesh used is the same as in Fig. 8.29. However, the load applied is σ_0 = 2000 units, and the parameters are different; they are shown in Table 8.6(b). Figure 8.31 shows computed strain vs. time plots for the three models. The trends in these plots are somewhat different than those in Fig. 8.30 for the von Mises criterion because the HISS δ_0-model allows for continuous yielding.

The parameters chosen in Examples 8.3 to 8.6 are arbitrary, as the examples are intended to show the capability of the models to simulate viscoelastic, viscoplastic creep or both. The parameters of given materials should be found

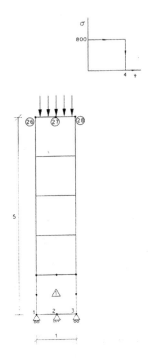

FIGURE 8.29
Creep analysis: von Mises and HISS models.

from laboratory stress–strain–time responses under uniaxial, triaxial, and/or shear loadings.

Example 8.7 Viscoplastic Model for Asphalt Concrete

Scarpas et al. (40) used the HISS δ_0-plasticity model (Chapter 7) with the Perzyna (evp) formulation for the characterization of the thermomechanical and rate-dependent behavior of asphalt concrete.

Figures 8.32 to 8.34 show laboratory uniaxial compression, tension, and incremental creep test results, respectively, from testing with 100-mm high and 100-mm-diameter specimens of asphalt-concrete mix of type 0/16 with 6% bitumen 80/100 (40). The tests were performed at different temperatures and deformation rates ($\dot{\delta}$) for the compressive and tensile loadings. For the creep behavior, the tests were performed at room temperature of 20°C. The creep response, Fig. 8.34(a), was divided in two parts; the first part up to about 12 seconds, in which for each increment of stress, the strains stabilize such that $\dot{\delta} \approx 0$ [Fig. 8.34(b)], and the second part from 12 to 17 seconds, during which the strains increase at approximately the constant rate, Fig. 8.34(c). After about 17 seconds, tertiary creep occurred.

The HISS δ_0-plasticity model was used in the context of the evp model to characterize the first part of the creep response, in which, for a given stress increment, the viscoplastic strains stabilized such that $F \approx 0$ (Fig. 8.3). For the

TABLE 8.6
Parameters for ve, vp, and vevp Overlay: Von Mises and HISS Plasticity

(a) von Mises: $\sigma_0 = 800$

	ve		vp		vevp	
	Unit 1	Unit 2	Unit 1	Unit 2	Unit 1	Unit 2
E	10^6	10^6	10^6		10^6	10^6
v	0.20	0.20	0.20		0.20	0.20
Γ	10^{-4}	10^{-4}	10^{-4}		10^{-4}	10^{-4}
N	1.0	1.0	1.0		1.0	1.0
σ_y	0.0	5000	750		0.0	1500
F_0	1.0	1.0	1.0		1.0	1.0

(b) HISS: $\sigma_0 = 2000$

	ve		vp		vevp	
	Unit 1	Unit 2	Unit 1	Unit 2	Unit 1	Unit 2
E		2×10^6	2×10^6		2×10^6	2×10^6
v		0.20	0.20		0.20	0.20
Γ		10^{-4}	10^{-4}		10^{-4}	10^{-4}
N		1.0	1.0		1.0	1.0
σ_y/F		F*	10,000	F*		0.0	F*
F_0		1.0	1.0	1.0		1.0	1.0

*$\gamma = 0.06784$, $\beta = 0.755$, $n = 5.237$, $a_1 = 0.46 \times 10^{-10}$, $\eta_1 = 0.826$, $3R = 50$.

second part in which the creep strains do not stabilize, the behavior was simulated by defining a creep initiation surface at the stress level when the first part stops. For subsequent stress (increments), the overstress, denoted by d in Fig. 8.35, allows the use of the evp model, in which the creep strain increases at a constant rate until the tertiary creep.

The hardening function, α, was expressed by Scarpas et al. (40) as a function of the dissipated energy, w_p, during the first part of the creep response as

$$\bar{\alpha} = \bar{\alpha}_0 e^{-kw_p} \qquad (7a)$$

where $\bar{\alpha}_0$ is the value of $\bar{\alpha}_0$ at 20°C up to which the response was considered to be elastic. Plots of yield surfaces and of α vs. w_p are shown in Fig. 8.36(a) and (b), respectively. The latter is similar to α expressed in terms of ξ_{vp}, Eq. (8.4b).

For fracture and degradation analysis, Scarpas et al. (40) expressed fracture energy given by the area under the stress (σ) vs. w (crack opening length) (Fig. 8.37), which was expressed as

$$w = \ell_c \varepsilon_{cr} \qquad (7b)$$

where ℓ_c is the characteristic length and ε_{cr} is the smeared crack strain. An alternative and general approach based on critical disturbance can be used for microcracking leading to fracture and degradation (Chapter 12). Here disturbance, D,

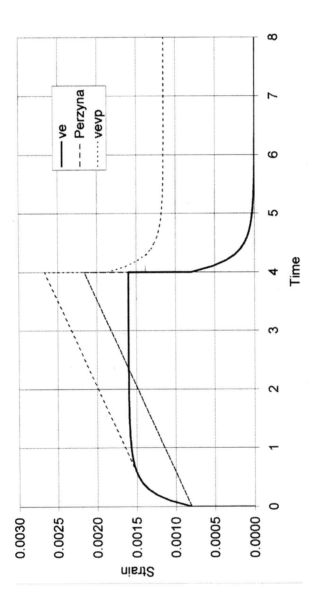

FIGURE 8.30
Predictions from overlay models: von Mises criterion.

Creep Behavior: Viscoelastic and Viscoplastic Models in DSC

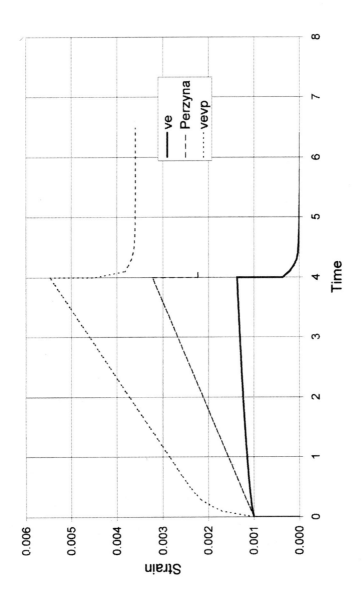

FIGURE 8.31
Predictions from overlay models: HISS δ_0-model.

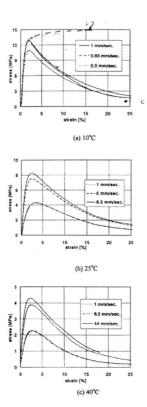

FIGURE 8.32
Uniaxial compression tests for asphalt concrete (40).

can be found from the test data (Figs. 8.32 and 8.33) by assuming linear elastic or elastoplastic (δ_0) model to characterize the RI behavior. The FA response can be characterized by using the asymptotic stress (c) (Fig. 8.32). Then D can be expressed as

$$D = D_u(T, \dot{\delta})\left[1 - e^{-A(T, \dot{\delta})\xi_{vp}^{Z(T, \dot{\delta})}}\right] \tag{7c}$$

where D_u, A, and Z are parameters dependent on temperature (T) and deformation rate ($\dot{\delta}$).

The parameters for the δ_0-plasticity and viscous response were determined from the test data (Figs. 8.32 to 8.34) by using procedures described in this chapter and Chapter 7. They are listed below (40).

$$\gamma = 0.089 \text{ (at } T = 20°C\text{)}, \beta = 0.442; n = 2.1$$
$$\bar{\alpha}_0 = 0.0865 \text{ (at } T = 20°C\text{)}, k = 2800$$
$$\Gamma = 1.5 \times 10^{-9} \text{ 1/sec}; N = 0.32$$

Creep Behavior: Viscoelastic and Viscoplastic Models in DSC

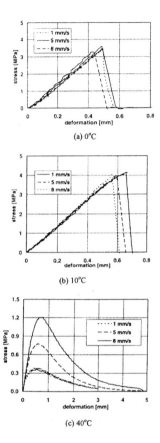

FIGURE 8.33
Uniaxial tension tests for asphalt concrete (40).

A thermoviscoplastic model was implemented in a three-dimensional, finite-element procedure (40), which was used to solve practical problems involving pavement structures in which permanent deformations (rutting), and microcracking and fracture are important design considerations.

Example 8.8 Overlay Model: Closed-Form Solution (38)

Consider two overlays 1 and 2 [Fig. 8.15(a)]. Let σ_0 be the applied (major principal) stress, σ_1. Also, let the thickness of the two overlays $t_1 = t_2 = 0.5$, and σ_1^1, σ_3^1, and σ_1^2, σ_3^2 be the major and minor principal stresses in overlays 1 and 2, respectively.

The stresses (Eq. 8.49) in the overlays are

$$\sum \sigma_1^i t_i = \sigma_1 \quad \text{or} \quad \sigma_1^1 \frac{1}{2} + \sigma_1^2 \frac{1}{2} = \sigma_1 \quad \text{or} \quad \sigma_1^1 + \sigma_1^2 = 2\sigma_0 \qquad (8a)$$

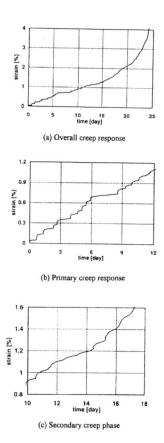

FIGURE 8.34
Creep response for asphalt concrete (40).

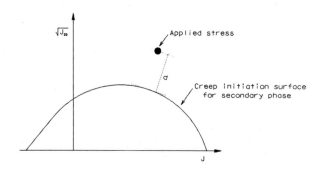

FIGURE 8.35
Secondary phase of creep due to overstress (40).

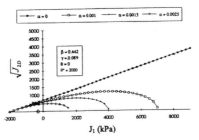

(a) Expanding hardening surfaces

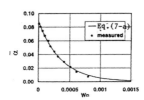

(b) Hardening as function of plastic work

FIGURE 8.36
Hardening for asphalt concrete (40).

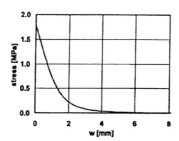

FIGURE 8.37
Fracture energy for asphalt concrete at $T = 25°C$ (40).

and

$$\sum \sigma_3^i t_i = \sigma_3 \quad \text{or} \quad \sigma_3^1 + \sigma_3^2 = 0 \tag{8b}$$

The strains in the overlays are

$$\varepsilon_1^1 = \varepsilon_{1e}^1 + \varepsilon_{1vp}^1, \quad \varepsilon_3^1 = \varepsilon_{3e}^1 + \varepsilon_{3vp}^1 \tag{8c}$$

and

$$\varepsilon_1^2 = \varepsilon_{1e}^2 + \varepsilon_{1vp}^2, \quad \varepsilon_3^2 = \varepsilon_{3e}^2 + \varepsilon_{3vp}^2$$

However, $\varepsilon_1^1 = \varepsilon_1^2$ and $\varepsilon_3^1 = \varepsilon_3^2$; therefore,

$$\varepsilon_1^1 = \varepsilon_1^2 = \varepsilon_1 \quad \text{and} \quad \varepsilon_3^1 = \varepsilon_3^2 = \varepsilon_3 \tag{8d}$$

For axisymmetric idealization and the linear elastic model:

$$\begin{aligned}\sigma_1^1 &= a_1 \varepsilon_{1e}^1 + 2b_1 \varepsilon_{3e}^1 \\ \sigma_3^1 &= (a_1 + b_1)\varepsilon_{3e}^1 + b_1 \varepsilon_{1e}^1\end{aligned} \tag{8e}$$

and

$$\begin{aligned}\sigma_1^2 &= a_2 \varepsilon_{1e}^2 + 2b_2 \varepsilon_{3e}^2 \\ \sigma_3^2 &= (a_2 + b_2)\varepsilon_{3e}^2 + b_2 \varepsilon_{1e}^2\end{aligned}$$

where

$$a_1 = \frac{E_1(1 - \nu_1)}{(1 + \nu_1)(1 - 2\nu_1)}, \quad b_1 = \frac{E_1 \nu_1}{(1 + \nu_1)(1 - 2\nu_1)}$$

$$a_2 = \frac{E_2(1 - \nu_2)}{(1 + \nu_2)(1 - 2\nu_2)}, \quad b_2 = \frac{E_2 \nu_2}{(1 + \nu_2)(1 - 2\nu_2)}$$

Substitution of Eqs. (8c), (8d), and (8e) in Eq. (8a), and simplifications, lead to

$$(a_1 + a_2)\varepsilon_1 + 2(b_1 + b_2)\varepsilon_3 = 2\sigma_0 + a_1 \varepsilon_{1vp}^1 + 2b_1 \varepsilon_{3vp}^1$$
$$+ a_2 \varepsilon_{1vp}^2 + 2b_2 \varepsilon_{3vp}^2 \tag{8f}$$

and

$$(b_1 + b_2)\varepsilon_1 + (a_1 + b_1 + a_2 + b_2)\varepsilon_3 = (a_1 + b_1)\varepsilon_{3vp}^1 + b_1 \varepsilon_{1vp}^1$$
$$+ (a_2 + b_2)\varepsilon_{3vp}^2 + b_2 \varepsilon_{3vp}^2$$

Equation (8f) can be used to solve for ε_1 and ε_3. Then from Eq. (8e), the expressions for stresses are found as

$$\begin{aligned}\sigma_1^1 &= a_1(\varepsilon_1 - \varepsilon_{1vp}^1) + 2b_1(\varepsilon_3 - \varepsilon_{3vp}^1) \\ \sigma_3^1 &= (a_1 + b_1)(\varepsilon_3 - \varepsilon_{3vp}^1) + b_1(\varepsilon_1 - \varepsilon_{1vp}^1) \\ \sigma_1^2 &= a_2(\varepsilon_1 - \varepsilon_{1vp}^2) + 2b_2(\varepsilon_3 - \varepsilon_{3vp}^2) \\ \sigma_3^2 &= (a_2 + b_2)(\varepsilon_3 - \varepsilon_{3vp}^2) + b_2(\varepsilon_1 - \varepsilon_{1vp}^1)\end{aligned} \tag{8g}$$

and the overall stresses in the overlay are

$$\sigma_1 = \sigma_1^1 t_1 + \sigma_1^2 t_2 = (\sigma_1^1 + \sigma_1^2)/2 = \sigma_0$$
$$\sigma_3 = \sigma_3^1 t_1 + \sigma_3^2 t_2 = (\sigma_3^1 + \sigma_3^2)/2 = 0$$
(8h)

Consider the yield function, F, in the Mohr–Coulomb criterion for the plastic response (Chapter 6)

$$F = -\frac{J_1}{3}\sin\phi + \sqrt{J_{2D}}\left(\cos\theta - \frac{\sin\phi\sin\theta}{\sqrt{3}}\right) - c\cos\theta = 0 \quad (8\text{ia})$$

Assume $\phi = 0$; then

$$F = \sqrt{J_{2D}}\cos\theta - c = 0 \quad (8\text{ib})$$

Now for σ_1 and $\sigma_2 = \sigma_3$, the deviatoric stresses, S, are given by

$$S_1 = \frac{2(\sigma_1 - \sigma_3)}{3}, \quad S_2 = S_3 = -\frac{\sigma_1 - \sigma_3}{3}$$

and

$$J_{2D} = \frac{1}{2}(S_1^2 + S_2^2 + S_3^2) = \frac{(\sigma_1 - \sigma_3)^2}{3} \quad (8\text{j})$$

Consider $\theta = 30°$ and $\cos\theta = \sqrt{3}/2$; hence

$$\frac{\partial F}{\partial J_{2D}} = \frac{\sqrt{3}}{4\sqrt{J_{2D}}}, \quad \frac{\partial J_{2D}}{\partial \sigma_1} = S_1, \quad \frac{\partial J_{2D}}{\partial \sigma_3} = S_3$$

$$\frac{\partial F}{\partial \sigma_1} = \frac{\partial F}{\partial J_{2D}} \cdot \frac{\partial J_{2D}}{\partial \sigma_1} = \frac{1}{2} \quad (8\text{k})$$

$$\frac{\partial F}{\partial \sigma_3} = \frac{\partial F}{\partial J_{2D}} \cdot \frac{\partial J_{2D}}{\partial \sigma_3} = -\frac{1}{4}$$

The incremental viscoplastic strains are now evaluated as

$$\Delta\varepsilon_{1vp} = \Gamma\langle F\rangle\frac{\partial F}{\partial \sigma_1}\cdot\Delta t = \frac{1}{2}\Gamma\left\langle\frac{\sigma_1 - \sigma_3}{2} - c\right\rangle$$

$$\Delta\varepsilon_{3vp} = \Gamma\langle F\rangle\frac{\partial F}{\partial \sigma_3}\cdot\Delta t = -\frac{1}{4}\Gamma\Delta t\left\langle\frac{\sigma_1 - \sigma_3}{2} - c\right\rangle$$
(8l)

The incremental viscoplastic strains in each overlay can be found by substituting the respective values of the fluidity parameters (Γ_1 and Γ_1) and cohesion (c_1 and c_2) in Eq. (8l).

A computer routine was prepared for the foregoing closed-form solution [Sharma (38)]. It was used to obtain the closed-form solution with the following parameters for a rock salt (7):

	Overlay 1	Overlay 2
Elastic modulus, E	4,700	15,400
Poisson's ratio, v	0.24	0.24
Fluidity parameter, Γ	2×10^{-4}/bar/day	1×10^{-5}/bar/day
Cohesion, c	0.0	55.0
Thickness	0.50	0.50

Applied stress, $\sigma_0 = 160$; time step, $\Delta t = 0.50$; 1 bar = 100 KN/m². The analysis can be continued for various time levels; for example, for 60 days a total number of increments = 121 are required. The finite-element computer procedure (37) was also used to solve the problem with the FE mesh [Fig. 8.38(b)]. Figure 8.39 shows computed strain vs. time from the closed-form solutions and the computer procedure for three different confining pressures. Both correlate very well and yield essentially the same results.

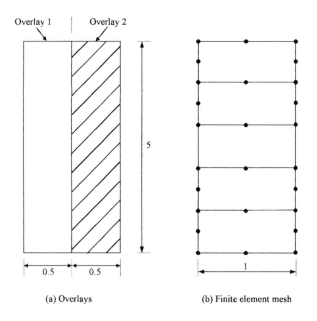

(a) Overlays

(b) Finite element mesh

FIGURE 8.38
Finite-element mesh for rock salt problem: closed-form solution with overlay model.

Creep Behavior: Viscoelastic and Viscoplastic Models in DSC

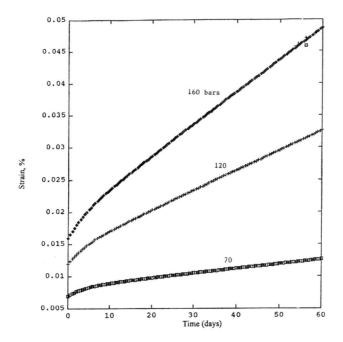

FIGURE 8.39
Closed-form and computer results for strain vs. time behavior.

Example 8.9 One-dimensional Elastoviscoplastic Model for Creep

Consider the solution in Eq. 8.24. Figure 8.40(a)–(c) show plots of strain vs. time for $H' = 10^5, 10^6,$ and 10^7 psi with $\Gamma = 10^{-10}$ to 10^{-2}. Here $\sigma_0 = 30{,}000$ psi, $\sigma_y = 10{,}000$ psi, and $E = 10^6$ psi. It can be seen that as the value of Γ increases, the results approach the solution as per inviscid plasticity.

Example 8.10 One-dimensional Elastoviscoplastic Model for Relaxation

Derive the relaxation solution for the one-dimensional case. Based on Eq. (8.2a), the viscoplastic rate is given by

$$\frac{d\varepsilon^{vp}(t)}{dt} = -\frac{d\varepsilon^e(t)}{dt} = -\frac{1}{E}\frac{d\sigma(t)}{dt} \tag{10a}$$

because the applied total strain is constant. The solution of Eq. (10a) is expressed as

$$\varepsilon^{vp}(t) = \frac{\sigma(t)}{dt} + C \tag{10b}$$

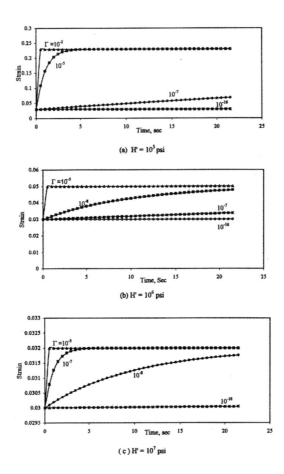

FIGURE 8.40
Creep predictions for different values of H' and Γ.

For the case of stress relaxation, consider that a strain (increment) ε_0 is applied at time $t = 0$. Stress at $t = 0$ is given by

$$\sigma_0 = E\varepsilon_0 \quad (10c)$$

and since the viscoplastic strain is zero at $t = 0$, we have

$$\frac{-\sigma(0)}{E} + C = 0 \quad (10d)$$

Therefore,

$$C = \varepsilon_0 \quad (10e)$$

Consider the linear hardening response; hence,

$$\bar{\sigma} = \sigma_y + H'\varepsilon^{vp} \quad (10f)$$

Creep Behavior: Viscoelastic and Viscoplastic Models in DSC

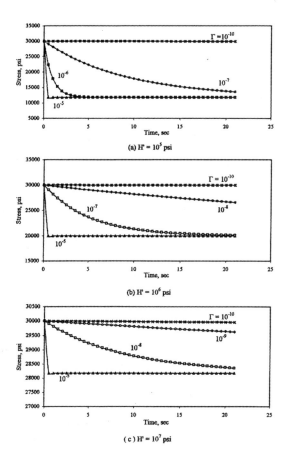

FIGURE 8.41
Stress relaxation predictions for different values of H' and Γ.

If the initial stress, σ_0, is greater than the yield stress (σ_y), stress relaxation will occur and the overstress will tend to zero. For this case, the basic equation is given by Eq. (8.19), which, after substitution of $d\varepsilon^{vp}/dt$ and ε^{vp} from Eqs. (10a) and (10b) leads to

$$\frac{\mu}{E}\frac{d\sigma(t)}{dt} + \left(1 + \frac{H'}{\varepsilon}\right)\sigma(t) = \sigma_y + H'\varepsilon(0) \qquad (10g)$$

The general solution for Eq. (10g) can be written as

$$\sigma(t) = \bar{C}e^{-t(H'+E)/\mu} + \frac{E}{H'+E}[\sigma_y + H\varepsilon(0)] \qquad (10h)$$

Using the initial condition for stress, $\bar{C}$ is obtained as

$$\bar{C} = \frac{E}{H'+E}[E\varepsilon(0) - \sigma_y] \qquad (10i)$$

Therefore, the solution for stress is given by

$$\sigma(t) = \frac{E}{H' + E}[E\varepsilon(0) - \sigma_y]e^{-t\Gamma(H'+E)} + \frac{E}{H' + E}[\sigma_y + H\varepsilon(0)] \quad (10j)$$

where $\mu = 1/\Gamma$. As $t \to \infty$,

$$\sigma(t) \to \frac{E}{H' + E}[\sigma_y + H\varepsilon(0)]$$

which corresponds to the inviscid plasticity solution.

Plots of σ vs. strain for $H' = 10^5, 10^6$, and 10^7 psi with $\Gamma = 10^{-5}$ to 10^{-10} are shown in Fig. 8.41. Here, $\varepsilon_0 = 0.03$, $E = 10^6$ psi, and $\sigma_y = 10{,}000$ psi.

References

1. Perzyna, P., "Fundamental Problems in Viscoplasticity," *Adv. in Appl. Mech.*, 9, 1966, 243–277.
2. Cormeau, I.C., "Viscoplasticity and Plasticity in Finite Element Method," *Ph.D. Dissertation*, Univ. College at Swansea, U.K., 1976.
3. Zienkiewicz, O.C. and Cormeau, I.C., "Viscoplasticity-Plasticity and Creep in Elastic Solids—A Unified Numerical Solution Approach," *Int. J. Num. Meth. in Eng.*, 8, 1974, 821–845.
4. Katona, M.G. and Mulert, M.A., "A Viscoplastic Cap Model for Soils and Rocks," Chap. 17 in *Mechanics of Engineering Materials*, C.S. Desai and R.H. Gallagher (Editors), John Wiley, Chichester, U.K., 1984, 335–349.
5. Baladi, G.Y. and Rohani, B., "Development of an Elastic Viscoplastic Constitutive Relationship for Earth Materials," Chap. 2 in *Mechanics of Engineering Materials*, C.S. Desai and R.H. Gallagher (Editors), John Wiley, Chichester, U.K., 1984, 23–43.
6. Kanchi, M.B., Zienkiewicz, O.C., and Owen, D.R.J., "The Viscoplastic Approach to Problems of Plasticity and Creep Involving Geometric Nonlinear Effects," *Int. J. Num. Methods in Eng.*, 12, 1978, 169–181.
7. Owen, D.R.J. and Hinton, E., *Finite Elements in Plasticity: Theory and Practice*, Pineridge Press, Swansea, U.K., 1980.
8. Desai, C.S. and Zhang, D., "Viscoplastic Model for Geologic Materials with Generalized Flow Rule," *Int. J. Num. Analyt. Methods Geomech.*, 11, 1987, 603–620.
9. Desai, C.S., Samtani, N.C., and Vulliet, L., "Constitutive Modelling and Analysis of Creeping Slopes," *J. Geotech. Eng.*, ASCE, 121, 1995, 43–56.
10. Samtani, N.C., Desai, C.S., and Vulliet, L., "An Interface Model to Describe Viscoplastic Behavior," *Int. J. Num. Analyt. Methods Geomech.*, 20, 1996, 231–252.
11. Chia, J. and Desai, C.S., "Constitutive Modeling of Thermomechanical Response of Materials in Semiconductor Devices with Emphasis on Interface Behavior," *Report to NSF*, Dept. of Civil Engng. and Engng. Mechs., University of Arizona, Tucson, AZ, 1994.

12. Desai, C.S., Chia, J., Kundu, T., and Prince, J., "Thermomechanical Response of Materials and Interfaces in Electronic Packaging, Part I: Unified Constitutive Model and Calibration; Part II: Unified Constitutive Models, Validation and Design," *J. of Electronic Packaging, ASME*, 119, 1997, 294–309.
13. Basaran, C., Desai, C.S., and Kundu, T., "Thermomechanical Finite Element Analysis of Problems in Electronic Packaging Using the Disturbed State Concept, Part I: Theory and Formulation; Part II: Verification and Application," *J. of Electronic Packaging, ASME*, 120, March 1998.
14. Chia, J. and Desai, C.S., "Constitutive Modelling of Thermoviscoplastic Response of Rock Salt," *Proc. 8th Int Conf of Computer Meth. & Adv. in Geomech.*, Balkema, Rotterdam, The Netherlands, 1994, 555–560.
15. Cristescu, N.D. and Hunsche, U., *Time Effects in Rock Mechanics*, John Wiley, Chichester, U.K., 1998.
16. Samtani, N.C. and Desai, C.S., "Constitutive Modeling and Finite Element Analysis of Slow Moving Landslides Using a Hierarchical Viscoplastic Material Model," *Report to NSF*, Dept. of Civil Eng. and Eng. Mechs., University of Arizona, Tucson, AZ, 1991.
17. Hughes, T.J.R. and Taylor, R.L., "Unconditionally Stable Algorithms for Quasi-static Elasto-visco-plastic Finite Element Analysis," *Computers and Structures*, 8, 1978, 19–43.
18. Dinis, L.M.S. and Owen, D.R.J., "Elastic-viscoplastic Analysis of Plates by the Finite Element Method," *Computers and Structures*, 8, 1978, 207–215.
19. Pan, T.Y., "Thermal Cyclic Induced Plastic Deformations in Solder Joints—Part I: Accumulated Deformations in Surface Mount Joints," *J. of Electronic Packaging, ASME*, 113, 1991, 8–15.
20. Duwez, P., "On the Plasticity of Crystals," *Physical Reviews*, 47, 6, 1935, 494–501.
21. Zienkiewicz, O.C., Nayak, G.C., and Owen, D.R.J., "Composites and Overlay Models in Numerical Analysis of Elasto-Plastic Continua," *Proc. Int. Symp. on Foundations of Plasticity*, Warsaw, Poland, 1972.
22. Pande, G.N., Owen, D.R.J., and Zienkiewicz, O.C., "Overlay Models in Time-Dependent Nonlinear Material Analysis," *Computer and Structures*, 7, 1977, 435–443.
23. Hardy, H.R., Kim, R., Stefanko, R., and Wang, Y.J., "Creep and Microseismic Activity in Geologic Materials," *Proc. 11th Symposium on Rock Mechanics* (AIME), 1970, 377–414.
24. Goodman, R.E., Introduction to Rock Mechanics, John Wiley & Sons, New York, 1980, 196–201.
25. Coleman, D., "On Thermomechanics, Strain Impulses and Viscoelasticity," *Archives for Rational Mech. and Analysis*, 17, 1964, 230–254.
26. Christensen, R.M., *Theory of Viscoelasticity: An Introduction*, Academic Press, London, 1982.
27. Schapery, R.A., "A Micromechanical Model for Nonlinear Viscoelastic Behavior of Particle-Reinforced Rubber with Disturbed Damage," *Eng. Fracture Mech.*, 25, 1986, 845–867.
28. Schapery, R.A., "A Theory of Mechanical Behavior of Elastic Media with Growing Damage and Other Changes in Structure," *J. Mech. Phys. Solids*, 28, 1990, 215–253.
29. Valanis, K.C., "A Theory of Viscoplasticity Without a Yield Surface, Part I," *Archives of Mechanics*, 23, 1971, 517–533.
30. Valanis, K.C., "An Endochromic Plasticity Theory for Concrete," *Mechanics of Materials*, 5, 1986, 227–295.

31. Krempl, E., "The Overstress Dependence of Inelastic Rate of Deformation Inferred from Transient Tests," *Mat. Sc. Res. Int.*, 1, 1, 1995, 3–10.
32. Krempl, E. and Gleason, J.M., "Isotropic Viscoplasticity Theory Based on Overstress (VBO): The Influence of the Direction of the Dynamic Recovery Theorem in the Growth Law of the Equilibrium Stress," *Int. J. of Plasticity*, 12, 6, 1996, 719–735.
33. Burke, P.M., "High Temperature Creep of Polycrystalline Sodium Chloride," *Doctoral Dissertation*, Stanford Univ., Stanford, CA, USA, 1968.
34. Yang, J.M., "Thermophysical Properties: Physical Properties Data for Rock Salt," *NBS 167*, Nat. Bureau of Stand., Washington, DC, USA, 1981, 205–221.
35. Kern, H. and Franke, J.H., "The Effect of Temperature on the Chemical and Mechanical Behavior of Carnallitic-Halite Rocks," *Proc. 1st Conf. Mech. Behavior of Salt*, Hardy, H. and Langer, M. (Editors), Pennsylvania State University, State College, PA, 1981, 181–191.
36. LeComte, P., "Creep in Rock Salt," *J. Geophysical Research*, 73, 1965, 469–484.
37. Desai, C.S., "Computer Procedure DSC-SST2D and User's Manuals," Tucson, AZ, 1999.
38. Sharma, K.G., Private communication, 1998.
39. Yamada, Y. and Iwata, K., "Finite Element Analysis of Thermo-Viscoelastic Problems," Seisan-Kenkyu, UDC 539.377, 24, 4, 1972, 41–46.
40. Scarpas, A., Al-Khoury, R., Van Gurp, C.A.P.M., and Erkens, S.M.J.G., "Finite Element Simulation of Damage Development in Asphalt Concrete Pavements," *Proc. 8th Int. Conf. on Asphalt Pavements*, Univ. of Washington, Seattle, WA, USA, 1997, 673–692.

9

The DSC for Saturated and Unsaturated Materials

CONTENTS
9.1 Brief Review 340
9.2 Fully Saturated Materials 340
9.3 Equations 341
9.4 Stress Equations 343
    9.4.1 Modified Form of Terzaghi's Equation 344
9.5 Incremental DSC Equations 346
9.6 Disturbance 348
    9.6.1 Stress–Strain Behavior 349
    9.6.2 Volumetric or Void Ratio 356
    9.6.3 Effective Stress Parameter 358
    9.6.4 Residual Flow Concept 359
9.7 HISS and DSC Models 361
    9.7.1 Plasticity δ_0- and δ_1-Models 362
9.8 Softening, Degradation, and Collapse 363
9.9 Material Parameters 363
    9.9.1 Disturbance 365
9.10 Examples 365
    9.10.1 Back Predictions 369
    9.10.2 Saturated Soil: δ_1-Model 373
    9.10.3 Partially Saturated Soil: δ_1-Model 374
    9.10.4 DSC Model 376

Porous materials can involve full saturation when all the pores in their solid particle matrix are filled with a fluid (or water). When the pores are only partially filled with the fluid, air or gas exists in the pore space not occupied by the fluid. Testing and modelling of the behavior of saturated porous materials have advanced more than those for the partially saturated or unsaturated materials. However, there has been a recent surge in research for the latter due to their importance in such areas as geoenvironmental engineering, mass transport, and other engineering systems. In the following, we consider the DSC model for both.

9.1 Brief Review

The literature on the modelling and mechanics of saturated and partially saturated porous materials is wide in scope. We do not intend to provide a comprehensive review here; however, selected works are stated. A number of early works and textbooks (1–6) address these problems. Biot's (7, 8) theory of coupled mechanical response of porous saturated materials has provided a basis for a number of formulations and implementation in computer (finite-element) procedures (9–15); Chapter 13 considers the computer implementation aspects. Terzaghi's (1) one-dimensional idealization for the behavior of saturated soils can be shown to be a special case of Biot's general three-dimensional formulation.

Although the research activity for the modelling of the behavior of partially saturated soils has increased recently, its importance for various engineering problems has been recognized for a long time. A number of investigations have proposed analytical and empirical models for saturated and expansive soils (16–33). Many recent works have considered the basic mechanisms of the behavior and have proposed analytical and numerical models, including the use of the critical-state, hierarchical single-surface (HISS) plasticity, disturbed state concept (DSC), coupled hydrothermomechanical aspects, and computer implementation (34–54).

In the context of fluid flow or seepage through porous media involving both saturated and unsaturated zones, Desai and co-workers (55–61) have proposed and used the residual flow procedure (RFP). The RFP has proven (62) equivalent to the variational inequality methods (63, 64) and shown to be an alternative procedure for the fixed-domain analysis (65). Similar procedures have been used for free surface flow in (66, 67), and for computer (finite-element) analysis of unsaturated media (68, 69).

9.2 Fully Saturated Materials

Terzaghi (1) proposed the effective stress principle for defining the relation between stresses and fluid or pore water pressure in one-dimensional saturated media:

$$\sigma^a = \bar{\sigma} + p = \bar{\sigma} + u_w \tag{9.1}$$

where σ^a is the total stress, $\bar{\sigma}$ is the effective stress carried by the solid skeleton through particle contacts, and p is the pore water pressure. Here we shall use u_w to denote the pore water pressure instead of p because u_w is commonly used in the literature and because p is often used to denote the mean (effective) stress = $\frac{J_1}{3}$. The derivation of Eq. (9.1) is based on a number of assumptions, e.g., the stresses and pore water pressure are defined over a nominal area, A, and the contact area between solid particles, A^s, is negligible during deformation. Thus, $\bar{\sigma}$ and u_w do not allow for the changing (increasing) contact areas during deformation (Chapter 2).

If the DSC formulation (Chapter 2) is used, the equilibrium of forces lead to

$$\sigma^a = (1-D)p^f + D\sigma^s \tag{9.2}$$

where D is disturbance, p^f is the fluid stress, which can be different from p in Eq. (9.1), and σ^s is the stress at contacts. Equation (9.2) allows for the calculation of σ^s and p^f as functions of D, which is expressed in terms of contact area, A^s (or void ratio, e); details are given in Chapter 2.

The theory developed by Biot (7, 8) is often used for the analysis of the coupled behavior of saturated media. It is also modified to characterize unsaturated media by introducing correction or residual (RFP) terms to allow for the partial saturation. We shall give details of Biot's equations and computer implementation in Chapter 13. Constitutive models for saturated and partially saturated media, and the equations that can be used for implementation in solution procedures, are discussed in this chapter. We define the quantities based on nominal area, as in Terzaghi's theory.

9.3 Equations

A symbolic representation of an element of partially saturated material is shown in Fig. 9.1. The responses are decomposed here into those of the solid skeleton, fluid (water), and gas (air). If the material is dry, the total force, F^a, is in equilibrium with the force, $\bar{F}$, in the solid skeleton:

$$F^a = \bar{F} = \bar{F}^i + \bar{F}^c \tag{9.3}$$

If the material is fully saturated, the force equilibrium gives

$$F^a = \bar{F}(u_w) + F^w(\bar{\sigma}) \tag{9.4a}$$

and

$$\sigma^a = \bar{\sigma}(u_w) + u_w(\bar{\sigma}) \tag{9.4b}$$

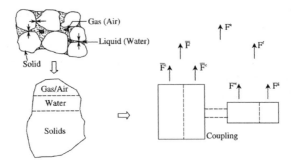

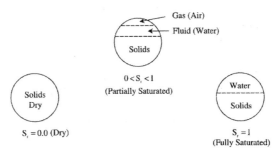

FIGURE 9.1
Dry and saturated materials.

where u_w and $\bar{\sigma}$ are the pore water pressure and effective stress in the solid skeleton, respectively. It is evident here that the responses of the solid skeleton and pore water are coupled; hence, $\bar{\sigma}$ and u_w are affected by each other.

For partially saturated materials, the equilibrium can be expressed as

$$F^a = \bar{F}(u_w, u_g) + F^w(\bar{\sigma}, u_g) + F^g(\bar{\sigma}, u_w) \qquad (9.5a)$$

or

$$F^a = \bar{F}(s) + F^w(\bar{\sigma}, s) + F^g(\bar{\sigma}, s) \qquad (9.5b)$$

$$\sigma^a = \bar{\sigma}(s) + u_w(\bar{\sigma}, s) + u_g(\bar{\sigma}, s) \qquad (9.5c)$$

where u_g is the pore gas or air pressure and $s = u_g - u_w$ is the matrix suction; if the gas happens to be air, $s = u_a - u_w$, where u_a is the pore air pressure.

Again, Eq. (9.5) indicates that the responses of the solid skeleton, pore water, and pore air are coupled.

In general, it would be necessary to define the constitutive equations for the solid skeleton and water and air phases from appropriate laboratory tests. Then, in a general solution (finite-element) procedure, coupled formulations involving displacements, pore water pressure, and pore air pressure are independent variables; in that case, all three will be evaluated during incremental procedures. However, it may become difficult to formulate such a coupled problem, particularly the constitutive equations for the three phases. Thus, approximate procedures can be developed. In a simple approximation, the response is considered to be uncoupled, in which the values of u_w and u_g are known and can be converted into equivalent "loads," as in the RFP.

In a somewhat more rigorous procedure, the stresses computed from displacements based on effective response can be used to evaluate relations that define u_w and/or u_g, which can then be used for the next step of incremental analysis for the effective response. Here the computed values of u_w and/or u_g can be converted into equivalent or residual vectors that are integrated in the coupled equations, e.g., based on Biot's theory (see Chapter 13).

9.4 Stress Equations

For the case of a partially saturated medium with air and water in the pores, the total stress tensor, σ_{ij}^a, can be expressed, based on Eq. (9.5), as

$$\sigma_{ij}^a = \bar{\sigma}_{ij} + u_a \delta_{ij} - f(u_a - u_w)\delta_{ij} \tag{9.6a}$$

$$= \bar{\sigma}_{ij} + [u_a - f(s)]\delta_{ij} \tag{9.6b}$$

where $\bar{\sigma}_{ij}$ is the effective stress tensor, and $f(s)$ is the function of air and pore water pressures. Figure 9.2 shows plots of expanding yield surfaces according to the HISS plasticity model (Chapter 7); the slope of the ultimate envelope (γ) can increase with suction. Here, the intercept on the J_1-axis can be derived from Eq. (9.6) as

$$\sigma_{ii}^a = \bar{\sigma}_{ii} + 3u_a - 3f(s) \tag{9.7a}$$

or

$$J_1^a = \bar{J}_1 + 3u_a - 3f(s) \tag{9.7b}$$

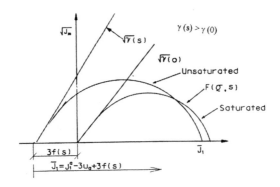

FIGURE 9.2
Expanding yield surfaces.

or

$$p^a = \bar{p} + u_a - f(s) \tag{9.7c}$$

where J_1^a and $\bar{J}_1$ are the first invariants of the total and effective stress tensors, respectively, and p^a and $\bar{p}$ are the total and effective mean pressures, respectively. The existence of suction modifies the response of the material and increases the magnitude of the material's tensile strength, which is proportional to the term $3f(s)$ and is called the *bonding stress*. If the material is fully saturated, Eq. (9.7) reduces to

$$J_1^a = \bar{J}_1 + 3u_w \tag{9.8a}$$

or

$$p^a = \bar{p} + u_w \tag{9.8b}$$

as in Terzaghi's effective stress equation.

9.4.1 Modified Form of Terzaghi's Equation

A simplified form of Eq. (9.7), as a modification of Terzaghi's one-dimensional effective stress equation for saturated materials, was proposed by Bishop (17) for partially saturated soils:

$$\sigma^a = \bar{\sigma} + u_a - \chi(u_a - u_w) \tag{9.9}$$

where χ is often called the weighting or effective stress parameter. In terms of effective stress $\bar{\sigma}$, Eq. (9.9) is written as

$$\bar{\sigma} = (\sigma^a - u_a) + \chi(u_a - u_w) \quad (9.10)$$

which, in terms of mean effective pressure, $\bar{p} = \bar{J}_1/3$, becomes

$$\bar{p} = (p^a - u_a) + \chi(s) \quad (9.11a)$$

$$\bar{p} = \bar{p}^a + \chi(s) \quad (9.11b)$$

where p^a is the total mean stress $= \bar{J}_1/3$, $\bar{p}^a$ is the total mean stress in excess of air pressure (u_a), which is often referred to as net mean stress, and s is the matrix suction or suction $= u_a - u_w$.

The parameter χ attains a value of unity for saturated materials when the relative saturation $S_r = 1$. Then Eq. (9.10) reduces to Eq. (9.1). For dry materials, $\chi = 0$ and $u_a = 0$; hence, Eq. (9.10) reduces to $\bar{\sigma} = \sigma^a$, indicating that the total or observed stress equals the effective or solid stress.

A plot of χ vs. S_r for different soils is shown in Fig. 9.3; the curves were adopted from different publications, which are detailed in (20). The validity of Eq. (9.10) has been questioned by some investigators (20, 22, 23) on the ground that the parameter χ may not be defined based only on an index property such as S_r, particularly when the material can experience plastic deformations, microcracking leading to softening (degradation) and collapse. Hence, we

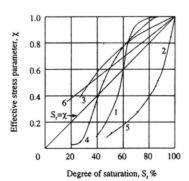

1. Compact boulder clay
2. Compacted shale
3. Silt 1
4. Silt 2
5. Silty clay
6. Theoretical

FIGURE 9.3
Relations between χ and S_r (20, which gives details of the origins of curves). Reproduced from Geotechnique, Institute of Civil Engineers, London, England.

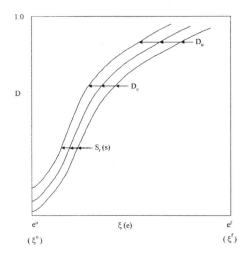

FIGURE 9.4
Disturbance vs. plastic strain trajectory (ξ) or void ratio (e).

propose the following generalization based on Eq. (9.6) as

$$\bar{\sigma} = (\sigma^a - u_a) + f(S_r, s, D)(u_a - u_w) \tag{9.12}$$

where D is the disturbance as a function of plastic strain trajectory (ξ), void ratio (e), or plastic work (w) (Chapter 3). If D is expressed as a function of S_r or s, and since the effective stress, $\bar{\sigma}$, is dependent on S_r or s, Eq. (9.12) can be written as

$$\bar{\sigma}(s) = \sigma^a(s) - u_a + D(\xi, e, w, s)(u_a - u_w) \tag{9.13}$$

Figure 9.4 shows a schematic of D vs. ξ or e, as functions of suction or saturation. Various definitions of D are given later.

9.5 Incremental DSC Equations

The microstructure (solid skeleton or matrix) experiences relative particle motions as affected by irreversible strains, air and the pore water pressures, and suction. Hence, the disturbance can be considered to occur in the solid skeleton and can be incorporated through the effective stress. Then for the saturated case, the observed effective stress, $\bar{\sigma}^a$ (in the solid skeleton), can be expressed as

$$\bar{\sigma}^a_{ij} = (1 - D)\bar{\sigma}^i_{ij} + D\bar{\sigma}^c_{ij} \tag{9.14a}$$

The DSC for Saturated and Unsaturated Materials

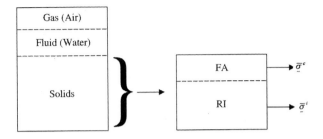

FIGURE 9.5
Mixture, and RI and FA parts in solids.

or

$$\bar{\sigma}^a = (1 - D)\bar{\sigma}^i + D\bar{\sigma}^c \tag{9.14b}$$

where $\bar{\sigma}^a$, $\bar{\sigma}^i$, and $\bar{\sigma}^c$ are the vectors of stresses in the observed, relative intact, and fully adjusted states in the solid skeleton (Fig. 9.5), respectively, and D is the disturbance in the solid skeleton, which can be defined in different ways (to be covered later). The incremental form of Eq. (9.14) can be written as

$$d\bar{\sigma}^a = (1 - D)d\bar{\sigma}^i + Dd\bar{\sigma}^c + dD(\bar{\sigma}^c - \bar{\sigma}^i) \tag{9.15a}$$

or

$$d\bar{\sigma}^a = (1 - D)\bar{C}^i d\bar{\varepsilon}^i + D\bar{C}^c d\bar{\varepsilon}^c + dD(\bar{\sigma}^c - \bar{\sigma}^i) \tag{9.15b}$$

where $\bar{C}^i$ and $\bar{C}^c$ are the constitutive matrices for the responses of the RI and FA parts of the solid skeleton, respectively; $\bar{\varepsilon}^i$ and $\bar{\varepsilon}^c$ are the strains in the RI and FA parts, respectively; and dD is the increment or rate of D.

In the case of partially saturated materials, Eq. (9.15) can be written as a function of s (or S_r) as

$$d\bar{\sigma}^a = [1 - D(s)]\bar{C}^i(s)d\bar{\varepsilon}^i + D(s)\bar{C}^c(s)d\bar{\varepsilon}^c + dD(s)(\bar{\sigma}^c - \bar{\sigma}^i) \tag{9.16}$$

Thus, the RI and FA responses, and D, are expressed as functions of suction (s) or the saturation ratio (S_r).

Equation (9.15) or (9.16) can be integrated to obtain the effective response. Then the pore water and air pressure or suction can be evaluated using the effective quantities if the total stress is known. This is usually possible in laboratory tests where both the total and effective response and suction are measured. For example, in the case of saturated materials [Eq. (9.8)], the pore water pressure can be evaluated as

$$u_w = \frac{J_1^a - \bar{J}_1}{3} = p^a - \bar{p} \qquad (9.17)$$

where J_1 ($= \sigma_1 + \sigma_2 + \sigma_3$) is the first invariant of the total stress tensor, σ_{ij}, measured in the laboratory, and $\bar{J}_1$ is the effective invariant computed from the integration of Eq. (9.15). Examples of prediction of laboratory behavior using Eq. (9.15) are given later in section 9.10.

In the case of partially saturated tests, the air pressure, u_a, can be computed from [Eq. (9.7)]

$$\bar{J}_1 = (J_1^a - 3u_a) + f(D)(u_a - u_w) \qquad (9.18)$$

where D is the disturbance (Fig. 9.4) and $s = u_a - u_w$ is the value of suction in the test.

In the case of general boundary-value problems, when coupled (e.g., Biot's) equations are solved incrementally, say, by using a finite-element procedure (Chapter 13), the values of u_w and u_a are computed as independent variables together with the displacements. Alternatively, a separate relation for u_a may be postulated; then the coupled formulation can involve only the displacements and pore water pressures, u_w.

9.6 Disturbance

Disturbance can be evaluated using laboratory test data such as stress–strain, volumetric (void ratio), effective (pore water pressure), or nondestructive properties. Then it can be expressed as

$$D(s, \bar{p}_0) = D_{u1}(s, \bar{p}_0)\left[1 - e^{-A_1(s, \bar{p}_0)\xi_D^{Z_1(s, \bar{p}_0)}}\right] \qquad (9.19a)$$

or

$$D(s, \bar{p}_0) = D_{u2}(s, \bar{p}_0)\left[1 - e^{-A_2(s, \bar{p}_0)w^{Z_1(s, \bar{p}_0)}}\right] \qquad (9.19b)$$

where s and $\bar{p}_0$ are the suction and initial effective mean pressure, respectively, for a given test, and ξ_D and w are the (deviatoric) plastic strain trajectory and plastic work, respectively. A number of possible ways to define the disturbance are described ahead. Although the disturbances from different measured quantities characterize the same phenomenon, their magnitudes and variations may not be the same. Hence, when used in the stress [Eq. (9.16)], D needs to be determined from appropriate correlations with respect to disturbances from different definitions.

9.6.1 Stress-Strain Behavior

Two of the possible ways for defining the disturbance with respect to the stress–strain response are indicated in Fig. 9.6. Figure 9.7 (a to f) shows the stress–strain volume change behavior of a partially saturated soil as reported by Cui and Delage (46) from triaxial tests under different confining pressures (σ_3) and suction (s). The test data show that under some combinations of σ_3

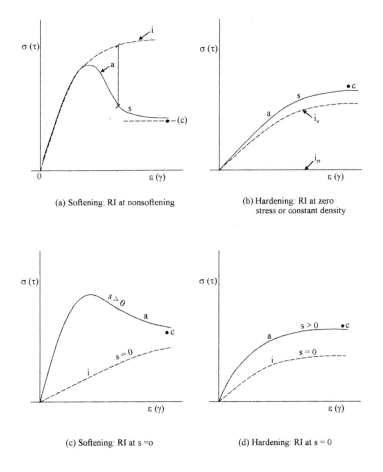

(a) Softening: RI at nonsoftening

(b) Hardening: RI at zero stress or constant density

(c) Softening: RI at s = 0

(d) Hardening: RI at s = 0

FIGURE 9.6
Definitions of disturbance.

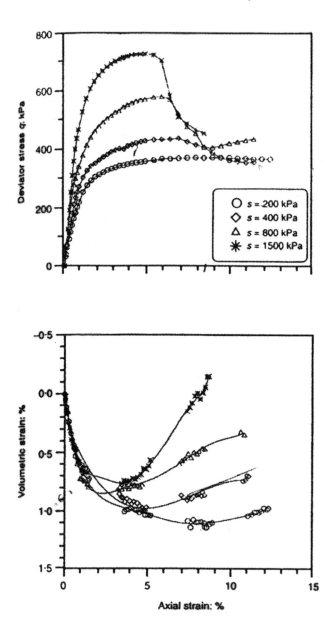

(a) $\sigma_3 = 50$ kPa

FIGURE 9.7(a)
Observed stress–strain responses with different confining pressures and suction (46).

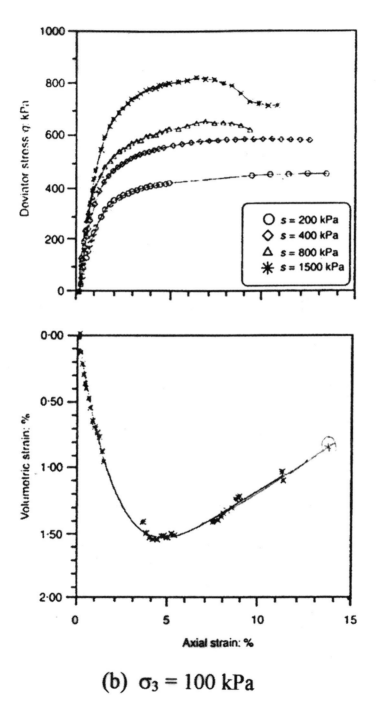

(b) $\sigma_3 = 100$ kPa

FIGURE 9.7(b)
(continued)

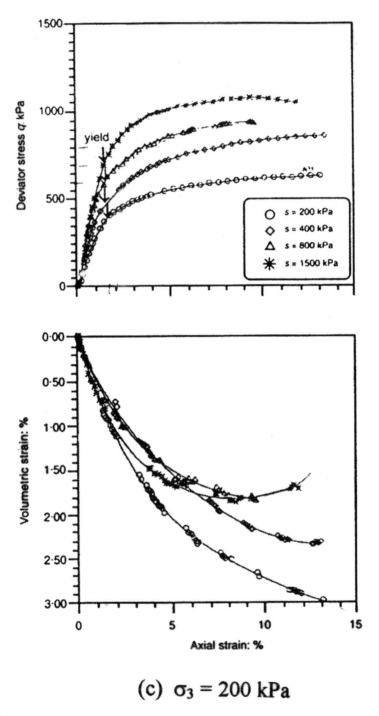

FIGURE 9.7(c) (continued)

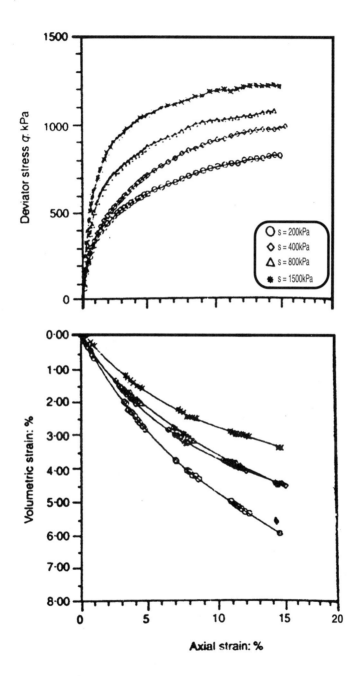

(d) $\sigma_3 = 400$ kPa

FIGURE 9.7(d)
(continued)

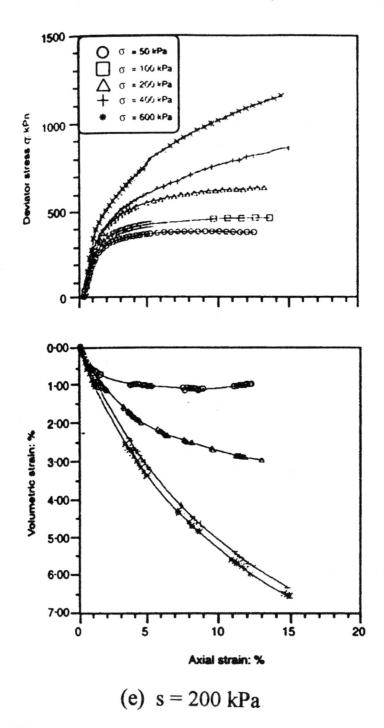

(e) s = 200 kPa

FIGURE 9.7(e)
(continued)

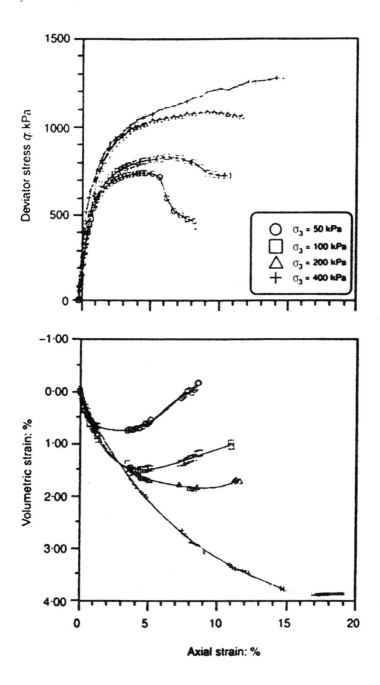

(f) s = 1500 kPa

FIGURE 9.7(f)
(continued)

and s, softening and dilative response are exhibited, while for other combinations, the response is mainly hardening and compactive.

The procedures for disturbance used for dense and loose soils (Chapter 3) can be used for partially saturated soils for test data at a given value of the suction. For the softening response, the RI behavior can be simulated as the hardening response (nonlinear elastic or elastoplastic) as the continuation of the prepeak curve. For the hardening behavior [Fig. 9.6(b)], the RI response can be characterized based on the response under constant volume or density, i_v. As a simplification, the zero stress (i_σ) can be adopted as the RI response. In both cases, the FA response (c) can be characterized based on the critical-state condition (c).

The choice of the RI and FA responses can depend on the behavior of a given material. For instance, the FA state can be defined based on the response at full saturation (i.e., zero suction), and the RI state on the response at (approximately) zero saturation (i.e., high value of suction).

Alternatively, the response under the zero suction (saturated) condition can be adopted as RI [Fig. 9.6(c) and (d)]. The FA response can be characterized as the critical-state behavior.

The disturbance function, D, can be expressed as

$$D = \frac{\sigma^i - \sigma^a}{\sigma^i - \sigma^c} \tag{9.20}$$

or

$$D = \frac{\tau^i - \tau^a}{\tau^i - \tau^c} \tag{9.21}$$

In the case of Fig. 9.6(a) and (b), the disturbance will vary from 0 to 1. Its value can increase from zero to greater than unity, and then decrease for Fig. 9.6(c), whereas from Fig. 9.6(d), it will vary from 0 to 1.

In the first alternative, softening will occur with positive D [Fig. 9.6(a)], while stiffening will occur with negative D [Fig. 9.6(b)]. Hence, it will be necessary to adopt appropriate values of quantities to identify softening and stiffening. For instance, with respect to the initial void ratio, $e > e^c$ will indicate stiffening, and for $e < e^c$, softening will occur. In the case of the second alternative, stiffening will occur with respect to the RI response at $s = 0$. Thus, disturbance with positive or negative sign will need to be used in Eq. (9.15).

9.6.2 Volumetric or Void Ratio

Disturbance can also be expressed in terms of the void ratio (e) or specific volume as

$$D_g = \frac{e^{max} - e^a}{e^{max} - e^{min}} \tag{9.22}$$

The DSC for Saturated and Unsaturated Materials

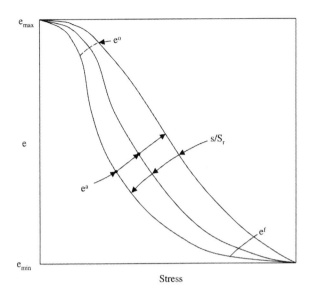

FIGURE 9.8
Schematic of void ratio (e) vs. stess.

where D_g denotes the "global" disturbance, e^{max} and e^{min} are the maximum and minimum void ratios, respectively, and e^a is the observed void ratio. Figure 9.8 shows a schematic plot of e vs. stress (e.g., mean net stress). The value of D_g varies from zero to unity. However, in practice, a porous material may have an initial void ratio $e^0 (<e^{max})$ before testing, and for a given loading, it may tend toward the final void ratio, e^f. In that case, the disturbance for the range of loading will start with an initial value of $D_0 (> 0)$ corresponding to e_0 and will tend toward the final disturbance, D_f or $D_u (< 1)$; thus, the disturbance will vary between D_0 and D_f (Fig. 9.8). If necessary and appropriate, the local disturbance, D_ℓ, for the given load can then be expressed as

$$D_\ell = \frac{e^0 - e^a}{e^0 - e^f} \tag{9.23}$$

which will vary from zero to unity for the given loading. It may be noted that the definition of D_g [Eq. (9.22)] can be expressed as

$$D_g = 1 - S_r \tag{9.24}$$

in which D_g depends on factors such as plastic strains and suction. Thus, D_g can be found from measurements of S_r during laboratory tests under different values of $\bar{p}_0$ and s. A physical interpretation of Eq. (9.24) is that degree of saturation can be considered a measure of disturbance.

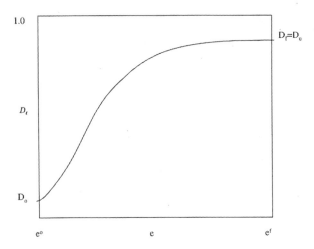

FIGURE 9.9
Disturbance vs. void ratio (e) between test range of e (e_0 to e_f).

9.6.3 Effective Stress Parameter

The foregoing definition of disturbance, which is applicable for the entire stress–strain behavior, can be considered analogous to the representation of the behavior of partially saturated material in terms of peak (failure) strength proposed by Khalili and Khabhaz (33). Here the shear strength, τ, is expressed as

$$\tau = c' + [(\sigma - u_a) + \chi(u_a - u_w)]\tan\phi' \qquad (9.25)$$

where c' is the effective cohesion, ϕ' is the effective angle of friction, and σ is the total stress. The drained shear strength of soil at full saturation, τ_0, is written as

$$\tau_0 = c' + (\sigma - u_a)\tan\phi' \qquad (9.26)$$

where, at saturation, $u_a = u_w$. The difference between τ and τ_0 can be considered as the contribution to the strength due to suction; therefore,

$$\tau - \tau_0 = \chi(u_a - u_w)\tan\phi' \qquad (9.27)$$

and

$$\chi = \frac{\tau - \tau_0}{(u_a - u_w)\tan\phi'} \qquad (9.28)$$

Figure 9.10 shows plots of shear strength vs. matric suction, s, reported by Gan et al. (70) at different initial void ratios. The disturbance at (peak)

The DSC for Saturated and Unsaturated Materials

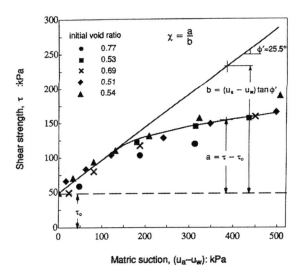

FIGURE 9.10
Shear strength vs. matric suction (70).

strength, D_p, can now be expressed using Eq. (9.27) as

$$D_p = \frac{(u_a - u_w)\tan\phi' - (\tau - \tau_0)}{(u_a - u_w)\tan\phi'} \quad (9.29a)$$

$$= 1 - \chi \quad (9.29b)$$

Figure 9.3 shows the variation of S_r with χ for different soils as reported in (20). Thus, for full saturation, $S_r = 1$ and $\chi = 1$; hence, $D_p = 0$. For dry condition, $S_r = 0$ and $\chi = 0$; therefore, $D_p = 1$. Thus, D_p, as plotted schematically in Fig. 9.11, shows that it varies from 0 to 1 as saturation varies from 1 to 0, i.e., from a fully saturated to a dry condition.

In Eq. (9.29), the parameter χ represents the change (increase) in strength from that in the fully saturated (RI) state to that in the partially saturated state. Thus, the parameter χ can be treated as a measure of the disturbance. For the entire stress–strain response, it will depend on factors such as plastic strain or work, initial pressure, and suction.

9.6.4 Residual Flow Concept

The idea of disturbance can be derived from the residual flow procedure (RFP) proposed by Desai and co-workers (55–61) for free surface seepage in two- and three-dimensional problems. Its application for uncoupled seepage and stress analysis is given by Li and Desai (58). In the RFP, a residual or correction fluid "load" is evaluated based on the difference between permeabilities or

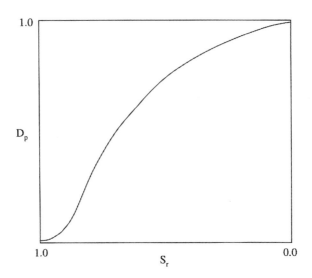

FIGURE 9.11
Plot of D_p vs. S_r.

conductivities or the degree of saturation, in the saturated and unsaturated conditions. The RFP concept can also be used to modify the coupled equations for saturated media, say, in Biot's theory, so as to allow for partial saturation.

Figure 9.12(a) shows a porous (soil) medium in which the material is saturated below the free surface (FS) and partially saturated above the FS. The fluid pressure (u_w) is positive in the saturated zone and negative (suction) in the partially saturated zone. Figure 9.12(b) and (c) show variations of the coefficient of permeability (k) and saturation with pressure, respectively. In the fully saturated state, the permeability is k_s, while $S_r = 1.0$. In the partially saturated regions, the permeability is $k_{us} < k_s$ and saturation $S_r < 1$. As the suction increases, the permeability and saturation decrease and approach asymptotically those in the dry state.

The disturbance, D, can be expressed as

$$D_k = \frac{k_s - k_{us}}{k_s - k_f} \tag{9.30}$$

and

$$D_s = \frac{1 - S_r}{1 - S_{rf}} \tag{9.31}$$

Initially, in the saturated or RI state, $k_{us} = k_s$ and $S_r = 1$; finally, $k_{us} = k_f$ and $S_r = S_{rf}$. Hence, D varies from 0 to 1. Thus, if the values of permeabilities or

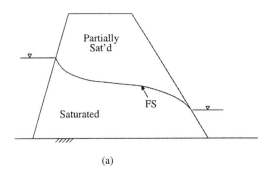

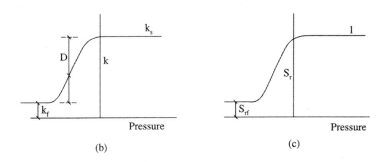

FIGURE 9.12
Residual flow procedure.

saturation are measured during deformation of a specimen, the values of D can be evaluated.

It is possible that the values and variations of D, according to the foregoing definitions, are not the same as those required in the constitutive equations [Eq. (9.16)]. Hence, it will be necessary to establish appropriate correlations between disturbances from different methods in order to define the D required in the DSC equations based on stresses [Eq. (9.16)].

9.7 HISS and DSC Models

The DSC and its hierarchical versions, e.g., HISS δ_0- and δ_1-plasticity models (Chapter 7), can be used to develop constitutive equations for saturated and partially saturated materials. A number of approximate and general formulations are possible. We describe some of them here for partially saturated materials. Models for saturated materials can be derived as special cases of those for the partially saturated case.

9.7.1 Plasticity δ_0- and δ_1-Models

If the effective response of the porous material (e.g., soil) is essentially the plastic yielding or hardening type [Fig. 9.7 (c, d, e)], the δ_0- and δ_1-versions in the HISS plasticity approach can be used by expressing the parameters as functions of suction (s) or degree of saturation, S_r.

The yield function, F, for the δ_0-model (Chapter 7) can now be written as

$$F = J_{2D} - [-\alpha(s)J_1^{n(s)} + \gamma(s)(J_1)^2][1 - \beta(s)S_r]^{-0.5} = 0 \qquad (9.32)$$

where the parameters are expressed in terms of suction, s. Here the effective stress quantities are nondimensionalized with respect to p_a; however, the overbar [as in Eq. (7.1)] is not shown, in order to avoid confusion with effective quantities.

If the behavior exhibits nonassociative properties, $\alpha(s)$ in Eq. (9.32) can be expressed as

$$\alpha_Q(s) = \alpha + \kappa(s)(\alpha_0 - \alpha)(1 - r_v) \qquad (9.33)$$

where κ is the nonassociative parameter, α_0 is the value of α at the end of initial (hydrostatic) loading, and r_v is the ratio of the volumetric plastic strains to the total plastic strain trajectory [Eq. (7.5), Chapter 7].

Figure 9.13 shows a schematic of F [Eq. (9.32)] in which the intercept $3R$ denotes the bonding (tensile) strength of the material given by

$$J_1^a = J_1 + 3R(s) \qquad (9.34)$$

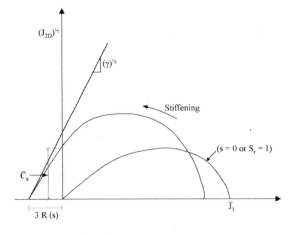

FIGURE 9.13
Schematic of yield surfaces and bonding strength.

The bonding strength will increase with suction, and the yield surface will move to the left as suction increases, starting with zero suction ($S_r = 1$). The parameters' dependence on suction needs to be defined on the basis of laboratory tests; an example is given subsequently.

The incremental effective stress equations [Eq. (9.16)] can be used to represent the behavior of partially saturated materials by using the elasticity and plasticity parameters.

A number of investigators have proposed and used elastic, thermoelastic, and elastoplastic models for partially saturated soils (37–41, 43–46, 48–53). The elastoplastic models based on the critical-state concept have been used commonly (41, 43, 45, 46, 50–52). The HISS and DSC models have been used recently (47, 54). The HISS model has been shown to include the critical-state and other plasticity (hardening) models as special cases (see also Chapter 7).

9.8 Softening, Degradation, and Collapse

Under certain combinations of suction (s) and initial (cell) pressure, $\bar{p}_0$, and density (ρ_0), a material may exhibit softening or degradation behavior [Fig. 9.7(a, b, f)]. As discussed in Chapter 10, the microstructural adjustment may entail threshold transition states, some of which lead to instability or collapse of the microstructure. The latter is identified by the critical value of disturbance, D_c, indicated in Fig. 9.4. Thus, the idea of critical disturbance can provide a means for the evaluation of instability or collapse in dry, saturated, or partially saturated materials; this is discussed in Chapter 13 and Appendix I.

The incremental constitutive equations [Eq. (9.16)] can be used to allow for the disturbance, which provides for the simulation of softening and collapse responses. Note that the general Eq. (9.16) includes plastic hardening response as a special case, when $D = 0$.

9.9 Material Parameters

The procedures for finding the parameters in the DSC/HISS models from laboratory tests are presented in Chapters 3 and 7. Here we briefly describe their evaluation for partially saturated soils.

Elastic. The modulus such as E and ν or G and K are found as average slopes of unloading curves and can be expressed as functions of suction, as necessary.

Ultimate Parameters. The parameters associated with the ultimate asymptotic stress–strain response are γ, β, and R. They are determined from test data for given saturation (S_r) or suction (s) and then expressed as

$$\gamma = \gamma(s) \qquad (9.35a)$$

$$\beta = \beta(s) \qquad (9.35b)$$

$$R = R(s) \qquad (9.35c)$$

Here γ represents the slope ($\sqrt{\gamma}$) of the ultimate envelope in $\sqrt{J_{2D}}$–J_1 space (Fig. 9.13), and β is related to the shape of F in σ_1–σ_2–σ_3 space (Chapter 7). The bonding strength R can be found approximately from

$$3R(s) = \frac{\bar{c}_a(s)}{\sqrt{\gamma(s)}} \qquad (9.35d)$$

where $\bar{c}_a$ is the (cohesion) intercept along the $\sqrt{J_{2D}}$-axis, when $J_1 = 0$ (Fig. 9.13) and is based on the assumption that the ultimate envelope is linear. If it is nonlinear, the slope to the initial part of the envelope can be used to define $\sqrt{\gamma}$.

Hardening Parameter. The hardening or growth function, α, in Eq. 9.32 can be expressed as

$$\alpha(s) = \frac{a_1(s)}{\xi^{\eta_1(s)}} \qquad (9.36)$$

where a_1 and η_1 are the hardening parameters dependent on suction, and ξ is the trajectory of plastic strains.

Phase Change Parameter. This parameter is associated with the state of volume change when transition from compaction to dilation occurs and the change in volumetric strains, $d\varepsilon_v$, vanishes (Fig. 9.7). It can be expressed as a function of suction as

$$n = n(s) \qquad (9.37)$$

For the FA response, the parameters can be defined on the basis of the constrained liquid or constrained liquid–solid assumption (Chapter 7). For the former, only the bulk or hydrostatic response is relevant, and if the behavior is represented by using the bulk modulus, K, it can be expressed as a function of suction as

$$K = K(s) \qquad (9.38)$$

The DSC for Saturated and Unsaturated Materials

For the liquid–solid assumption, the critical-state concept can be used. Here the critical responses can be expressed using the following equations (Chapter 4):

$$\sqrt{J^c_{2D}} = \overline{m}(s) J^c_1 \tag{9.39a}$$

and

$$e^c = e^c_0 - \lambda(s)\ln\left(\frac{J^c_1}{3p_a}\right) \tag{9.39b}$$

where the slopes $\overline{m}$ and λ are both expressed as functions of suction.

9.9.1 Disturbance

For a set of tests under given suction, s, Eq. (9.19a) can be written as

$$\ln A + Z \ln(\xi_D) = \ln\left[-\ln\left(\frac{D_u - D}{D_u}\right)\right] = D^* \tag{9.40}$$

Plots of $\ln \xi_D$ and D^* lead to the values of A and Z (Chapter 3). D_u is obtained based on ultimate or residual values (of stress) (Fig. 9.6); its value can also often be adopted as 1 or close to 1 (≈ 0.95). Then A, Z, and D_u can be expressed as

$$A = A(s) \tag{9.41a}$$

$$Z = Z(s) \tag{9.41b}$$

$$D_u = D_u(s) \tag{9.41c}$$

9.10 Examples

We now consider examples of application of the HISS and DSC models for the behavior of saturated and partially saturated materials. Two examples of the use of the HISS plasticity models for saturated clay and sand are presented as Examples 7.7 and 7.8 in Chapter 7. Additional examples are given here.

TABLE 9.1

Material Parameters for the DSC Model:
Ottawa Sand (72, 73)

	Parameter	Value
Elastic	E	193,000 kPa
	ν	0.380
Ultimate		
	γ	0.123
	β	0.00
Phase Change	n	2.45
Hardening	h_1	0.845
	h_2	0.0215
FA (Critical state)	$\bar{m}$	0.150
	λ	0.020
	e_0^c	0.601
Disturbance	A	4.22
	Z	0.43
	D_u	0.99

Example 9.1 Saturated Sand

A series of laboratory undrained cyclic (frequency = 0.10 Hz) to multiaxial tests were performed on saturated Ottawa sand with cubic specimens (10 × 10 × 10 cm) under different initial confining pressures $\bar{\sigma}_0 = \sigma_3 = 69, 138,$ and 207 kPa and at relative density $D_r = 60\%$. Figures 9.14 to 9.16 show test data in terms of applied cyclic stress, $\sigma_1 - \sigma_3 = \sigma_d$ vs. time, axial strain (ε_1) vs. time, pore water pressure u_w vs. time, and shear stress (σ_d) vs. axial strain (ε_1) for the three values of $\bar{\sigma}_0$, respectively (71–73). These data were used to find the parameters. The (average) values of parameters obtained using the procedures in Chapters 3 and 7 are shown in Table 9.1; brief details are given below (74).

Elastic Moduli. Figure 9.17 shows typical data in terms of τ_{oct} vs. ε_1 and $\varepsilon_2 = \varepsilon_3$ from quasi-static or one-way (loading, unloading, reloading) CTC ($\sigma_1 > \sigma_2 = \sigma_3$) tests, for $\bar{\sigma}_0 = 207$ kPa. The values of E and ν were found from

$$E = \frac{3}{\sqrt{2}} E_1 \tag{1a}$$

$$\nu = \frac{2E_1}{E_2 + E_3} \tag{1b}$$

where E_i ($i = 1, 2, 3$) are the slopes shown in Fig. 9.17.

Ultimate. Figure 9.18 shows a schematic in terms of J_1 vs. $\sqrt{J_{2D}}$ in which the ultimate and critical-state or phase change envelopes are shown. The ultimate parameters, γ and β, were found using the procedure given in Chapter 7.

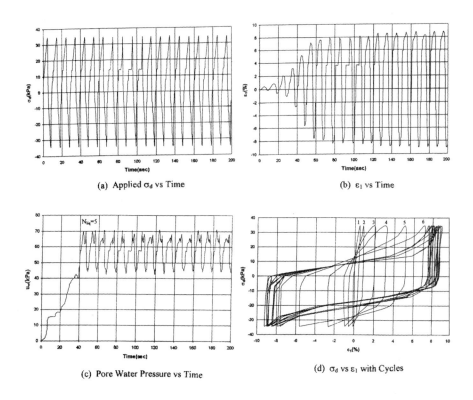

FIGURE 9.14
Observed behavior of sand: $\bar{\sigma}_0 = 69$ kPa (71, 72).

Phase Change. Figure 9.19 shows a plot of J_1 vs. $\sqrt{J_{2D}}$ from the three tests. The value of n was found using the following equation (see Chapter 7):

$$\frac{J_{1a}}{J_{1m}} = \left(\frac{2}{n}\right)^{\frac{1}{n-2}} \tag{1c}$$

where J_{1a} and J_{1m} are shown in Fig. 9.19. An average value of $n = 2.45$ was found.

Hardening. The following equation, expressed in terms of the deviatoric plastic strain trajectory, ξ_D, was used to find the parameters h_1 and h_2:

$$\alpha = \frac{h_1}{\xi_D^{h_2}} \tag{1d}$$

which is the special case of Eq. (9.36). Figure 9.20 shows plots of $\ln \xi_D$ vs. $\ln \alpha$ used to find the average values of h_1 and h_2 (Table 9.1).

Fully Adjusted State. Figure 9.21 shows plots of J_1^c vs. $\sqrt{J_{2D}^c}$ [Eq. (9.39a)] and e^c vs. $\ln(J_1^c/3p_a)$ [Eq. (9.39b)], used to find the values of $\overline{m}$ and λ.

Disturbance. Figure 9.22 shows plots of effective stress $\bar{\sigma}$ vs. the number of cycles (N) for $\bar{\sigma}_0 = 69, 138, 207$ kPa, respectively; here the observed pore water pressures were used to evaluate the effective stress at peak values of

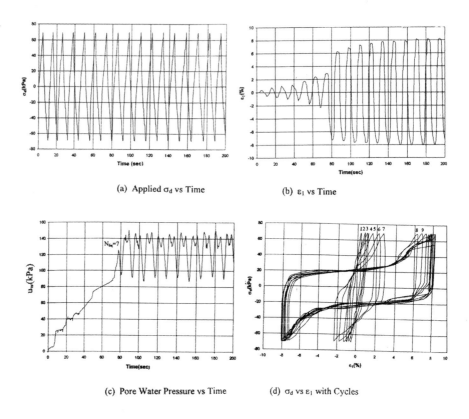

FIGURE 9.15
Observed behavior of sand: $\bar{\sigma} = 138$ kPa (71, 72).

cyclic stress (Figs. 9.14 to 9.16). Note that the ultimate asymptotic value of the effective stress is zero; hence, $D_u \approx 0.99$ was used. The disturbance was found using the following equation:

$$D = \frac{\bar{\sigma}^i - \bar{\sigma}^a}{\bar{\sigma}^i - \bar{\sigma}^c} \tag{1e}$$

where $\bar{\sigma}^i$ is the RI effective stress computed using the δ_0-model to simulate the first cycle response without cyclic degradation (73). The disturbance function was expressed as

$$D = D_u(1 - e^{-A\xi_D^Z}) \tag{1f}$$

Equations (1e) and (1f) were used to obtain plots of $\ln(\xi_D)$ vs. $\ln[-\ln(1 - \frac{D}{0.99})]$ as shown in Fig. 9.23, which provided the values of A and Z. Plots of disturbance vs. $\xi_D(N)$ are shown in Fig. 9.24. Here, the test data values refer to Eq. (1e), while the function values refer to Eq. (1f) in which the parameters A, Z, and D_u (Table 9.1) were used.

The DSC for Saturated and Unsaturated Materials 369

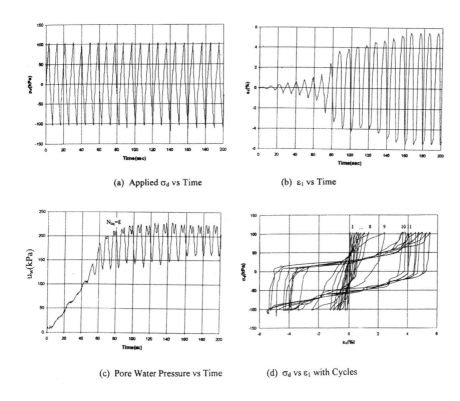

FIGURE 9.16
Observed behavior of sand: $\bar{\sigma}$ = 207 kPa (71, 72).

9.10.1 Back Predictions

The dynamic finite-element procedure (74–76 and Chapter 13) was used to predict the behavior of the test specimens. Figure 9.25 shows the finite-element mesh and loading with frequency = 0.1 Hz. The latter involved applying the confining stress incrementally and then applying the cyclic, σ_d, stress incrementally with different amplitudes (Figs. 9.14 to 9.16), as it was done in the laboratory tests.

Figures 9.26 to 9.28 show comparisons between predicted and observed effective stress ($\bar{\sigma}$) and excess pore water pressure vs. time (or number of cycles) for $\bar{\sigma}_0$ = 69, 138, and 207 kPa, respectively. The pore water pressures were computed using Eq. (9.17).

In the above-mentioned back predictions, the parameters used were those for the tests used to find them, the relative density of which was D_r = 60%. In order to validate the model further, the average parameters (Table 9.1) were used to predict an *independent* test, and not used to find the parameters. This test involved $\bar{\sigma}_0$ = 69 kPa and relative density D_r = 40%. Figure 9.29 shows comparisons between predictions and test data in terms of the effective stress

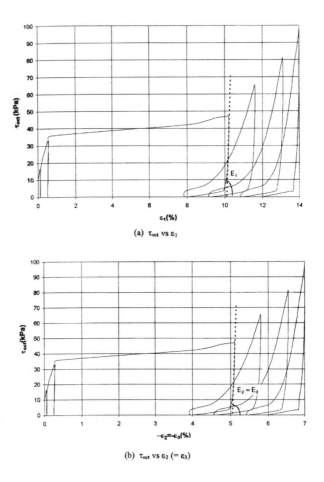

FIGURE 9.17
τ_{oct} vs. ε_i ($i = 1, 2, 3$) curves for CTC test: $\bar{\sigma}_0 = 207$ kPa (71, 72).

($\bar{\sigma}$) and excess pore water pressure (u_w). The model also provides good predictions for the independent tests (74).

Liquefaction. The DSC provides a basis for the evaluation of instability and liquefaction. Analysis and prediction of liquefaction for the Ottawa sand behavior are given in Chapter 12.

Example 9.2 Saturated Clay

The DSC model was used to predict the laboratory undrained behavior of a saturated marine clay (77, 78). These predictions for typical tests using the DSC model appear in Chapter 7. Here we present predictions of typical (triaxial) tests for the clay by using a finite-element procedure in which the DSC model was implemented (75). The DSC model is used for the virgin stress–strain response (hardening, peak, and softening), while separate and

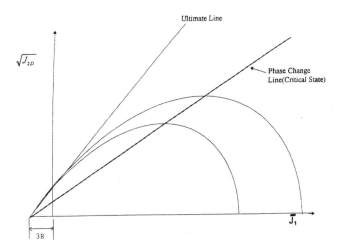

FIGURE 9.18
Ultimate and phase change envelopes (72).

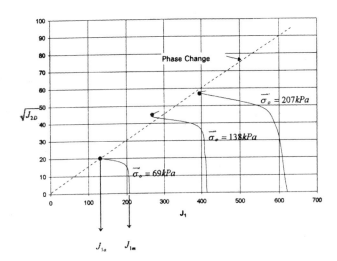

FIGURE 9.19
Phase change envelope in $\sqrt{J_{2D}} - J_1$ space: determination of n (72).

simplified procedures are used to simulate unloading and reloading; details of these procedures are given in Chapter 13.

The material parameters for the clay are given in Table 9.2. The procedures for finding the DSC parameters are given in Chapters 3 and 7. Figure 9.30 indicates the procedure for finding the unloading and reloading parameters included in Table 9.2 (see Chapter 13).

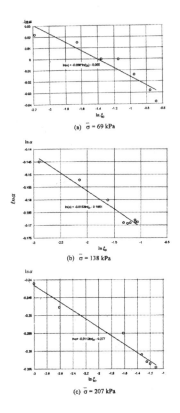

FIGURE 9.20
Plots for determining of hardening parameters (72).

The DSC model with the unloading–reloading procedures was used to predict the conventional triaxial compression (CTC) ($\sigma_1 > \sigma_2 = \sigma_3$) behavior with initial confining pressure $\bar{\sigma}_0 = 276$ kPa (40 psi) and initial void ratio $e_0 = 0.8614$. The FE analyses were obtained by using 1-, 4-, and 16-element meshes for triaxial specimens, with boundary conditions similar to those in Fig. 9.25. The confining stress ($\bar{\sigma}$) was first applied in increments, and then the stress difference ($\sigma_1 - \sigma_3$) was applied incrementally. The three meshes, as expected, provided essentially the same results (76).

Figures 9.31(a) and (b) show comparisons between predicted and observed stress ($\sigma_1 - \sigma_3$) vs. strain (ε_1) and pore water pressure [Eq. (9.17)] vs. ε_1 responses for the above test. The correlations, including the unloading–reloading responses, are considered satisfactory.

Example 9.3 Saturated and Partially Saturated Sandy Silt

A series of laboratory triaxial tests were performed on saturated and partially saturated silt under different initial confinements ($p_0 = \sigma_3 = 200, 400,$ and 600 kPa) and different values of matrix suction ($u_a - u_w$) (54, 79, 80). The HISS δ_1-plasticity and DSC models (47) were used to predict the behavior

The DSC for Saturated and Unsaturated Materials

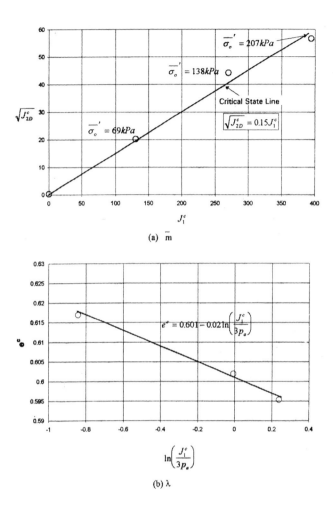

FIGURE 9.21
Determination of parameters $\bar{m}$ and λ (72).

of saturated and unsaturated specimens. The model parameters were found from the triaxial tests and are shown in Table 9.3.

9.10.2 Saturated Soil: δ_1-Model

Figure 9.32 shows comparisons between predictions from the δ_1-model and test data for two typical values of $\bar{\sigma}_3 = 200$, 400, and 600 kPa. The stress–strain responses in terms of τ_{oct} vs. ε_1, $\varepsilon_2 = \varepsilon_3$ correlate well, except in the strain-softening zone. The volumetric response compares well in the early regions; however, the correlation is not good in the later regions. For improved correlation, it would be appropriate to use the DSC model, as discussed ahead.

TABLE 9.2
Parameters for DSC Model: Marine Clay (76–78)

	Parameter	Value
Elastic	E	10,350 kPa (1500 psi)
	ν	0.35
Ultimate	γ	0.047
	β	0.0
Phase change	n	2.8
Bonding stress (normally consolidated clay)	R	0.0
Hardening	h_1 Eq. (7.11b),	0.0001
	h_2 Chapter 7	0.780
	h_3	0.0
	h_4	0.0
FA (critical state)	$\overline{m}$	0.0694
	λ	0.1692
	e_0^c	0.9033
Disturbance	A	1.73
	Z	0.309
	D_u	0.75
Unloading/reloading (see Chapter 13)	E^b	34,500 kPa
	E^e	3,450 kPa
	ε_1^p	0.005

TABLE 9.3
Parameters for HISS δ_1- Model for Saturated Sandy Silt (54, 79, 80)

Parameter	Symbol	Value
Young's modulus	E	145 MPa
Poisson's ratio	ν	0.40
Ultimate	γ	0.08
	β	0.58
Hardening	a_1	0.005
	η_1	0.36
Phase change	n	2.45
Bonding stress	R	0.0
Nonassociative	κ	0.50

9.10.3 Partially Saturated Soil: δ_1-Model

In the tests for the unsaturated soil, the parameters γ, R, a_1, and η_1 were expressed as functions of the suction (s) (54). Figures 9.33 (a)–(d) show variations of these parameters for $s = 0$, 50, 100, and 200 kPa. They show that the slope γ of the ultimate envelope, the bonding stress R, and the phase change parameter n increase with suction, while the hardening parameter a_1 decreases

The DSC for Saturated and Unsaturated Materials

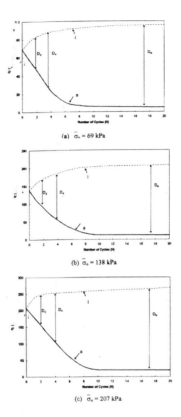

FIGURE 9.22
Effective stress vs. cycles (N): determination of disturbance (72).

with suction. The other parameters (Table 9.3) were assumed to be constant. The δ_1-model, with the variations of parameters in Fig. 9.33 expressed as linear or nonlinear functions of suction, was used to predict the behavior of drained triaxial compression tests.

The following three tests involved the same initial net mean pressure, $p^a - u_a = 400$ kPa (54):

C1—saturated soil: $\bar{\sigma}_3 = 400$ kPa, $s = 0$ kPa
C2—small suction: $\bar{\sigma}_3 = 450$ kPa, $s = 50$ kPa
C3—high suction: $\bar{\sigma}_3 = 600$ kPa, $s = 200$ kPa

Two additional tests involved average suction:

C4—average suction: $\sigma_3 = 600$ kPa, $s = 100$ kPa
C5—average suction: $\sigma_3 = 400$ kPa, $s = 100$ kPa

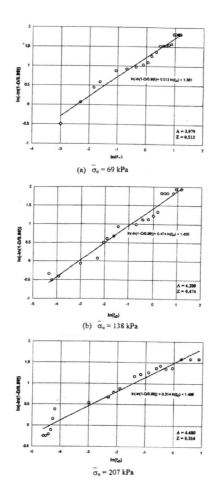

FIGURE 9.23
Determination of disturbance parameters A and Z (72).

Figure 9.34 shows comparisons between predictions and experimental behavior using the δ_1-model for tests C1, C2, and C3, where the foregoing parameters were used as functions of suction. The correlations, particularly for the stress–strain behavior, are very good for the saturated (C1) and small saturation (C2) cases when the behavior is essentially of the plastic yielding type. At higher suction (C3), the test data show softening behavior, and the δ_1-model provides good correlation up to about the peak stress. The correlations for volumetric responses for the three cases are not as good as the stress–strain responses, but are considered satisfactory.

9.10.4 DSC Model

In order to simulate the softening response, it is appropriate to use the DSC model. The disturbance, D, was evaluated based on the test data involving

The DSC for Saturated and Unsaturated Materials

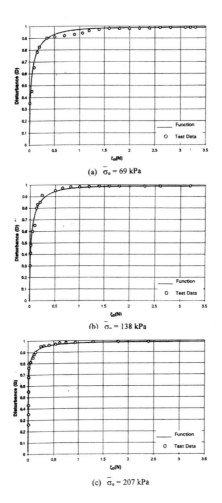

FIGURE 9.24
Plots of disturbance (D) vs. deviatoric plastic strain trajectory (ξ_D) (72).

softening, by using the following equation:

$$D = \frac{\sqrt{J_{2D}^i} - \sqrt{J_{2D}^a}}{\sqrt{J_{2D}^i} - \sqrt{J_{2D}^c}} \tag{3a}$$

Here the results for the shear stress measure $\sqrt{J_{2D}}$ vs. ε_1 are used, where i, a, and c denote RI, observed, and FA states, respectively. Now D is expressed in terms of the trajectory of the deviatoric plastic strain, ξ_D (Chapter 3):

$$D = D_u(1 - e^{-A\xi_D^Z}) \tag{3b}$$

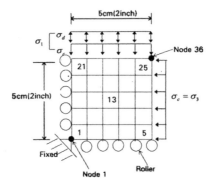

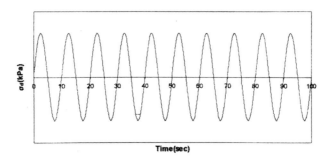

(b) Applied Stress: $\sigma_d = \sigma_1 - \sigma_3$

FIGURE 9.25
Finite-element mesh and applied stress for test specimen (72, 74).

The use of Eqs. (3a) and (3b) provides the values of D_u, A, and Z. Plots of $D^* = \ln[-\ln(1 - D_u)]$ vs. $\ln \xi_D$ [Eq. (3b)], based on tests C3 and C4, are shown in Fig. 9.35 (54); they provide the values of A and Z.

Tests C3 ($p^a - u_a = 500$ kPa and $s = 200$ kPa) and C4 ($p^a - u_a = 500$ kPa and $s = 200$ kPa) exhibited softening response. The parameter $D_u = 0.85$ was assumed to be independent of suction. The parameters for the two tests are as follows (54):

Test	A	Z	D_u
C3: $s = 200$ kPa	4.5×10^{-5}	3.90	0.85
C4: $s = 100$ kPa	1.4×10^{-10}	8.10	0.85

Figure 9.36 shows the comparison between the observed disturbance [Eq. (3a)] and the computed disturbance [Eq. (3b)] for the two tests and indicates very good correlations.

The DSC for Saturated and Unsaturated Materials

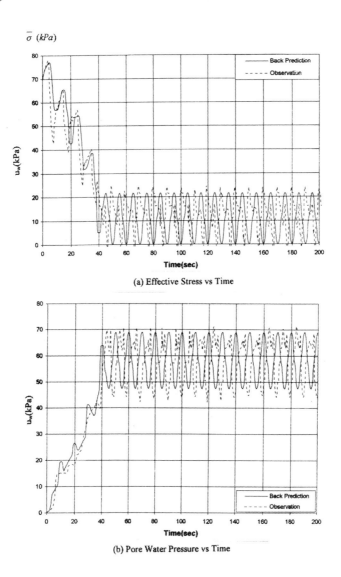

FIGURE 9.26
Observed and predicted behavior: $D_r = 60\%$; $\bar{\sigma}_0 = 69$ kPa (74) (with permission).

Figure 9.37 shows comparisons between prediction by the (RI) δ_1 and DSC models and the data for the two tests C3 and C4. It can be seen that the DSC model simulates the softening response well and overall provides improved correlations compared to those by the δ_1-plastic hardening model.

The foregoing predictions of tests involved the use of the parameters obtained by using given tests. An independent test (not used to find those parameters) was also predicted. Figure 9.38 shows the predictions and test data for the drained triaxial compression test (C5) with $\sigma_3 = 300$ kPa and $s = 100$ kPa. The DSC model provides a good simulation of the independent test.

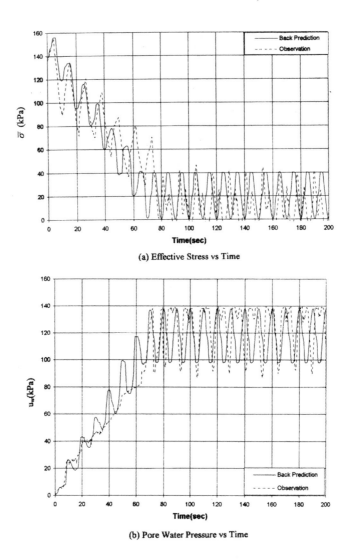

FIGURE 9.27
Observed and predicted behavior: $D_r = 60\%$; $\bar{\sigma}_0 = 138$ kPa (74) (with permission).

References

1. Terzaghi, K., "The Shear Resistance and Saturated Soils," *Proc. 1st Int. Conf. Soil Mech. and Found. Eng.*, 1, 1936, 54–56.
2. Terzaghi, K., *Theoretical Soil Mechanics*, John Wiley, New York, 1943.
3. DeWiest, R. J. M. (Editor), *Flow Through Porous Media*, Academic Press, New York, 1969.

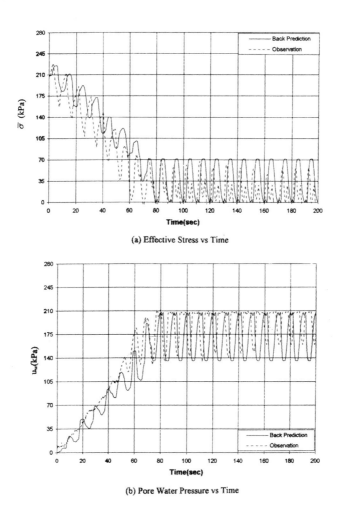

FIGURE 9.28
Observed and predicted behavior: D_r = 60%; $\bar{\sigma}_0$ = 207 kPa (74) (with permission).

4. Bear, J., *Dynamics of Fluids in Porous Media*, American Elsevier, New York, 1972.
5. Lewis, R. W. and Schrefler, B. A., *The Finite Element Method in the Static and Dynamic Deformation and Consolidation of Porous Media*, John Wiley and Sons, New York, 1998.
6. Coussy, O., *Mechanics of Porous Continua*, John Wiley, New York, 1995.
7. Biot, M. A., "General Theory of Three Dimensional Consolidation," *J. of Appl. Physics*, 12, 1941, 155–164.
8. Biot, M. A., "Theory of Elasticity and Consolidation for a Porous Anisotropic Solid," *J. of Appl. Physics*, 26, 1955, 182–185.
9. Sandhu, R. S. and Wilson, E. L., "Finite Element Analysis of Flow in Saturated Elastic Media," *J. of Eng. Mech.*, ASCE, 95, 1969, 641–652.
10. Ghaboussi, J. and Wilson, E. L., "Flow of Compressible Fluid in Porous Elastic Media," *Int. J. Num. Meth. in Eng.*, 5, 1973, 419–442.

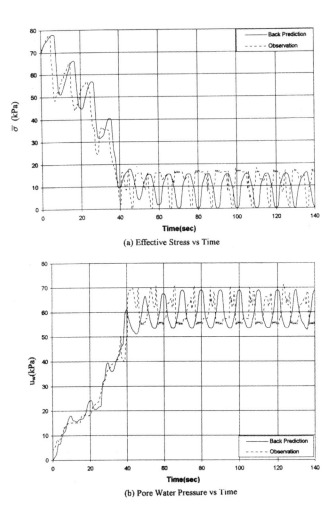

FIGURE 9.29
Observed and predicted behavior for independent test: $D_r = 40\%$; $\bar{\sigma}_0 = 69$ kPa (72, 74).

11. Desai, C. S. and Saxena, S. K., "Consolidation of Layered Anisotropic Foundations," *Int. J. Num. Analyt. Methods in Geomech.*, 1, 1977, 5–23.
12. Booker, J. R. and Savvidou, C., "Consolidation Around a Point Heat Source," *Int. J. Numer. Analyt. Methods Geomech.*, 9, 1985, 173–184.
13. Desai, C. S. and Galagoda, H. M., "Earthquake Analysis with Generalized Plasticity Models for Saturated Soils," *Int. J. Earthquake Eng. and Struct. Dyn.*, 18, 6, 1989, 903–919.
14. Desai, C. S., Wathugala, G. W., and Matlock, H., "Constitutive Model for Cyclic Behavior of Cohesive Soils II: Application," *J. of Geotech. Eng.*, ASCE, 119, 4, 1993, 730–748.
15. Desai, C. S., Shao, C., and Park, I. J., "Disturbed State Modeling of Cyclic Behavior of Soils and Interfaces in Dynamics Soil-structure Interaction," *Proc. 5th Int. Conf. on Computer Methods and Advances in Geomech.*, Wuhan, China; Balkema, Rotterdam, The Netherlands, 1997.

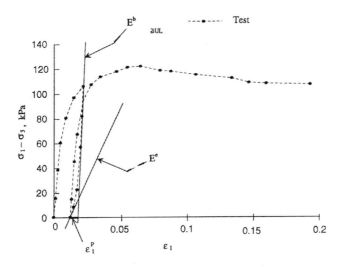

FIGURE 9.30
Determination of unloading and reloading parameters (76).

16. Aitchison, G. D. and Donald, I. B., "Some Preliminary Studies of Unsaturated Solids," *Proc. Second Australia-New Zealand Conf. on Soil Mech. and Found. Eng.*, 1956, 192–199.
17. Bishop, A. W., "The Principle of Effective Stress," *Tecknish Ukebad* 106, 1959, 859–863.
18. Aitchison, G. D., "Relationships of Moisture Stress and Effective Stress Function in Unsaturated Soils," *Proc. Conf. On Pore Pressure and Suction in Soils,* Butterworth, London, 1960, 47–52.
19. Bishop, A. W. and Donald, I. B., "The Experimental Study of Partly Saturated Soil in the Triaxial Apparatus," *Proc. 5th Int. Conf. On Soils Mech.*, Paris, 1, 1961, 13–21.
20. Jennings, J. E. B. and Burland, J. B., "Limitations of the Use of Effective Stresses in Partially Saturated Soils," *Geotechnique*, 12, 2, 1962, 125–144.
21. Bishop, A. W. and Blight, G. E., "Some Aspects of Effective Stress in Saturated and Partly Saturated Soils," *Geotechnique*, 13, 3, 1963, 177–197.
22. Aitchison, G. D., "Soil Properties, Shear Strength and Consolidation," *Proc. 6th, Int. Conf. on Soil Mech. and Found. Eng.*, 3, 1965, 319–321.
23. Maytyas, E. L. and Rathakrishna, H. S., "Volume Change Characteristics of Partially Saturated Soils," *Geotechnique*, 18, 4, 1968, 432–448.
24. Barden, R. J., Madedor, A. O., and Sides, G. R., "Volume Characteristics of Unsaturated Clays," *J. Soil Mech. Fdn. Div.*, ASCE, 95, 1, 1969, 33–51.
25. Fredlund, D. G., "Appropriate Concepts and Technology for Unsaturated Soils," *Can. Geotech. J.*, 16, 1979, 121–139.
26. Yong, R. N., Japp, R. D., and How, G., "Shear Strength of Partially Saturated Clays," *Proc. 4th Asian Reg. Conf. Soil Mech. Fdn. Eng.*, Bangkok, 2, 12, 1971, 183–187.
27. Kassif, G. and Ben Shalom, A., "Experimental Relationship between Swell Pressure and Suction," *Geotechnique*, 21, 3, 1971, 255.

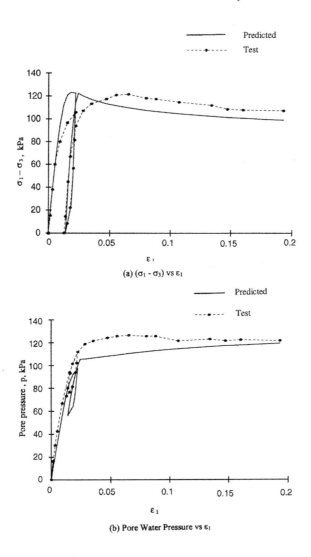

FIGURE 9.31
Comparison of predicted and observed stress–strain behavior: CTC test, $\bar{\sigma}_0 = 276$ kPa (40 psi): $e_0 = 0.8614$ (76).

28. Richards, B. G., "Moisture Flow and Equilibria in Unsaturated Soils," in *Soil Mechanics: New Horizons*, I. K. Lee (Editor), Newness, Butterworths, London, 1974, 113–157.
29. Lytton, R. L., "Foundation in Expansive Soils," in *Numerical Methods in Geotechnical Engineering*, C. S. Desai, and J. T. Christian (Eds.), McGraw-Hill, New York, 1977, 427–457.
30. Lloret, A. and Alonso, E. E., "Consolidation of Unsaturated Soil Including Swelling and Collapse Behavior," *Geotechnique*, 30, 4, 1980, 449–477.
31. Komornik, A., Livneh, M., and Smut, S., "Shear Strength and Swelling of Clays under Suction," *Proc. 4th Int. Conf. Expansive Soils*, Denver, 1, 1980, 206–226.

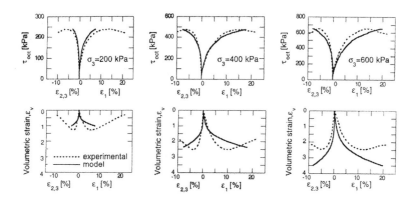

FIGURE 9.32
Comparison between experimental results and numerical predictions with the δ_1-model: $\bar{\sigma}_3$ = 200, 400, 600 kPa (54). (Reprinted with permission from A.A. Balkema.)

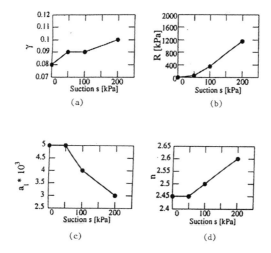

FIGURE 9.33
Variations of γ, R, n, and a_1 with suction (54). (Reprinted with permission from A.A. Balkema.)

32. Escario, V. and Saez, J., "The Shear Strength of Partly Saturated Soils," *Geotechnique*, 36, 3, 1986, 453–456.
33. Khalili, N. and Khabhaz, M. H., "A Unique Relationship for χ for Shear Strength Determination of Unsaturated Soils," *Geotechnique*, 48, 5, 1998, 681–688.
34. Coleman, J. D., "Stress Strain Relations for Partly Saturated Soil, Correspondence," *Geotechnique*, 12, 4, 1962, 348–350.
35. Fredlund, D. G. and Morgenstern, N. R., "Constitutive Relations for Volume Change in Unsaturated Soil," *Can. Geotech. J.* 13, 3, 1976, 261–276.
36. Leo, C. J. and Booker, J. R., "A Boundary Element Method for Analysis of Contaminant Transport in Porous Media I: Homogeneous Porous Media, II: Non-Homogeneous Porous Media," *Int. J. Num. Analyt. Methods in Geomech.*, 23, 14, 1999, 1681–1699, 1701–1715.

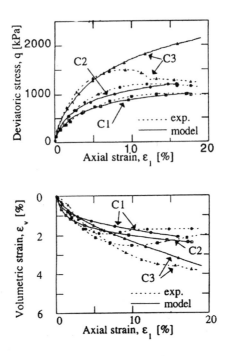

FIGURE 9.34
Comparison of predicted and observed behavior: δ_1-model (54, 80). (Reprinted with permission from A.A. Balkema.)

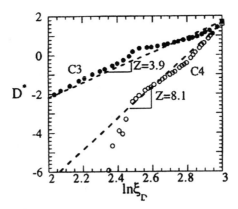

FIGURE 9.35
Determination of disturbance parameters A and Z (54, 79). (Reprinted with permission from A.A. Balkema.)

37. Chang, C. S. and Duncan, J. M., "Consolidation Analysis for Partly Saturated Clay by using an Elastic-Plastic Effective Stress-Strain Model," *Int. J. Num. and Analyt. Methods in Geomech.*, 7, 1983, 39–55.
38. Geraminegad, M. and Saxena, S. K., "A Coupled Thermoelastic Model for Saturated-unsaturated Porous Media," *Geotechnique*, 36, 4, 1986, 536–550.

The DSC for Saturated and Unsaturated Materials

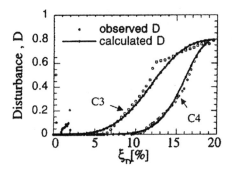

FIGURE 9.36
Comparison between observed and calculated disturbances (54, 79). (Reprinted with permission from A.A. Balkema.)

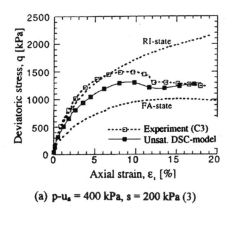

(a) $p-u_a = 400$ kPa, $s = 200$ kPa (3)

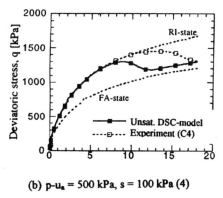

(b) $p-u_a = 500$ kPa, $s = 100$ kPa (4)

FIGURE 9.37
Comparisons between experimental data and predictions (54). Reprinted with permission from A.A. Balkema.)

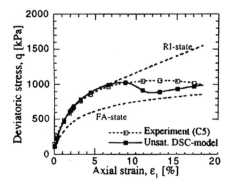

FIGURE 9.38
Comparison between experimental data and predictions: independent test (5), $p - u_a = 300$ kPa, $s = 100$ kPa (54). Reprinted with permission from A.A. Balkema.)

39. Desai, C. S. and Katti, D. R., "Contitutive Modelling with Extension to Expansive Soils," *Proc. 6th Int. Conf. on Expansive Soils*, New Delhi, India, Dec. 1987.
40. Schrefler, B. A. and Simon, L., "A United Approach to the Analysis of Saturated-Unsaturated Elasto-Plastic Porous Media," *Proc. 7 Int. Conf. on Computer Method and Adv. in Geomech.* G. Swoboda (Editor), Innsbruck, Austria, Balkema, Rotterdam, The Netherlands, 1988, 205–212.
41. Alonso, E. E., Gens, A., and Josa, A., "A Constitutive Model for Partially Saturated Soils," *Geotechnique*, 40, 3, 1990, 405–430.
42. Biarez, J. and Hicher, P. Y., *Elementary Mechanics of Soil Behavior, Vol. 1: Saturated Remolded Soils*, Balkema, Rotterdam, The Netherlands, 1994.
43. Modaressi, A. and Abou-Bekr, N., "A Unified Approach to Model the Behavior of Saturated and Unsaturated Soils," *Proc. 9th Int. Conf. Computer Methods and Adv. Geomech.*, Balkema, Rotterdam, The Netherlands, 1994.
44. Thomas, H. R. and He, Y., "Analysis of Coupled Heat, Moisture and Air Transfer in Unsaturated Soil," *Geotechnique*, 45, 4, 1995, 677–689.
45. Khalili, N. and Loret, B., "An Elastoplastic Model for the Behavior of Unsaturated Soils," *Proc., 48th Asia-Pacific Conf. on Comp. Mech.*, Singapore, December 1999.
46. Cui, Y. J. and Delage, P., "Yielding and Plastic Behavior of Unsaturated Compacted Silt," *Geotechnique*, 46, 2, 1996, 291–311.
47. Desai, C. S., Vulliet, L., Laloui, L., and Geiser, F., "Disturbed State Concept for Constitutive Modeling of Partially Saturated Porous Media," Report, Laboratoire de Mecanique des Sols, École Polytechnique Federale de Lausanne, Laussane, Switzerland, July 1996.
48. Hueckel, T. and Peano, A. (Editors) "Thermomechanics of Clays and Clay Barrier," *Special Issue, Eng. Geology*, 41, 1996, 1–318.
49. Selvadurai, A. P. S. (Editor), "Mechanics of Poroelastic Media," Kluwer Academic Publishers, The Netherlands, 1996.
50. Wheeler, S. J., "Inclusion of Specific Water Volume within an Elasto-plastic Model for Unsaturated Soil," *Can. Geotech.*, 33, 1996, 42–57.
51. Pietruszczak, S. and Pande, G. N., "Constitutive Behavior for Partially Saturated Soils Containing Gas Inclusions," *J. of Geotech. Eng., ESCE*, 122, 1, 1996, 50–59.

52. Bolzon, G., Schrefler, B. A., and Zienkiewicz, O. C., "Elasto-plastic Soil Constitutive Laws Generalized to Partially Saturated States," *Geotechnique,* 46, 2, 1996, 279–289.
53. Smith, D. W., Rowe, R. K., and Booker, J. R., "The Analysis of Pollutant Migration Through Soil with Linear Hereditary Time-Dependent Sorption," *Int. J. Num. Analyt. Methods in Geomech.,* 17, 4, 1993, 255–274.
54. Geiser, F., Laloui, L., Vulliet, L., and Desai, C.S., "Disturbed State Concept for Partially Saturated Soils," *Proc., 6th Intl. Symp. on Num. Models in Geomechanics,* Montreal, Pietrusczka, S. and Pande, G.N., (Eds.), Canada, Balkema, Netherlands, July, 1997.
55. Desai, C. S. and Sherman, W. C., "Unconfined Transient Seepage in Sloping Banks," *J. of Soil Mech. and Found. Div.,* ASCE, 97, 2, 1971, 357–373.
56. Desai, C. S., "Finite Element Residual Schemes for Unconfined Flow," *Int. J. Num. Meth. Eng.,* 10, 1976, 1415–1418.
57. Desai., C. S. and Li, G. C., "A Residual Flow Procedure and Application for Free Surface Flow in Porous Media," *Int. J. Adv. in Water Resources,* 6, 1983, 27–35.
58. Li, G. C. and Desai, C. S., "Stress and Seepage Analysis of Earth Dams," *J. of Geotech. Eng. Div.,* ASCE, 109, 7, 1983, 947–960.
59. Sugio, S. and Desai, C. S., "Residual Flow Procedure for Salt Water Intrusion in Unconfined Aquifers," *Int. J. Num. Meth. Eng.,* 24, 1987, 1439–1450.
60. Desai, C. S. and Baseghi, B., "Theory and Verification of Residual Flow Procedure for 3-D Free Surface Seepage," *Int. J. Adv. in Water Resources,* 11, 1988, 195–203.
61. Baseghi, B. and Desai, C. S., "Laboratory Verification of the Residual Flow Procedure for 3-D Free Surface Flow," *J. of Water Resources Research,* 26, 2, 1990, 259–272.
62. Westbrook, D. R., "Analysis of Inequality and Residual Flow Procedures and an Iterative Scheme for Free Surface Seepage," *Int. J. Num. Meth. Eng.,* 21, 1985, 1791–1802.
63. Baiocchi, C., "Free Boundary Problems in the Theory of Fluid Flow Through Porous Media," *Proc. Int. Congr. Math.,* Vol. II, Vancouver, Canada, 1974, 237–243.
64. Alt, H. W., "Numerical Solution of Steady-state Porous Flow Free Surface Boundary Problems," *Numer. Math.* 36, 1980, 73–98.
65. Bruch, J. C., "Fixed Domain Methods for Free and Moving Boundary Flows in Porous Media," *J. of Transport in Porous Media,* 6, 1991, 627–649.
66. Cathie, D. N. and Dungar, D. N., "The Influence of the Pressure-Permeability Relationship on the Stability of Rock-Filled Dam," in *Criteria and Assumptions for Numerical Analysis of Dams,* Naylor, D. J., Stagg, K. G., and Zienkiewicz, O. C. (Editors), Univ. of Wales, Swansea, U.K., 1975.
67. Bathe, K. J. and Khoshgoftaar, M. R., "Finite Element Free Surface Seepage Analysis without Mesh Iteration," *Int. J. Num. Analyt. Meth. Geomech.,* 3, 1979, 13–22.
68. Ng, A. K. L. and Small, J. C., "Coupled Finite Element Analysis for Unsaturated Soils," *Int. J. Num. Analyt. Meth. in Geomech.,* Submitted, 1998.
69. Kim, J. M. and Parizek, R. R., "Three-Dimensional Finite Element Modeling for Consolidation due to Groundwater Withdrawal in a Desaturating Anisotropic Aquifer System," *Int. J. Num. Analyt. Meth. in Geomechanics,* 23, 6, 1999, 549–571.

70. Gan, J. K. M., Fredlund, D. G., and Rahandjo, H., "Determination of the Shear Strength Parameters of an Unsaturated Soil Using the Direct Shear Test," *Canadian Geotech. J.*, 25, 3, 1988, 500–510.
71. Gyi, M. M., "Multaxial Cyclic Testing of Saturated Ottawa Soil," *M.S. Thesis*, Dept. of CEEM, Univ. of Arizona, AZ, USA, 1996.
72. Park, I. J. and Desai, C. S., "Disturbed State Modeling for Dynamic and Liquefaction Analysis," *Report*, Dept. of CEEM, Univ. of Arizona, AZ, USA, 1997.
73. Desai, C. S., Park, I. J., and Shao, C., "Fundamental yet Simplified Model for Liquefaction Instability," *Int. J. Num. Analyt. Meth. in Geomech.*, 22, 7, 1998, 721–748.
74. Park, I. J. and Desai, C. S., "Cyclic Behavior and Liquefaction of Sand Using Disturbed State Concept," *J. Geotech. Geoenv. Eng.*, ASCE, 126, 9, 2000.
75. Desai, C. S., "Computer Code DSC-DYN2D: Theoretical Background, User's Manual and Examples," Tucson, AZ, 2000.
76. Shao, C. and Desai, C. S., "Implementation of DSC Model for Dynamic Analysis of Soil-structure Interaction Problems," *Report*, Dept. of CEEM, Univ. of Arizona, AZ, USA, 1998.
77. Katti, D. R. and Desai, C. S., "Modeling Including Associated Testing of Cohesive Soil Using Disturbed State Concept," *Report to NSF*, Dept. of CEEM, Univ. of Arizona, AZ, USA, 1991.
78. Katti, D. R. and Desai, C. S., "Modeling and Testing of Cohesive Soil Using the Disturbed State Concept," *J. of Eng. Mech.*, ASCE, 121, 5, 1995, 648–658.
79. Geiser, F., Laloui, L., and Vulliet, L., "Constitutive Modelling of Unsaturated Sandy Silt," *Proc. 9th Int. Conf. on Computer Meth. and Adv. in Geomech.*, Wuhan, China, Balkema, The Netherlands, 1997.
80. Geiser, F., "Testing and Constitutive Modeling of Saturated and Partially Saturated Soils," *Ph.D. Dissertation*, Institute des Sols, Roches et Fondations, École Polytechnique Federale De Lausanne, Lausanne, Switzerland, 1999.

10

The DSC for Structured and Stiffened Materials

CONTENTS
10.1 Definition of Disturbance 392
10.2 Structured Soils 396
      10.2.1 Validations 399
10.3 Dislocations, Softening, and Stiffening 401
      10.3.1 Formulation of DSC Model 402
      10.3.2 Validation 403
      10.3.3 Disturbance and Dislocation Density 405
10.4 Reinforced and Jointed Systems 407
      10.4.1 Equivalent Composite 407
      10.4.2 Individual Solid and Joint Elements 409
      10.4.3 Validation for Test Results with HISS Model and
             FE Method 413
      10.4.4 Jointed Rock 414
10.5 Rest Periods: Unloading 415

In this chapter, we consider use of the DSC for materials that are referred to as *structured* and those that experience stiffening or healing during mechanical and thermal loading, and/or chemical effects. The term "structured" is relative. For our purposes, it can refer to a material that possesses a micro-or macro-structural matrix or particle arrangement that is different relative to, or is superimposed on, that of its basic or reference state. In this sense, many materials can be classified as structured. An anisotropic material can be considered as structured with respect to its isotropic state. A material that exhibits nonassociative plastic response caused by friction possesses a different structure compared to its state that involves associative response. A softening or degrading material has a structure that is different from its state that exhibits nonsoftening or hardening response.

Figure 10.1 depicts responses of a soil in its overconsolidated (OC) and normally consolidated (NC) states (1). The NC state can result due to gradual deposition of soil particles under increasing (normal) load, while the OC state results if the material has undergone loading and unloading such that

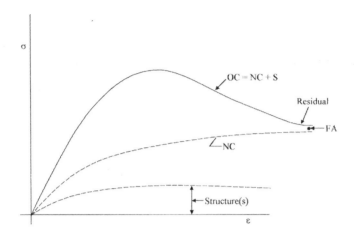

FIGURE 10.1
Overconsolidation structure (OC) with respect to normal consolidation (NC).

the current stress is lower than the previous one. As a result of the OC, the material's microstructure changes, which can entail formations of preferred directions and additional interparticle bonds (Chapter 2). Under loading, an OC soil usually exhibits a stiffer response compared to the soil's response in its basic NC state. As the load is increased, the peak stress is reached, after which the material softens or degrades. The softening is caused by the annihilation of the OC structure and bonds, which may initiate before or at the peak stress. On further deformations beyond the peak, the response enters the residual state, and in the limit, the FA (or the critical state) is approached. In other words, the limiting state of an OC material would be the same as that for the NC material.

10.1 Definition of Disturbance

In the DSC, the disturbance (D) for structured materials can be defined in different ways. Figure 10.2(a) shows a schematic of disturbance, in which D_b denotes disturbance relevant to the basic structure, e.g., NC or reconstituted state. The latter can be obtained when a natural (OC) material is broken down, say, by pulverization, and then it is remolded (or reconstituted) with the natural moisture content.

The overall disturbance, $\bar{D}$, is expressed as

$$\bar{D} = D_b + D_s \tag{10.1}$$

where D_b denotes disturbance for the basic (e.g., NC) state, and D_s is the disturbance associated with the structure. A special form of Eq. (10.1) can be

The DSC for Structured and Stiffened Materials

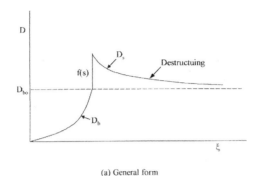

(a) General form

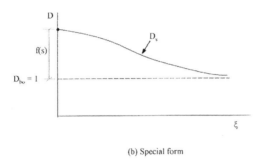

(b) Special form

FIGURE 10.2
Disturbance due to structure.

written as

$$\bar{D} = D_{ub}\left(1 - e^{-A_b \xi^{Z_b}}\right) + f(s)e^{-A_s \xi^{Z_s}} \qquad (10.2a)$$

in which the first term represents D_b, the function $f(s)$ denotes the change (increase or decrease) in disturbance due to the structure, ξ is the (deviatoric) plastic strain trajectory, and D_u, A, and Z are the material parameters. Equation (10.2a) can be simplified such that the response is considered only for the behavior of the structured state. Then, it is specialized as [Fig. 10.2(b)]

$$\bar{D} = f(s)e^{-A_s \xi^{Z_s}} \qquad (10.2b)$$

Alternatively, the disturbance of the material in the basic state, D_b, can be adopted as unity; then

$$\bar{D} = 1 + f(s)e^{-A_s \xi^{Z_s}} \qquad (10.2c)$$

Figure 10.2 shows that the structure imposed on the basic state is annihilated gradually as the load is applied, and in the limit, it approaches the basic structure, D_{b0}. An example of the characterization of the compressive behavior of a structured soil obtained by using the simplified Eq. (10.2c) is given subsequently.

Figure 10.3(a) shows the response of a structured material that exhibits hardening, peak, and softening. After the critical or threshold plastic strains, $\varepsilon_t(\xi_t)$, the internal microstructure of the material can modify due to the strengthening of bonds as a result of different influences. In the case of dislocated silicon under thermomechanical loading, such an influence can arise due to the *locking* of dislocations with impurities such as oxygen and nitrogen, in the presence of specific temperatures.

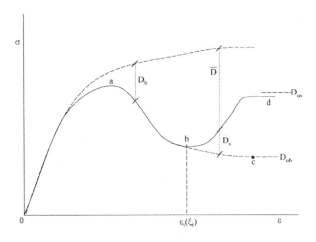

(a) Stress – strain response

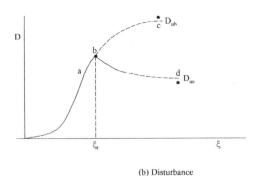

(b) Disturbance

FIGURE 10.3
Softening and stiffening responses.

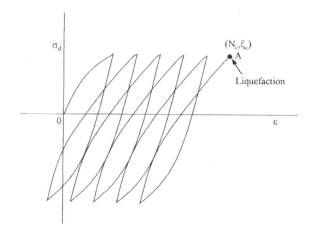

(a) Cyclic behavior under constant shear stress amplitude

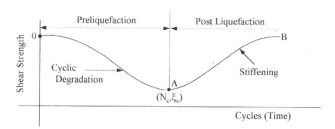

(b) Shear Strength vs. Cycles (Time)

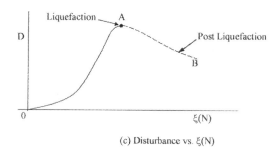

(c) Disturbance vs. $\xi(N)$

FIGURE 10.4
Liquefaction and postliquefaction behavior of saturated sand.

A saturated soil (sand) under cyclic (earthquake) loading can first experience degradation due to the increase in pore water pressures, i.e., decrease in the effective stress, leading to instability and liquefaction at the critical disturbance (D_c) corresponding to plastic strains (ξ_c) after loading cycles (N_c); see Fig. 10.4(a) and (b). Then, in the postliquefaction region, the porous material can experience drainage with a resultant decrease in the pore water pressure (2):

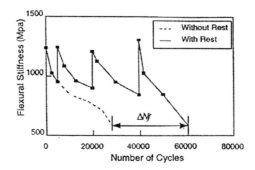

FIGURE 10.5
Flexural stiffness of asphalt concrete with and without rest periods (4). 2000©John Wiley & Sons Ltd. Reproduced with permission.

This can lead to an increase in the strength as the effective stress increases. Hence, the disturbance decreases in the postliquefaction zone (3) [Fig. 10.4(c)].

An asphaltic concrete (used in pavements) is subjected to vehicular traffic load during days, while it can experience healing or increase in stiffness due to rest periods (during nights) (4). Such healing can be attributed to chemical effects as well as unloading, which can result in a decrease in plastic strains and microcracking. The specific contributions of mechanical unloading and chemical effect are not known precisely. Figure 10.5 shows the effect of rest periods on the flexural stiffness of an asphalt obtained from flexural fatigue tests on beams, with and without rest periods (4). The figure also shows that the cycles to fatigue failure (ΔN_f) increase with rest periods.

In Chapter 2 we discussed the response of a porous chalk that exhibits the behavior in Fig. 2.10 similar to that in Fig. 10.3. In the case of composite or reinforced materials, the disturbance can be defined relative to the responses of the component materials (Chapter 2).

Examples of the formulation of the DSC for a structured soil, a dislocated silicon with oxygen impurities, a reinforced soil and a jointed rock, and healing or stiffening due to unloading are discussed in the remainder of this chapter.

10.2 Structured Soils

Materials such as soils can develop a structure due to natural deposition processes caused by geological history and by artificial methods such as application of mechanical load. Such a structure can be defined with respect to that of the soil in its reconstituted or remolded state, which can be considered to represent its basic state before the forces causing the change in the structure were imposed.

In the DSC, the behavior of the reconstituted soil can be treated as the fully adjusted state, and then the effect of the structure developed can be introduced as the change in disturbance with respect to the disturbance at the reconstituted state. Then the modified disturbance, $\overline{D}$, to allow for the structure, which can result in stiffening with respect to the FA state, can be expressed using Eq. (10.2).

We describe an example of the application of the DSC for naturally or artificially structured soil. The formulation presented here has been reported by Liu et al. (5). The incremental strain equations for the DSC [Eq. (4.46), Chapter 4] are given by

$$d\underline{\varepsilon}^a = (1 - D_\varepsilon)d\underline{\varepsilon}^i + D_\varepsilon d\underline{\varepsilon}^c + dD_\varepsilon(\underline{\varepsilon}^c - \underline{\varepsilon}^i) \tag{10.3a}$$

where $D\varepsilon$ is the disturbance function. Liu et al. (5) assumed that the RI represents "zero strain state," i.e., it is characterized as a perfectly rigid material. Hence, $d\underline{\varepsilon}^i = \underline{0}$, and Eq. (10.3a) is simplified as

$$d\underline{\varepsilon}^a = D_\varepsilon d\underline{\varepsilon}^c + dD_\varepsilon \underline{\varepsilon}^c = d(D_\varepsilon \underline{\varepsilon}^c) \tag{10.3b}$$

The disturbance function for the structured soil is expressed as

$$\overline{D}_\varepsilon = D_\varepsilon + \left(\frac{\partial D_\varepsilon}{\partial \underline{\varepsilon}^c}\right)^T \underline{\varepsilon}^c \tag{10.4}$$

Hence, the observed strain is given by

$$d\underline{\varepsilon}^a = \overline{D}_\varepsilon d\underline{\varepsilon}^c \tag{10.5}$$

The disturbance function can be decomposed in two parts as

$$d\varepsilon_v^a = \overline{D}_{\varepsilon v} d\varepsilon_v^c \tag{10.6a}$$

and

$$d\varepsilon_d^a = \overline{D}_{\varepsilon d} d\varepsilon_d^c \tag{10.6b}$$

where $\overline{D}_{\varepsilon v}$ and $\overline{D}_{\varepsilon d}$ are disturbance functions related to volumetric and deviatoric behavior, respectively, and

$$\varepsilon_v = \varepsilon_1 + \varepsilon_2 + \varepsilon_3 \tag{10.7a}$$

$$\varepsilon_d = \frac{\sqrt{2}}{3}\sqrt{(\varepsilon_1 - \varepsilon_2)^2 + (\varepsilon_2 - \varepsilon_3)^2 + (\varepsilon_3 - \varepsilon_1)^2} \tag{10.7b}$$

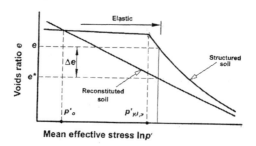

FIGURE 10.6
Schematic of compression behavior of reconstituted and structured soils (5). ©John Wiley & Sons Ltd. Reproduced with permission.

Here, ε_i ($i = 1, 2, 3$) are the principal strains. Thus, the DSC can be formulated for the combined volumetric and deviatoric response using Eq. (10.6). Liu et al. (5) considered the compressive behavior of structured soil, based on Eq. (10.6a).

Example 10.1 Compression Behavior of a Structured Soil

Figure 10.6 shows a schematic of the compression behavior of structured and reconstituted soils expressed in terms of void ratio (e) and ln p', where p' is the mean effective pressure. As the pressure is increased on the structured soil, it experiences a process of *destructuring* during which the soil's structure (bonds) is annihilated and, in the limit, the soil approaches the reconstituted state, at high pressures.

It was assumed that when the pressure p' was below the pressure $p'_{y,i}$, when yielding occurred, the behavior was elastic, and there was no disturbance. In other words, disturbance and destructuring occur only due to plastic yielding beyond the pressure $p'_{y,i}$.

The yielding virgin response beyond $p'_{y,i}$ was defined using the DSC model, in which the disturbance, $\bar{D}_{\varepsilon v}$ [Eq. (10.6a)] was defined based on the observed (compression) behavior of a number of structured soils as [Eq. (10.2c)]

$$\bar{D}_{\varepsilon v} = 1 + b\left(\frac{p'_{y,i}}{p'}\right) \qquad (10.8)$$

where b is the disturbance index, to represent the structure with reference to that of the reconstituted soil. In Eq. (10.8), the value of D_b in Eq. (10.2) is assumed to be unity relevant to the reconstituted state. A plot of $\bar{D}_{\varepsilon v}$ is shown in Fig. 10.2(b), where $b = f(s)$.

Liu et al. (5) presented various versions of the DSC model for the compressive behavior: (1) for the case when the (experimental) compression data are available in terms of mean effective pressure and volumetric response, and (2) for the case when the data are available in terms of effective vertical or axial stress (σ'_v) from one-dimensional compression tests. These formulations were based on the critical-state concept (5–7), and the resulting equations are given below:

Case 1:

$$d\varepsilon_v^a = \begin{cases} \dfrac{\kappa^*}{1+e}\dfrac{dp'}{p} & \text{for } p' < p'_y \\ \dfrac{\lambda^*}{1+e}\left[1 + b\left(\dfrac{p'_{y,i}}{p'}\right)\right]\left(\dfrac{dp'}{p}\right) & \text{for } p' \geq p'_y \end{cases} \quad (10.9)$$

Case 2:

$$d\varepsilon_v^a = \begin{cases} \dfrac{\kappa_v^*}{1+e}\left(\dfrac{d\sigma'_v}{\sigma'_v}\right) & \text{for } \sigma'_v < \sigma'_{vy} \\ \dfrac{\lambda_v^*}{1+e}\left[1 + b_v\left(\dfrac{\sigma'_{vy,i}}{\sigma'_v}\right)\right]\left(\dfrac{d\sigma'_v}{\sigma'_v}\right) & \text{for } \sigma'_v \geq p'_{vy} \end{cases} \quad (10.10)$$

where λ^* and λ_v^* denote slopes of a virgin curve (Chapter 3), and κ^* and κ_v^* denote the slope of an unloading curve on an $e - \ln p'$ plot.

10.2.1 Validations

The DSC model was used to predict compression behavior of seven different soils (5). Typical results for two soils are included here; Table 10.1 shows their properties.

For soil 1, the test data were available (8) in terms of mean effective stress, p'. Then, Eq. (10.9) was used for the back predictions. For soil 2, the test data were available (9) in terms of effective vertical stress (σ'_v); hence, Eq. (10.10) was used for the back predictions.

TABLE 10.1
Details of Compression Tests and Parameters

Soil	Reference	$p'_{y,i}$ (kPa)	b	κ^*	λ^*	Comments
1. Mexico City clay	Terzaghi (8)	92.0	1.65	—	—	Very high void ratio; $e - \ln p'$ nonlinear
2. Leda clay	Quigley and Thompson (9)	185.0	4.10	κ_v^* 0.02	λ_v^* 0.16	$e - \ln \sigma'_v$ linear

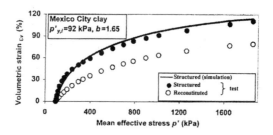

FIGURE 10.7
Comparisons for one-dimensional compression tests on Mexico City clay (5).

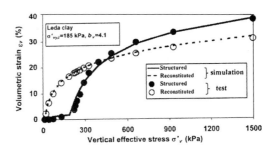

FIGURE 10.8
Comparisons for one-dimensional compression tests on Leda clay (5).

The values of the initial yield stress, $p'_{y,i}$, were determined from the compression curves plotted in the $e - \ln p'_v$ coordinate system. The value of the disturbance index, b, was evaluated as the best fit for the predictions using Eq. (10.9) or (10.10) and the test data for virgin compressions behavior. The parameters, λ^* and κ^*, were found using the test data for the reconstituted soil. When the $e - \ln \sigma'_v$ response was linear (soil 2), behavior for both the structured and reconstituted states was simulated. When the $e - \ln p'_v$ response was nonlinear (soil 1), only the structured soil behavior was predicted; Figure 10.7 shows comparisons between predictions and test data for soil 1 (Mexico City clay). Figure 10.8 shows similar comparisons for the Leda clay (soil 2), in which predictions are shown for both the structured and reconstituted soils.

These results and comparisons for other soils (5) show that the DSC model provides very good predictions for both the structured and reconstituted states for under compressive (volumetric) behavior. The DSC model can be developed for the shear response [Eq. (10.6b)]. Then, it would be possible to simulate the combined compressive and shear behavior of structured soils by using the DSC.

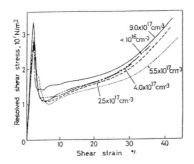

(a) Dislocation Free Crystals: T = 900°C and $\dot{\varepsilon} = 1.1 \times 10^{-4} s^{-1(2)}$

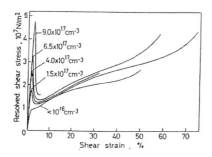

(b) Dislocation Density: 1×10^6 cm^{-2}: T = 800°C and $\dot{\varepsilon} = 1.1 \times 10^{-4} s^{-1}$

FIGURE 10.9
Stress–strain behavior of silicon doped with different oxygen content (11) (with permission).

10.3 Dislocations, Softening, and Stiffening

The mechanical behavior of silicon crystals as affected by factors such as dislocation density (N_0), temperature (T), strain rate ($\dot{\varepsilon}$), and impurities such as oxygen and nitrogen may exhibit responses as depicted in Fig. 10.3 (10, 11).

Example 10.2 Dislocated Silicon with Impurities

Figure 10.9 shows stress–strain data for silicon crystals under different dislocations, temperature, and strain rates reported by Yonenaga et al. (11). It can be seen that the peak or upper yield stress was not affected significantly for the dislocation free crystals [Fig. 10.9(a)], while it shows significant increase for the dislocation, $N_0 = 1 \times 10^6$ cm^{-2}, as the oxygen concentration increases. Such an effect is attributed to the locking of dislocations by oxygen atoms dissolved in silica crystals because the dislocations become immobile due to impurity locking, while the crystal is kept at elevated temperature under no applied stress (11).

The prepeak region of the stress–strain response is usually elastoplastic with plastic yielding or hardening, during which the yield stresses increase with plastic deformations (Chapter 7); Fig. 10.9. There occurs a reduction in the stress after the peak stress, often referred to as softening or degradation, which results due to the initiation of microcracking (sometimes before the peak stress), and its growth due to the coalescence of microcracks. For some materials, microstructural instability occurs at critical locations or disturbance, D_c, and failure may subsequently occur near the ultimate condition, D_u. In the case of the silicon with impurities, however, the softening may continue up to the point b in Fig. 10.3(a), and then under the specific combination of N_0, T, and $\dot{\varepsilon}$, and after the threshold plastic strains (ξ_t), the material exhibits stiffening or healing response. Often, the stiffening is higher for greater values of the dislocation density (N_0).

10.3.1 Formulation of DSC Model

Figure 10.3 shows a schematic of the response as a combination of hardening–softening–stiffening behavior. The response, 0–a–b–c, is the basic hardening–softening behavior, which has been characterized before (Chapter 4), using Eq. (4.1) as

$$d\underline{\sigma}^a = (1 - D)\underline{C}^i \, d\underline{\varepsilon}^i + D\underline{C}^c \, d\underline{\varepsilon}^c + dD(\underline{\sigma}^c - \underline{\sigma}^i) \qquad (10.11)$$

In this case, the disturbance, D, increases with (plastic) deformation, shown as 0–a–b, Fig. 10.3(b).

In the case of the stiffening response beyond point b, the material exhibits higher stress-carrying capacity and is assumed to approach an asymptotic value (near d). Then, during stiffening, the disturbance decreases, as depicted by b–d in Fig. 10.3(b).

For the combined hardening–softening–stiffening response, the modified disturbance function, $\bar{D}$, in Eq. (10.2) is expressed as (12)

$$\bar{D} = D_{ub}[\{1 - \exp(-A_b \xi_D^{Z_b})\} - \bar{\lambda}\{1 - \exp(-A_s \xi_D^{Z_s})\}] \qquad (10.12)$$

where D_{ub}, A_b, and Z_b are parameters for the basic softening behavior; $\bar{\lambda}$, A_s and Z_s are parameters related to the stiffening response; and $\bar{\lambda}$ is given by

$$\bar{\lambda} = \begin{cases} 0 & \text{if } \xi_D \leq \xi_t \\ \dfrac{D_{us}}{D_{ub}} & \text{if } \xi_D > \xi_t \end{cases} \qquad (10.13)$$

where D_{us} is the ultimate disturbance corresponding to the limiting stiffening response, and $\bar{\xi}_t$ is the threshold value of the deviatoric plastic strain trajectory when the stiffening initiates.

With the modified disturbance, $\bar{D}$, the DSC incremental equations are modified as

$$d\underset{\sim}{\sigma}^a = (1 - \bar{D})\underline{C}^i\,d\underset{\sim}{\varepsilon}^i + D\underline{C}^c\,d\underset{\sim}{\varepsilon}^c + d\bar{D}(\underset{\sim}{\sigma}^c - \underset{\sim}{\sigma}^i) \quad (10.14)$$

10.3.2 Validation

Applications of the DSC model for the basic softening behavior exhibited by silicon crystal (ribbons) with relatively low dislocation densities ($N_0 \approx 2 \times 10^3$ cm^{-2}) and low oxygen concentration ($C_0 \approx 10^{14}$ atoms/cm^3) (10) are given in Example 7.16, Chapter 7; this material does not exhibit stiffening behavior. Here, we present application of the modified DSC equation (10.14) for characterizing the behavior of dislocated silicon with high levels of dislocation densities (N_0) and oxygen concentration (C_0), which exhibits the stiffening response. Figure 10.9 shows laboratory test results for such silicon doped with oxygen as reported by Yonenaga et al. (11).

The stress–strain data (Fig. 10.9) were used to find the parameters for each individual curve for the dislocation-free ($N_0 = 0$) and dislocation ($N_0 = 1 \times 10^{16}$ cm^{-2}) conditions. The elastic modulus, E (slope at the origin), was found for each stress–strain curve. Its average value for $T = 800°C$ was found to be 2.45×10^9 Pa, and for $T = 900°C$, it was found to be 1.67×10^9 Pa. The RI response was assumed to be linear elastic.

The disturbance parameters were found using procedures given in Chapters 4 and 7 and are shown in Tables 10.2(a) and (b) for $T = 900°C$ and $800°C$, respectively. The parameters were expressed as functions of the oxygen concentration, C_0, using the following relations:

$$A_b(C_0) = A_{br}\left(\frac{C_0}{C_r}\right)^{n_1}$$

$$Z_b(C_0) = Z_{br}\left(\frac{C_0}{C_r}\right)^{n_2}$$

$$A_s(C_0) = A_{sr}\left(\frac{C_0}{C_r}\right)^{n_3}$$

$$Z_s(C_0) = Z_{sr}\left(\frac{C_0}{C_r}\right)^{n_4}$$

$$(10.15)$$

where the subscript r denotes reference concentration, $C_0 = 1 \times 10^7$ cm^{-3}; accordingly, e.g., A_{br} is the value of A_b at $C_0 = 1 \times 10^7$ cm^{-3}. Table 10.2(c) shows the parameters in Eq. (10.15). The parameters obtained from the test with $C_0 = 4 \times 10^{17}$ cm^{-3} were not used in the "average" relations [Eq. (10.15)]; this test was used as an independent validation.

TABLE 10.2(a)
Parameters for $N_0 = 0$ at $T = 900°C$

$C_0 \times 10^{17}$ cm^{-3}	D_{ub}	D_{us}	$\bar{\lambda}$	A_b	A_s	Z_b	Z_s
2.5	0.5909	0.6558	1.1099	23.9657	−3.6464	1.1196	2.0844
4.0	0.3174	0.6397	2.0156	15.0097	−2.1515	0.7941	1.7377
5.5	0.4815	0.6667	1.3846	7.4037	−1.9060	0.6109	1.6889
9.0	0.4661	0.6260	2.0156	3.5548	−1.2710	0.4148	1.4809

TABLE 10.2(b)
Parameters for $N_0 = 1 \times 10^{16}$ cm^{-2} at $T = 800°C$

$C_0 \times 10^{17}$ cm^{-3}	D_{ub}	D_{us}	$\bar{\lambda}$	A_b	A_s	Z_b	Z_s
1.5	0.4497	0.7512	0.5987	4.0748	−7.2044	0.5637	1.9819
4.0	0.5629	0.4545	1.2385	8.3227	−3.7178	0.6800	1.8443
6.5	0.6624	0.8112	0.7672	11.8467	−2.6796	0.7462	1.7797
9.0	0.7017	—	—	15.0097	−2.1515	0.7941	1.7377

TABLE 10.2(c)
Parameters in Eq. (10.15)

Temperature (°C)	A_{br}	A_{sr}	Z_{br}	Z_{sr}	η_1	η_2	η_3	η_4
800	4.0785	7.2044	0.5638	1.9820	0.7272	−0.6745	0.1912	−0.07341
900	51.2980	5.5514	1.6577	2.3889	−1.4898	−0.8228	0.7683	−0.2669

The laboratory curves were predicted by integrating Eq. (10.14), in which the RI behavior, $\underset{\sim}{C}^i$, was defined based on linear elastic response, and the FA response, $\underset{\sim}{C}^c$, was based on the bulk modulus (K). The disturbance function $\bar{D}$ in Eq. (10.12) was used with the parameters shown in Table 10.2.

Figures 10.10(a) and (b) show comparisons between predictions and test data for two individual responses: (1) $N_0 = 0.0$, $C_0 = 2.5 \times 10^{17}$ cm^{-3}, $T = 900°C$, and (2) $N_0 = 1 \times 10^6$ cm^{-2}, $C_0 = 1.5 \times 10^{17}$ cm^{-3}, $T = 800°C$, respectively. The parameters used for these predictions were relevant to the individual stress–strain data.

Figures 10.11(a) and (b) show comparisons between predictions and test data for two independent tests: (1) $N_0 = 0.0$, $C_0 = 4 \times 10^{17}$ cm^{-3}, $T = 900°C$, and (2) $N_0 = 1 \times 10^6$ cm^{-2}, $C_0 = 4 \times 10^{17}$ cm^{-3}, $T = 800°C$. Here the average parameters were used to predict independent tests that were not used in finding the parameters. It can be seen from Figs. 10.10 and 10.11 that the DSC model provides very good predictions of the behavior of dislocated silicon doped with oxygen, as affected by C_0, N_0, and T.

The DSC for Structured and Stiffened Materials

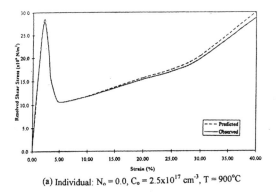

(a) Individual: $N_o = 0.0$, $C_o = 2.5 \times 10^{17}$ cm^{-3}, $T = 900°C$

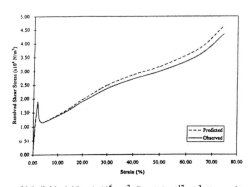

(b) Individual: $N_o = 1 \times 10^6$ cm^{-2}, $C_o = 1.5 \times 10^{17}$ cm^{-3}, $T = 800°C$

FIGURE 10.10
Comparisons between observed and predicted individual responses (12) (with permission).

10.3.3 Disturbance and Dislocation Density

It was also shown by Desai et al. (12) that disturbance in the DSC and dislocation density can be correlated. The latter is given as [Eq. (3.25a)], Chapter 3 (10, 12)

$$\dot{N}_m = \bar{K} N_m k_0 (\sqrt{J_{2D}^a} - \lambda \sqrt{N_m})^{p+r} e^{-Q/kT'} \quad (10.16a)$$

where k is Boltzmann's constant (= 8.617×10^{-5} evK); $\lambda\sqrt{N_m}$ is the back stress, Q is the Peierl's energy (= 2.17 ev), $\bar{K}$, p, and r are material constants (= 3.1×10^{-4}, 1.1, and 1.0, respectively), $k_0 = B_0/\tau_0$, B_0 is mobility (= 4.30×10^4 m/s), T' is absolute temperature, $\tau_0 = 10^7$ N/m^2, $\sqrt{J_{2D}^a}$ denotes the observed stress, the overdot denotes time derivative, and m denotes any point. By using the stress-based disturbance [Eq. (3.7), Chapter 3], Eq. (10.16a) can be

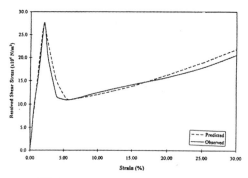

(a) Independent: $N_o = 0.0$, $C_o = 4 \times 10^{17}$ cm^{-3}, T = 900°C

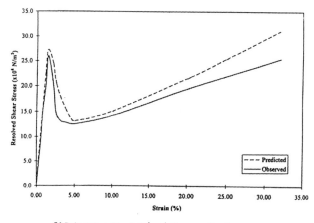

(b) Independent: $N_o = 1 \times 10^6$ cm^{-2}, $C_o = 4 \times 10^{17}$ cm^{-3} and T = 800°C

FIGURE 10.11
Comparisons between observed and predicted independent responses (12) (with permission).

expressed as

$$\dot{N}_m = \bar{K} N_m k_0 [(1-D)\sqrt{J^i_{2D}} - \lambda\sqrt{N_m})]^{p+\gamma} e^{-Q/kT'} \quad (10.16b)$$

where $\sqrt{J^i_{2D}}$ denotes the RI shear stress.

Figure 10.12 shows the plot of the dislocation density vs. axial strain for the test (10) in Fig. 7.44, Chapter 7, for the curve $T = 1100°C$, $\dot{\varepsilon} = 4.8 \times 10^{-4}$ s^{-1}, and $N_0 = 1.8 \times 10^5$ cm^{-2}. This plot is obtained by integrating Eqs. (10.16a) and (10.16b); here the values of disturbance at various points are used in Eq. (10.16b) for the DSC predictions. It can be seen that both Eqs. (10.16a) and

The DSC for Structured and Stiffened Materials

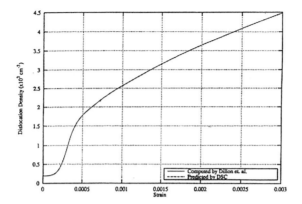

FIGURE 10.12
Comparison of dislocation density vs. strain with Eq. (10.16): independent tests (12) (with permission).

(10.16b) show essentially the same results, implying that the disturbance and dislocation densities bear a correlation.

10.4 Reinforced and Jointed Systems

Materials are often reinforced to enhance their deformation and strength properties, e.g., reinforced concrete, reinforced earth, and composites (metallic, ceramic, etc.). Some materials involve joints or interfaces that exist before the load is applied or are induced during loading. Structure-foundation systems and jointed rocks are examples of the former, whereas the latter can occur in any material that experiences microcracking leading to macro- or finite-sized cracks. In the case of joints, the material is usually weaker than the parent or solid material parts; see Fig. 10.13.

Reinforced and jointed material systems can be characterized using the DSC in two ways: (1) by expressing the observed behavior of the equivalent reinforced or jointed composite in terms of the behavior of the solid and the joint (Fig. 10.13), or (2) by treating the body as made of solid, interfaces or joints, and then introducing their responses individually, say, in the finite-element analysis.

10.4.1 Equivalent Composite

In the case of the equivalent composite, the observed or actual behavior of the composite can be expressed as (Chapter 2)

$$d\underline{\sigma}_m^a = (1 - \bar{D}_m)d\underline{\sigma}_s^a + \bar{D}_m \, d\underline{\sigma}_j^a + d\bar{D}_m(\underline{\sigma}_j^a - \underline{\sigma}_s^s) \qquad (10.17)$$

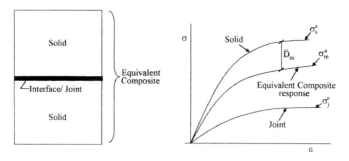

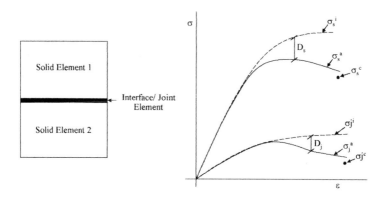

FIGURE 10.13
Representations for composite material systems.

where σ_m^a, σ_s^a, and σ_j^a denote the observed stresses in the composite, solid, and joint/interface, respectively [Fig. 10.13(a)]. Then it will be necessary to conduct laboratory tests on specimens of the solid and joint to measure and define their observed behavior. The observed behavior of the solid, σ_s^a, and that of the joint, σ_j^a, can be expressed in terms of the responses of their respective RI and FA states:

$$d\sigma_s^a = (1 - D_s)d\sigma_s^i + D_s\, d\sigma_s^c + dD_s(\sigma_s^c - \sigma_s^i) \tag{10.18a}$$

and

$$d\sigma_j^a = (1 - D_j)d\sigma_j^i + D_j\, d\sigma_j^c + dD_j(\sigma_j^c - \sigma_j^i) \tag{10.18b}$$

where σ_s^i and σ_j^i are the RI stresses for the solid and joint, respectively, σ_s^c and σ_j^c are the FA stresses for the solid and joint, respectively, and D_s and D_j are disturbances for the solid and joint behavior, respectively [Fig. 10.13(b)].

With the foregoing formulation in the FE procedure, the elements in the mesh will be considered as a composite containing a joint or interface with surrounding elements [Fig. 10.13(a)]. The equations for such a composite element can be expressed as

$$\underset{\sim}{k}_m d\underset{\sim}{q}_m = d\underset{\sim}{Q}_m \qquad (10.19)$$

where the stiffness

$$\underset{\sim}{k}_m = \int \underset{\sim}{B}^T \underset{\sim}{C}^a_m \underset{\sim}{B} \, dV \qquad (10.20)$$

where $d\underset{\sim}{Q}_m$ denotes the applied load, and $\underset{\sim}{C}^a_m$ is the constitutive matrix for the composite element, which is expressed in Eq. (10.17).

10.4.2 Individual Solid and Joint Elements

In this approach, the structure is divided into solid and joint or interface elements. Then the behavior of each is characterized using the DSC model. This method is similar to the standard FE method in which solid and joint regions are treated as individual elements with definition of properties for material (solid or joint) for each element.

Example 10.3 Reinforced Earth and Jointed Rock

We present an example of reinforced earth in which both preceding approaches were used (13). Figure 10.14 shows the finite-element mesh for a laboratory triaxial test with cylindrical specimen reinforced by a single layer of nonwoven geotextile (R1NW) at the midheight of the specimen. Tests were also performed with two layers of the nonwoven geotextile placed at one third and two thirds of the specimen's height (R2NW), and two layers of woven geotextile at one third and two thirds height (R2W). A number of triaxial tests were also performed for the natural sand specimens, without reinforcement. The soil was Enmore sand procured from the coastal area in India near the City of Chennai (old Madras). The physical properties of the sand are as follows:

Specific gravity = 2.64
Uniformity coefficient = 1.63
Median size, D_{50} = 0.60 mm
Maximum dry density = 18.0 KN/m^3
Minimum dry density = 16.0 KN/m^3

TABLE 10.3(a)

Material Parameters for Natural and Reinforced Soils

Parameter	Natural Soil	Reinforced Soil (R1NW)	Reinforced Soil (R2NW)	Reinforced Soil (R2W)
Elastic constants				
k	600.00	500.00	230.00	690.00
n'	0.95	0.96	1.06	0.90
ν	0.34	0.37	0.36	0.34
Ultimate parameters				
γ	0.071	0.072	0.088	0.089
β	0.610	0.687	0.727	0.667
Phase change parameter, n	2.54	2.98	3.07	2.90
Hardening parameters				
a_1	0.366×10^{-3}	0.405×10^{-5}	0.397×10^{-4}	0.273×10^{-3}
η_1	0.711	1.611	1.327	0.721
Nonassociative parameter, κ	0.228	0.276	0.242	0.257

From Ref. 13, ©John Wiley & Sons Ltd. Reproduced with permission.

TABLE 10.3(b)

Material Parameters of Interface Between Soil and Nonwoven and Woven Geotextiles

Parameter		Nonwoven Geotextile	Woven Geotextile
Elastic constants ($k\,N/m^2/m$)	K_s	3,000	8,000
	K_n	60,000	90,000
Ultimate parameters	γ	0.722	0.656
Phase change parameter	n	2.84	2.50
Hardening parameters	a_1	0.098	0.253
	η_1	0.675	0.241
Nonassociative parameter	κ	0.875	0.802

See Chapter 11 for details of interfaces.

TABLE 10.3(c)

Parameters for Geotextile Reinforcement

Properties	Nonwoven Geotextile	Woven Geotextile
Material/color	Polypropylene/white	Polypropylene/white
Thickness (mm)	2.8	0.64
Stiffness modulus ($k\,N/m$)	23.13	660
Yield strength ($k\,N/m$)	11.65	19.93

The tests were performed under three confining pressures in the range of 100 to 300 kPa, with different stress paths, compression −CTC ($S_r = 1.00$) and extension − RTE ($S_r = -1.0$); see Chapter 7, Fig. 7.13.

The DSC for Structured and Stiffened Materials

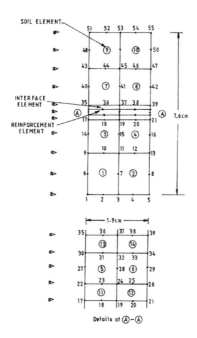

FIGURE 10.14
Finite-element discretization of reinforced soil sample (13). ©John Wiley & Sons Ltd. Reproduced with permission.

Tests for the sand-reinforcement interfaces were conducted by using a test box of 6.0 × 6.0 cm size, in which the lower half of the box contained a steel block over which the geotextile was wrapped and fixed. The upper half of the box was filled with dry sand. Normal stresses of 50, 100, and 100 kPa were used with shear displacement rate of 0.25 mm/min.

The elastic moduli and the yield strength of the geotextile were determined from tension tests conducted by using a Universal Testing Machine (13).

The HISS δ_0- and δ_1-plasticity models (Chapter 7) were used to characterize the behavior of the sand and interface (Chapter 11), while the woven and nonwoven reinforcements were modelled using the von Mises yield criterion. Table 10.3 (a), (b), and (c) shows the HISS parameters for the natural and reinforced sand, interface, and the parameters for the geotextiles, respectively. In the case of the sand, the elastic modulus, E, was expressed as a function of the confining stress, σ_3, as

$$E = kp_a \left(\frac{\sigma_3}{p_a}\right)^{n'} \qquad (10.21)$$

where k and n' are parameters and p_a is the atmospheric pressure constant.

Analysis The observed hardening behavior of the natural and reinforced soil with woven geotextile is compared in Fig. 10.15 (a) and (b) for the woven

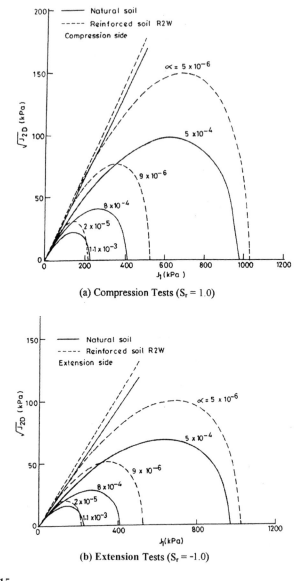

FIGURE 10.15
Comparisons for yield behavior of natural and reinforced soil (R2W) (13) ©John Wiley & Sons Ltd. Reproduced with permission.

geotextile. Figure 10.16 shows similar comparisons for nonwoven geotextiles. It can be seen that the reinforced soils exhibit lower values of the hardening or growth function, α; from Eq. (7.2), Chapter 7, the lower the value of α, the higher the value of plastic strains, which signifies increased yielding. Thus, reinforcement causes higher levels of yielding, and an increase in the number of layers of reinforcement increases yielding.

The DSC for Structured and Stiffened Materials

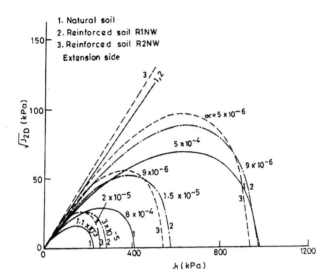

FIGURE 10.16
Comparisons of yield behavior for natural soil and reinforced soil, R1NW and R2NW: extension test ($S_r = -1.0$)(13). ©John Wiley & Sons Ltd. Reproduced with permission.

Figure 10.17 shows yielding behavior for the interfaces with nonwoven and woven reinforcements. It indicates that the growth function, α, is lower for nonwoven geotextile; i.e., the yielding is higher for the nonwoven geotextile compared to that for the woven geotextile.

10.4.3 Validation for Test Results with HISS Model and FE Method

The triaxial stress–strain–volume change behavior was predicted using two methods: SPM and FEM. In the SPM method, the specimen was treated as an equivalent composite element. The predictions were obtained by integrating the incremental constitutive equations [Eq. (10.17)], in which the HISS parameters for the composite reinforced soil were used; see Table 10.3.

In the FEM, the specimen was considered to be made of three components: natural soil, reinforcement (geotextile), and interface between the soil and reinforcement. The properties of individual components were provided in the FEM (Table 10.3).

The validations were obtained by using (average) parameters for the tests which were used for finding the parameters (Group A). Then validations were obtained for independent tests *not* used for finding the parameters (Group B). Only typical results are given below.

Figures 10.18 and 10.19 show comparisons between predictions and test data for the reinforced specimen (R2W) for the reduced triaxial extension (RTE) and triaxial compression (TC) stress paths (Fig. 7.13), respectively, for Group A and Group B analyses. It can be seen that both the SPM and FEM predictions correlate well with the test behavior.

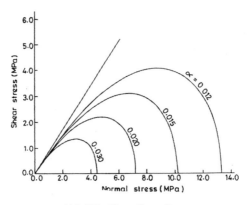

(a) Soil-Non-Woven Geotextile

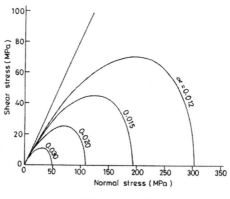

(b) Soil-Woven Geotextile

FIGURE 10.17
Plots of yield surfaces for soil-reinforcement interfaces (13). ©John Wiley & Sons Ltd. Reproduced with permission.

10.4.4 Jointed Rock

The FEM approach, in which each component is treated individually, was applied to analyze the (laboratory) behavior of jointed rock masses using the DSC model (14). It was found that the DSC model provided very good correlation with the laboratory behavior of a rock mass with three inclined joints.

The foregoing results show that the special version, the HISS plasticity model of the DSC, provides satisfactory simulation of the behavior of the reinforced soil. Both the equivalent composite and the component approaches can be used for characterizing the behavior of a reinforced composite. The DSC model including softening or degradation would also provide satisfactory results, as was reported by Chia et al. (14).

Characterization of soils and interfaces, and FE predictions for a Tensar (plastic) reinforced retaining wall using the DSC, are given in Chapter 13.

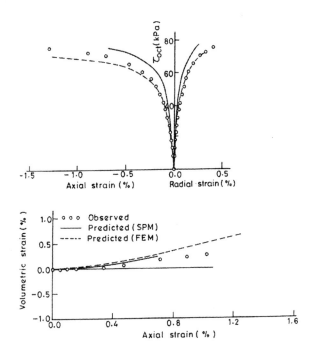

FIGURE 10.18
Comparisons for stress–strain–volume change response of reinforced soil R2W: RTE path, σ_3 = 200 kPa (Group A) (13). ©John Wiley & Sons Ltd. Reproduced with permission.

10.5 Rest Periods: Unloading

During rest periods and under unloading, some materials may experience healing or strengthening due to interparticle rebonding under chemical effects and/or reduction in microcracks; the latter, however, will depend on the stress state (compressive or tensile) existing in the material before unloading took place. It is difficult to identify the effect of chemical rebonding. We now give two examples of microstructural changes due to unloading.

We first consider the example of the behavior under uniaxial tensile loading for a flat specimen of a fiber-reinforced ceramic composite (15, and Example 5.6, Chapter 5).

Figure 10.20 shows the plots of crack density, C_d vs. $\sqrt{J_{2D}}$, and disturbance, D_v vs. $\sqrt{J_{2D}}$, under a cycle of (virgin) loading, unloading, and reloading. The value of C_d is computed using Eq. (6b), Example 5.6 of Chapter 5, and that for disturbance, D_v [Eq. (3.9), Chapter 3], is computed using measured (average) ultrasonic P-wave velocity ($\bar{V}$); see Fig. 10.21 (15).

It can be seen from Fig. 10.20 that during the virgin loading, the crack density and disturbance increase with shear stress, $\sqrt{J_{2D}}$. During unloading both decrease because, as the tensile stress is removed, the microcracks close with

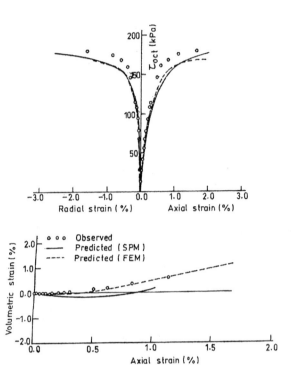

FIGURE 10.19
Comparisons for stress–strain–volume change response of reinforced soil R2W: TC path, $\sigma_3 = 200$ kPa (Group B) (13). ©John Wiley & Sons Ltd. Reproduced with permission.

a consequent reduction in the disturbance. Upon reloading, both C_d and D_v increase again; at the end of reloading, the virgin loading response resumes. Thus, during unloading, as the disturbance decreases, the material exhibits a healing or stiffer response [Eq. (10.14)].

On the other hand, if the material is under compressive stress state when unloading occurs, the situation can be different. Figure 10.22 illustrates this phenomenon for the behavior of a cemented sand tested under multiaxial compressive loading, e.g., the CTC stress path ($\sigma_1 > \sigma_2 = \sigma_3$) (Fig. 7.13, Chapter 7). Figure 10.23 shows the plot of measured $\bar{V}$ vs. $\sqrt{J_{2D}}$ for the cemented sand used to compute D_v.

During the virgin loading, both C_d and D_v (Fig. 10.22) increase. Under unloading, both increase because as the compressive stress is removed, the microcracks can experience opening, resulting in an increase in the disturbance. During reloading, C_d and D_v first decrease due to the coalescence of microcracks, and then they increase. Thus, mechanical healing during unloading is not present, except during a part of the reloading.

Hence, the existence of healing or stiffening due to mechanical unloading depends on the stress state. Healing due to chemical effects can occur simultaneously during unloading, but requires additional studies and research for its definition and quantification.

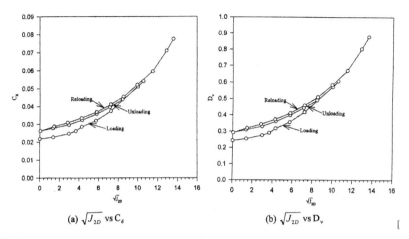

FIGURE 10.20
Variations of D_v and C_d for fiber ceramic (15) (with permission).

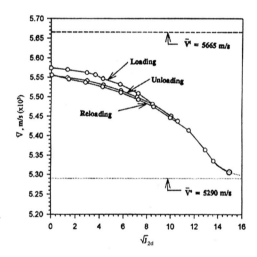

FIGURE 10.21
Average ultrasonic P-wave velocity for fiber-reinforced ceramic (15) (with permission from Elsevier Science).

References

1. Desai, C.S., "Consistent Finite Element Technique for Worksoftening Behavior," *Proc. Int. Conf. on Comp. Meth. in Nonlinear Mech.*, J.T. Oden (Editor), Univ. of Texas, Austin, TX, 1974.
2. Vaid, Y.P. and Thomas, J., "Liquefaction and Post Liquefaction Behavior of Sand," *J. of Geotech. Eng., ASCE*, 121, 2, 1995, 163–173.
3. Desai, C.S., Park, I.J., and Shao, C., "Fundamental Yet Simplified Model for Liquefaction Instability," *Int. J. Num. Analyt. Meth. Geomech.*, 22, 1998, 721–748.

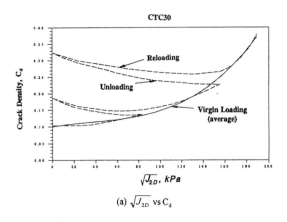

(a) $\sqrt{J_{2D}}$ vs C_d

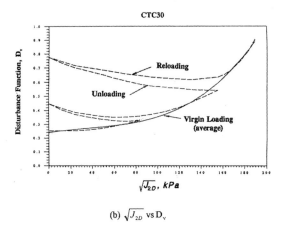

(b) $\sqrt{J_{2D}}$ vs D_v

FIGURE 10.22
Variations of D_v and C_d for CTC30-cemented sand (15) (with permission from Elsevier Science)..

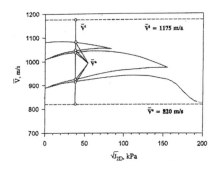

FIGURE 10.23
Average ultrasonic P-wave velocity for cemented sand (15) (with permission from Elsevier Science)..

4. Kim, Y.R., Lee, H.J., Kim, Y., and Little, D.N., "Mechanistic Evaluation of Fatigue Damage Growth and Healing of Asphalt Concrete: Laboratory and Field Experiments," *Proc. 8th Int. Conf. on Asphalt Pavements,* Seattle, WA, USA, 1987, 1089–1108.
5. Liu, M.D., Carter, J.P., Desai, C.S., and Xu, K.J., "Analysis of the Compression of Structured Soils Using the Disturbed State Concept," *Int. J. Num. Analyt. Methods in Geomech.,* 24, 8, 2000, 723–735.
6. Scofield, A.N. and Wroth, C.P., *Critical State Soil Mechanics,* McGraw Hill, London, U.K., 1968.
7. Liu, M.D. and Carter, J.P., "On the Volumetric Deformation of Reconstituted Soils," *Int. J. Num. Analyt. Meth. Geomech.,* 24, 2, 2000, 101–133.
8. Terzaghi, K., "Fifty Years of Subsoil Exploration," *Proc. 3rd Int. Conf. on Soil Mech. and Found. Eng.,* 1953, 227–238.
9. Quigley, R.M. and Thompson, C.D., "The Fabric of Anisotropic Consolidated Sensitive Marine Clay," *Canadian Geotech. J.,* 3, 1966, 61–73.
10. Dillon, G.W., Tsai, C.T., and DeAnglis, R.J., "Dislocation Dynamics During the Growth of Silicon Ribon," *J. of Appl. Physics,* 60, 5, 1986, 1784–1792.
11. Yonenaga, I., Sumino, K., and Hoshi, J., "Mechanical Strength of Silicon Crystals as a Function of the Oxygen Concentration," *J. of Appl. Physics,* 56, 8, 1984, 2346–2350.
12. Desai, C.S., Dishongh, T.J., and Deneke, P., "Disturbed State Constitutive Model for Thermomechanical Behavior of Dislocated Silicon with Impurities," *J. of Appl. Physics,* 84, 11, 1998, 5977–5984.
13. Varadarajan, A., Sharma, K.G., and Soni, K.M., "Constitutive Modelling of a Reinforced Soil Using Hierarchical Model," *Int. J. Num. Analyt. Meth. Geomech.,* 23,3, 1999, 217–241.
14. Chia, J., Chern, J.C., and Lin, C.C., "DSC Finite Element Modeling of Progressive Failure in Jointed Rock," *Proc. 9th Int. Conf. on Computer Meth. and Advances in Geomech.,* Yuan, J.X., (Editor), Balkema, Rotterdam, 1997.
15. Desai, C.S. and Toth, J., "Disturbed State Constitutive Modeling Based on Stress–Strain and Nondestructive Behavior," *Int. J. Solids and Struct.,* 33, 11, 1996, 1619–1650.

11

The DSC for Interfaces and Joints

CONTENTS

11.1 General Problem 422
11.2 Review 422
    11.2.1 Comments 424
11.3 Thin-Layer Interface Model 425
11.4 Disturbed State Concept 427
    11.4.1 Relative Intact Behavior 427
    11.4.2 Stress-Displacement Equations 428
    11.4.3 Elastoplastic Models 430
    11.4.4 Fully Adjusted State 432
    11.4.5 FA as the Critical State 432
11.5 Disturbance Function 435
11.6 Incremental Equations 437
    11.6.1 Specializations 439
    11.6.2 Alternative DSC Equations..................................... 440
11.7 Determination of Parameters 441
    11.7.1 Mathematical and Physical Characteristics of the DSC 447
    11.7.2 Regularization and Penalty 448
11.8 Testing 448
11.9 Examples 449
    11.9.1 Creep Behavior 464
    11.9.2 Testing 464
11.10 Computer Implementation 472

The subject of contacts between two similar or dissimilar materials has been studied for a long time. The classical friction laws by Coulomb and Amontons considered the case of dry friction between (rigid) bodies. Subsequently, the pursuit for understanding the complex response at contacts as affected by many practical factors beyond those considered in the classical laws has continued.

    In engineering systems contacts are often referred to as interfaces or joints between two materials. Such discontinuities may preexist or are induced in a deforming material system. Existence of discontinuities in a material body whose parts around and beyond the discontinuities may be continuous can lead the overall system to be discontinuous. Hence, the theories based on continuum mechanics may not be applicable to characterize its response. As described in Chapter 2, an initially continuous body may experience internal microcracking during loading. Up to a certain extent of microcracks, the material may still be treated as continuous. However, the microcracks often coalesce and lead to zones of macrolevel fractures. As a result, the material may no longer be treated as a continuum. In fact, such finite-fracture discontinuities may represent interfaces or joints and need to be treated as such.

11.1 General Problem

Contacts can occur in metals, ceramics, concrete, rocks and soils, composites, and structure-medium (soil or rock) combinations. In mechanical and aerospace engineering, particularly in machinery and aircraft structures, metal-to-metal contacts are common, and behavior is often treated under the subject of *tribology*, involving motions and resulting wear, tear, and degradation. In civil engineering, such contacts refer to interfaces in structures and geologic foundations, between concrete and reinforcement, and joints in rock masses. Examples of contacts in the electronic industry can involve interfaces and joints in electronic assemblies such as chip-substrate systems.

Although a number of models have been proposed for the contact problem, its behavior has not yet been fully understood and characterized, particularly when we consider the number of significant factors that influence the behavior. In the following, we provide a brief review of various models before describing the rationale for the development and use of the models based on the DSC.

11.2 Review

The subjects of contacts, friction, interfaces and joints involve extensive research and publications. Comprehensive reviews are available in (1–8). The objective here is to provide a limited review, particularly with respect to constitutive models leading to the development of the DSC.

Amontons' (9) and Coulomb's law of friction can be considered to be the first formal constitutive model to describe the response of two bodies in contact [Fig. 11.1(a)]. The law is expressed as

$$F = \mu N \qquad (11.1)$$

where F is the tangential or shear force parallel to the plane of contact, μ is the coefficient of friction, and N is the force normal to the contact with

The DSC for Interfaces and Joints

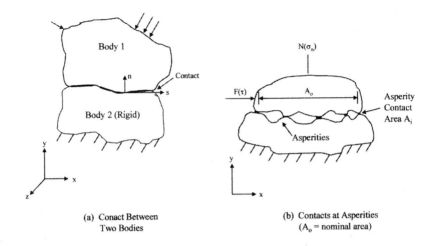

FIGURE 11.1
Schematic of contact and asperities.

nominal area A_0. The above law is valid for dry friction between rigid bodies, assumes μ to be constant, and provides a pointwise description leading to definition of *gross* sliding of one body relative to the other. In reality, however, the contact involves mated or nonmated asperities [Fig. 11.1(b)] and the bodies in contact can be deformable. As a result, the pointwise or local definition is not sufficient, and consideration needs to be given to nonlocal effects due to factors such as nonuniform properties and lack of complete contact. The latter would require that the contact area $\bar{A}$ be smaller than the nominal area, A_0, and its effect on the behavior needs to be considered.

For including elastic component of response, Archard (10) proposed the following law:

$$F = \mu N^m \tag{11.2}$$

where m is the exponent whose value varies between 2/3 and 1; for purely elastic behavior, $m = 2/3$, and for ductile contacts $m = 1$.

Oden and Pires (11) have discussed the limitations of the Coulomb friction law, in particular its local and pointwise character, and presented variational principles by treating the materials as elastic. The law in Eq. (11.1) was modified as

$$F = A_r s \tag{11.3a}$$

where s is the average shear strength of the interface and A_r is the weighted actual contact area, given by

$$A_r = A_1 + A_2 \ldots = \frac{N_1}{p_0} + \frac{N_2}{p_0} + \ldots = \frac{N}{p_0} \tag{11.3b}$$

where $A_1, A_2 \ldots$ are the contact areas of individual deforming asperities, $N_1, N_2 \ldots$ are the normal loads on the asperities, and p_0 is the approximately constant local plastic yield pressure. Substitution of A_r in Eq. (11.3a) gives

$$F = \frac{s}{p_0}N = \mu N \tag{11.3c}$$

Here $s/p_0 = \mu$. Thus, F is defined over the weighted area, A_r.

In the general areas of friction and contact mechanisms, models based on strength, limit equilibrium, elastic and classical elastoplastic theories have been presented in (1, 2, 4, 6–8, 12–18). In the context of interfaces in structure-medium (soil) and joints in rock, models based on similar and advanced concepts have been presented in (3, 5, 19–42).

11.2.1 Comments

As indicated earlier, a number of significant factors influence the behavior at interface or joint. They include nonlocal considerations, particularly with respect to the normal behavior, nonlinear effects involving elastic, plastic and creep strains, microcracking, degradation and softening, stiffening or healing, existence of filler materials (oxides, gouge, etc.), fluids and type of loading (static, cyclic, environmental, etc.). Although the foregoing available models allow for strength, elastic, and limited plasticity response, they usually do not allow for factors such as continuous yielding or hardening, microcracking and softening leading to postpeak degradation, stiffening, and viscous (time) effects. The objective here is to present the unified DSC model that allows for these factors, in addition to those included in previous models. It is also noted that many of the previous models can be derived as special cases of the DSC.

It is felt that the interface behavior needs to be treated as a problem in constitutive modelling in which the foregoing factors are integrated. This is in contrast to some of the previous models, which have treated contact or interface behavior by introducing constraints (kinematic and/or force) to allow for the effects of special characteristics such as relative motions (sliding, debonding, etc.).

Another important issue is the implementation of the models in solution (computer) procedures with due consideration to factors such as robustness, accuracy, and stability of the numerical predictions. This aspect is handled by using the concept of the *thin-layer* element, in which the interface zone is simulated as a thin zone of finite thickness, t (36). In the finite-element procedure, the interface zone is treated as a regular element whose constitutive behavior (with the DSC) is defined based on appropriate laboratory tests using special shear testing devices.

Models based on the DSC that include elastic, plastic, and creep deformations, microcracking, and (asperity) degradation response have been presented in (42–51). The DSC approach allows for the nonlocal effects and characteristic dimension and hence leads to computations that are unaffected by spurious mesh dependence (52–54). It may be mentioned that these models allow realistic characterization of the normal response discussed before (11, 12, 36–39, 41) and the coupled shear and normal response. Particular attention is given here to the (laboratory) measurements of the normal response (55), which is discussed subsequently under testing.

The following descriptions include the DSC models for interfaces and joints, laboratory testing for the calibration of material parameters, and implementation of the models in numerical procedures using the thin-layer element. The latter is described first, as the DSC is formulated with respect to the thin-layer simulation of interfaces and joints.

11.3 Thin-Layer Interface Model

An interface can involve a number of configurations. The interface between two metallic bodies can be considered *clean* in the sense that there is no third material between them. Then the smeared interface zone can entail different levels of roughness defined by the surface (microlevel) asperities [Fig. 11.2(a)]. The contact can be very rough to medium-rough to smooth, depending on the characteristics (height, length) of irregular asperities. It may be noted that even a "smooth" contact involves asperities at different levels (macro, micro, etc.); hence, an *ideally* smooth surface is essentially hypothetical.

It may happen that when two bodies made of different materials are in contact, like steel (pile) and soft clays [Fig. 11.2(b)], there exists a *finite "smeared"* zone between the two that behaves as an interface [Fig. 11.2(e)].

In the case of rock joints, it can happen that the contact is filled with a third (e.g., gouge) material that acts as a *bulk* interface [Fig. 11.2(c)]. Similar situations occur in the case of chip-substrate system joined together, say, by solders. Here, the filled zone can be treated as the bulk interface [Fig. 11.2(d)]. It may be noted, however, that in Fig. 11.2(b), (c), and (d), other interfaces occur between the material in the interface and the bodies in contact, which may involve additional considerations such as diffused layers, intermetallics, and surface effects. These are not considered at this time.

The objective herein is to present a model for the bulk interface or joint that can provide a realistic simulation of relative motions between the two bodies. The interface zone is referred to as thin-layer (36, 56) with thickness, t, that can be treated as equivalent or smeared zone between two materials [Fig. 11.2(e)]. If characterized appropriately, it can provide, in a weighted sense, model(s) for the response at the interface. Even in the case of clean contact [Fig. 11.2(a)], it can be possible to develop the dimension (thickness) of the equivalent "smeared" zone, as affected by the asperities, to represent the interface behavior. Thus, in all cases in Fig. 11.2, it is assumed that the interface zone can be represented by an equivalent dimension with thickness, t [Fig. 11.2(e)].

Then the vital question arises as to how to determine the thickness, t. This can be difficult. One of the direct ways would be to perform tests in which the

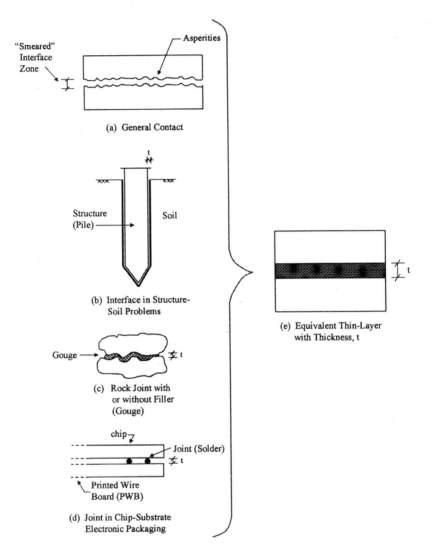

FIGURE 11.2
Examples of interfaces and joints.

deformation behavior at the interface is measured, and its significant dimension (t) is determined. Both nondestructive and mechanical testing are possible. In the case of nondestructive tests such as X-ray computerized tomography and acoustic methods, it can be possible to measure the dimensions of the influence zones around the asperities. It is also possible to perform integrated numerical (finite-element) predictions and (laboratory) observations to develop empirical criteria for defining the thickness, t (36, 56). We shall discuss these aspects later.

The DSC for Interfaces and Joints

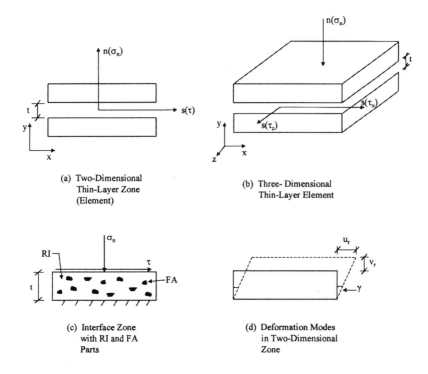

FIGURE 11.3
Thin-layer element and material parts in RI and FA states.

11.4 Disturbed State Concept

Now, as the interface is represented by an equivalent thickness, t [Fig. 11.3(a, b)], it can be treated as a deforming material element that is composed of the relative intact (RI) and fully adjusted (FA) parts. The task now is to define RI and FA behavior for the interface zone.

11.4.1 Relative Intact Behavior

The RI behavior can be represented by using theories such as linear (nonlinear) elasticity, elastoplasticity (δ_0-model for interfaces and joints), thermoelastoplasticity, and thermoviscoplasticity (Chapters 3–10).

If the (linear) theory of elasticity is used, the RI behavior can be simulated using two moduli, shear stiffness, k_s, and normal stiffness, k_n (Fig. 11.4). The values of these moduli can depend on such factors as the (initial) normal stress (σ_{n0}) and roughness (R_0). Thus,

$$k_s = k_s(\sigma_{n0}, R) \tag{11.4a}$$

$$k_n = k_n(\sigma_{n0}, R) \tag{11.4b}$$

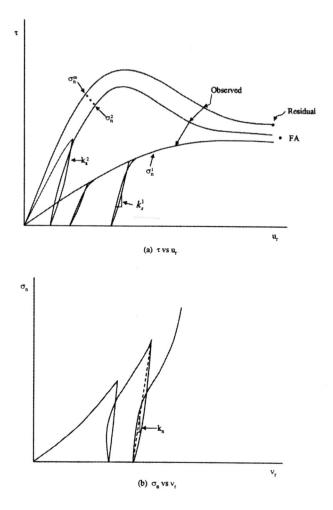

FIGURE 11.4
Schematics of shear and normal behavior.

11.4.2 Stress-Displacement Equations

For the two-dimensional idealization [Fig. 11.3(c)], the incremental equations for the RI behavior with the nonlinear piecewise elastic response are expressed as

$$\left\{ \begin{array}{c} d\tau \\ d\sigma_n \end{array} \right\}^i = \begin{bmatrix} k_{ss}^t & k_{sn}^t \\ k_{ns}^t & k_{nn}^t \end{bmatrix}^i \left\{ \begin{array}{c} du_r \\ dv_r \end{array} \right\}^i \quad (11.5a)$$

where τ and σ_n are the shear and normal stresses, respectively, the superscript t denotes tangent shear and normal stiffnesses, u_r and v_r are the relative shear

The DSC for Interfaces and Joints

and normal displacements, respectively, i denotes RI or continuum response, and d denotes an increment. If it is assumed that the elastic shear and normal responses are uncoupled, Eq. (11.5a) will reduce to

$$\left\{\begin{array}{c} d\tau \\ d\sigma_n \end{array}\right\}^i = \begin{bmatrix} k_s^t & 0 \\ 0 & k_n^t \end{bmatrix} \left\{\begin{array}{c} du_r \\ dv_r \end{array}\right\} \tag{11.5b}$$

or

$$d\underline{\sigma}^i = \bar{C}_j^i\, d\underline{\bar{u}}^i \tag{11.5c}$$

where $d\underline{\sigma}^T = [d\tau\ d\sigma_n]$, $d\underline{\bar{u}}^T = [du_r\ dv_r]$, and $\bar{C}_j$ is the interface or joint tangent constitutive matrix.

As discussed earlier, the net contact area involves contacts at the asperities between the two bodies and is usually smaller than the nominal or total area of the contact, A_0. In order to account for the nonlocal effects, it is necessary to introduce weighting functions. In the DSC, such weighting is introduced through the disturbance function, D (see later). However, based on the nominal area, the shear and normal stress are first defined as

$$\tau = \frac{F}{A_0} \tag{11.6a}$$

$$\sigma_n = \frac{N}{A_0} \tag{11.6b}$$

With the thin-layer element, the stiffness moduli can be expressed approximately as

$$k_s t \approx G \tag{11.7a}$$

$$k_n t \approx E \tag{11.7b}$$

where G and E are the equivalent shear and elastic moduli for the interface, respectively.

Now, the relative shear displacement, u_r, is expressed as [Fig. 11.3(d)]

$$u_r \approx \gamma t \tag{11.8a}$$

Therefore,

$$\gamma \approx \frac{u_r}{t} \tag{11.8b}$$

and

$$v_r \approx \varepsilon_n t \tag{11.9a}$$

Hence,

$$\varepsilon_n \approx \frac{v_r}{t} \tag{11.9b}$$

where γ and ε_n are the shear and normal strains, respectively. Then, Eq. (11.5) can be written as

$$\begin{Bmatrix} d\tau \\ d\sigma_n \end{Bmatrix}^i = \begin{bmatrix} G_t & 0 \\ 0 & E_t \end{bmatrix}^i \begin{Bmatrix} d\gamma \\ d\varepsilon_n \end{Bmatrix}^i \tag{11.10a}$$

or

$$d\underline{\sigma}^i = \underline{C}_j^i \, d\underline{\varepsilon}^i \tag{11.10b}$$

Here, tangent values of G and E are obtained from laboratory shear and normal behavior (Fig. 11.4) using Eq. (11.7), and $\underline{\varepsilon} = [\gamma \; \varepsilon_n]$ is the vector of interface strains.

11.4.3 Elastoplastic Models

The RI behavior can be simulated using an elastic or elastoplastic model. Here, the δ_0-version of the HISS plasticity family (Chapter 7) is considered.

It is shown (42, 46, 47) that for the two-dimensional interface (Fig. 11.3), the yield function, F (Eq. 7.1, Chapter 7) can be specialized as

$$F = \tau^2 + \alpha \sigma_n^{*n} - \gamma \sigma_n^{*2} \tag{11.11}$$

where $\sigma_n^* = \sigma_n + R$, R is the intercept along $-ve \; \sigma_n$ (Fig. 11.5) and is related to adhesion, c_0, the stress quantities are nondimensionalized with respect to p_a, n and γ are the phase change and ultimate parameters, and α is the hardening or growth function:

$$\alpha = \frac{a}{\xi^b} \tag{11.12}$$

where a and b are the hardening parameters and ξ is the trajectory of irreversible or plastic shear and normal displacements (strains):

$$\xi = \int (du_r^p \, du_r^p + dv_r^p \, dv_r^p)^{1/2} = \xi_D + \xi_v \tag{11.13}$$

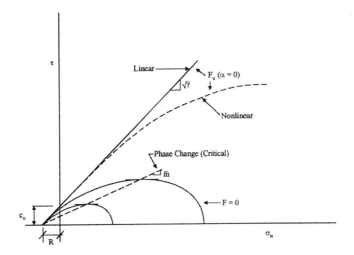

FIGURE 11.5
Yield and ultimate surfaces.

Here, u_r^p and v_r^p are the relative plastic shear and normal displacements, respectively. The total values of relative shear (u_r) and normal (v_r) displacements can be expressed as (44)

$$u_r = u^e + \bar{u}^p + u^s \qquad (11.14a)$$

$$v_r = v^e + \bar{v}^p + v^s \qquad (11.14b)$$

where u^e and v^e are the elastic deformations of the contact asperities, $\bar{u}^p$ and $\bar{v}^p$ are the plastic deformations of the asperities, and u^s and v^s are the slip (irrecoverable) displacements, respectively. As the last two are difficult to separate and are irrecoverable, they are combined as u^p and v^p as

$$u_r = u^e + u^p \qquad (11.15a)$$

$$v_r = v^e + v^p \qquad (11.15b)$$

Then, u^p and v^p are used in the modelling.

As in the case of solids (Chapter 7), the surfaces given by F in Eq. (11.11) plot as continuous yield surfaces, which approach the ultimate yield surface, F_u, when $\alpha = 0$. A schematic of F in the $\tau - \sigma_n$ stress space is shown in Fig. 11.5. Here, the function F will change with α, depending on the roughness of the interface (43, 44).

Very often, the ultimate yield surface or envelope that represents the asymptotic stress states to the observed stress–strain ($\tau - u_r$) curves (Fig. 11.4) will be an (average) straight line with slope equal to $\sqrt{\gamma}$ (Fig. 11.5) [Eq. (11.11)]. If the envelope is curved, F can be expressed as (44)

$$F = \tau^2 + \alpha \sigma_n^{*n} - \gamma \sigma_n^{*q} = 0 \ldots \tag{11.16}$$

where q is a parameter; its value equals 2 for the straight-line envelope. As noted before, $\alpha = 0$ at the ultimate, hence, Eq. (11.16) gives

$$\gamma = \frac{\tau^2}{\sigma_n^{*q}} \tag{11.17a}$$

which, for $q = 2$, leads to

$$\gamma = \frac{\tau^2}{\sigma^{*2}} \tag{11.17b}$$

that is,

$$\sqrt{\gamma} = \frac{\tau}{\sigma_n^{*}} \tag{11.17c}$$

For the linear ultimate envelope, R can be found as

$$R = \frac{c_0}{\sqrt{\gamma}} \tag{11.17d}$$

11.4.4 Fully Adjusted State

As in the case of solids, the material in the FA state can be characterized such that it (1) has no strength at all, that is, it can carry no shear or normal stress, or (2) is a *constrained liquid* that can carry normal stress but no shear stress, or (3) is at the *critical state* at which it can continue to carry shear stress reached up to that state for given (initial) normal stress with no change in volume or normal displacement. The first two conditions are straightforward and have been discussed before (Chapter 3). The idea of a critical state for joints and interfaces is presented below, based on experimental and analytical studies.

11.4.5 FA as the Critical State

The idea of the critical state for the behavior of joints and interfaces can be based on their observed response. Figure 11.6 shows schematics of τ vs. u_r, and normal displacement during shear, v_r vs. u_r for typical interface or joint. Both initially smooth and rough joints for a given σ_n approach an invariant

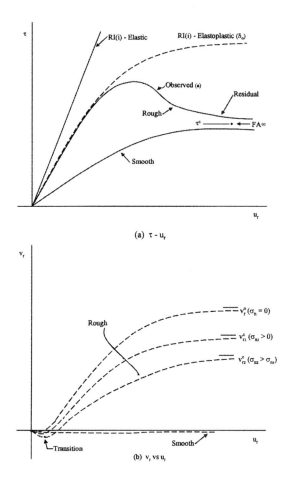

FIGURE 11.6
RI and FA states during shear.

shear stress, τ^c, and normal displacement, v_r^c, through compaction, and compaction followed by dilation [Fig. 11.6(b)]. Based on similar results, Archard (10) proposed the following expression for critical shear stress and normal stress, σ_n:

$$\tau^c = c_0 + c_1 \sigma_n^{(c)c_2} \tag{11.18}$$

where c_0 is the critical value of τ^c when $\sigma_n = 0$ (related to the adhesive strength), σ_n^c is the normal stress at the critical state, and c_1 and c_2 are parameters for the critical state. Equation (11.18) is similar to the critical-state line equation for normally consolidated soils [i.e., relevant to the second term in Eq. (11.18)]:

$$\sqrt{J_{2D}^c} = \overline{m} J_1^c \tag{11.19}$$

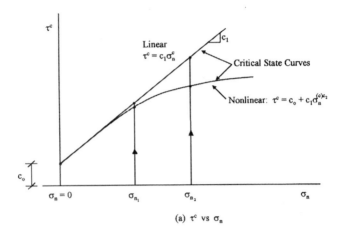

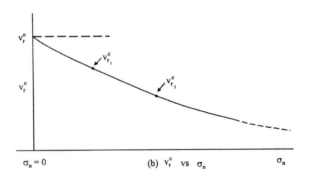

FIGURE 11.7
Behavior at critical (FA) state.

(see Chapter 7). Here, c_1 is analogous to $\bar{m}$ and c_2 is analogous to the power of J_1^c, which is unity. Indeed, Eq. (11.18) can represent a critical-state curve [Fig. 11.7(a)] instead of straight line (Chap. 7) and includes the interface adhesion or cohesion (c_0).

Based on shear test on joints, Schneider (57, 58) reported that during shear loading, a joint under given normal stress would approach a critical value of normal displacement (dilation), v_r^c [Fig. 11.6(b)], and its relation with σ_n is given by

$$v_r^c = v_r^0 e^{-\bar{\lambda}\sigma_n} \tag{11.20}$$

where v_r^0 is the critical or ultimate dilation when $\sigma_n = 0$ [Fig. 11.6(b)] and $\bar{\lambda}$ is a material parameter. As shown in Fig. 11.6(b), for each σ_n there occurs an associated value of critical normal displacement, v_{r1}^c, v_{r2}^c, and so on. Equation (11.20) is similar to the void ratio–pressure relationship in the critical-state concept (Chapter 7).

$$e^c = e_0^c - \lambda \ln(p) \tag{11.21}$$

where e^c is the critical void ratio for given mean pressure, $p = J_1/3 = (\sigma_1 + \sigma_2 + \sigma_3)/3$, λ is a parameter, and e_0^c is the void ratio corresponding to $p =$ unity.

Thus, as can be seen in Fig. 11.7(b), for a given value of σ_n, the interface or joint approaches the critical (or FA) state at which the shear stress, τ^c, and the normal displacement, v_r^c, remain the same, and shear displacements (u_r) continue under constant τ^c.

Then Eqs. (11.18) and (11.20) can be used to characterize the behavior of parts in the joint or interface zone that are in the FA state. Now we consider the issue of the disturbance function, D.

11.5 Disturbance Function

The disturbance function, D, can be defined based on the shear stress (τ) vs. relative shear displacement (u_r) and/or relative normal (v_r) vs. u_r curves (Fig. 11.8). In the case of the former [Fig. 11.8(a)]

$$D_r = \frac{\tau^i - \tau^a}{\tau^i - \tau^c} \tag{11.22a}$$

and for the latter [Fig. 11.8(b)]

$$D_v = \frac{v_r^i - v_r^a}{v_r^i - v_r^c} \tag{11.22b}$$

For saturated interfaces, the effective normal stress ($\bar{\sigma}_n = \sigma - p$) data can be used to define the disturbance [Fig. 11.8(c)]:

$$D_{\sigma_n} = \frac{\bar{\sigma}_n^i - \bar{\sigma}_n^a}{\bar{\sigma}_n^i - \bar{\sigma}_n^c} \tag{11.22c}$$

Details for the evaluation of RI normal stress, $\bar{\sigma}_n^i$, are given in Chapter 9.

As discussed in Chapter 7, for cyclic behavior, disturbance as a function of cycles, N (time), is calculated based on the RI and peak values of the quantities, e.g. [Fig. 11.8(d)];

$$D(N) = \frac{\tau^i - \tau^p}{\tau^i - \tau^c} \tag{11.22d}$$

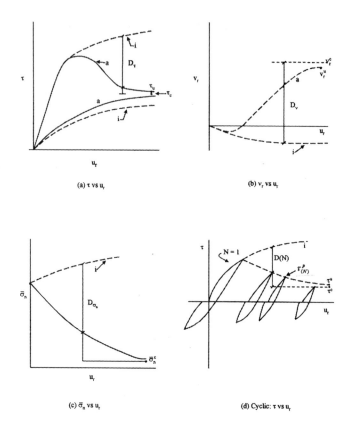

FIGURE 11.8
Disturbance based on different tests.

The disturbances, Eq. (11.22), can be expressed in terms of plastic relative displacement trajectory, ξ,

$$D = D_u(1 - e^{-A\xi^Z}) \qquad (11.23a)$$

where ξ is composed of both the deviatoric and normal components [Eq. (11.13)]. It can be also expressed in terms of the normal (ξ_v) or deviatoric (ξ_D) component:

$$D_n = D_{nu}\left(1 - e^{-A_n \xi_v^{Z_n}}\right) \qquad (11.23b)$$

and

$$D_\tau = D_{\tau u}\left(1 - e^{-A_\tau \xi_D^{Z_\tau}}\right) \qquad (11.23c)$$

The DSC for Interfaces and Joints

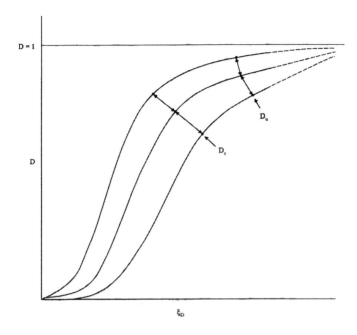

FIGURE 11.9
Disturbance vs. deviatoric displacement (strain) trajectory.

where D_u is the ultimate value of D at the residual which can be adopted to be unity, as all responses asymptotically tend to $D = 1$, and A and Z are parameters. For the shear behavior, they will be A_τ and Z_τ, and for volumetric (normal displacement) they will be A_n and Z_n. The value of D_u corresponding to the residual state can be calculated using Fig. 11.8; a schematic of D vs. ξ_D is shown in Fig. 11.9.

11.6 Incremental Equations

The incremental DSC equations for solids are given by (Chapter 7)

$$d\sigma_{ij}^a = (1 - D)C_{ijk\ell}^i d\varepsilon_{k\ell}^i + DC_{ijk\ell}^c d\varepsilon_{k\ell}^c + dD(\sigma_{ij}^c - \sigma_{ij}^i) \qquad (11.24a)$$

or

$$d\underset{\sim}{\sigma} = (1 - D)\underset{\sim}{C}^i \, d\underset{\sim}{\varepsilon}^i + D\underset{\sim}{C}^c \, d\underset{\sim}{\varepsilon}^c + dD(\underset{\sim}{\sigma}^c - \underset{\sim}{\sigma}^i) \qquad (11.24b)$$

The interface can be treated as three-dimensional. Here the formulation is presented by assuming the two-dimensional idealization (Fig. 11.3), where only the shear stress ($\sigma_{12} = \tau$) and the normal stress ($\sigma_{22} = \sigma_n$) are significant (42, 44, 56), and Eq. (11.24) specializes to

$$d\tau^a = (1 - D)d\tau^i + Dd\tau^c + dD(\tau^c - \tau^i) \qquad (11.25a)$$

$$d\sigma_n^a = (1 - D)d\sigma_n^i + Dd\sigma_n^c + dD(\sigma_n^c - \sigma_n^i) \qquad (11.25b)$$

or

$$\begin{Bmatrix} d\tau^a \\ d\sigma_n^a \end{Bmatrix} = (1 - D)\bar{\underline{C}}^i \begin{Bmatrix} du_r^i \\ dv_r^i \end{Bmatrix} + D\bar{\underline{C}}^c \begin{Bmatrix} du_r^c \\ dv_r^c \end{Bmatrix} + dD \begin{Bmatrix} \tau_R \\ \sigma_{nR} \end{Bmatrix} \qquad (11.25c)$$

where $\bar{\underline{C}}^i$ is the RI constitutive matrix, $\bar{\underline{C}}^c$ is the FA constitutive matrix, and $\tau_R = \tau^c - \tau^i$ and $\sigma_{nR} = \sigma_n^c - \sigma_n^i$ are the relative stresses. We assumed the same D for the shear and normal responses.

If $D = 0$, Eq. (11.25c) reduces to RI equations as

$$\begin{Bmatrix} d\tau^i \\ d\sigma_n^i \end{Bmatrix} = \bar{\underline{C}}^i \begin{Bmatrix} du_r^i \\ dv_r^i \end{Bmatrix} \qquad (11.26)$$

For (linear) elastic RI response, $\bar{\underline{C}}^i$ will be uncoupled and composed of shear and normal stiffnesses, k_s and k_n. For elastoplastic characterization, $\bar{\underline{C}}^i$ matrix will be coupled, and symmetric or nonsymmetric depending on the associative or nonassociative model used.

Now, assume that the FA displacement can be expressed as

$$d\underline{U}^c = (1 + \bar{\alpha})d\underline{U}^i \qquad (11.27a)$$

where $\underline{U}^{(c)T} = [du_r^c, dv_r^c]$ and $\underline{U}^{(i)T} = [du_r^i, dv_r^i]$, and $\bar{\alpha}$ is the relative motion parameter (Chapter 4). Also, dD can be derived as (Chapter 4)

$$dD = \underline{R}^T d\underline{U}^i \qquad (11.28)$$

Then, Eq. (11.25) can be written as

$$d\underline{\sigma}^a = [(1 - D)\underline{C}^i + D\underline{C}^c(1 + \bar{\alpha}) + \underline{R}^T \bar{\tau}_R]d\underline{U}^i \qquad (11.29a)$$

or

$$d\underline{\sigma}^a = \underline{C}^{DSC} d\underline{U}^i \qquad (11.29b)$$

where $d\underset{\sim}{\sigma}^{(a)T} = [d\tau^a d\sigma_n^a]$ and $\bar{\tau}_R^T = [\tau_R \sigma_{nR}]$. Here matrix $\underset{\sim}{C}^{DSC}$ will be coupled and nonsymmetric. Analysis with Eq. (11.29) can be complex and require special iterative schemes. Hence, as a simplified approximation, $\bar{\alpha}$ can be assumed as zero.

11.6.1 Specializations

If the FA is assumed to carry no stress at all, as in the case of the continuum damage model (59), $\bar{\underset{\sim}{C}}^c = \underset{\sim}{0}$ and $\tau^c = 0$ and Eq. (11.25) reduces to

$$d\underset{\sim}{\sigma}^a = (1-D)\bar{\underset{\sim}{C}}^i d\underset{\sim}{u}^i - dD\underset{\sim}{\sigma}^i \qquad (11.30)$$

where $\underset{\sim}{\sigma}^{(i)T} = [\tau^i \ \sigma_n^i]$. Such a model is not appropriate, as it does not include the interaction between the RI and FA parts. As a result, it would include spurious mesh-dependent effects (Chapter 13).

If it is assumed that the FA part can carry normal stress but no shear stress like a *constrained liquid* (Chapter 4), Eq. (11.25) would reduce to

$$\begin{aligned} d\tau^a &= (1-D)d\tau^i + D(0) + dD(0-\tau^i) \\ d\sigma_n^a &= (1-D)d\sigma_n^i + D\ d\sigma_n^c + dD(\sigma_n^c - \sigma_n^i) \end{aligned} \qquad (11.31a)$$

or

$$d\underset{\sim}{\sigma}^a = \bar{\underset{\sim}{C}}^i d\underset{\sim}{\sigma}^i + Dd\underset{\sim}{\sigma}^c + dD\ \bar{\underset{\sim}{\sigma}}_r \qquad (11.31b)$$

where $\bar{\underset{\sim}{C}}^i$ is the constitutive matrix for the RI (elastic, elastoplastic-δ_0, etc.) material, $d\underset{\sim}{\sigma}^{(c)T} = [0 \ d\sigma_n^c]$ and $\bar{\underset{\sim}{\sigma}}_r^T = [-\tau^i \ \sigma_n^c - \sigma_n^i]$.

Now, by using Eq. (11.18), we have

$$d\tau^c = c_1 c_2 \cdot \sigma_n^{(c)^{(c_2-1)}} d\sigma_n^c \qquad (11.31c)$$

If the relation between τ^c and σ_n is linear, $c_2 = 1$ and

$$d\tau^c = c_1 d\sigma_n^c \qquad (11.31d)$$

where c_1 is the slope of the average straight line (Fig. 11.7).

If the FA part is assumed to be at the critical state, $d\tau^c$, $d\sigma_n^c$ and τ^c, σ_n^c need to be evaluated based on Eqs. (11.18) and (11.20). This would require an iterative analysis. To simplify the formulation, it can be assumed that the normal stresses are equal, i.e., $d\sigma_n^i = d\sigma_n^c$. Then, using Eq. (11.31d), Eq. (11.25) can be expressed as

$$\begin{Bmatrix} d\tau^a \\ d\sigma_n^a \end{Bmatrix} = (1-D) \begin{Bmatrix} d\tau^i \\ d\sigma_n^i \end{Bmatrix} + D \begin{Bmatrix} c_1 d\sigma_n^i \\ d\sigma_n^i \end{Bmatrix} + dD \begin{Bmatrix} \tau_R \\ 0 \end{Bmatrix} \qquad (11.32)$$

11.6.2 Alternative DSC Equations

It can be appropriate to treat the interface zone as "solid" material with (small) finite thickness (Fig. 11.3). In that case, the formulation can be obtained by using the elastoplastic (RI) and critical-state (FA) models for interface treated as a solid material (Chapter 7).

The stress σ_x (Fig. 11.3) can be assumed to be negligible, and the formulation can be obtained in terms of the shear stress, τ, and normal stress ($\sigma_n = \sigma_y$) (56), with corresponding shear and normal strains, γ and ε_n, respectively [Eqs. (11.8) and (11.9)]. Furthermore, the disturbance, D, can be expressed (separately) in terms of shear and normal responses as in Eqs. (11.23b and c).

Now Eq. (11.25) can be written as

$$\begin{Bmatrix} d\tau^a \\ d\sigma_n^a \end{Bmatrix} = \begin{Bmatrix} (1-D_r)d\tau^i \\ (1-D_n)d\sigma_n^i \end{Bmatrix} + \begin{Bmatrix} D_r \, d\tau^c \\ D_n \, d\sigma_n^c \end{Bmatrix} + \begin{Bmatrix} dD_r(\tau^c - \tau^i) \\ dD_n(\sigma_n^c - \sigma_n^i) \end{Bmatrix} \qquad (11.33)$$

Then $d\sigma_n^c$ and $d\tau^c$ can be obtained using the critical state, Eqs. (11.18) and (11.21), as

$$d\sigma_n^c = \sigma_n^c \frac{1+e_0}{\lambda} d\varepsilon_n^c \qquad (11.34a)$$

$$d\tau^c = c_1 d\sigma_n^c \qquad (11.34b)$$

where e_0 is the initial void ratio of the interface material and the linear relation is assumed between τ^c and σ_n^c (Fig. 11.7). The equations can be simplified by assuming that $d\sigma_n^i = d\sigma_n^c$, $\sigma_n^i = \sigma_n^c$, and $D = D_r = D_n$:

$$\begin{Bmatrix} d\tau^a \\ d\sigma_n^a \end{Bmatrix} = (1-D) \begin{Bmatrix} d\tau^i \\ d\sigma_n^i \end{Bmatrix} + D \begin{Bmatrix} c_1 \, d\tau^i \\ d^i \, \sigma_n^i \end{Bmatrix} + \begin{Bmatrix} dD \, (\tau^c - \tau^i) \\ 0 \end{Bmatrix} \qquad (11.35)$$

When the interface zone is treated as the "solid" thin-layer element, we need parameters relevant to the DSC/HISS models for solids. They can be determined from interface shear tests by using appropriate assumptions and

The DSC for Interfaces and Joints

transformations, which are given below (here, the subscripts j and s denote quantities relevant to interface and solid, respectively):

Assumptions:	$\tau_j \approx \sqrt{J_{2D(s)}}$; $\sigma_{nj} = p_j = J_{1s}/3$
Elastic:	for E and G, use Eq. (11.7)
Plasticity:	$\sqrt{\gamma_s} \approx \sqrt{\gamma_j}/3.0$
	$R_s = \left(\dfrac{c_a}{c_0}\right) \cdot R_j$
	n from Eq. (11.37) below
	a and b from Eq. (11.12), with $\xi_s = \dfrac{1}{t}\xi_j$
Critical state:	$\bar{m}_s = \bar{m}_j/3.0$
	λ_s from the slope of the $\ln e$ vs. $\ln p_j$ plot
	e_0^c = void ratio, Eq. (11.21)
Disturbance:	D_u, A, and Z, Eq. (11.22)

11.7 Determination of Parameters

The DSC model with the elastoplastic δ_0-version for the RI response for interfaces involves the following constants:

Model	Constants	Comment
Elastic	k_s, k_n	
Plasticity (RI)	a, b, n, and γ	Straight ultimate envelope
	a, b, n, γ, and q	Curved ultimate envelope
Critical state (FA)	None	Zero strength
	k_n	Normal strength
	c_0, c_1, c_2; v_r^0, $\bar{\lambda}$	Critical state
Disturbance	D_u, A, Z	If $D = D_\tau = D_n$

The procedures for finding the above parameters are essentially similar to that for solid materials (Chapter 7). They are described briefly below.

Elastic. The shear stiffness, k_s, and normal stiffness, k_n, are found as (average) unloading slopes of τ vs. u_r and σ_n vs. v_r curves (Fig. 11.10), respectively. If they are considered to be variable in a nonlinear elastic model, they can be expressed as functions of factors such as τ and σ_n.

Plasticity. The parameter γ is determined from the slope of the ultimate envelope in τ–σ_n space [Fig. 11.11(a)]. The value of q for the curved envelope is found from (here the superscript * is dropped for convenience)

$$\tau_u = \sqrt{\gamma}\sigma_n^{q/2} \qquad (11.36a)$$

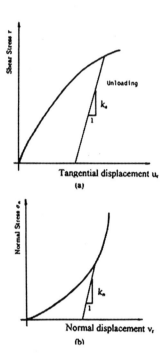

FIGURE 11.10
Shear and normal stiffnesses: (a) shear, k_s; (b) normal, k_n.

where τ_u is the ultimate shear stress ($\alpha = 0$). The value of q is found by writing Eq. (11.36a) as

$$\ln \tau_u = \ln \sqrt{\gamma} + \frac{q}{2} \ln \sigma_n \qquad (11.36b)$$

Then the plot of $\ln \tau_u$ vs. $\ln \sigma_n$ provides the value of q from the slope of the average line [Fig. 11.11(b)]. This and subsequent figures for other parameters are relevant to test data for various joints (44), e.g., Example 11.2.

The phase or state change parameter, n, is evaluated on the basis of the state of stress at the transition point [Fig. 11.6(b)], where the normal displacement is zero or changes from compression to dilation. Then, based on the substitution of α from Eq. 11.16 in the expression for $\partial F/\partial \sigma_n$, the equation for n is derived as

$$n = \frac{q}{1 - \dfrac{\bar{\tau}^2}{\gamma \bar{\sigma}_n^q}} \qquad (11.37)$$

where $\bar{\tau}$ and $\bar{\sigma}_n$ are the stresses at the transition point. In the case of hard materials, the compactive response may be very small. In that case, the

The DSC for Interfaces and Joints

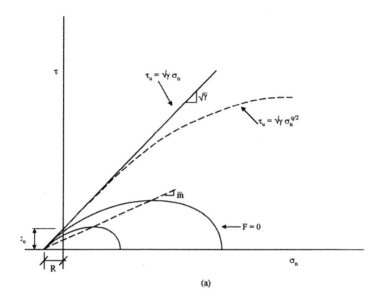

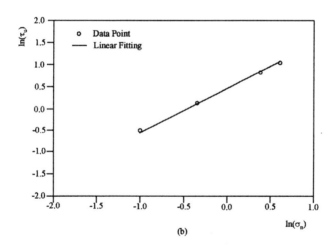

FIGURE 11.11
Determination of ultimate and critical parameters: (a) plot for γ; and (b) plot for q (44, 64).

hardening parameters a and b [Eq. (11.12)] and n can be found simultaneously by first writing Eq. (11.16) as

$$\alpha \sigma_n^n = \gamma \sigma_n^a - \tau^2 = \Delta \tag{11.38a}$$

and then from Eq. (11.12a)

$$\ln a - b \ln \xi + n \ln \sigma_n = \ln \Delta \tag{11.38b}$$

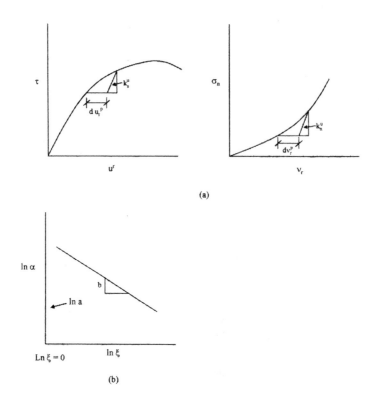

FIGURE 11.12
Determination of hardening parameters.

where ξ is computed at various points on the $\tau - u_r$ and $\sigma_n - v_r$ curves [Fig. 11.12(a)] for stresses (τ and σ_n) at those points. Then Eq. (11.38b) is written for various points, and a least-square fit solution provides values of a, b, and n.

The values of a and b can also be found by writing Eq. (11.12) as

$$\ln a - b \ln \xi = \ln \alpha \tag{11.39}$$

The values of ξ are computed at various points on the curves [Fig. 11.12(a)] and those of α for those points are obtained from Eq. (11.16). Then a plot of $\ln \alpha$ vs. $\ln \xi$ provides values of a and b as intercept when $\ln \xi = 0$, and the slope of the average curve [Fig. 11.12(b)]. The parameter b denotes rate of hardening, and a is the value of α when $\xi = 1$, i.e., at a high value of the plastic displacement trajectory. The value of a is usually small and needs to be determined as precisely as possible.

The critical-state parameters are found by writing Eq. (11.18) as

$$\ln(\tau^c - \tau_0) = \ln c_1 + c_2 \ln \sigma_n^c \tag{11.40}$$

The DSC for Interfaces and Joints

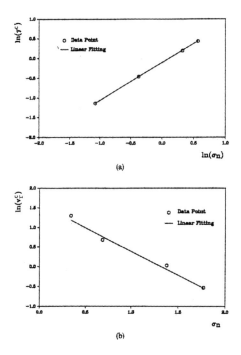

FIGURE 11.13
Typical plots for critical parameters for rock joint type A: (a) plot for c_1 and c_2; and (b) plot for v_r^0 and $\bar{\lambda}$ (44). ©John Wiley & Sons Ltd. Reproduced with permission.

Here, the value of c_0 is obtained from a plot of τ^c vs. σ_n when $\sigma_n = 0$ (Fig. 11.7). Then the plot of $\ln(\tau^c)$ vs. $\ln(\sigma_n)$ [Fig. 11.13(a)] provides the values of c_1 and c_2 as the intercept when $\ln \sigma_n = 0$ and the slope of the average line, respectively.

The parameters in Eq. (11.20) are found by writing it as

$$\ln v_r^c = \ln v_r^0 - \bar{\lambda}\sigma_n \qquad (11.41)$$

The plot in Fig. 11.13(b) of $\ln(v_r^c)$ vs. σ_n provides values of v_r^0 and $\bar{\lambda}$ as intercept when $\sigma_n = 0$ and average slope, respectively.

The disturbance parameters (with $D_u = 1$) are found by writing Eq. (11.23) as

$$\ln[-\ln(1-D)] = \ln A + Z \ln \xi_D \qquad (11.42)$$

The values of D at different points on the observed data (Fig. 11.8) are found using corresponding curves. For example, for $\tau - u_r$ curves, D is found from

$$D_r = \frac{\tau^i - \tau^a}{\tau^i - \tau^c} \qquad (11.42a)$$

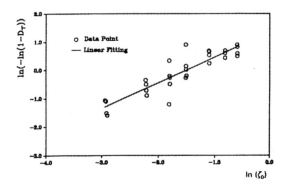

FIGURE 11.14
Determination of disturbance parameters (A, Z); assume $D_u \approx 1.0$ (44). ©John Wiley & Sons Ltd. Reproduced with permission.

The values of ξ_D are found corresponding to the points chosen. Then the plot of $\ln[-\ln(1 - D_r)]$ vs. $\ln \xi_D$ (Fig. 11.14) provides the values of A and Z as the intercept when $\ln \xi_D = 0$ and the average slope, respectively.

For the case of cyclic loading involving degradation [Fig. 11.8(d)], the disturbance can be expressed in terms of measured shear stress or effective stress with number of cycles (N). In the case of shear stress, D can be expressed as

$$D(N) = \frac{\tau^i - \tau^p}{\tau^i - \tau^c} \tag{11.42b}$$

where τ^p is the peak shear stress at given cycles, and τ^c is the residual shear stress. Then, Eq. (11.23) is expressed in terms of $\xi_D(N)$, which is the deviatoric plastic displacement trajectory corresponding to peaks at given cycles. The RI shear stresses (τ^i) are determined based on the assumption that the continuation of the first cycle response, in Fig. 11.8(d), characterized using a linear elastic or elastoplastic (δ_0) or other model, can provide the RI response by integrating:

$$d\underline{\sigma}^i = \underline{C}^i \, d\underline{u}^i \tag{11.43}$$

where $\overline{\underline{C}}^i$ is the RI constitutive matrix in which the parameters are found from the first cycle response. For linear elasticity, the unloading or initial slope can provide the elastic moduli. For the plasticity (δ_0) model, the ultimate parameters can be found from quasistatic tests augmented by hardening parameters from the initial part of the first cycle response.

For stress-controlled tests under saturated conditions in which pore water pressures are measured under cyclic loading, D can be found based on the effective normal stress ($\bar{\sigma}_n$) vs. N relation [Fig. 11.8(c)]. The RI values of $\bar{\sigma}_n^{(i)}$ are obtained by treating the first cycle $\tau - u_r$ response as RI. Then

$$D(N) = \frac{\bar{\sigma}_n^{(i)} - \bar{\sigma}_n^{(a)}}{\bar{\sigma}_n^{(i)} - \bar{\sigma}_n^{(c)}} \quad (11.44a)$$

where $\bar{\sigma}_n^{(a)}$ is the observed stress given by

$$\bar{\sigma}_n^{(a)} = \sigma_n^{(a)} - p \quad (11.44b)$$

where $\sigma_n^{(a)}$ is the total normal stress and p is the measured pore water pressure.

11.7.1 Mathematical and Physical Characteristics of the DSC

In modelling interfaces and joints, it is necessary to account for the nonlocal consideration with appropriate attention to the normal behavior. Furthermore, the microcrack interaction in the interface zone as affected by deformations of asperities needs to be included. Also, as the material in the interface is not a continuum and involves discontinuities due to microcracks and fractures, characteristic dimension should be a part of the model (Chapter 12). The DSC model allows for these factors implicitly within its framework (52–54).

Assuming that the same disturbances apply for the shear and normal behavior, the total observed stresses, τ^a and σ_n^a, in the DSC are given by

$$\tau^a = (1 - D)\tau^i + D\tau^c \quad (11.45a)$$

$$\sigma_n^a = (1 - D)\sigma_n^i + D\sigma_n^c \quad (11.45b)$$

where the disturbance is given by (Chapter 4)

$$D = \frac{A^c}{A} \quad (11.46)$$

and A^c represents the area of distributed fully adjusted parts due to the internal microstructural self-adjustment leading to microcracking and degradation (of asperities). That is, the observed stresses are expressed in terms of the weighted value of the distributed FA parts in the interface zone (Fig. 11.3). Thus, the effect of the area modified by the deformations in the asperities is included in the description of the stresses. This approach is considered similar to that in Eq. (11.3b) (11).

The incremental Eq. (11.25) includes the effect of the deformation characteristics of the FA (or microcracked) zones because of the existence of the second term and the first part of the third term. Note that in the continuum damage model, Eq. (11.30), the interaction due to the response of the damaged part is not included. Thus, the DSC model allows for the interaction between the mechanisms in the RA and FA parts. Furthermore, the third term in Eq. (11.25), which represents the relative stress (σ_r), provides for interactive (relative) motions between the RI and FA parts.

In the description of D in Eq. (11.23), the parameters D_u, A, and Z can be expressed as functions of the size of the (test) specimen or roughness or mean particle size of material at the interface (52–54). Also, D_u is given by

$$D = \frac{V^c}{V} \approx \frac{tA^c}{tA} \tag{11.47}$$

where V^c is the critical or unique volume (dilation) to which a given interface zone would tend. It can be a function of (initial) normal stress or roughness (particle size). In other words, D_u can be considered to act as the characteristic dimension. Thus, the model can allow for the characteristic dimension through D_u.

11.7.2 Regularization and Penalty

The thin-layer element (Fig. 11.3) is assigned the thickness t, which can be evaluated based on analytical and/or experimental considerations (36, 56, 60). The effect of (deformation of) neighboring bodies can also be included in the formulation (56). The inclusion of thickness can be interpreted as a mathematical scheme to provide a *penalty* leading to the regularization of the unilateral contact (56, 61, 62). Consider the normal stress (σ_n) given by

$$\begin{aligned}\sigma_n &\approx E_n \varepsilon_n \\ &\approx E_n \frac{v_r}{t} \\ &\approx k_n v_r\end{aligned} \tag{11.48}$$

where v_r is the normal relative displacement assuming no penetration of body 1 into body 2 (Fig. 11.3), and ε_n is the normal strain. If $t \to 0$, body 2 tends to "rigid" behavior. It has been shown that the thickness t acts as a penalty parameter and regularizes the system (62).

11.8 Testing

Laboratory or field testing are essential for the determination of parameters and calibration of constitutive models. In the context of interfaces and joints, a number of laboratory test devices have been used to measure shear and normal responses. They include direct shear, torsion and ring shear, and simple shear devices; comprehensive reviews of these testing devices are given elsewhere (5, 42, 49).

The testing here is identified as static, quasistatic or one-way, and cyclic or two-way. The static test involves monotonic loading with increasing (shear) stress. The quasistatic or one-way loading involves loading–unloading and reloading cycles; however, the stress does not change sign. The cyclic or two-way case involves cycles of loading–unloading and reloading that include change in the sign of the stress.

A brief description of the cyclic multi degree-of-freedom (CYMDOF) shear device (49, 63) with pore pressure measurements is given. Figure 11.15(a), (b), and (c) show an overall view, cross section, and rings for a simple shear deformation mode, respectively. The CYMDOF device allows testing under static, quasistatic, and cyclic load and displacement-controlled loading, dry and saturated interfaces, direct and simple shear modes of deformation, and measurements of shear, normal and pore water pressure responses. It allows for three degrees of freedom, normal, translational, and rotational. Maximum load is about 130 KN (30 Kips) and displacement amplitudes of about 2.00 cm (0.80 in) and ± 1.30 cm (± 0.50 in.) in the vertical and horizontal directions, respectively. It can permit simulation of a wide range of interfaces (soil-structure, rock joints, metal–metal, soil–geosynthetics, etc.).

11.9 Examples

The DSC and its hierarchical (elastic, nonlinear elastic, d_0-plasticity, etc.) versions have been used to characterize a wide range of dry and fluid saturated interfaces and joints under static, quasistatic, and cyclic loading. Table 11.1 shows details of these applications.

TABLE 11.1
Review of Interface/Joint Models

Model and Details	References
Nonlinear elastic models for shear and normal response under quasistatic or one-way loading for dry interfaces	3, 5, 28, 30, 36, 37, 38, 39, 41
Elastoplastic (δ_0/δ_1) models under quasistatic loading for dry interfaces and joints	42, 43, 44, 45, 46, 74, 75
DSC and elastoplastic models under quasistatic and cyclic loading for dry interfaces	43, 44, 46, 51, 69, 70
DSC and elastoplastic models under quasistatic and cycling loading for saturated interfaces	49, 50
Viscoplastic model under quasistatic loading for dry and wet interfaces	47, 48
DSC model for thermomechanical behavior of joints (Pb/Sn solders) under cyclic loading	77–79

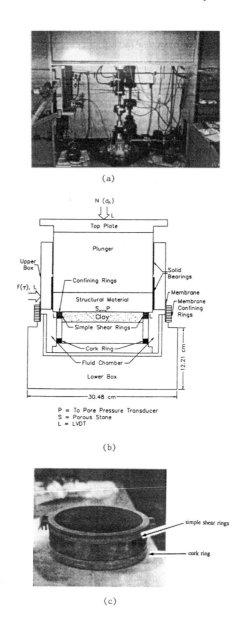

FIGURE 11.15
Details of CYMDOF-P device: (a) overall view; (b) cross section of specimen; and (c) rings for simple shear mode (49, 63, 73).

It is found that for more realistic simulation of interface behavior, in general, it is necessary to include factors such as irreversible deformations, microcracking, and degradation at interfaces in addition to the elastic response. Hence, in the upcoming examples, major attention is given to the

TABLE 11.2

Constants for Back Predictions. Concrete–Concrete Interface
(a) $\alpha = 5°$; and (b) $\alpha = 9°$ (64)

		$\alpha = 5°$	$\alpha = 9°$
Elastic constants	k_n	6.2 MPa/mm	6.2 MPa/mm
	k_s	31.0 MPa/mm	31.0 MPa/mm
Yield function F	a	0.0035	0.0085
	b	0.9	0.9
	n	2.1	2.1
	γ	0.0936	0.303
	q	1.79	1.98
Critical state	c_1	0.48	0.81
	c_2	0.895	0.99
	v_r^o	0.81	0.81
	$\bar{\lambda}$	0.0	0.0
Disturbed state	A_r	1.15	3.29
Eq. (11.23)	Z_r	0.37	0.65
	A_n	0.85	0.85
	Z_n	1.85	1.85

DSC and its specialized versions such as δ_0-associative plasticity models in the HISS approach.

Example 11.1   Concrete–Concrete Interface

Figure 11.16(a) shows a schematic of interface or joint between concrete with different teeth angles α (= 5, 7, 9 degrees) that were tested using the CYM-DOF device under different normal stresses (42, 55). The parameters for the DSC model for $\alpha = 5°$ and $9°$ (Table 11.2) were obtained as average values from tests with $\alpha = 5°$ and $9°$ (44, 64). Figure 11.16(b) and (c) show comparisons between back predictions and observed test results for $\alpha = 7°$. This is an independent validation, as it was not used for finding the average constants (Table 11.2).

Example 11.2   Rock Joints

Schneider (57, 58) reported a series of shear tests on rock joints, Type C, A, and B, indicating smooth to rough conditions. Type C was joint in a limestone and is considered to be smooth; Type A was joint in a granite and is considered to be medium-rough; and Type B was a joint in sandstone, which was rougher. Table 11.3 shows material parameters for the three joint types.

Figure 11.17 shows comparisons between the $\tau - u_r$ and v_r vs. u_r relations for the three joint types from smooth (C) to intermediate (A) to rough (B) joints. The v_r vs. u_r plots are shown only for typical values of $\sigma_n = 0.61, 1.38$, and 1.29 MPa, respectively.

Bandis et al. (65) reported shear tests for simulated rock joints with different normal stresses and roughness defined by JRC (joint roughness coefficient).

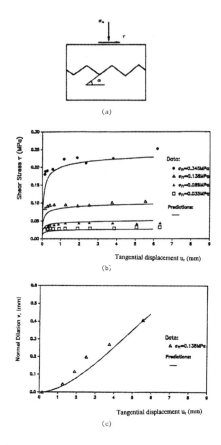

FIGURE 11.16
Comparisons of predictions and observations for concrete–concrete interface, teeth angle = 7°; (a) τ vs. u_r, (b) v_r vs. u_r (44, 64). ©John Wiley & Sons Ltd. Reproduced with permission.

Details of the DSC modelling and validation for these test data are given elsewhere (44, 64).

Example 11.3 Rock–Pile Interface

Williams (66) reported laboratory direct shear tests on interface between pile and a weak rock. In contrast to the foregoing tests where the normal stress was kept constant, the normal stiffness (k_n) was kept constant. The normal stiffness was directly proportional to the normal displacement (dilation). The tests were performed under different (constant) normal stiffnesses and initial normal stress, σ_n^0. The material parameters for the DSC model were determined using a number of tests. Their average values (Table 11.4) were used to perform independent back prediction as described ahead.

Figure 11.18 shows comparisons between back predictions and observations in terms of τ vs. u_r, v_r vs. u_r, σ_n vs. u_r, and τ vs. σ_n for $\sigma_n^0 = 15.0$ kPa and

The DSC for Interfaces and Joints

TABLE 11.3

Constants for Back Predictions. Rock Joints: (a) Type C — Smooth, (b) Type A — Medium Rough, and Type B — Rough (64)

		Type C	Type A	Type B
Elastic constants	k_n	4.2 MPa/mm	5.6 MPa/mm	5.0 MPa/mm
	k_s	21.0 MPa/mm	28.0 MPa/mm	25.0 MPa/mm
Yield function F	a	0.0082	0.0656	0.124
	b	1.05	1.42	1.21
	n	2.10	2.31	2.23
	γ	1.91	3.24	3.68
	q	1.98	1.91	1.97
Critical state	c_1	0.82	0.90	0.89
	c_2	0.99	0.955	0.985
	v_r^0	3.80	4.94	7.52
	$\bar{\lambda}$	0.873	1.21	1.069
Disturbed state	A_τ	2.97	3.87	6.27
	Z_τ	0.30	0.36	0.90
	A_n	4.18	3.88	7.99
	Z_n	1.64	0.90	1.52

TABLE 11.4

Constants for Back Predictions. Rock-Pile Interface: Test SM4 (64)

Elastic constants	k_n	1.0 MPa/mm
	k_s	5.0 MPa/mm
Yield function F	a	0.1
	b	4.5
	n	3.0
	γ	42.5
	q	1.78
Critical state	c_1	1.27
	c_2	0.89
	v_r^0	2.11
	$\bar{\lambda}$	0.0
Disturbed state	A_τ	5.16
	Z_τ	0.96
	A_n	5.23
	Z_n	1.19

$k_n = 36.0$ kPa/mm. This test was not used for finding the parameters (Table 11.4); hence, it is an *independent* validation.

The constant normal stiffness test is relatively complex and exhibits different behavior compared to the constant normal stress tests. Here, the normal stress increases [Fig. 11.18(d)], and the mobilized friction angle ϕ_m (tan $\phi_m = \tau/\sigma_n$) decreases and tends to the residual value. This is because, as the shear stress increases, the normal stress increases faster, and the ratio τ/σ_n decreases with shearing. This phenomenon, before the peak stress, is termed as *weakening* in contrast to softening. In the case of the latter, the shear stress after the

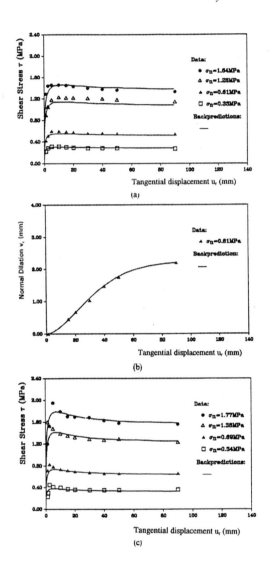

FIGURE 11.17
Comparisons of predictions and observations for rock joints: (a) τ vs. u_r, type C joint; (b) v_r vs. u_r, type C joint; (c) τ vs. u_r, type A joint; (d) v_r vs. u_r type A joint; (e) τ vs. u_r, type B joint; (f) v_r vs. u_r type B joint (44, 64).

peak drops. In contrast, in the constant normal stress tests, ϕ_m increases up to the peak and then decreases.

Example 11.4 Dry Sand–Concrete Interfaces: Static and Cyclic Behavior

A series of laboratory tests using a simple shear device have been reported by Uesugi et al. (67, 68) for sand (Toyoura)–concrete and sand–steel interfaces.

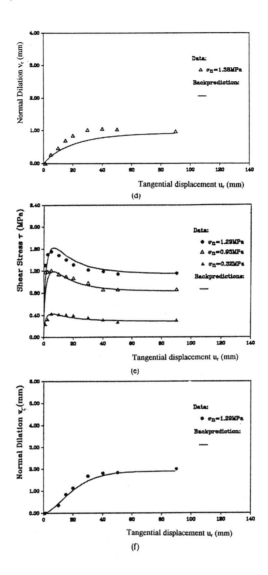

FIGURE 11.17
(continued)

The tests were performed under different surface roughness (R), initial normal stress (σ_n), and relative density (D_r). Details of the DSC model and associated plasticity (RI) model are given in (43). Here the effect of roughness of the interface was included in the ultimate parameters and disturbance parameters for the static and cyclic tests. The roughness was defined as (43, 67, 68)

$$R = \frac{R_n}{R_n^{Cr}} \quad \text{if } R_n^{Cr} > R_n$$
$$R = 1 \quad \text{if } R_n^{Cr} \leq R_n$$

(11.49a)

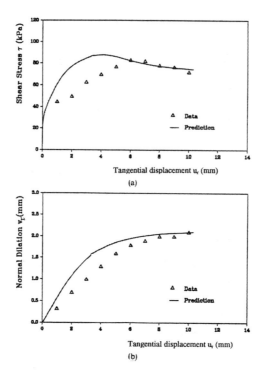

FIGURE 11.18
Comparisons of predictions and observations for rock–pile interface, independent test: $\sigma_n^0 = 15$ kPa, $k_n = 36.0$ kPa/mm; (a) τ vs. u_r; (b) v_r vs. u_r; (c) σ_n vs. u_r; (d) τ vs. σ_n (44, 64).

where R_n is the normalized roughness:

$$R_n = \frac{R_{\max}(L)}{D_{50}} \qquad (11.49b)$$

$R_{\max}$ is the relative height between the highest peak and lowest trough along a surface (asperity) profile over the length L, which can be adopted depending on roughness, e.g., $L = D_{50}$ or 0.2 mm, D_{50} is the mean diameter of sand particles, and R_n^{Cr} is the critical normalized roughness beyond which the shear failure would occur in soil rather than at the interface.

The effect of roughness (R) on the hardening and disturbance behavior was incorporated by expressing various quantities in terms of R. The hardening function, α, in Eq. (11.11) was expressed as

$$\alpha = \gamma \exp(-\alpha_1 \xi_v) \left(\frac{\xi_D^* - \xi_D}{\xi_D^*} \right)^{b_1} \quad \text{for } \xi_D < \xi_D^* \qquad (11.50a)$$

$$\alpha = 0 \quad \text{for } \xi_D \geq \xi_D^* \qquad (11.50b)$$

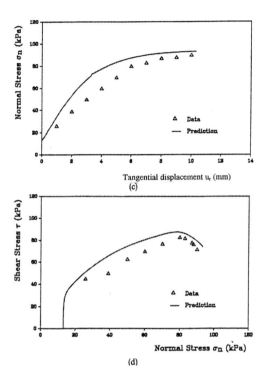

FIGURE 11.18
(continued)

where ξ_D^* is the value of ξ_D at the peak shear stress, τ_p, and a_1 and b_1 are hardening parameters. For the nonassociative δ_1 model, the hardening function, α_Q (Eq. 11.57b below), was expressed as

$$\alpha_Q = \alpha + \alpha_{ph}\left(1 - \frac{\alpha}{\alpha_i}\right)\left[1 - \kappa\left(1 - \frac{D}{D_u}\right)\right] \qquad (11.51a)$$

where α_{ph} is the value of α at the phase change point, κ is the nonassociative parameter, and

$$D = 0 \quad \text{for } \xi_D < \xi_D^* \qquad (11.51b)$$

$$D = D_u[1 - \exp(-A(\xi_D - \xi_D^*)^2)] \quad \text{for } \xi_D \geq \xi_D^* \qquad (11.51c)$$

The ratios τ_p/σ_n, τ_r/σ_n, and ξ_D^* were expressed as functions of R:

$$\mu_p = \sqrt{\gamma} = \frac{\tau_p}{\sigma_n} = \frac{1}{R}(\mu_{p1} + \mu_{p2}R) \qquad (11.52a)$$

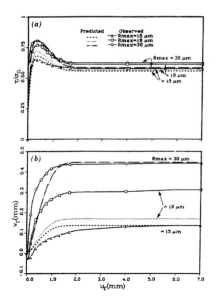

FIGURE 11.19
Comparisons for dry sand (Toyoura)–concrete interface, static loading: $\sigma_n = 98$ kPa, $D_r = 90\%$ (43).

$$\mu_0 = \frac{\tau_r}{\sigma_n} = \frac{1}{R}(\mu_{01} + \mu_{02}R) \qquad (11.52b)$$

$$\xi_D^* = (\xi_{D1}^* + \xi_{D2}^* R) \qquad (11.52c)$$

where $\mu_{p1}, \mu_{p2}, \mu_{01}, \mu_{02}, \xi_{D1}^*$, and ξ_{D2}^* are material parameters, τ_r is the residual shear stress, and $\bar{R}$ denotes the roundness of sand particles. Equations (11.52a and b) are used in Eq. (11.50), and Eq. (11.52c) introduces the effect of R on the disturbance through D_u in Eq. (11.51c).

The DSC parameters were found from tests reported in (67, 68); their values are given in (43).

Figure 11.19 shows comparisons between back predictions and static test data for different roughness for $\sigma_n = 98$ kPa and $D_r = 90\%$, for concrete–Toyoura sand interface.

Figure 11.20 shows comparisons for cyclic tests for $\sigma_n = 98$ kPa, $D_r = 90\%$, and $R_{max} = 23$ μm, in terms of τ/σ_n vs. u_r and v_r vs. u_r, with number cycles ($N = 1, 2, \cdots, 15$).

Example 11.5 Dry Sand-Steel Interfaces: Static Two- and Three-Dimensional Behavior

Fakharian and Evgin (69) used the DSC with HISS δ_1-plasticity model for the RI behavior and constrained-liquid simulation for the FA response, i.e., the

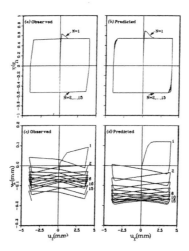

FIGURE 11.20
Comparisons for sand (Toyoura)–steel interface, cyclic loading: $\sigma_n = 98$ kPa, $D_r = 90\%$, $R_{max} = 23$ μm (43).

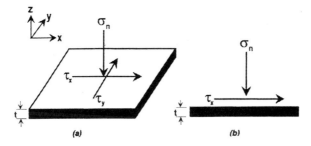

FIGURE 11.21
Stresses acting on an interface: (a) 3-D, (b) 2-D (69). ©John Wiley & Sons Ltd. Reproduced with permission.

material parts in the FA state cannot carry shear stress. They calibrated the model on the basis of laboratory tests for interfaces between a medium silica sand with relative density, $D_r = 88\%$, and steel (plate) by using the test device described in (70). The sand-blasting technique was used to obtain different surface roughnesses, R_{max} (Eq. 11.49b) = 4, 15, and 25 μm, which cover the range from smooth to rough conditions.

The DSC model used was the same as in Example 11.4. The material parameters were found from a series of two-dimensional laboratory tests depicted in Fig. 11.21; details of parameters and the test device are given in (69, 70).

Figure 11.22 shows comparisons between back predictions and test data for the two-dimensional test in which the shear displacement was applied with constant normal stress $\sigma_n = 100$ kPa and sand density $D_r = 88\%$. The comparisons involve $R_{max} = 3.6$ and 25 μm, while the results for $R_{max} = 15$ μm

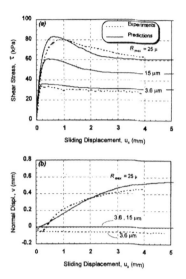

FIGURE 11.22
Comparisons between predictions and test data for 2-D tests: σ_n = 100 kPa, D_r = 88%. For R_{max} = 15, only predictions are shown (69). ©John Wiley & Sons Ltd. Reproduced with permission.

involved only predictions. It can be seen that the model provides satisfactory predictions of hardening, dilative, peak, and softening responses. For the case of the smooth interface (R_{max} = 3.6 μm), the predictions show essentially zero normal displacement, while the test data show nonzero compression.

For the three-dimensional case, the yield function [Eq. (11.11)] was modified as (69)

$$F = \tau_x^2 + \tau_y^2 + \alpha\sigma_n^n - \gamma\sigma_n^2 = 0 \tag{11.53}$$

In the three-dimensional tests, the shear stress τ_y = 0, 20, 40, and 60 kPa was applied with σ_n = 100 kPa (Fig. 11.21). Then the interface was sheared in the x-direction by increasing τ_x with constant values of τ_y and σ_n. The stress and displacement paths followed in the tests are shown in Fig. 11.23.

Figure 11.24 shows comparisons between the predictions and test data for shear stress, τ_x, versus shear displacement, u_x, and normal displacement, v, versus u_x for the three-dimensional tests. Figure 11.23 shows comparisons between the predicted and observed stress paths. The parameters from the two-dimensional tests were used for the three-dimensional analysis; hence, this is considered to be an *independent* prediction. Overall, the predictions compare well with the observed behavior, particularly because this is an independent validation. Details of tests and validations for tests under constant stiffness condition (Example 11.3) are also reported by Fakharian and Evgin (69).

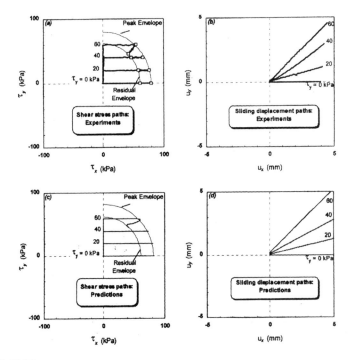

FIGURE 11.23
Comparisons between predictions and test data for stress and displacement paths for 3-D tests: $\sigma_n = 100$ kPa (69). ©John Wiley & Sons Ltd. Reproduced with permission.

Example 11.6   Soil–Rock Interface: Viscoplastic Model

A smeared interface zone occurs between the moving mass of soil and bed rock in the case of creeping slopes and landslides (47, 48); Fig. 11.25(a) shows a schematic of such an interface. The mechanical behavior of the interface zone can involve creep or viscous deformations in addition to elastic and plastic deformations.

A viscoplastic model for the interface zone has been developed, verified, and used (47, 48) to characterize the interface zone at the site of the Villarbeney landslide in Switzerland (71). Brief details of the model are given ahead.

The incremental elastoplastic equations for the HISS (δ_0/δ_1) models can be written as

$$d\underset{\sim}{\sigma} = \overline{\underset{\sim}{C}}^{ep} d\underset{\sim}{U} \tag{11.54a}$$

where $d\underset{\sim}{\sigma}^T = [d\tau \ d\sigma_n]$, $d\underset{\sim}{U} = [du_r \ dv_r]$, and $\overline{\underset{\sim}{C}}^{ep}$ is the elastoplastic matrix derived using the yield function, F [Eq. (11.11)], in the context of consistency condition, $dF = 0$, and the normality rule

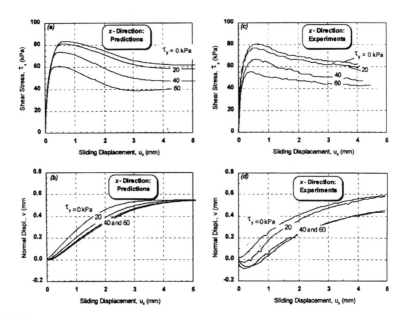

FIGURE 11.24
Comparisons between predictions and test data for different values of τ_y in 3-D tests: $\sigma_n = 100$ kPa, $R_{max} = 25$ μm, $D_r = 88\%$ (69). ©John Wiley & Sons Ltd. Reproduced with permission.

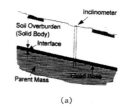

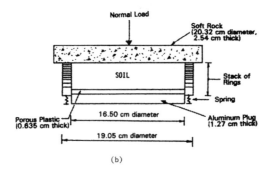

FIGURE 11.25
(a) Schematic of interface in creeping slope; (b) cross section of interface test specimen (47, 48). ©John Wiley & Sons Ltd. Reproduced with permission.

$$d\underset{\sim}{U}^p = \lambda \frac{\partial Q}{\partial \underset{\sim}{\sigma}} \tag{11.54b}$$

where $d\underset{\sim}{U}^{(p)T} = [du_r^p \ dv_r^p]$, λ is the scalar proportionality factor, and Q is the plastic potential function. In the case of associative plasticity, $Q \equiv F$ (Chapter 7).

Perzyna's (72) theory of viscoplasticity is used to characterize the predominant plastic deformations, i.e., the viscoelastic component is assumed to be small. Accordingly, the viscoplastic strain rate is given by (see Chapter 8)

$$\dot{\underset{\sim}{\varepsilon}}^{vp} = \Gamma < \phi\left(\frac{F}{F_0}\right) > \frac{\partial Q}{\partial \underset{\sim}{\sigma}} \tag{11.55a}$$

or

$$= \Gamma \left\langle \phi \frac{F}{F_0} \right\rangle \begin{Bmatrix} \dfrac{\partial Q}{\partial \tau} \\ \dfrac{\partial Q}{\partial \sigma_n} \end{Bmatrix} \tag{11.55b}$$

where Q is the scalar flow (plastic potential) function, Γ is the fluidity parameter, F_0 is the normalizing constant (e.g., yield stress, atmospheric pressure), and the angle bracket $\langle \ \rangle$ has the following meaning:

$$\phi(F/F_0) = \begin{cases} \phi(F/F_0) & \text{if } F/F_0 > 0 \\ 0 & \text{if } F/F_0 \leq 0 \end{cases} \tag{11.56}$$

Nonassociative Response. For nonassociative behavior, the plastic potential function, Q, is written as

$$Q = \tau^2 + \alpha_Q \sigma_n^{*n} - \gamma_n^{*2} \tag{11.57a}$$

and the hardening function, α_Q, is given by

$$\alpha_Q = \alpha + \kappa(\alpha_0 - \alpha)(1 - r_v) \tag{11.57b}$$

where α_0 is the value of α at the initiation of the nonassociative response, r_v is the ratio of the plastic normal displacement trajectory (ξ_v) to the total trajectory, ξ [Eq. (11.13)], and κ is the nonassociative parameter (Chapter 7). For the case of smooth interfaces, dilation is not significant, and the following expression can be used for α_Q:

$$\alpha_Q = \alpha + \alpha_{ph}\left[1 - \frac{\alpha}{\alpha_0}\right]^\theta \tag{11.58}$$

where $\alpha_{ph} = (2/n)\gamma\sigma_n^{2-n}$ and $\alpha_0 = \gamma\sigma_n^{2-n}$ are the values of α (in Eq. 11.11) at the transition point (Fig. 11.6) and at the start of the shear loading (i.e., at the end of the normal loading), respectively, and θ is the nonassociative parameter. The hardening function α in Eq. (11.11) used here is given by

$$\alpha = \frac{\gamma e^{-a\xi_v^{vp}}}{(1+\xi_D^{vp})^b} \tag{11.59}$$

where ξ_v^{vp} and ξ_D^{vp} are the trajectories of volumetric and deviatoric viscoplastic strains (displacements), respectively, and a and b are hardening parameters.

11.9.1 Creep Behavior

The parameters Γ and ϕ define the viscous or creep response. The latter is given by

$$\phi\left(\frac{F}{F_0}\right) = \left(\frac{F}{F_0}\right)^N \tag{11.60}$$

where N is the material parameter.

11.9.2 Testing

A series of laboratory tests were performed by using the CYMDOF, triaxial, and direct shear devices (47, 48, 71). They include both quasistatic or one-way (loading–unloading–reloading) and creep tests. The drained tests using the CYMDOF device involved interface simulation [Fig. 11.25(b)] and were conducted under different normal stresses, σ_n = 103, 207, and 345 kPa, with different amplitudes of shear displacements, u_r^a = 0.19, 0.64, and 1.27 cm. The parameters for the viscoplastic model are shown in Table 11.5.

Figures 11.26 and 11.27 show comparisons between back predictions and observations for a typical static test and for the creep tests, respectively.

Example 11.7 Saturated Clay–Steel Interfaces: Cyclic Behavior

The CYMDOF-P device (Fig. 11.15) was used to perform a number of cyclic undrained (two-way) tests for saturated marine clay–steel interfaces, under different normal stresses and amplitudes of displacements (49, 73). The DSC constitutive model was formulated based on the elastoplastic (δ_0) model to characterize the RI response, the critical state for the FA response, and the disturbance was found based on cyclic degradation data [Fig. 11.28(a)].

TABLE 11.5

Parameters for Interface Model in Creeping Slope (47, 48)

Parameters	Symbol	Value
Elastic		
Normal stiffness	k_n	8×10^6 kPa/cm
Shear stiffness	k_s	2800 kPa/cm
Plastic		
Ultimate	γ	0.24
Transition	n	2.04
Hardening	a	143.0
Hardening	b	10.0
Nonassociative	κ	0.57
Viscous		
Fluidity	Γ	0.057/min
Exponent	N	3.15

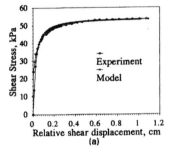
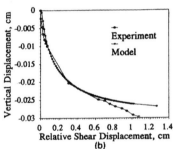

FIGURE 11.26
Comparisons between predictions and observations for interface test at $\sigma_n = 103$ kPa (47, 48).

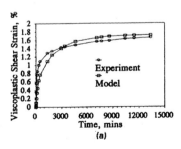
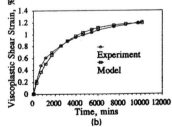

FIGURE 11.27
Comparisons between prediction and observation for creep tests, stress ratio = 0.6: (a) preconsolidation stress = 200 kPa; (b) preconsolidation stress = 400 kPa (47, 48).

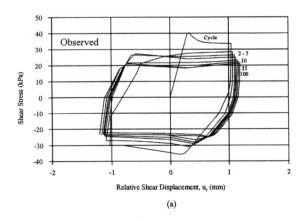

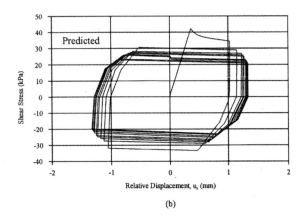

FIGURE 11.28
(a) Observed cyclic $\tau - u_r$ curves and (b) predicted curves for different cycles: $\sigma_n = 138$ kPa, $u_r^a = 1.5$ mm (49, 73).

The alternative form of disturbance function [Eq. (3.15b), Chapter 3], was used as

$$D = D_u\left[1 - \left\{1 + \left(\frac{\xi}{h}\right)^w\right\}^{-s}\right] \quad (11.58)$$

where $D_u \approx 1$ and h, w, and s are parameters found from the laboratory tests.

The incremental constitutive relations for the monotonic loading and reverse loading were based on the alternative formulation in Eq. (11.33), while the unloading response was assumed to be bilinearly elastic, i.e., the unloading modulus was bilinear, as indicated in Fig. 11.28.

The material parameters were found from test data with different values of σ_n (= 69 and 138 kPa) and displacement amplitudes, u_r^a (= 0.5 and 1.5 mm).

TABLE 11.6
Parameters for Cyclic Test, Clay–Steel Interface, for $u_r^a = 1.5$ mm (73)

Parameters		Average	$\sigma_n = 69$ kPa	$\sigma_n = 138$ kPa	$\sigma_n = 207$ kPa
Intact State					
Elastic	E	4,400	2,400	4,300	6,500
	ν	0.42	0.41	0.42	0.42
Plastic	n	2.1	2.1	2.1	2.1
	$\bar{\gamma}$	3.622	2.822	2.88	5.222
	a	2.63	2.701	2.586	2.601
	b	0.067	0.024	0.087	0.09
Critical State	$\bar{m}$	0.37	0.408	0.356	0.346
	λ	0.298	0.298	0.298	0.298
	e_0^c	1.359	1.359	1.359	1.359
	e_0		1.69	1.49	1.37
Disturbance Function					
Shear	h_τ	3.76	4.153	9.411	0.156
Eq. (11.58)	w_τ	0.64	0.916	0.498	1.035
	s_τ	2	2	2	2
Normal	h_n	8.941	17.62	10.62	0.791
Eq. (11.58)	w_n	0.745	0.825	0.766	0.703
	s_n	2	2	2	2

Typical parameters for the test with $\sigma_n = 138$ kPa and $u_r^a = 1.5$ mm are shown in Table 11.6.

Figures 11.28 and 11.29 show typical comparisons between predicted results and observed data in terms of τ vs. u_r and pore water pressure vs. time (cycles) for the test with $\sigma_n = 138$ kPa and $u_r^a = 1.5$ mm.

Sand–Steel Interfaces: Quasistatic and Cyclic Behavior. A series of laboratory tests were performed for dry and (partially) saturated sand (Ottawa)–steel interfaces by using the CYMDOF-P device (50). The tests were performed under different normal stresses, σ_n (69, 138, 207 kPa), amplitudes of displacements, u_r^a (3.5, 4.3, 1.27 mm) and rates of loading (1.27, 2.54, 36.58 mm/min). Calibration of the DSC model and its validation are given in (50).

Example 11.8 Sand–Geosynthetic Interfaces: Static Response

Laboratory static direct tests were performed (51) to measure the interface response between a sand and geogrid (geosynthetic reinforcement) under different normal stresses, σ_n (26.8, 53.6, 80.4, and 214.4 kPa).

The DSC model was formulated based on the test data with Eq. (11.32). The RI behavior was assumed to be elastoplastic (δ_0-model), and the FA response was simulated using the critical state. Table 11.7 shows the material parame-

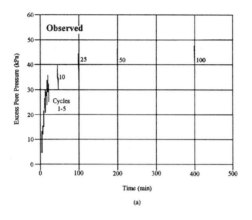

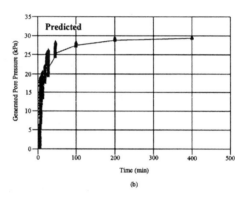

FIGURE 11.29
(a) Observed excess pore water pressure curves and (b) predicted curves for different cycles: $\sigma_n = 138$ kPa, $u_r^a = 1.5$ mm (49, 73).

ters. Figure 11.30 shows typical comparisons between predictions and observed data for the test with $\sigma_n = 26.8$ kPa.

Example 11.9   Aluminum Shaft–Sand Interfaces

El-Sakhawy and Edil (74, 75) reported a series of laboratory interface tests between aluminum shaft and a sand (dry, medium-dense portage) using the test device and instrumentation shown in Fig. 11.31. The measurements involved shear and normal stresses and displacements at the interface between soil and the shaft. In order to measure the interface normal stresses, the aluminum shaft was instrumented with two 90° strain gauge rosettes mounted inside the shaft at two locations, 152 mm (6.0 in.) apart. Two surface roughnesses were considered: smooth with $R_{max} = 5.72$ μm (0.225 μ in.) and rough with $R_{max} = 4.27$ mm (0.168 in.) [Eq. (11.49)].

Tests were performed under two conditions using the cell pressure and flexible donut-shaped pressure gauges (Fig. 11.31). In the constant normal

The DSC for Interfaces and Joints

TABLE 11.7
Material Constants for Geosythetic Interface (51)

Elastic parameters	k_n	550 kPa/mm
	k_s	88.83 kPa/mm
Intact state	γ	95.7
	n	1.4
	q	1.395
	a_1	96.9324
	η_1	0.01965
Critical state	c_1	0.456
	v_r^0	−0.2635
	λ	0.0444
Disturbance	A_τ	0.0896
	Z_τ	2.2935
	A_n	1.1158
	Z_n	1.17
	D_u	0.90

Note: The parameters are obtained from nondimensionalized quantities, with respect to p_a.
©John Wiley & Sons Ltd. Reproduced with permission.

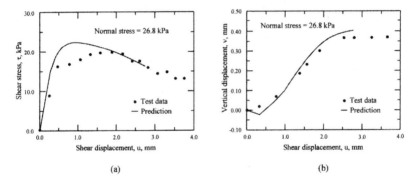

FIGURE 11.30
(a) Observed and predicted τ vs. u_r and (b) v_r vs. u_r curves, one-way loading for sand–geosynthetic interface: $\sigma_n = 26.8$ kPa (51). ©John Wiley & Sons Ltd. Reproduced with permission.

stress tests, independent cell fluid and flexible-top boundary bag pressures provided constant horizontal (radial) and vertical normal stresses. In the constant volume test, the cell fluid and the fluid in the flexible-top boundary bag were sealed after the application of the initial horizontal and vertical normal stresses (75).

The interface model based on HISS δ_1-plasticity model (42, 47) [Eq. (11.57)] was used to characterize the behavior of interfaces. The parameters obtained from the test data are shown in Table 11.8 (74); they were found from tests under three normal stresses: $\sigma_n = 34.5, 69.0,$ and 103.5 kPa.

TABLE 11.8

Parameters for Shaft–Sand Interfaces (75)

Parameter		Smooth interface	Rough interface
Elastic	k_s	460 kPa/mm	1600 kPa/mm
	k_n	5400 kPa/mm	5400 kPa/mm
Plastic (average)	γ	0.246	1.25
	n	2.05	2.13
	a	0.0096	0.088
	b	0.517	0.415
	θ	1.124	—
	κ	—	1.120

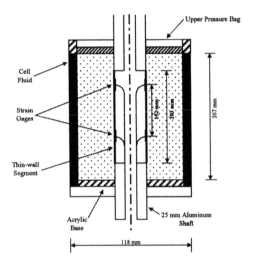

FIGURE 11.31
Instrumented shaft and sand (75).

Figure 11.32(a and b) show comparisons between the laboratory shear tests and nonassociative model predictions for the smooth and rough interfaces, respectively, under constant stress tests. It can be seen that the correlations are very good. It was reported (75) that the predictions for normal displacements showed consistent trends but did not provide as good correlations.

Example 11.10 Solder Joint in Electronic Chip-Substrate Systems: Thermomechanical Behavior

In electronic packaging and semiconductor systems (76) (Fig. 11.33), the joining materials, e.g., solders (Pb/Sn) whose thickness is small, of the order 200 μm, can be treated as bulk interface zones with thickness t. Development and

The DSC for Interfaces and Joints

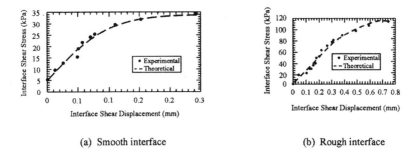

FIGURE 11.32
Comparison between model predictions and test behavior for constant stress tests (75).

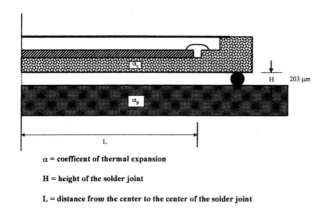

FIGURE 11.33
Leadless ceramic chip carrier (LCCC) (76).

application of the DSC model with δ_0-plasticity model for the RI response for solders are given in Chapter 7. Development, validations, and computer predictions for various chip-substrate systems are given in (77–79) and in Chapter 13.

Example 11.11   Homework Problems
Derive the matrix $\underset{\sim}{C}^i$ in Eq. (11.25) by assuming that the RI behavior is simulated as linear elastic with k_s and k_n, and as elastoplastic with F in Eq. (11.11). Then by assuming that the strains (displacements) in the RI and FA parts are equal, derive the special form of Eq. (11.25). Further, assume that the FA behavior is simulated such that it cannot carry any stress at all, i.e., τ^c in Eq. (11.22a) is zero. Then by using the parameters in Table 11.2, backpredict the responses for $\sigma = 0.138$ MPa (Fig. 11.16).

Hint: You may prepare a computer routine for the integration of the specialized form of Eq. (11.25). The predictions you obtain may not be the same as in Fig. 11.16 because the FA response is characterized using the critical-state equations. This problem is designed mainly to illustrate the procedure involved in back predictions.

11.10 Computer Implementation

Details of the implementation of the DSC models are given in Chapter 13. The thin-layer interface/joint element is formulated in the same manner as the solid (two- or three-dimensional) element. The stiffness matrix for the interface element usually has the same number of degrees of freedom as the solid element; hence, it is assembled directly with the solid elements.

References

1. Bowden, F.P. and Tabor, D., *The Friction and Lubrication of Solids,* Part I, Clarendon Press, Oxford, 1950.
2. Rabinowicz, E., *Friction and Wear of Materials,* John Wiley, New York, 1965.
3. Desai, C.S. and Christian, J.T. (Eds.), *Numerical Methods in Geotechnical Engineering,* McGraw-Hill, New York, 1977.
4. Dawson, D., *The History of Tribology,* Longman, London, 1978.
5. Desai, C.S., "Behavior of Interfaces Between Structural and Geologic Media," A state-of-the-art paper, *Proc., Int. Conf. on Recent Advances in Geotech. Earthquake Eng. and Soil Dynamics,* St. Louis, MO, 1981.
6. Kikuchi, N. and Oden, J.T., *Contact Problems in Elasticity,* Siam, Philadelphia, 1981.
7. Suh, N.P. and Sih, H.C., "The Genesis of Friction," *Wear,* 69, 1981, 91–114.
8. Madakson, P.B., "The Frictional Behavior of Materials," *Wear,* 87, 1983, 191–206.
9. Amontons, G., "De La Resistance Causee les Machines," *Memories de l'Academie Royale,* A. Chez Gerard Kuyper, Amsterdam, 1699, 257–282.
10. Archard, J.I., "Elastic Deformation and the Laws of Friction," *Proc., Roy. Soc.,* London, Series A, Vol. 243, 1959, 190–205.
11. Oden, J.T. and Pires, T.B., "Nonlocal and Nonlinear Friction Laws and Variational Principles for Contact Problems in Elasticity," *J. of Appl. Mech.,* Vol. 50, 1983, 67–75.
12. Tolstoi, D.M., "Significance of Normal Degree-of-Freedom and Natural Normal Vibrations in Contact Friction," *Wear,* 10, 1967, 199–213.
13. Seguchi, Y., Shindo, A., Tomita, Y., and Sunohara, M., "Sliding Rule of Friction in Plastic Forming Metal," *Computational Methods in Nonlinear Mechanics* (Oden, J.T., et al., Editors), The Univ. of Texas, Austin, 1974, 683–692.
14. Fredriksson, B., "Finite Element Solution of Surface Nonlinearities in Structural Mechanics with Special Emphasis to Contact and Fracture Problems," *Computer and Structures,* Vol. 6, 1976, 281–290.
15. Herrmann, L.R., "Finite Element Analysis of Contact Problems," *J. of Eng. Mech.,* ASCE, 104, 1978, 1043–1057.

16. Michalowski, R. and Mroz, Z., "Associated and Non-associated Sliding Rules in Contact-Friction Problems," *Arch. Mech.*, 30, 1978, 259–276.
17. Kikuchi, N. and Song, Y.S., "Penalty/Finite-Element Approximations of a Class of Unilateral Problems in Linear Elasticity," *Appl. Math.*, XXXIX, 1981, 1–22.
18. Cheng, J.H. and Kikuchi, N., "An Incremental Constitutive Relation of Unilateral Contact Friction for Large Deformation Analysis," *J. of Appl. Mech.*, ASME Trans. Paper No. 85-APM-24, 1985.
19. Jaeger, J.C., "The Frictional Properties of Joints in Rock," *Pure Appl. Geophys.*, 43, 1959, 148–158.
20. Patton, F.D., "Multiple Modelling of Shear Failure in Rock," *Proc., 1st Cong. Rock Mech.*, Lisbon, 1, 1966, 509–513.
21. Goodman, R.E., Taylor, R.L., and Brekke, T.L., "A Model for the Mechanics of Jointed Rock," *J. of Soil Mech. & Found. Eng.*, ASCE, 94, 1968, 637–659.
22. Zienkiewicz, O.C., Best, B., Dullage, C., and Stag, K.G., "Analysis of Nonlinear Problems in Rock Mechanics with Particular Reference to Jointed Rock Systems," *Proc., 2nd Int. Conf. of Rock Mechanics*, Belgrade, Yugoslavia, 3, 1970, 501–509.
23. Ladanyi, B. and Archmbault, G., "Simulation of Shear Behavior of a Jointed Rock Mass," *Proc., 11th O.S. Symp. Rock Mech.*, Berkeley, CA, 1970, 105–125.
24. Isenberg, J., Lee, L., and Agbabian, M.S., "Response of Structures to Combined Blast Effects," *J. Transp. Eng.*, ASCE, 99, 1973, 887–908.
25. Ghaboussi, J., Wilson, E.L. and Isenberg, J., "Finite Element for Rock Joints and Interfaces," *J. of Soil Mech. and Found. Eng.*, ASCE, 99, 1973, 833–848.
26. Barton, N.R., "A Review of a New Shear Strength Criterion for Rock Joints," *Eng. Geology*, 7, 1973, 287–332.
27. Hsu-Jun, K., "Nonlinear Analysis of the Mechanical Properties of Joint and Weak Interaction in Rock," *Proc., 3rd Int. Conf. Num. Meth. in Geomech.*, Aachen, W. Germany, 1973, 523–532.
28. Desai, C.S., "Numerical Design-Analysis of Piles in Sands," *J. Geotech. Eng.*, ASCE, 100, 1974, 613–625.
29. Kausel, E., Roesset, J.M., and Christian, J.T., "Nonlinear Behavior in Soil-Structure Interaction," *J. of Geotech. Eng.*, ASCE, 102, 12, 1976.
30. Desai, C.S., "Soil-Structure Interaction and Simulation Problems," Chap. 7 in *Finite Elements in Geomechanics*, G. Gudehus (Editor), Wiley, Chichester, U.K., 1977.
31. Wolf, J.P., "Seismic Response Due to Travelling Wave Including Soil-Structure Interaction with Base Mat Uplift," *Earthquake Eng. & Struct. Dyn.*, 5, 1977, 337–363.
32. Lightner, J.G. and Desai, C.S., "Improved Numerical Procedures for Soil-Structure Interaction Including Simulation of Construction Sequences," *Report No. VPI-E-79-32*, Dept. of Civil Eng., Virginia Tech., Blacksburg, VA, 1979.
33. Toki, K.T., Sato, T., and Muira, F., "Separation and Sliding Between Soil and Structure During Strong Ground Motion," *Earthquake Eng. & Struct. Dyn.*, 9, 1981, 263–277.
34. Katona, M.G., "A Simple Contact-Friction Interface Element with Applications to Buried Culverts," *Int. J. Num. Analyt. Methods in Geomech.*, 7, 3, 1983, 371–384.
35. Vaughan, D.K. and Isenberg, J., "Nonlinear Rocking Response of Model Containment Structures," *Earthquake Eng. & Struct. Dyn.*, 11, 1983, 275–296.

36. Desai, C.S., Zaman, M.M., Lightner, J.G., and Siriwardane, H.J., "Thin-Layer Element for Interfaces and Joints," *Int. J. Num. Analyt. Meth. in Geomech.*, 8, 1, 1984, 19–43.
37. Zaman, M.M., Desai, C.S., and Drumm, E.C., "An Interface Model for Dynamic Soil Structure Interaction," *J. Geotech. Eng. Div., ASCE*, 110, 9, 1984, 1257–1273.
38. Drumm, E.C. and Desai, C.S., "Determination of Parameters for a Model for Cyclic Behavior of Interfaces," *J. of Earthquake Eng. & Struct. Dynamics*, 14, 1, 1986, 1–18.
39. Desai, C.S., Drumm, E.C., and Zaman, M.M., "Cyclic Testing and Modelling of Interfaces," *J. Geotech. Eng., ASCE*, 111, 6, 1985, 793–815.
40. Plesha, M.E., "Constitutive Models for Rock Discontinuities with Dilatancy and Surface Degradation," *Int. J. Num. Analyt. Methods in Geomech.*, 11, 1987, 345-362.
41. Desai, C.S. and Nagaraj, B.K., "Modeling for Cyclic Normal and Shear Behavior of Interfaces," *J. Eng. Mech., ASCE*, 114, 1988, 1198–1217.
42. Desai, C.S. and Fishman, K.L., "Plasticity Based Constitutive Model with Associated Testing for Joints," *Int. J. Rock Mech. & Min. Sc.*, 4, 28, 1991, 15–26.
43. Navayogarajah, N., Desai, C.S., and Kiousis, P.D., "Hierarchical Single Surface Model for Static and Cyclic Behavior of Interfaces," *J. of Eng. Mech., ASCE*, 118, 5, 1992, 990–1011.
44. Desai, C.S. and Ma, Y., "Modelling of Joints and Interfaces Using the Disturbed State Concept," *Int. J. Num. Analyt. Meth. Geomech.*, 16, 1992, 623–653.
45. Desai, C.S., "Hierarchical Single Surface and the Disturbed State Constitutive Models with Emphasis on Geotechnical Applications," Chap. 5 in *Geotechnical Engineering*, K.R. Saxena (Editor), Oxford & IBM Publishing Co., New Delhi, India, 1994.
46. Desai, C.S., "Constitutive Modelling Using the Disturbed State as Microstructure Self-Adjustment," Chap. 8 in *Continuum Models for Materials with Microstructures*, H.B. Mühlhaus (Editor), John Wiley, Chichester, U.K., 1995.
47. Desai, C.S., Samtani, N.C., and Vulliet, L., "Constitutive Modelling and Analysis of Creeping Slopes," *J. Geotech. Eng., ASCE*, 121, 1, 1995, 43–56.
48. Samtani, N.C., Desai, C.S., and Vulliet, L., "An Interface Model to Describe Viscoplastic Behavior," *Int. J. Num. Analty. Meth. Geomech.*, 20, 1996, 231–252.
49. Desai, C.S. and Rigby, D.B., "Cyclic Interface and Joint Shear Device Including Pore Pressure Effects," *J. of Geotech. & Geoenv. Eng., ASCE*, 123, 6, 1997, 568–579.
50. Alanazy, A. and Desai, C.S., "Testing and Modelling of Sand-Steel Interfaces Under Static and Cyclic Loading," *Report*, Dept. of Civil Engng. and Engng. Mechs., The Univ. of Arizona, Tucson, AZ, 1996.
51. Pal, S. and Wathugala, G.W., "Disturbed State Model for Sand-Geosynthetic Interfaces and Application to Pull-out Tests," *Int. J. Num. Analyt. Meth. Geomech.*, 23, 15, 1999, 1873–1892.
52. Desai, C.S. and Toth, J., "Disturbed State Constitutive Modeling Based on Stress-Strain and Nondestructive Behavior," *Int. J. Solids and Struct.*, 33, 11, 1996, 1619–1650.
53. Desai, C.S., Basaran, C. and Zhang, W., "Numerical Algorithms and Mesh Dependence in the Disturbed State Concept," *Int. J. Num. Meth. Eng.*, 40, 1997, 3059–3083.
54. Desai, C.S. and Zhang, W., "Computational Aspects of Disturbed State Constitutive Models," *Int. J. Comp. Meth. in Appl. Mech. & Eng.*, 151, 1988, 361–376.

55. Fishman, K.L. and Desai, C.S., "Measurements of Normal Deformations in Joints During Shear Using Inductance Devices," *Geotech. Testing J.*, 12, 1989, 297–301.
56. Sharma, K.G. and Desai, C.S., "An Analysis and Implementation of Thin-Layer Element for Interfaces and Joints," *J. of Eng. Mech., ASCE*, 118, 1992, 545–569.
57. Schneider, H.J., "Rock Friction – A Laboratory Investigation," *Proc., 3rd Cong. Int. Soc. Rock Mech.*, Denver, CO, 2, Part A, 1974, 311–315.
58. Schneider, H.J., "The Friction and Deformation Behavior of Rock Joint," *Rock Mech.*, 8, 1976, 169–184.
59. Kachanov, L.M., *Introduction to Continuum Damage Mechanics*, Martinus Nijhuft Publ., Dordrecht, The Netherlands, 1986.
60. Pande, G.N. and Sharma, K.G., "On Joint/Interface Elements and Associated Problems of Numerical Ill-Conditioning," *Int. J. Num. Analyt. Meth. in Geomech.*, 3, 1979, 293–300.
61. Hohberg, J.M. and Schweiger, H.F., "On the Penalty Behaviour of Thin-Layer Elements," *Proc., Num. Models in Geomech.*, Swansea, U.K., 1992.
62. Kikuchi, N. and Song, Y.J., "Penalty/Finite-Element Approximations of a Class of Unilateral Problems in Linear Elasticity," *Quart. Appl. Mathematics*, 39, 1, 1981, 1–22.
63. Desai, C.S., "Cyclic Multi-Degree-of-Freedom Shear Device," *Report No. 8-36*, Dept. of Civil Eng., Virginia Tech, Blacksburg, VA, 1980.
64. Ma, Y. and Desai, C.S., "Constitutive Modeling of Joints and Interfaces by Using Disturbed State Concept," *Report to NSF and AFOSR*, Dept. of Civil Eng. and Eng. Mechs., The Univ. of Arizona, Tucson, AZ, 1990.
65. Bandis, S., Lumsden, A.C., and Barton, N.R., "Experimental Studies of Scale Effects on the Shear Behaviour of Rock Joints," *Int. J. Rock Mech. and Min. Sc.*, 18, 1981, 1–21.
66. Williams, A.F., "The Design and Performance of Piles into Weak Rock," *Ph.D. Dissertation*, Dept. of Civil Eng., Monash Univ., Melbourne, 1980.
67. Uesugi, M. and Kishida, H., "Influential Factor of Friction Between Steel and Dry Sands," *Soils and Foundations*, 26, 2, 1986, 33–46.
68. Uesugi, M. and Kishida, H., "Frictional Resistance at Yielding Between Dry Sand and Mild Steel," *Soils and Foundations*, 26, 4, 1986, 139–149.
69. Fakharian, K. and Evgin, E., "Elasto-plastic Modelling of Stress-Path Dependent Behaviour of Interfaces," *Int. J. Num. Analyt. Meth. Geomech.*, 24, 2, 2000, 183–1999.
70. Fakharian, K. and Evgin, E., "An Automated Apparatus for Three-Dimensional Monotonic and Cyclic Testing of Interfaces," *Geotech. Testing J., ASTM*, 19, 1, 1996, 22–31.
71. Vulliet, L., "Modelisation des Pents Naturelles en Mouvement," *These No. 635*, École Polytechnique Federale de Lausanne, Lausanne, Switzerland, 1986.
72. Perzyna, P., "Fundamental Problems in Viscoplasticity," *Adv. Appl. Mech.*, 9, 1966, 243–377.
73. Rigby, D.B. and Desai, C.S., "Modelling of Interface Behavior Using the Cyclic Multi-Degree-of-Freedom Shear Device with Pore Water Pressures," *Report to NSF*, Dept. of Civil Eng. and Eng. Mechs., The Univ. of Arizona, Tucson, AZ, 1996.
74. El Sakhawy, N.R., "Experimental and Numerical Study of the Pile Shaft–Sand Interface," *Ph.D. Dissertation*, Dept. of Civil Eng., Univ. of Wisconsin, Madison, WI, USA, 1991.

75. El Sakhawy, N.R. and Edil, T.B., "Behavior of Shaft–Sand Interface from Local Measurements," *Transportation Research Record 1548,* Transp. Res. Board, Washington, DC, 1996, 74–80.
76. Hall, P.M. and Sherry, W.M., "Materials, Structures and Mechanics of Joints for Surface-Mount Microelectronics Technology," *Proc., 3rd Int. Conf. on Techniques de Connexion en Electronique,* Welding Society, Miami, FL, 1986, 18–20.
77. Desai, C.S., Chia, J., Kundu, T. and Prince, J.L., "Thermomechanical Response of Materials and Interfaces in Electronic Packaging: Part I - Unified Constitutive Models and Calibration; Part II - Unified Constitutive Models, Validation and Design," *J. of Elect. Packaging, ASME,* 119, 4, December 1997, 294–309.
78. Basaran, C., Desai, C.S., and Kundu, T., "Thermomechanical Finite Element Analysis of Problems in Electronic Packaging Using the Disturbed State Concept - Part I: Theory and Formulations; Part II: Verification and Application," *J. of Elect. Packaging, ASME,* 120, 1998.
79. Desai, C.S., Basaran, C., Dishongh, T., and Prince, J., "Thermomechanical Analysis in Electronic Packaging with Unified Constitutive Model for Materials and Joints," *Part B: Advanced Packaging, IEEE Trans.,* 21, 1, February 1998.

12

Microstructure: Localization and Instability

CONTENTS

12.1 Microstructure 478
12.2 Wellposedness 482
12.3 Localization 482
      12.3.1 Nonlocality and Characteristic Dimension 483
12.4 Regularization and Nonlocal Models 484
      12.4.1 Microcrack Interaction Models 485
      12.4.2 Rate-Dependent Models 486
      12.4.3 Continuum Damage Model 486
      12.4.4 Models for Nonlocal Effects 487
12.5 Nonlocal Continuum 488
      12.5.1 Strain and Energy Based Models 489
      12.5.2 Gradient Enrichment of Continuum Models 489
      12.5.3 Cosserat Continuum 491
12.6 Stability 492
12.7 Disturbed State Concept: Nonlocality, Microcrack Interaction, Characteristic Dimension, Mesh Dependence, and Instability 494
      12.7.1 Approximate Decoupled DSC 500
      12.7.2 Instability Through Disturbance 502
      12.7.3 Stability Analysis of DSC 503
      12.7.4 Stability Condition for One-Dimensional DSC Model 503
12.8 Examples 507
      12.8.1 Instability Based on Critical Dissipated Energy and Disturbance 522

In this chapter, we consider the

(a) characteristics of deforming materials resulting from microstructural self-adjustment caused by microcracking and/or relative particle motions,

(b) nonlocality due to development of discontinuous and nonhomogeneous deformations resulting in localization, and the need of including nonlocal effects in constitutive models,

(c) review of various nonlocal models,
(d) spurious or pathological mesh dependence,
(e) instability at local and global levels leading to threshold transitions in the microstructure such as contraction to dilation, peak and critical conditions like liquefaction and fatigue failure,
(f) discussion of the capabilities of the DSC to allow for localization and characteristic dimension and instability, including comparison of the DSC with other models, and
(g) examples to illustrate the capabilities of the DSC.

12.1 Microstructure

A material can be considered to be a matrix or structure made of (clusters of) particles connected through friction, and/or bonding at the contacts, Fig. 12.1. The matrix framework is composed of solid particles and pore spaces. For dry materials, the pore spaces are empty, and usually contain air. For wet materials, the pores are partially or fully saturated with a liquid; e.g., water. In the case of partial saturation, the part of the pore space not occupied by the liquid, is filled with air (or gas), Chapter 9.

Under load, the material's matrix deforms due to particle motions that can involve deformations without sliding, particle sliding, rotation and movement towards or away from each other, Chapter 2. As a result, the particles may experience instantaneous and locally unstable motions at the microlevel. Such motions can occur in the material element at random locations, Fig. 12.2(a). The macrolevel or global response of the finite-sized test specimen, however, shows average or observed response that is "smooth", without oscillations. As the deformations grow, the (clusters of) particles experiencing the local instability

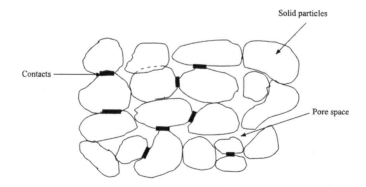

FIGURE 12.1
Material matrix and particle contacts.

Microstructure: Localization and Instability

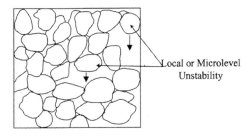

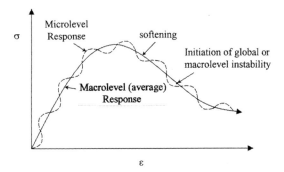

(a) Local and Global Instabilities

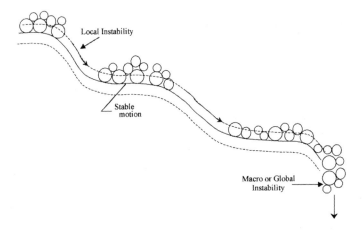

(b) Symbolic representation of local and global instabilities

FIGURE 12.2
Deformation of microstructure or matrix of material.

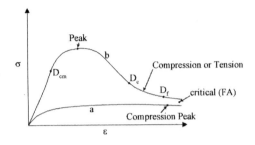
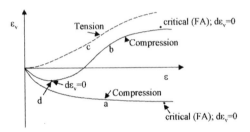

FIGURE 12.3
Schematic of stress–strain and volumetric responses for loose (a) and dense (b) materials in compression, and material in tension.

may reach critical magnitudes, and global or macrolevel instability, failure or collapse may initiate near the end (softening) region. Subsequently, complete failure can occur with the collapse of the entire specimen.

Figure 12.2(b) shows an analogy of motions of material particles over a series of undulating slopes involving (horizontal) plateaus. A particle or cluster of particles can experience instantaneous "fall" or instability when it is at the end of a plateau. However, overall stable motion can occur on the plateau. The macrolevel response of a material element composed of a large number (millions) of such particle (clusters) may, however, maintain global stability until the time the extents of the particles (clusters) experiencing local instability reach a critical threshold value. Then global or macro instability can occur.

Figure 12.3 shows schematics of stress-strain and volume change behavior for materials under compressive and tensile loading. Some materials, e.g., metals, may exhibit nonlinear behavior marked (a) and very little or no volume change, under both compression and tension.

Other materials like geologic (granular) may exhibit different behavior under initially loose (a) and dense conditions (b). In the case of the former, the volume change behavior is compactive and approaches the critical state ($d\varepsilon_v = 0$) asymptotically. On the other hand, the dense material may first compact, then dilate, and then approach the state when $d\varepsilon_v = 0$. Under tension, a material

Microstructure: Localization and Instability 481

(like concrete) may exhibit softening (b) response, where the volume change is essentially dilative (increase), marked (c).

Threshold transitions occur at states when the extents of the accumulation of local motions reach certain values. Examples of threshold transitions are the peak stress, transition from contractive to dilative state when $d\varepsilon_v = 0$ (point d), the later asymptotic state when $d\varepsilon_v = 0$, the peak or earlier stress states at which microcracking initiates (D_{cm}), and the state at which the material experiences initiation of instability (D_c), and enters the region leading to failure (D_f), Fig. 12.3.

From the viewpoints of global instability at the specimen level, the peak stress, and the states, D_c and D_f, are usually relevant. Localization may initiate at the peak stress, and the tangent modulus or tangent constitutive matrix, $\underline{C}$, vanishes; i.e., $|\underline{C}| \to 0$, Fig. 12.4. In the case of softening, the stiffness assumes a negative definite value from peak to D_c and then approaches zero in the ultimate

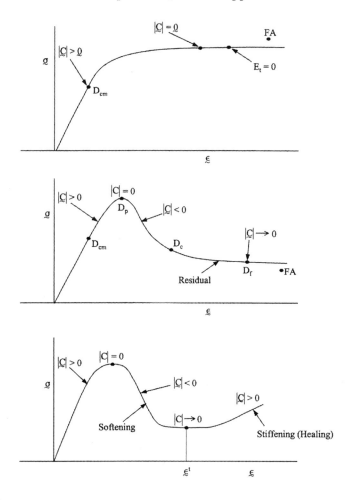

FIGURE 12.4
Threshold transition states and instability.

or residual zone. In the case of healing or stiffening, the material exhibits stiffer response after the threshold state, (ε^t), after which $|\underset{\sim}{C}| > 0$.

One of the reasons for material instability can be due to the frictional effects, which result in the constitutive matrix, $\underset{\sim}{C}$, to be nonsymmetric. This can occur in the case of both compressive and tensile loading. The lack of symmetry can be sufficient to lead to the loss of material stability at certain states during deformation. Thus, material instabilities can occur due to both decohesion (particle separation) in the case of tensile loading and the frictional behavior. The former can also occur in materials such as concrete and rocks under low confining pressures. The friction case can involve slip processes in metals, soils and concrete under higher confining pressures (1–15).

12.2 Wellposedness

An initially continuous material may experience discontinuous deformations due to microcracking and relative particle motions. Characterization of such materials requires considerations beyond those for continuous materials defined by using the theories of continuum mechanics.

As the discontinuities grow, the continuum material gradually transforms into a discontinuum material; in the limit, the entire material can approach the discontinuum state. As discussed in Chapters 2 and 3, the limiting or asymptotic state, which represents "total" failure, is usually not measurable in engineering practice; the engineering failure occurs before the limiting state is reached. Such a state is treated, approximately, as the FA state in the DSC.

The initiation of discontinuities involving localization (strain, damage or disturbance) represents loss of *wellposedness* of the initial or boundary value problem. Such a critical state entails a change in the type of the differential problem. For instance, it may result in the loss of *ellipticity* of the equilibrium differential equations. Maintenance of ellipticity is considered to be a condition for wellposedness of the boundary value problem (10–12).

12.3 Localization

In some materials, discontinuous deformations often localize in narrow zones, while the material in the neighboring regions may load or unload and remain in its intact or continuum state (10–13, 16–23). Factors such as material type and composition, boundary conditions and kind of applied loads influence the localization phenomenon. A material can fail due to the formation and growth of microcracking (disturbance) at randomly distributed locations. This type of localization is often called Mode-I localization (24–26) and is usually observed in rocks, alloys, and composite media.

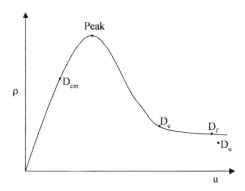

FIGURE 12.5
Microcracking and localization.

Localization may result in motions along preferred directions, called *shear bands*, and can lead to excessive deformations and failure. This type of localization is called Mode-II localization, and is often observed in metals, geomaterials and concrete (in compression) (24–26).

Localization usually initiates at the peak load (stress), Fig. 12.5; however, microcracking leading to localization may initiate before peak (D_{cm}).

Beyond the peak, a material may experience softening or degradation in its load-carrying capacity, but still continue to carry reduced load compared to the peak load while growth and coalescence of microcracks continue during the falling region of the curve, Fig. 12.5. Finally, the microcrack growth can lead to a macrocrack or fracture at point C, where critical disturbance, D_c, occurs. Further deformations lead to engineering "failure" at D_f. In Fig. 12.5, D_u is the ultimate disturbance.

If the disturbance (or damage) causing strain-softening is characterized by using the continuum approach, it becomes necessary to include in the constitutive model a characteristic dimension that identifies and provides limit to the zone of localization. In other words, the zone of disturbance is not allowed to localize to zero volume (11, 12, 27). At the macrolevel, therefore, tensile or shear strains evolve and grow within a zone of finite thickness or width, which is related to characteristic dimension. It was reported that the width for concrete was about 2.7 times the maximum aggregate size. For geological media, the shear band width is about 10 to 30 times the average grain diameter (26–28).

12.3.1 Nonlocality and Characteristic Dimension

Nonlocality can be explained, in a simple and philosophical way, as the requirement that the behavior at a point in a material should include the effect of the (discontinuous) responses in the neighboring zones. This requirement is rooted in the interlinkedness that pervades physical systems, of which a material is a subset. Thus, effects such as microcracking and particle motions leading to disturbance (damage or strengthening) should include coupled influences from

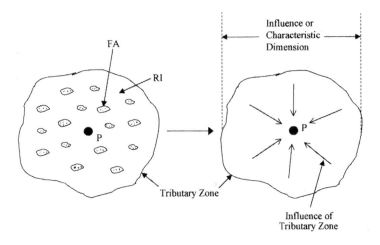

FIGURE 12.6
Influence of tributary zone around a point.

tributary zones, Fig. 12.6, in the regions adjacent to the point (P). In the case of computer (FE) analysis, the effect of disturbance in the adjoining element should be included in the response of a given finite element. In other words, the nonlocal effects should be included in the stiffness matrix assembly beyond that given by the approximation (10–15, 29, 30) functions for the unknown (displacement).

Characteristic Dimension. The issue of characteristic dimension is an integral part in the nonlocal considerations. It defines the connection between the behavior of neighboring zones in the vicinity of a point, and the extent or length of the influence. It acts as a localization limiter. The characteristic dimension is dependent on such factors as the particle shape and size, material fabric and size of heterogeneities. A constitutive model that can account for the nonlocal effects should include the characteristic dimension in its framework.

Nonlocality is often allowed for by defining average or weighted values of quantities (stress, strain, disturbance) at a point by including the tributary regions. Here, it is considered that the impending motions at points of contact between particles in solid deformable bodies, or at an interface or joint between (two) deformable bodies, will occur when the stress at a point reaches a value proportional to the weighted value of mean pressure in a solid or normal stress at an interface. An example of such nonlocal formulations for friction laws is presented in (31).

12.4 Regularization and Nonlocal Models

When classical continuum and damage models are used to define response of materials experiencing discontinuous deformations and localization, disturbance (damage) leading to degradation or softening can occur without

Microstructure: Localization and Instability

energy dissipation as the (FE) mesh is refined (10–13, 29, 30, 32). In other words, continuum models that admit only a pointwise or local description of quantities (strains, stresses, damage, disturbance) are not capable to handle discontinuous deformations. A consequence of their use in numerical (FE) calculations is that the computer solutions are severely mesh dependent, as failure is predicted without energy dissipation (10–12). Such mesh dependence, which is beyond the traditional effect of mesh refinement (33, 34), is called *pathological* or *spurious* mesh dependence.

To overcome the foregoing difficulties, it becomes necessary to regularize the problem such that the strain field is "regular" during softening. Various schemes such as homogenization, weighted averaging of quantities, and introduction of enrichments and constraints are used for the regularization. We first present a review of these schemes, and then describe how the DSC allows for nonlocality, characteristic dimension and regularization.

12.4.1 Microcrack Interaction Models

One of the ways to introduce nonlocality in constitutive models is to allow for the interaction or coupling between the undamaged (RI) and microcracked (FA) material parts in the deforming material element. A number of studies have considered superposition of the microcrack kinematics on the continuum model. One such model is proposed by Bazant (13), in which the weighted average stresses are incorporated in the continuum damage model. The incremental stress, $d\underline{\sigma}$, is expressed as

$$d\underline{\sigma} = \underline{C}^e d\underline{\varepsilon} - d\underline{\bar{\sigma}} \tag{12.1}$$

where $\underline{C}^e$ is the elasticity matrix for the undamaged material, $d\underline{\varepsilon}$ is the incremental strain vector, and $d\underline{\bar{\sigma}}$ is the nonlocal plastic or inelastic incremental stress vector, given by

$$d\underline{\bar{\sigma}}(x) = \int \alpha(\underline{x},\underline{\xi}) d\underline{\sigma}(\underline{\xi}) dV(\underline{\xi}) \tag{12.2}$$

Here, V is the volume of the body, $\underline{x}, \underline{\xi}$ are the coordinate vectors, (α, ξ) is the nonlocal weight function, and $d\underline{\sigma}(\underline{\xi})$ is the incremental plastic stress vector. Details of the derivation of the model are given by Bazant (13). Here we make the following observations:

Figure 12.7 shows the schematic of the model for an incremental loading step. Assuming that the cracks are "frozen" or "glued", the stress increment due to the strain increment, $d\underline{\varepsilon}$, is represented by the line segment $\overline{13}$ with the stiffness, E, for the undamaged material. Thus, this step can be considered to represent the response of the material as if it is a continuum or in the RI state (as in the DSC). In the second step, the cracks are unfrozen, which results in transmission of stresses across the relaxed cracks.

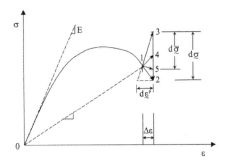

FIGURE 12.7
Local and nonlocal inelastic stress increments during loading step (13).

If there is no growth or closing of cracks, the unfreezing of cracks under given stress or displacement increment would cause the stress to drop from point 3 to point 4, Fig. 12.7, along the secant line $\overline{01}$. Effect of such a change in stress is calculated by applying equal but opposite stress corresponding to the drop. When cracks propagate and new cracks nucleate, there occurs a greater stress drop, which is given by the segment $\overline{32}$. The combined effect of unfreezing, propagation and nucleation of cracks is to decrease the stress from the continuum (undamaged) or RI stress at point 3 to that corresponding to the observed or actual stress at point 2.

Thus, the effect of the microcrack response and its coupling with the behavior of the undamaged part is superimposed on the response of the undamaged (or RI) part. In the DSC, such effect of the microcracks on the FA part is incorporated implicitly in the model, Eq. (4.1), through the terms, $D\underline{\sigma}^c$ and $dD(\underline{\sigma}^c - \underline{\sigma}^i)$ in which $\underline{\sigma}^c$ denotes the stresses carried by the microcracked or FA part.

12.4.2 Rate-Dependent Models

Regularization can be provided by incorporating deformation-rate dependence in the constitutive equations. The viscoplastic models, e.g., Perzyna (Chapter 8), possess the regularization attribute as the energy dissipation remains finite during plastic flow for dynamic loading (19). Here, the viscosity (Γ) that defines rate dependence is related to the internal length scale (35,36). The regularization property of the viscoplastic model has not been proven for quasi-static loadings (32). However, with proper choice of the spatial mesh and time-step size, the regularization effect can be achieved for quasistatic loadings (36).

12.4.3 Continuum Damage Model

The development and use of continuum and nonlocal damage models have been available in many publications including various text books (37–46). Kachanov (38) defined the damage variable, ω, as the ratio of the area A^d of the damaged, fractured or "lost" part of the material element to its original area, A, i.e.,

$$\omega = \frac{A^d}{A} \tag{12.3}$$

Then the actual, observed or nominal stress, σ^a, at a point, is given by

$$\sigma^a = (1 - \omega)\sigma^i \tag{12.4}$$

where σ^i is the stress at a point in the undamaged (or relative intact) part.

Equation (12.4) implies that the actual stress at a point is affected by damage (ω) which represents the effect of the lost or fractured part of the material. However, it ignores the deformation in the damaged part and its coupled influence on the actual behavior; this is because, once a part is damaged, it is treated like a "void" which can carry no stress at all.

In reality, however, the damage may be treated as a "void" only at or near failure when "finite" cracks and separation occur. In general, however, microcracked or damaged parts can deform, carry stresses, and interact with the undamaged parts so as to lead to the actual response. Thus, it is not realistic to ignore the effect of microcracks and damaged parts on the behavior of the material element. The damage model, Eq. (12.4), which assumes that the deformation in the damaged part has no effect on the coupled actual response, is often called the *local* damage model. As the effect of deformation in the neighboring zone of a point is not included in the local models, they suffer from the spurious mesh dependence.

12.4.4 Models for Nonlocal Effects

Recognition of the influence of the discontinuous nature of deforming materials and nonlocal effects has led to the development of a number of nonlocal models. They include enrichment of continuum models by using various schemes such as imposition of microcrack interaction (12, 13, 45), and gradient (10,12,47–54) and Cosserat theories (10, 12, 55–59). Here, brief descriptions are presented together with the analysis of DSC in accounting for the nonlocal effects.

A symbolic representation of nonlocal models is shown in Fig. 12.8. The crux of the approach is to allow for the coupling or interaction between the interconnected zones or clusters of continuum and discontinuous or microcracked zones in a deforming material element. This leads to inclusion of nonlocality and connection through the characteristic dimension between the continuum and microcracked (or fully adjusted) zones or clusters. It may be noted that almost all the available models are still based on the continuum approach, which is modified by the enrichments and special

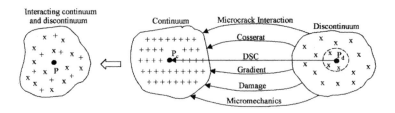

FIGURE 12.8
Symbolic representation of nonlocal models.

schemes. Thus, behavior, $B(P)$, at a point, P, in the mixture of continuum (or RI) and discontinuum (or FA) is expressed through a combination of the behavior of the continuum (or RI) parts, $B(P_c)$, and that of the discontinuous (or FA) parts, $B(P_d)$

$$B(P) \Leftarrow B(P_c) \ddagger B(P_d) \qquad (12.5)$$

where the symbol, $\ddagger$, represents the coupled integration of the behavior of the two parts.

12.5 Nonlocal Continuum

In the context of the theory of elasticity, the concept of nonlocal continuum was introduced by Eringen (37), Kröner (60) and others, the basic idea being that the stress at a point should depend not only upon the deformation (strain) at that point, but also on the deformations in the neighboring tributary zones.

Various investigators have used the nonlocal concept in the context of continuum damage models; these include nonlocal damage models with microcrack interaction (12,13,45). The effect of deformation in the tributary zone is often obtained by using weighted average quantities such as strain and stress. For example, weighted strain ($\bar{\varepsilon}$) is given by

$$\bar{\varepsilon}(x) = \frac{1}{V_r(x)} \int \psi(x - s)\varepsilon(s)dV \qquad (12.6)$$

where V is the volume of the body or structure, $V_r(x)$ is the representative volume at point x, $\psi(x - s)$ is the weighting function, and $\varepsilon(s)$ is the (equivalent) strain at a point. The weighted average for energy release rate at a point x, $\bar{E}(x)$, is expressed as

$$\bar{E}(x) = \frac{1}{V_r(x)} \int \psi(x-s) E(s) dV \tag{12.7}$$

where $E(s)$ is the energy release rate at a point s in the material.

12.5.1 Strain and Energy Based Models

Nonlocal damage models are then developed by using weighted strain ($\bar{\varepsilon}$) or ($\bar{E}$) in the definition of damage, ω (45), e.g.,

$$\omega_\varepsilon = f_\varepsilon(\bar{\varepsilon}) \tag{12.8a}$$

$$\omega_e = f_e(\bar{E}) \tag{12.8b}$$

Thus, the use of the weighted average quantities $\bar{\varepsilon}$, $\bar{E}$, etc., in the constitutive equation and in the finite element calculation allows for the effect of the state or deformation in the neighboring region on the behavior at a point.

12.5.2 Gradient Enrichment of Continuum Models

In the gradient enriched models, the nonlocality is introduced by adding gradient terms in the continuum model. For example, in the case of plasticity models, the yield function, F, is enhanced by introducing the gradients of plastic equivalent strain, $\dot{\varepsilon}^p$ (10, 44, 50–52)

$$\dot{\bar{\varepsilon}}^p = \dot{\varepsilon}^p + a_1 \nabla \dot{\varepsilon}^p + a_2 \nabla^2 \dot{\varepsilon}^p + \cdots \tag{12.9}$$

where $\dot{\bar{\varepsilon}}^p$ is the nonlocal equivalent plastic rate, $\dot{\varepsilon}^p$ is the local plastic strain rate, and a_1 and a_2 are parameters that depend on the weight function and dimension of the problem. If isotropy is considered, the odd derivatives cancel and the yield function, F, is expressed as

$$F = F(\sigma, \varepsilon^p, \nabla^2 \varepsilon^p) \tag{12.10}$$

where $\varepsilon^p = \int \dot{\varepsilon}^p dt$, and t is time. Note that in the local and classical plasticity, the gradient term, $\nabla^2 \varepsilon^p$ is not included. The gradient terms can be considered to introduce nonlocal effects in the zones (in the neighborhood of a point) in the microstructural deformations involving discontinuities. Discussion of the effect of gradient terms are presented in (7, 16, 17, 47, 48, 61).

De Borst et al. (10) have presented the gradient enhanced plasticity model with the Drucker-Prager yield criterion, which, with the gradient terms, Eq. (12.10), is given by

$$F = \sqrt{3J_{2D}} + \alpha^* p - \bar{\sigma}(\varepsilon^p, \nabla^2 \varepsilon^p) \quad (12.11)$$

where J_{2D} is the second invariant of the deviatoric stress tensor, S_{ij}, $p = J_1/3$ is the hydrostatic or mean pressure, J_1 is the first invariant of the stress tensor, σ_{ij}, α^* is the friction parameter, and $\bar{\sigma}$ denotes the cohesion in the material.

By considering the plastic scalar multiplier, λ, in the normality rule as an independent variable, in addition to displacements, $\underline{u} = [u \ v \ w]^T$, as unknowns, the finite element equations are derived as (10)

$$\begin{bmatrix} \underline{k}_{qq} & \underline{k}_{q\lambda} \\ \underline{k}_{q\lambda}^T & \underline{k}_{\lambda\lambda} \end{bmatrix} \begin{Bmatrix} \dot{\underline{q}} \\ \dot{\underline{\lambda}} \end{Bmatrix} = \begin{Bmatrix} \dot{\underline{Q}} \\ 0 \end{Bmatrix} \quad (12.12a)$$

where

$$\underline{k}_{qq} = \int_V \underline{B}^T \underline{C}^e \underline{B} \, dV$$

is the traditional stiffness matrix from the displacement formulation,

$$\underline{k}_{q\lambda} = -\int \underline{B}^T \underline{C}^e \underline{n} \underline{h}^T dV \quad (12.12b)$$

$$\underline{k}_{\lambda\lambda} = \int_V (h + \underline{n}^T \underline{C}^e \underline{n}) \underline{h} \, \underline{h}^T - ch \, \underline{p}^T) dV \quad (12.12c)$$

$\dot{\underline{q}}$ is the nodal (rate) displacement vector, $\dot{\underline{\lambda}}$ nodal plastic multiplier (rate) vector, $\dot{\underline{Q}}$ is the applied external load (rate) vector, $\underline{h} = [h_1 \cdots h_n]$ is the vector of interpolation functions for the plastic multiplier (λ), $\underline{p}^T = [\nabla^2 h_1, \cdots \nabla^2 h_n]$, and c is given by

$$c = \bar{c}\sqrt{1 + 2\alpha^2/9} \quad (12.13a)$$

and

$$\bar{c} = \frac{\partial F}{\partial (\nabla^2 \varepsilon^p)} \quad (12.13b)$$

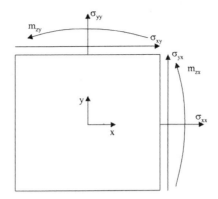

FIGURE 12.9
Material element in Cosserat model.

The matrix, $k_{\lambda\lambda}$, is nonsymmetric, which would require a nonsymmetric equation solver and may entail computational difficulties. Often, various schemes (10) are used to solve the equations approximately by considering only the symmetric part, k_{qq}. One way is to move the nonsymmetric part as a load vector on the right-hand side as a load vector (15, 62–65). Details of the computational schemes in the DSC are given later and in Chapter 13.

12.5.3 Cosserat Continuum

In the classical strength of materials approach, say, for the two-dimensional case, the state of stress is defined by two normal (σ_{xx}, σ_{yy}) and one shear ($\sigma_{xy} = \sigma_{yx}$) stresses. In the Cosserat theory (10, 12, 55–59), additional static and kinematic variables are introduced, Fig. 12.9. Hence, the displacement, strain and stress vector are augmented as follows:

$$\underline{u}^T = [u \; v \; \omega_z]^T \tag{12.14a}$$

$$\underline{\varepsilon}^T = [\varepsilon_{xx} \; \varepsilon_{yy} \; \varepsilon_{zz} \; \varepsilon_{xy} \; \varepsilon_{yx} \; \kappa_{xz}\ell \; \kappa_{yz}\cdot\ell] \tag{12.14b}$$

$$\underline{\sigma}^T = [\sigma_{xx} \; \sigma_{yy} \; \sigma_{zz} \; \sigma_{xy} \; \sigma_{yz} \; m_{xz}/\ell \; m_{yz}/\ell]^T \tag{12.14c}$$

where m_{zx} and m_{zy} are the coupled stresses, $\kappa_{zx} = \partial\omega_z/\partial x$, $\kappa_{zy} = \partial\omega_z/\partial y$ are the micro-curvatures, and ω_z is the micro-rotation about the z-axis. The internal length scale, ℓ, is introduced so that the strains and stresses have the same dimensions.

It may be noted that the length scale (or characteristic dimension) is introduced directly in the basic (elastic) formulation and provides the regularization for the nonlocal effect. On the other hand, the regularization effect in the previous gradient enrichment is introduced only in the plastic regime. In other words, in the case of the Cosserat model, the yield function, F, is defined as in classical plasticity in terms of $\underset{\sim}{\sigma}$ and $\underset{\sim}{\varepsilon}^p$ only.

Furthermore, the Cosserat theory introduces micro-rotation, ω_z, as an additional unknown; hence, in the finite element formulation, there occur three independent generalized displacements u, v and ω_z. Details of the Cosserat model and its implementation in computational (finite element) procedures are given in a number of publications (12, 59).

12.6 Stability

The cause of localization involving discontinuous deformations is considered to be material instabilities in its microstructure. Stability has been defined differently by different investigators (1–6, 8–10, 13). However, the basic notion is that if a material experiences undefined or infinite changes under finite (small) loading (or disturbance), instability can be considered to have occurred. In this sense, when the determinant of the constituitive matrix, $|\underset{\sim}{C}| \to \underset{\sim}{0}$ (or for the system stiffness matrix, $|\underset{\sim}{K}| \to \underset{\sim}{0}$), large changes in motion under small changes in load can occur, causing instability.

According to Hill (1), instability occurs when the following stability criterion is violated:

$$\dot{\varepsilon}^T \dot{\sigma} > 0 \tag{12.15}$$

where ε and σ denote strain and stress, respectively, and the overdot denotes rate (or increment). When the slope of average or homogenized (uniaxial) compression or tension curve becomes negative, strain softening occurs, Fig. 12.4. Average or homogenized refers to the fact that the response at the macrolevel is obtained by using stress as the ratio of force to the original area, and strain as the ratio of the change in displacement to the original dimension of the specimen. It does not take into account what happens at the microlevel affected by factors such as local instability, initial (micro) flaws, and resulting nonhomogeneities in deformations. As indicated before, such an approach may not provide consistent simulation of the response which involves nonhomogeneous or discontinuous states of deformations and does not allow for what happens in the neighborhood of a material point. Then it becomes necessary to superimpose external enrichments such as microcrack interaction and gradient or Cosserat theories so as to "regularize" the problem. In the DSC, such regularizing is included internally in the model. These aspects are discussed later.

Equation (12.15) can be expressed as

$$\dot{\underset{\sim}{\varepsilon}}^T \underset{\sim}{C} \dot{\underset{\sim}{\varepsilon}} > 0 \tag{12.16}$$

Microstructure: Localization and Instability

where $\underset{\sim}{\varepsilon}$ is the strain vector for multidimensional behavior, and $\underset{\sim}{C}$ is given by

$$\underset{\sim}{\dot{\sigma}} = \underset{\sim}{C}\underset{\sim}{\dot{\varepsilon}} \tag{12.17}$$

where $\underset{\sim}{\sigma}$ is the stress vector. As discussed earlier, the limiting case of the inequality, Eq. (12.16) can be replaced by an equality to show the initiation of material instability. Hence, the loss of positive-definiteness of the tangent constitutive (stiffness) matrix, $\underset{\sim}{C}$ is expressed as (10)

$$\det(\underset{\sim}{C} + \underset{\sim}{C}^T) = 0 \tag{12.18}$$

Structural Instability. A structure is made of material elements with volume, V, and, according to Eq. (12-16), the stability condition can be expressed as

$$\int_V \underset{\sim}{\varepsilon}^T \underset{\sim}{\dot{\sigma}} dV > 0 \tag{12.19a}$$

or

$$\int_V (\underset{\sim}{B}^T \underset{\sim}{C} \underset{\sim}{B}) dV \cdot \dot{q} > 0 \tag{12.19b}$$

or

$$\underset{\sim}{k} \cdot \dot{q} > 0 \tag{12.19c}$$

where q is the vector of (nodal) displacements, $\underset{\sim}{B}$ is the strain-displacement transformation matrix given by (33, 34)

$$\underset{\sim}{\dot{\varepsilon}} = \underset{\sim}{B}\dot{q} \tag{12.20a}$$

and $\underset{\sim}{k}$ is the tangent stiffness matrix:

$$\underset{\sim}{k} = \sum \int_V \underset{\sim}{B}^T \underset{\sim}{C} \underset{\sim}{B} \, dV \tag{12.20b}$$

Hence, for structural instability, we can write the limiting condition as

$$\det(\underset{\sim}{K} + \underset{\sim}{K}^T) = 0 \tag{12.21}$$

Thus, when the tangent constitutive matrix, $\underset{\sim}{C}$, loses positive-definiteness, the structural stiffness matrix can also lose positive-definiteness. Many investigators have studied material and structural instabilities and resulting localization and phenomena using the above and other formulations (1–13).

Although the stiffness matrix becomes negative-definite in the softening zone, it is possible to devise computational algorithms to achieve convergent solutions. For example, the incremental calculations can be obtained by ensuring convergence along the fixed applied or computed strain increment. Also, in the DSC, the incremental calculations are performed for the RI behavior, and the observed stresses are calculated for the same computed strains in the RI and observed states. Then the observed stresses are calculated using Eq. (4.1) through an iterative procedure based on disturbance from the observed (stress-strain) response. These procedures are described in Chapter 13.

12.7 Disturbed State Concept: Nonlocality, Microcrack Interaction, Characteristic Dimension, Mesh Dependence, and Instability

Now, we discuss the DSC and how the foregoing aspects such as nonlocality, localization, characteristic dimension, and avoidance of the spurious mesh dependence are incorporated in it; a thermodynamical analysis of the DSC is presented in Appendix 12.1. Then, the microstructural instability in the DSC is considered. Examples of theoretical derivations and computer solutions to illustrate these capabilities of the DSC are presented subsequently.

The DSC incremental equations, Eq. (4.1), can be expressed as

$$d\underline{\sigma}^a = d\underline{\sigma}^i + D(d\underline{\sigma}^c - d\underline{\sigma}^i) + dD(\underline{\sigma}^c - \underline{\sigma}^i) \tag{12.22a}$$

or

$$d\underline{\sigma}^a = d\underline{\sigma}^i + W_1 d\underline{\sigma}^r + W_2 \underline{\sigma}^r \tag{12.22b}$$

where $\underline{\sigma}^r$ is the relative stress in the material element, Fig. 12.10(a), and W_1 and W_2 are considered to be weighting functions, Fig. 12.10(b). The RI (local) stress, $d\underline{\sigma}^i$, at a point in the material element can be considered to be modified due to the effect of the relative stress, $\underline{\sigma}^r$, and relative stress increment, $d\underline{\sigma}^r$, Fig. 12.10(c). The equivalent length (or volume or area) of the material element, d_{ch}, that influences the observed stress, $d\underline{\sigma}^a$, can be written proportional to $(D + dD)$ as [Fig. 12.10(c)]

$$d_{ch} \alpha (D + dD) \tag{12.23}$$

Microstructure: Localization and Instability

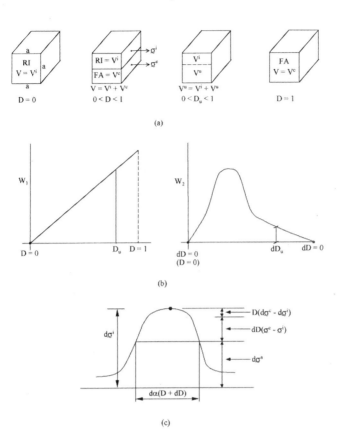

FIGURE 12.10
Averaging and weighting function in DSC.

The dimension, d_{ch}, occurs in the range of 0 to 1; however, its value is limited by the ultimate disturbance, D_u; therefore,

$$(d_{ch})_u \, \alpha \, (D_u + dD_u) \tag{12.24}$$

Equation (12.22) is similar to Eq. (12.1) for the microcrack interaction nonlocal damage model (13), Fig. 12.7. The term $W_1 d\sigma^i + W_2 d\sigma^r$ in Eq. (12.22) is similar to the nonlocal (weighted) stress increment, $d\bar{\sigma}^c$ in Eq. (12.2). However, in the DSC, the influence of the stress (strain) in the microcracked or FA region (σ^c) is included implicitly in Eq. (12.22); in other words, the coupling or interaction between the RI and FA responses is included through the relative stress terms. As a consequence, in the DSC, it is not necessary to superimpose the microcrack interaction effects externally. Thus, the averaging of the observed stress, $d\sigma^a$, Eq. (12.22), allows for the localization and the resulting nonlocal effects.

For the classical continuum damage model, Eq. (12.22) reduces to (38)

$$d\underline{\sigma}^a = d\underline{\sigma}^i + W_1(-d\underline{\sigma}^i) + W_2(-\underline{\sigma}^i) \qquad (12.22c)$$

because the damaged or cracked zone cannot carry any stress (i.e., $\underline{\sigma}^c = d\underline{\sigma}^c = 0$). Thus, the classical model does not include the coupling effect of the microcracked (or damage) zone in the formulation. As the observed stress, $d\underline{\sigma}^a$, is computed only on the basis of the stresses in the undamaged (or RI) part, the nonlocal effect due to microcrack interaction is not included in the classical model. As a result, the classical damage model suffers from spurious or pathological mesh dependence.

As discussed in Chapter 2, the ultimate disturbance, D_u, represents the invariant volume (V^u/V) (or density) a deforming loose or dense material approaches irrespective of its initial density. It can be dependent on factors such as initial mean pressure (p_0), (mean) particle dimension (d_m) and characteristic or size ratio (ℓ/L), where ℓ is the characteristic dimension or width of material test specimen, and L is the length of the structure or height of the material test specimen (10, 20, 29). Hence,

$$D_u = D_u(p_0, d_m, \ell/L) \qquad (12.25)$$

Thus, the characteristic dimension $(d_{ch})_u$ is dependent on the ultimate disturbance, D_u, which is an invariant parameter. It denotes the limit of the influence of the localization zone, d_{ch}, [Eq. (12.23)] around a point.

An interpretation of the characteristic dimension can be advanced based on deformation of a particulate (solid) material or interface (joint). Figure 12.11 shows schematics of the deformation of a deformable solid square element (dimension a × a) and interface between (two) deforming materials. As discussed (Chapter 2), an initially loose granular material may compact continuously, whereas a dense material may first compact and then dilate. The motions of particles develop along preferred directions, with inclination, i. In the ultimate state, the material approaches the unique volume, V^u (or area A^u), irrespective of its initial density, (ρ_i), Fig. 12.11(a); the inclination approaches an ultimate value (i^u) corresponding to A^u.

A smooth interface may exhibit continuous compactive normal displacement (v_r), while a rough interface may first compact and then dilate along direction denoted by (i) (Chapter 11), which is proportional to dimensions of deforming asperities. The asperities are gradually annihilated with shear loading, and irrespective of the initial roughness (R_i); it approaches the unique value of normal displacement, v_r^u, or asperity, i^u, in the ultimate region.

As mentioned earlier, the ultimate values, A^u and v_r^u, are dependent on the mean pressure (p_0) or normal stress (σ_0), mean particle size, d_m in the solid or on the surface of the interface, and the ratio (ℓ/L). In the case of the solid, ℓ/L can represent the size or characteristic ratio, and in the case of interface, it can

Microstructure: Localization and Instability

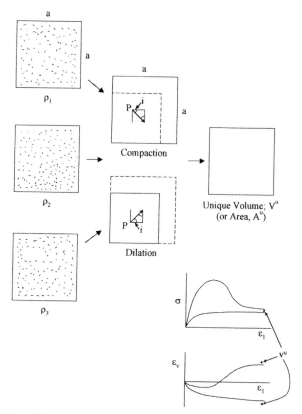

(a) Unique volume, V^u, for Solid, Irrespective of initial density (ρ) for given p_o, d_m and ℓ/L

FIGURE 12.11
Deformation in solid and interface material elements.

represent the ratio of thickness (t) of the interface zone (or asperity height, h) to the width, B, of the interface (Chapter 11). Then the limiting values of the zones of influence (localization) can be expressed as
Solid:

$$(d_{ch})_s = \frac{a^u}{\tan i} \qquad (12.26a)$$

Interface:

$$(d_{ch})_j = \frac{v_r^u}{\tan i} \qquad (12.26b)$$

where a^u is the ultimate dimension in A^u. Thus, in Eq. (12.26) the characteristic

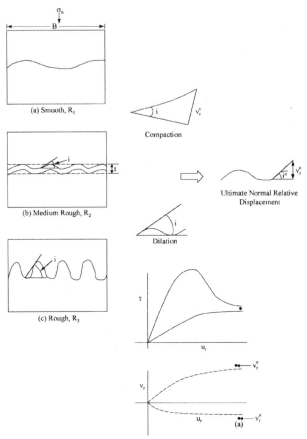

(b) Unique Normal Displacement, v_r^u, for Interface, Irrespective of Roughness (R) for Given σ_n

FIGURE 12.11
(continued)

length is proportional to disturbance, D_u, as D_u for solids $= A^u/A$, and for interface, $D_u = v_r^u/v_r^0$, where $v_r^0 =$ ultimate dilation when $\sigma_n = 0$, Fig. 11.6(b).
Furthermore, consider the following form of Eq. (12.22):

$$d\underset{\sim}{\sigma}^a = (1-D)\underset{\sim}{C}^i\,d\underset{\sim}{\varepsilon}^i + D\underset{\sim}{C}^c \cdot d\underset{\sim}{\varepsilon}^c + dD(\underset{\sim}{\sigma}^c - \underset{\sim}{\sigma}^i) \quad (12.27)$$

Assume that the incremental analysis is performed at the same level of strain increment, $d\underset{\sim}{\varepsilon}^i = d\underset{\sim}{\varepsilon}^a = d\underset{\sim}{\varepsilon}^c$, Fig. 12.12(a) (Chapter 13). Then, Eq. (12.27) can be expressed as

$$d\underset{\sim}{\sigma}^a = [(1-D)\underset{\sim}{C}^i + D\underset{\sim}{C}^c + \underset{\sim}{R}^T(\underset{\sim}{\sigma}^c - \underset{\sim}{\sigma}^i)]d\underset{\sim}{\varepsilon}^i = \underset{\sim}{C}^{DSC} \cdot d\underset{\sim}{\varepsilon}^i \quad (12.28)$$

Microstructure: Localization and Instability

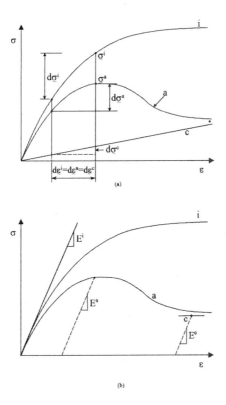

FIGURE 12.12
Incremental iterative analysis at constant strain level.

Here, D and $\underset{\sim}{R}$ (dD), Eq. (4.14), Chapter 4, are functions of D_u, A, Z, ξ and $d\xi$, e.g.,

$$\underset{\sim}{R}^T = \underset{\sim}{R}^T(D_u, A, Z, \xi, d\xi) \tag{12.29}$$

The constitutive matrix, $\underset{\sim}{C}^{DSC}$ in Eq. (12.28), includes the effects of both the plastic strains (ξ) and gradient of plastic strains ($d\xi$). Thus, the DSC model allows for the gradients of plastic strains in addition to the strains, and can be considered to allow for nonlocal effects similar to those provided by the gradient-enriched models, Eq. (12.10) (10). Additional derivations to show the inclusion of gradient effects in the DSC are given in (29).

Now, consider the simple one-dimensional (uniaxial) form of Eq. (12.27), Fig. 12.12(b):

$$d\sigma^a = (1 - D)E^i d\varepsilon^i + D \cdot E^c d\varepsilon^c + dD(\sigma^c - \sigma^i) \tag{12.30}$$

The computed value of $d\sigma^a$ is limited with respect to the ultimate disturbance, D_u:

$$(d\sigma^a)_u = (1 - D_u)E^i d\varepsilon^i + D_u E^c d\varepsilon^c + dD_u(\sigma^c - \sigma^i)_u \qquad (12.31)$$

As $D_u < 1$, $E^i = E^c \neq 0$, $d\varepsilon^i = d\varepsilon^c \neq 0$, $dD_u \neq 0$ and $(\sigma^c - \sigma^i)_u \neq 0$, the computed value of $(d\sigma^a)_u \neq 0$. Hence, at localization corresponding to D_u, there are always nonzero values of $d\sigma^a$ and $d\varepsilon^i$ $(=d\varepsilon^a)$. In other words, localization occurs at nonzero energy dissipation.

In view of the above, the DSC allows for nonlocal effects, microcrack interaction and characteristic dimension. As a result, it avoids spurious mesh dependence. Examples to illustrate these capabilities are given subsequently.

12.7.1 Approximate Decoupled DSC

The foregoing considerations apply when the coupled DSC, Eq. (12.28), leads to the following (FE) equations:

$$K^{DSC} \cdot dq^i = dQ \qquad (12.32)$$

where $K^{DSC} = \int_V B^T C^{DSC} B \, dV$ is the stiffness matrix, dq^i is the incremental (RI) displacement vector and dQ is the applied load vector. As noted before, K^{DSC} is nonsymmetric and becomes negative definite after the peak. Convergent computer results are obtained during incremental analysis at the same levels of strains ($d\varepsilon^a = d\varepsilon^i = d\varepsilon^c$). Then the observed stress is found for Eq. (12.28). Details are given in Chapter 13.

As discussed in Chapter 13, an approximate and decoupled but efficient procedure can be developed and used. Here, the FE equations are first solved only for the RI response:

$$K^i \cdot dq^i = dQ^i \qquad (12.33)$$

where $K^i = \int_V B^T C^{ep} B dV$, C^{ep} is the RI elastoplastic constitutive matrix. The increment RI stress as $d\sigma^i$ is found as

$$d\sigma^i = C^{ep} \cdot d\varepsilon^i \qquad (12.34)$$

and the FA stress is defined based on given assumptions, such as the critical state (Chapters 3 and 4):

$$d\sigma^c = C^c \cdot d\varepsilon^i \qquad (12.35)$$

Then the observed stress, $d\sigma^a$, is found by using Eq. (12.28) through an iterative procedure in which the disturbance based on the stress-strain model from the observed behavior is used as a convergence parameter (Chapter 13), Fig. 12.13.

Microstructure: Localization and Instability

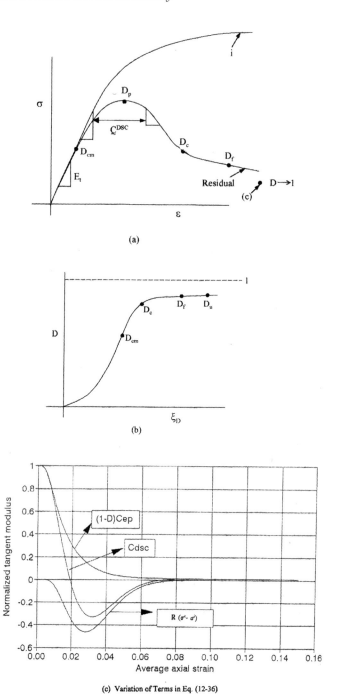

FIGURE 12.13
Approximate decoupled analysis and critical disturbances.

For the foregoing uncoupled procedure, insofar as the FE computer analysis is concerned, there arise no questions of nonlocality and spurious mesh dependence, because the computer solution is based on the continuum plastic hardening response, Fig. 12.13. However, the calculation of $d\underset{\sim}{\sigma}^a$ does involve averaging and the effect of coupling between the RI and RI responses.

12.7.2 Instability Through Disturbance

The limiting case, $|\underset{\sim}{C}| = 0$, Eq. (12.18), results because the states of stress, strain and disturbance (damage) that define the changing (nonlinear) response reach certain critical values. For instance, in the case of nonlinear elastic behavior, the tangent elastic modulus, E_t, approaches zero at the peak, and leads to the limiting condition.

In the DSC, the disturbance, D, is involved in the definition of the tangent matrix, $\underset{\sim}{C}^{DSC}$, Eq. (12.28). Hence, the examination of the variation of disturbance during deformation can provide the conditions in Eqs. (12.18) and (12.21). For instance, when the disturbance reaches the value of D_{cm}, microcracking can initiate; when it reaches the value of D_p at the peak, localization may initiate; when it reaches the critical value, D_c, material instability (failure, liquefaction) can initiate; and when it reaches the value of D_f, engineering failure or fracture occurs, Fig. 12.13 (b).

Assuming that the incremental iterative calculations are performed at the same level of strains, i.e., $d\underset{\sim}{\varepsilon}^i = d\underset{\sim}{\varepsilon}^a = d\underset{\sim}{\varepsilon}^c$, the $\underset{\sim}{C}^{DSC}$ matrix is given by

$$\underset{\sim}{C}^{DSC} = (1 - D)\underset{\sim}{C}^i + D\underset{\sim}{C}^c + \underset{\sim}{R}(\underset{\sim}{\sigma}^c - \underset{\sim}{\sigma}^i) \qquad (12.36)$$

Figure 12.13(c) shows schematic variations of the first two terms, the last term and $\underset{\sim}{C}^{DSC}$ in Eq. (12.36) (29). When $D = 0$, $\underset{\sim}{C}^{DSC} = \underset{\sim}{C}^i$, i.e., the initial matrix is the same as that for the RI (e.g., elastoplastic) material. The matrix $\underset{\sim}{C}^{DSC} = 0$ at the peak, and then it becomes negative (definite) during the falling (softening) region of the stress-strain response. During the softening region when $\underset{\sim}{C}^{DSC} \to 0$, initiation of instability or failure occurs.

If the FA material does not carry any stress as in the classical damage model (38), i.e., $\underset{\sim}{C}^c = 0$, as $D \to 1$, $\underset{\sim}{C}^{DSC} \to o$ as $dD \to o$. However, if the FA material can carry stress, i.e., $\underset{\sim}{C}^c \neq \underset{\sim}{0}$, $\underset{\sim}{C}^{DSC}$ will be positive with small magnitude depending on the properties of the FA material. In that case, microcracking leading to fracture and instability, which initiates when $D = D_c$, will grow and the final failure will occur when $D = D_f$, Fig. 12.13(a), due to large magnitudes of the FA zones.

Hence, the initiation of microcracking and of failure (instability) and fracture can be identified on the basis of threshold values of disturbance such as D_{cm} and D_c, which are determined from appropriate laboratory tests.

The foregoing states of the $\underset{\sim}{C}^{DSC}$ matrix at the material level will be reflected at the structural system level, in stiffness matrix, $\underset{\sim}{K}^{DSC}$, Eq. (12.21):

$$\underset{\sim}{K}^{DSC} d\underset{\sim}{q}^i = d\underset{\sim}{Q} \qquad (12.37a)$$

where

$$K^{DSC} = \int_v B^T C^{DSC} B \, dV \qquad (12.37b)$$

which will become negative definite after the peak, and at $D = D_c$, it will indicate structural instability, as $|K^{DSC}| \to 0$.

Now, we present a mathematical derivation of instability in the DSC.

12.7.3 Stability Analysis of DSC

Mathematical stability analyses of various constitutive models with the enrichments such as Cosserat and gradient theories are presented in a number of publications (10–12). The stability analysis of the DSC for the one-dimensional specialization together with simple examples is presented below. First, we present an analytical stability condition for the one-dimensional DSC model, Davis (66). Then stability analyses for the DSC using computer finite element method are presented.

12.7.4 Stability Condition for One-Dimensional DSC Model

Consider the one-dimensional problem, Fig. 12.14. The one-dimensional rate (incremental) constitutive, evolution, equilibrium and strain-displacement equations are given by

$$\dot{\sigma}^a = (1 - D)\dot{\sigma}^i + D\dot{\sigma}^c + \dot{D}(\sigma^c - \sigma^i) \qquad (12.38a)$$

$$\dot{\sigma}^i = C^i \cdot \dot{\varepsilon}^i \qquad (12.38b)$$

$$\dot{\sigma}^c = C^c \cdot \dot{\varepsilon}^c \qquad (12.38c)$$

$$\dot{D} = \alpha|\dot{\varepsilon}| \qquad (12.38d)$$

$$\frac{\partial \sigma}{\partial x} = \rho \ddot{u} \qquad (12.38e)$$

$$\frac{\partial u}{\partial x} = \dot{\varepsilon} \qquad (12.38f)$$

FIGURE 12.14
One-dimensional idealization: stability analysis (66).

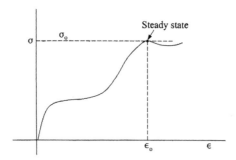

FIGURE 12.15
Perturbation from steady state: stability analysis.

where the superscript a, i, and c denote observed, RI, and FA states, the overdot denotes rate, α is a scalar parameter, ρ is the density, u is the displacement [Fig. (12.14)], ε is the observed strain, and it is assumed that the strains in the RI and FA parts are equal.

Let σ_o^a, τ_o^i, τ_o^c, D_o, ε_o, and u_o relate to a homogeneous steady state condition. Consider a perturbation from the steady state, Fig. 12.15, with the resulting quantities as

$$\begin{aligned}
\sigma^a &= \sigma_o^a + \hat{\sigma}^a \cdot e^{i(kx+\omega t)} \\
\sigma^i &= \sigma_o^i + \hat{\sigma}^i \cdot e^{i(kx+\omega t)} \\
\sigma^c &= \sigma_o^c + \hat{\sigma}^c \cdot e^{i(kx+\omega t)} \\
D &= D_o + \hat{D} \cdot e^{i(kx+\omega t)} \\
\varepsilon &= \varepsilon_o + \hat{\varepsilon} e^{i(kx+\omega t)} \\
u &= u_o + \hat{u} e^{i(kx+\omega t)}
\end{aligned} \qquad (12.39)$$

where k is the material parameter, ω is the frequency, t is time, and o denotes the steady state condition. Now the rate of various quantities in Eq. (12.39) and their derivatives with respect to x can be defined; for example:

$$\dot{\sigma}^a = \hat{\sigma}^a i \omega e^{i(kx+\omega t)} \qquad (12.40a)$$

$$\frac{\partial \sigma^a}{\partial x} = \hat{\sigma}^a i k e^{i(kx+\omega t)} \qquad (12.40b)$$

and so on.

Now use of Eq. (12.39) in Eq. (12.38a) leads to

$$\hat{\sigma}^a i \omega = (1 - D_o)\hat{\sigma} i \omega + D_o \hat{\sigma}^c \cdot i \omega + (\sigma_o^c - \sigma_o^i)\hat{D} i \omega \qquad (12.41a)$$

or

$$\hat{\sigma}^a = (1 - D)\hat{\sigma}^i + D_o\hat{\sigma}^c + \hat{D}(\sigma_c^c - \sigma_o^i) \qquad (12.41b)$$

Similarly, Eqs. (12.38b), (12.38c) and (12.38d) become

$$\hat{\sigma}^i = C^i\hat{\varepsilon} \qquad (12.42a)$$

$$\hat{\sigma}^c = C^c \cdot \hat{\varepsilon} \qquad (12.42b)$$

and

$$\hat{D} = \alpha\hat{\varepsilon}, \quad \text{if } \hat{\varepsilon}\omega e^{i(kx+\omega t)} > 0 \qquad (12.42c)$$

$$\hat{D} = -\alpha\hat{\varepsilon}, \quad \text{if } \hat{\varepsilon}\omega e^{i(kx+\omega t)} < 0 \qquad (12.42d)$$

Now, the equilibrium and strain-displacement relations, Eqs. (12.38e) and (12.38f) become

$$k\hat{\sigma}^a = \omega\rho\hat{u} \qquad (12.43a)$$

$$k\hat{u} = \omega\hat{\varepsilon} \qquad (12.43b)$$

Equations (12.41) to (12.43) with Eq. (12.38a) can be written in the matrix notation as

$$\begin{bmatrix} 1 & D_o-1 & -D_o & \sigma_o^i-\sigma_o^c & 0 & 0 \\ 0 & 1 & 0 & 0 & -C^i & 0 \\ 0 & 0 & 1 & 0 & -C^c & 0 \\ 0 & 0 & 0 & 1 & -\alpha & 0 \\ k & 0 & 0 & 0 & 0 & -\omega\rho \\ 0 & 0 & 0 & 0 & -\omega & k \end{bmatrix} \begin{Bmatrix} \hat{\sigma}^a \\ \hat{\sigma}^i \\ \hat{\sigma}^c \\ \hat{D} \\ \hat{\varepsilon} \\ \hat{u} \end{Bmatrix} = \begin{Bmatrix} 0 \\ 0 \\ 0 \\ 0 \\ 0 \\ 0 \end{Bmatrix} \qquad (12.44)$$

By setting the determinant of the matrix in Eq. (12.44) to zero, we obtain

$$-\rho\omega^2 - k^2[\alpha(\sigma_o^i - \sigma_o^c) - D_oC^c + C^i(D_o - 1)] = 0 \qquad (12.45a)$$

or

$$\frac{\omega^2}{k^2} = \frac{C^i(1 - D_o) + C^cD_o - \alpha(\sigma_o^i - \sigma_o^c)}{\rho} \qquad (12.45b)$$

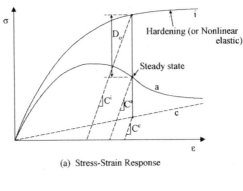

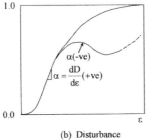

(a) Stress-Strain Response

(b) Disturbance

FIGURE 12.16
Stress-strain response: softening and stiffening.

If Eq. (12.42d) is used, we have

$$\frac{\omega^2}{k^2} = \frac{C^i(1 - D_o) + C^c \cdot D_o - (-\alpha)(\sigma_o^i - \sigma_o^c)}{\rho} \tag{12.46}$$

In Eq. (12.45), ω/k is the wave velocity, $C^i(1 - D_o)$ and $C^c D_o$ are the tangent moduli or slopes at the steady state, Fig. 12.16, and $\alpha(\sigma_o^i - \sigma_o^c)$ denotes the reduction or increase in the effective tangent modulus corresponding to the first two terms, for positive or negative α, respectively. Assume that $\sigma_o^i - \sigma_o^c$ is always positive, Fig. 12.16; a positive value of α denotes increase in the disturbance for the softening response, while a negative value denotes decrease in disturbance for stiffening or healing, Fig. 12.16.

Now, to investigate stability, we examine the condition when the term on the right-hand side, Eq. (12.45), becomes negative. Consider the possibility with respect to Eq. (12.42c):

$$C^i(1 - D_o) + C^c D_o - \alpha(\sigma_o^i - \sigma_c^c) \le 0 \tag{12.47}$$

Now, for the RI behavior (say, linear or nonlinear elastic or plastic hardening, Fig. 12.16), and as $0 \leq D \geq 1$, we have

$$\sigma_o^i - \sigma_o^c \geq \frac{1}{\alpha}[C^i(1 - D_o) + C^c D_o] \tag{12.48}$$

which represents the instability criterion. Now, with $\alpha = dD/d\varepsilon$ and $C^c = \frac{d\sigma^i}{d\varepsilon}$ and $C^c = \frac{d\sigma^c}{d\varepsilon}$, Eq. (12.48) becomes

$$\sigma_o^i - \sigma_o^c \geq \frac{d\varepsilon}{dD}\left[\frac{d\sigma^i}{d\varepsilon}(1 - D_o) + \frac{d\sigma^c}{d\varepsilon}D_o\right] \tag{12.49}$$

With Eq. (12.42d), we have for instability criterion

$$C^i = (1 - D_o) + C^c \cdot D_o + \alpha(\sigma_o^i - \sigma_o^c) \leq 0 \tag{12.50a}$$

$$\sigma_o^i - \sigma_o^c \leq \frac{1}{\alpha}[C^i(1 - D_o) + C^c D_o] \tag{12.50b}$$

12.8 Examples

Now we consider examples to illustrate various characteristics of the DSC, namely, spurious mesh dependence, localization and instability.

Example 12.1 Spurious Mesh Dependence

A number of computer analyses using the finite element procedures described in Chapter 13 were performed to study the issue of spurious mesh dependence with the DSC. Results from two of the studies are described below.

The first study involved computer solutions based on the general DSC equations in which the constitutive matrix $\underline{C}^{DSC}$, Eq. (12.28), is nonsymmetric and becomes negative definite after the peak and during the softening region.

Figure 12.17(a) shows the finite element mesh for a concrete specimen tested under multiaxial loading conditions (67). Four different meshes with 4, 16, 64 and 256 elements (8-noded isoparametric) were used.

The specimen was subjected to prescribed compressive displacement loading with increment = 0.05 mm on the top surface. The test specimen involved no confining stress; hence, the plane-strain idealization was used as an approximation. In the incremental analysis, strains were averaged over the Gauss points for the quarter of the specimen (29).

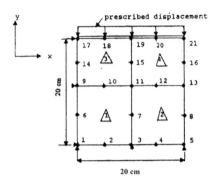

(a) Finite Element Mesh

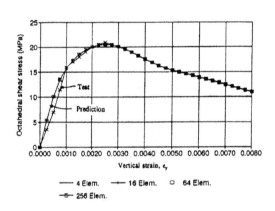

(b) τ_{oct} vs ε_y

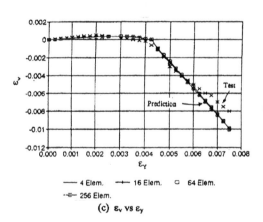

(c) ε_v vs ε_y

FIGURE 12.17
Comparisons of finite element prediction with laboratory test results for concrete (29). ©John Wiley & Sons Ltd. Reproduced with permission.

Microstructure: Localization and Instability

The material of the specimen was concrete and was characterized as elastoplastic (δ_0-model) with disturbance (Chapters 4 and 7). The parameters used are given below (29, 62, 64):

$$E = 37.00 \text{ GPa}, \quad \nu = 0.25;$$

$$\gamma = 0.0678, \quad \beta = 0.755, \quad n = 5.237$$

$$a_1 = 4.61 \times 10^{-11}, \quad \eta = 0.8262;$$

$$A = 688.00, \quad Z = 1.502, \quad D_u = 0.875.$$

Figures 12.17(b) and (c) show comparisons between the computed octahedral stress, τ_{oct}, vs. axial strain, ε_y, and volumetric strain, ε_v, vs. ε_y and the observed data from the laboratory tests. It can be seen that the results involve essentially no spurious mesh sensitivity, and that the predictions compare well with the laboratory observations.

The second study involved the approximate uncoupled FE procedure in which the incremental analysis, Eq. (12.33), is performed only for the RI response with C^{ep} as the constitutive matrix, and observed response is then computed by using Eq. (12.28).

The finite element analysis for the second study was performed for tension and compression loading on a rectangular zone made of Pb/Sn (40/60) solder. Three meshes with 16, 64, and 256 elements were used; Fig. 12.18 shows typical mesh with 8-noded isoparametric elements (68).

The solder material was characterized using the DSC model in which the RI material was modelled as elastoplastic (HISS-δ_0) model (Chapter 7), and the FA material was assumed to carry hydrostatic stress. The elastic, plastic and disturbance parameters used are given below:

$$E = 15.7 \text{ GPa}, \quad \nu = 0.40;$$

$$\gamma = 0.00081, \quad \beta = 0.00, \quad n = 2.10$$

$$a_1 = 0.78 \times 10^{-5}, \quad \eta = 0.46, \quad R = 208 \text{ MPa};$$

$$A = 0.102, \quad Z = 0.676, \quad D_u = 0.90$$

The load was applied in increments. For the compression case, a total load of 36.5 MPa was applied in 50, 100 and 100 steps with increments = 0.60, 0.05 and .015 MPa, respectively. For the tension case, the total load of 33.5 MPa was applied in 50, 100 and 280 steps with increments = 0.40, 0.10 and 0.0125 MPa.

Figure 12.19 shows computed axial stress (σ_y) vs. axial stress (ε_y) at point A, Fig. 12.18, for both the observed (a) and RI (i) responses; similar plots were also obtained at other points (B to F) marked in Fig. 12.18. It can be seen that

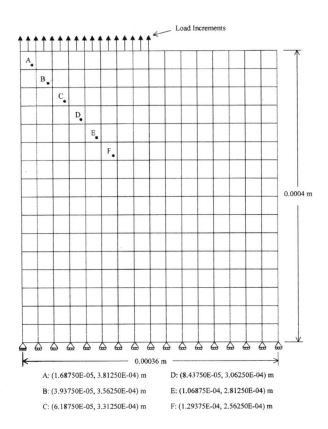

FIGURE 12.18
Finite element mesh with 256 elements: tensile (and compressive) loadings (68).

the results are independent of the mesh; i.e., there is no spurious mesh dependence. It may be noted that the stresses and strains are computed in the FE (DSC) analysis at specific points such as A, without averaging (over Gauss) points.

Example 12.2 Localization

As discussed earlier, strain localization occurs in a deforming material, particularly during the softening zone, Fig. 12.2(a). Differences of opinion have been expressed regarding the existence and location of possible slip lines through the study of the acoustic tensor (10, 69). For example, Zienkiewicz and Huang (70) have stated that ".... Acoustic tensor is a very unreliable indicator and most of the reported difficulties result simply from using an incorrect finite element approximation in the solution of plasticity problems involved." The DSC model allows for localization and can also be used for adaptive mesh refinement (see Chapter 13).

To illustrate the computational and localization capabilities of the DSC, a number of problems have been solved; they include one-dimensional bar

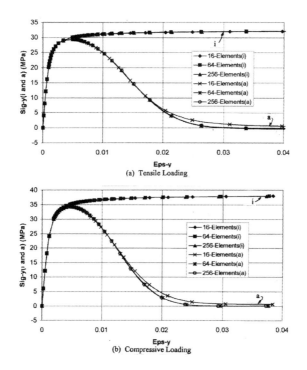

FIGURE 12.19
Computed axial stress (σ_y) vs. axial strain (ε_y): relative intact (*i*) and observed (*a*) responses for point A.

with imperfection, chip-substrate systems in electronic packaging and initiation of liquefaction instability due to localization in saturated porous media (Chapters 9 and 13). Here, description of the bar problem is presented.

A one-dimensional bar with a central imperfection and subjected to tensile loading is shown in Fig. 12.20 (10, 29); the middle zone of 10 mm was assigned tensile strength reduced by 10 percent. The same problem has been solved by other investigators, e.g., de Borst et al. (10), who studied the localization performance of their gradient-type regularization method. Figure 12.20(b) shows three finite element meshes with 20, 40, and 80 eight-noded isoparametric elements. The computer analysis involved the nonsymmetric, $\underset{\sim}{C}^{DSC}$, matrix, Eq. (12.28).

The material properties for the bar are given below:

$$E = 20{,}000 \text{ MPa}, \quad \sigma_t \text{ (tensile strength)} = 2.0 \text{ MPa};$$

$$\gamma = 1.1 \times 10^{-3}, \quad \beta = 0.0, \quad n = 2.10;$$

$$a_1 = 5.24, \quad \eta_1 = 1.1 \times 10^{-10}$$

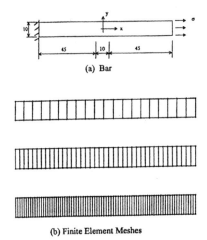

FIGURE 12.20
Tension bar with imperfection. ©John Wiley & Sons Ltd. Reproduced with permission.

Disturbance

Case	ℓ/L	A	Z	D_u
1	0.10	500	2.37	0.50
2	0.05	530	1.35	0.60
3	0.025	550	0.31	0.75

The elastic parameter (E) and σ_t were the same as those used by de Borst, et al. (10). The parameters for the plasticity (δ_0) model and disturbance were evaluated based on the stress-strain curves given in (10). A small non-linearity with plasticity response was considered. The three curves corresponding to different values of the ratio of the characteristic length (ℓ) to the length of the bar (L) were smoothed and extended to approach residual stresses of about 1.00, 0.80 and 0.50 MPa, respectively. The disturbance parameters were found to be dependent of the ratio, ℓ/L, as shown above.

Figure 12.21 shows computed results using the DSC in terms of strain (ε) vs. length, and stress (σ) vs. displacement (u) for three meshes with 20, 40 and 80, 8-noded isoparametric elements, Fig. 12.20(b), and for $\ell/L = 0.05$. Figure 12.21(a) shows consistent convergence for strains with the meshes. With the gradient model, the strain approached the value of about 1.0×10^{-3} with 160 elements (10); the DSC model shows similar value with 80 elements.

The expression for the width of localization zone, w, is given by (10):

$$w = 2\pi\ell \qquad (12.51)$$

The localization width, w, predicted by the DSC model, Fig. 12.21(a), is about 33 mm. This gives the value of ℓ from Eq. (12.51) as 5.25 mm, which compares well with $\ell = 5.00$ mm used in (10). It may be mentioned that the

Microstructure: Localization and Instability

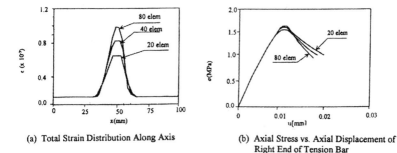

(a) Total Strain Distribution Along Axis

(b) Axial Stress vs. Axial Displacement of Right End of Tension Bar

FIGURE 12.21
Computed results using DSC for $\ell/L = 0.05$ (29). ©John Wiley & Sons Ltd. Reproduced with permission.

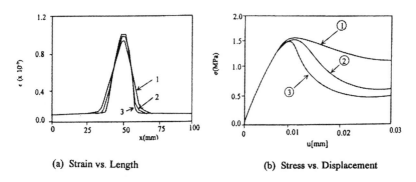

(a) Strain vs. Length

(b) Stress vs. Displacement

FIGURE 12.22
Computed results for 80-element mesh for three ℓ/L ratios: (1) 0.10, (2) 0.05, (3) 0.025 (29). ©John Wiley & Sons Ltd. Reproduced with permission.

DSC does not involve ℓ as a (direct) parameter; hence, the comparison is meant as an indirect validation. The computed results in terms of stress vs displacements, Fig. 12.21(b), also show trends similar to those presented by de Borst et al. (10).

Computed DSC results for $\ell/L = 0.10, 0.05$, and 0.025 for strain vs. length and stress vs. displacement for the 80-element mesh, Fig. 12.20(b), are shown in Fig. 12.22. The localization width, w, increases with ℓ/L, Fig. 12.22(a), indicating more brittle response for smaller values of ℓ. The values of ℓ using Eq. (12.51) for the three ratios are found to be 3.2, 5.25 and 7.00 mm, while those used in (10) were 2.50, 5.00 and 10.00 mm, respectively. The overall magnitudes of stress and trends in Fig. 12.22(b) are similar to those from the gradient theory (10). However, there are some differences due to reasons such as: (1) the δ_0-plasticity model with disturbance is used in the DSC, while de Borst et al. (10) used an elastoplastic (Drucker-Prager) model with the gradient enrichments, (2) the smoothing and digitization of stress-strain curves, and (3) in the DSC, the plastic strains are developed from the beginning, while in the Drucker-Prager model, they develop only after the (peak) stress is reached.

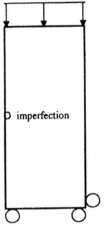

FIGURE 12.23
Plate with imperfection (10, 30). Reprinted with permission from Elsevier Science.

Localization behavior of the DSC is further illustrated in Chapter 13 with respect to other problems such as chip-substrate in electronic packaging.

Example 12.3 Localization and Mesh Dependence

Localization in a biaxial plate specimen with an imperfection, Fig. 12.23, was analyzed by de Borst et al. (10) using gradient enrichment with the Drucker-Prager plasticity model. The imperfection was simulated by reducing its yield strength by 10% compared to that for the plate material. de Borst et al. used four-node elements with a bilinear displacement field and a Hermitian (bicubic) interpolation for the plastic multiplier (λ), Eq. (12.12). The plate was subjected to incremental displacement = 0.0001 m at the top nodes.

The DSC model was used here with the RI response characterized as elastoplastic with the HISS-δ_0 plasticity model (30). The material parameters used are given below:

Elastic: $E = 1192$ MPa, $\nu = 0.49$;
Plasticity: $\gamma = 0.001$, $\beta = 0.00$, $n = 2.10$;
   $a_1 = 10^{-10}$, $\eta_1 = 0.20$
Disturbance: $A = 530$, $Z = 1.35$, $D_u = 0.60$

In the DSC analysis, two mesh layout were used: 6 × 18 and 13 × 36 with 8-noded quadrilateral elements.

Figures 12.24(a) and 12.24(b) show computed deformation patterns at the peak and after the peak for the two mesh layouts, respectively. The deformations, localization and shear band patterns from the DSC are consistent and

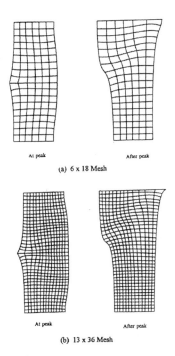

FIGURE 12.24
Deformation patterns in imperfect bar for 13 × 36 mesh (30). Reprinted with permission from Elsevier Science.

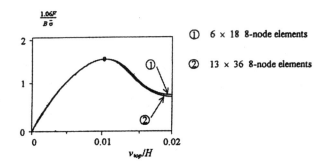

FIGURE 12.25
Load-displacement curves for 6 × 18 and 13 × 36 meshes (30). Reprinted with permission from Elsevier Science.

show essentially the same trends as in the analysis by de Borst, et al. (10); some differences between the two results can be due to factors such as the different material model and element orders used in the present analysis. Figure 12.25 shows load displacement curves from the two mesh layouts; they indicate no significant mesh dependence. In Fig. 12.25, F is the load induced at the top, B is the width of the plate, $\bar{\sigma}$ is the observed or actual stress, v_{top} is the displacement at the top and H is the height of the plate.

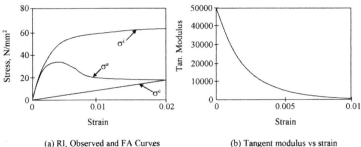

FIGURE 12.26
Stress-strain behavior and variation of tangent modulus.

These results show that the DSC model provides consistent and satisfactory localization and shear banding and avoids the spurious mesh dependence without the need for external enrichments.

Example 12.4 Instability in the DSC

The first example involves analysis of instability using Eq. (12.49) for given numerical stress-strain data. Consider the observed stress-strain response, Fig. 12.26(a). The RI response is simulated as nonlinear using hyperbolic function as

$$\sigma^i = \frac{\varepsilon}{a + b\varepsilon} \qquad (12.52a)$$

where $a = 1/E_i$ and $b = 1/\sigma_u$ are parameters, E_i is the initial modulus, σ_u is the ultimate asymptotic stress, σ^i is the RI stress and ε is the strain. The values of a and b for the RI curve, Fig. 12.26(a), are obtained as 2.33×10^{-5} and 0.0133, respectively (Chapter 5). The FA response is simulated as linear elastic given by

$$\sigma^c = 850 \cdot \varepsilon \qquad (12.52b)$$

where σ^c is the stress in the FA material.
The instability criterion, Eq. (12.49), is expressed as

$$\sigma_o^i - \sigma_o^c \geq \frac{1}{\alpha}\left[\frac{d\sigma^i}{d\varepsilon}(1 - D_o) + \frac{d\sigma^c}{d\varepsilon}D_o\right] \qquad (12.53)$$

where α is computed as

$$\alpha = \frac{dD}{d\varepsilon} \approx \frac{D_o^{m+1} - D_o^m}{\varepsilon_o^{m+1} - \varepsilon_o^m} \qquad (12.54)$$

TABLE 12.1

Instability Calculations

ε	σ^a	σ^i	σ^c	D	$dD/d\varepsilon$	$\sigma^i - \sigma^c$	R.H.S.*
0.00021	8.05	8.05	0.18	5.40×10^{-8}	0.0043	7.87	80×10^5
.00054	17.71	17.71	0.46	4.2×10^{-6}	115.55	17.25	215.50
.00092	25.00	25.90	0.78	0.100	280.27	25.12	60.10
0.00129	29.00	31.88	1.10	0.144	146.69	30.78	77.73
0.00175	33.00	37.57	1.49	0.222	93.61	36.08	103.82
0.0020	35.00	41.00	1.70	0.250	151.40	39.30	50.00
.0030	34.00	47.47	2.55	0.440	148.29	44.92	21.24
.0040	33.00	52.29	3.40	0.546	90.03	48.89	20.98
.0060	29.00	58.20	5.10	0.709	65.07	53.10	10.03
0.0080	24.00	61.68	6.80	0.843	72.37	54.88	3.11
0.010	18.00	63.98	8.50	0.978	33.00	55.48	0.754
0.020	18.00	69.13	17.00	0.981	24.63	52.13	2.22

$$^*\text{R.H.S.} = \frac{1}{(dD/d\varepsilon)}\left[\frac{d\sigma^i}{d\varepsilon}(1-D) + \frac{d\sigma^c}{d\varepsilon}D\right]$$

where m is the incremental step. Disturbance, D, is expressed as a function of strain, ε, and can be computed by using the stress-strain response as

$$D = \frac{\sigma_o^i - \sigma_o^a}{\sigma_o^i - \sigma_o^c} \tag{12.55}$$

where σ_o^a is the measured stress at a (steady) state. Table 12.1 shows the calculations of various quantities.

For the strain (ε) levels, before about 0.003, stability is indicated as per Eq. (12.53). At and after $\varepsilon = 0.003$, the values of the left-hand side in Eq. (12.53) become greater than those of the right-hand side (RHS), indicating instability. The instability occurs at strain levels beyond the peak stress, σ^a, of about 35 N/mm², in the softening zone at disturbance $D \approx 0.45$. Figure 12.26(b) shows a plot of the RI tangent modulus $(d\sigma^i/d\varepsilon)$ vs. strain.

Liquefaction Instability The second example involves instability that indicates liquefaction in saturated (soil) media. Conventionally, liquefaction is considered to initiate when the excess pore fluid (water) pressure that increases during (cyclic) loading equals the initial effective pressure (Chapter 9). In the DSC, the instability leading to initiation of liquefaction occurs when the microstructure reaches the critical threshold state at the critical disturbance, D_c (Chapters 9 and 13).

Figure 12.27 shows the dimensions of a (cubical) test specimen idealized as axisymmetric. The specimen is first subjected to initial effective confining stress, $\bar{\sigma}_o$, and then deviatoric (shear) stress, $\sigma_1 - \bar{\sigma}_o$, where σ_1 is the axial stress.

The specimens of saturated Ottawa sand at relative density ($D_r = 60\%$) were tested under three different initial confining stresses, $\bar{\sigma}_o = 69$, 138 and 207 kPa (71–73). The laboratory test behavior in terms of applied deviatoric stress vs. time, strain vs. time and excess pore water pressure vs. time, as well as plots of effective stress, $\bar{\sigma}$, $\bar{\sigma} = \sigma^t - p$, where σ^t is the total stress, and p is the

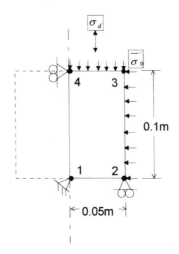

FIGURE 12.27
Finite element mesh and loading for triaxial test specimen (72). ©John Wiley & Sons Ltd. Reproduced with permission.

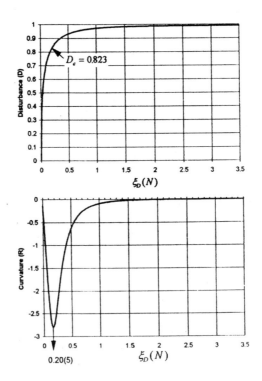

FIGURE 12.28
Disturbance and curvature vs. deviatoric plastic strain trajectory as function of cycles, N: $\bar{\sigma}_o$ = 69 kPa. ©John Wiley & Sons Ltd. Reproduced with permission.

Microstructure: Localization and Instability

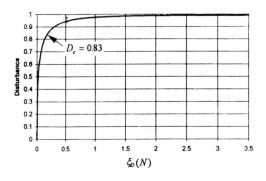

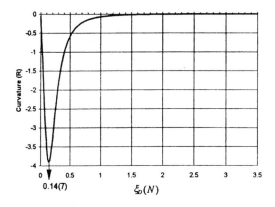

FIGURE 12.29
Disturbance and curvature vs. deviatoric plastic strain trajectory as function of cycles, N: $\bar{\sigma}_o$ = 138 kPa. ©John Wiley & Sons Ltd. Reproduced with permission.

pore water pressure are given in Chapter 9. Disturbance, D, is expressed as

$$D = \frac{\bar{\sigma}^i - \bar{\sigma}^a}{\bar{\sigma}^i - \bar{\sigma}^c} \qquad (12.56)$$

where i, a and c denote RI, observed and FA states, respectively.
The (average) material parameters for the sand are given by

E = 193.00 MPa, ν = 0.38;
γ = 1.713, β = 0.0, n = 2.45;
a_1 = 0.845, η_1 = 0.0215;
$\bar{m}$ = 0.20, λ = 0.019, e_o = 0.593;
A = 4.22, Z = 0.43, D_u = 0.99.

Figures 12.28 to 12.30 show plots of disturbance vs deviatoric plastic strain trajectory, ξ_D (or number of cycles, N) and the curvature (D'') vs. $\xi_D(N)$ for the three confining pressures (72). Here,

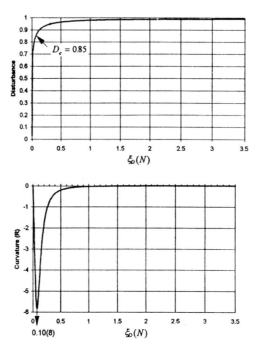

FIGURE 12.30
Disturbance and curvature vs. deviatoric plastic strain trajectory as function of cycles, N: $\bar{\sigma}_o = 207$ kPa. ©John Wiley & Sons Ltd. Reproduced with permission.

$$D'' = \frac{d^2 D}{d\xi_D^2} = AZ\xi_D^{z-2} \exp(-A\xi_D^z)(Z - 1 - AZ\xi_D^z) \tag{12.57}$$

The critical disturbance, D_c, occurs when D'' is the minimum; it can also be found approximately at the intersection of tangents to the earlier and later (saturation) parts of the disturbance curve. The values of D_c for the three confining pressures are found to be 0.82, 0.830 and 0.850, respectively, with corresponding cycles 5, 7 and 8. These cycles compare well with those observed in the tests (Chapter 9) when the pore pressure becomes approximately equal to the initial confining pressure. Thus, the critical disturbance, D_c, identifies the initiation of instability leading to liquefaction.

Finite element analyses with one-element mesh, Fig. 12.27, were performed in which the initial confining pressures were first applied, followed by incremental increase of $\sigma_1 - \bar{\sigma}_o$. Figure 12.31 shows variations of the trace of the constitutive matrix, $\|C_{ii}^{DSC}\|$ with ξ_D (as the function of N), for $\bar{\sigma} = 69$, 138 and 207 kPa. The trace, which represents the sum of eigen values, first decreases and then assumes a minimum value. Thereafter, it stabilizes to a small value as per Eq. (12.28) in the ultimate region when D approaches $D_u \approx 0.99$, and

Microstructure: Localization and Instability

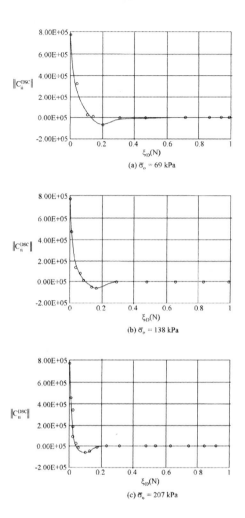

FIGURE 12.31
Variation of $\left\|C_{ii}^{CDC}\right\|$ for different $\bar{\sigma}_o$. ©John Wiley & Sons Ltd. Reproduced with permission.

$dD \approx 0$. The trace assumes minimum values at $\xi_D \approx 0.20$, 0.15 and 0.10, respectively, for the three confining pressures. These values correspond closely with the values at the initiation of liquefaction corresponding to $D_c \approx$ 0.82, 0.830 and 0.850 in Figs. 12.28 to 12.30 at cycles to liquefaction, $N_c = 5, 7$, and 8, respectively. Similar results were obtained with respect to plots of the trace of the stiffness matrix, $\left\|K_{ii}^{DSC}\right\|$ vs. ξ_D. Thus, the mathematical properties of the constitutive and stiffnesses matrices in the DSC correlate well with liquefaction instability identified through critical values of disturbance and the experimental response of the sand.

12.8.1 Instability Based on Critical Dissipated Energy and Disturbance

The microstructural instability (liquefaction) can be identified on the basis of critical dissipated energy and disturbance. Desai (73) has presented detailed analysis, correlations between energy and disturbance approaches, advantages of the DSC and application for the prediction of laboratory and field (Port Island, Kobe, Japan earthquake) behavior of different saturated sands. Some details are presented in Appendix I.

Appendix I: Thermodynamical Analysis of the DSC

Consider the weighted average observed or actual stress, $\underline{\sigma}^a$, as

$$\underline{\sigma}^a = (1-D)\underline{\sigma}^i + D\underline{\sigma}^c \tag{12A.1}$$

with the corresponding strain, $\underline{\varepsilon}^a$, which can be different from the RI ($\underline{\varepsilon}^i$) and ($\underline{\varepsilon}^c$) strains.

Now, from the thermodynamical considerations, the state of the material can be characterized by its free energy density (ψ^a) in the absence of thermal effects. The free energy, $\rho\psi^a$, is given by

$$\rho\psi^a = \frac{1}{2}\underline{\sigma}^a \underline{\varepsilon}^a \tag{12A.2}$$

where ρ is the mass density, and the stress is given by

$$\underline{\sigma}^a = \frac{\partial(e\psi^a)}{\partial \underline{\varepsilon}^a} \tag{12A.3}$$

Substitution of $\underline{\sigma}^a$ from Eq. (12A.1) into Eq. (12A.2) gives

$$\rho\psi^a = \frac{1}{2}[(1-D)\underline{\sigma}^i + D\underline{\sigma}^c]^T \cdot \underline{\varepsilon}^a \tag{12A.4}$$

or

$$\rho\psi^a = \frac{1}{2}[(1-D)\underline{C}^i\underline{\varepsilon}^i + D\underline{C}^c\underline{\varepsilon}^c]^T \cdot \underline{\varepsilon}^a$$

$$= \frac{1}{2}[(1-D)\underline{\varepsilon}^{iT}\underline{C}^i\underline{\varepsilon}^i + D\underline{\varepsilon}^{cT} \cdot \underline{C}^c\underline{\varepsilon}^c] \tag{12A.5}$$

As the process of the growth of disturbance involves energy dissipation, differentiation of Eq. (12A.5) leads to the energy dissipation rate, ϕ, as

$$\phi = -\frac{\partial(\rho\psi^a)}{\partial t} = -\frac{\partial(\rho\psi^a)}{\partial D} \cdot \frac{\partial D}{\partial t}$$
$$= \bar{y}\dot{D} \tag{12A.6}$$

where (43)

$$\bar{y} = -\frac{\partial(\rho\psi^a)}{\partial D} = \frac{1}{2}\underline{\varepsilon}^{aT}\underline{C}^i\underline{\varepsilon}^i - \frac{1}{2}\underline{\varepsilon}^{aT}\underline{C}^c\underline{\varepsilon}^c$$
$$= \rho\psi^i - \rho\psi^c \tag{12A.6}$$

where $\rho\psi^i$ and $\rho\psi^c$ are the energies for the RI and FA parts, respectively. Now, ϕ is given by

$$\phi = (\rho\psi^i - \rho\psi^c)\dot{D} \tag{12A.7}$$

$\rho\psi^i$ and $\rho\psi^c$ are positive definite functions of $\underline{\varepsilon}^a$. Also, from physical considerations, $\rho\psi^i > \rho\psi^c$ in a softening or degrading material because the energy in the RI state is greater (or equal) to that in the FA state. Furthermore, $\dot{D}$ increases with time. Hence,

$$\phi \geq 0 \tag{12A.8}$$

which satisfies the Clausius-Duhem inequality as per the second law of thermodynamics (1, 9, 11, 15, 43), which will also apply for the case when $\underline{\varepsilon}^a = \underline{\varepsilon}^i = \underline{\varepsilon}^c$.

References

1. Hill, R., "A General Theory of Uniqueness and Stability in Elastic-Plastic Solids," *J. Mech. Phys. Solids*, 6, 1958, 236–249.
2. Drucker, D.C., "A Definition of Stable Inelastic Materials," *J. Appl. Mech.*, 26, 1959, 101–106.
3. Koiter, W.T., "On the Thermodynamics Background of Elastic Stability Theory," in *Problems of Hydrodynamics and Continuum Mechanics*, SIAM, Philadelphia, 1969, 423–433.
4. Rudnicki, J.W. and Rice, J.R., "Conditions for the Localization of Deformation in Pressure-Sensitive Dilatant Solids," *J. Mech. Phys. of Solids*, 23, 1975, 371–394.

5. Mailer, G. and Hueckel, T., "Nonassociated and Coupled Flow Rules of Elastoplasticity for Rock-like Materials," *Int. J. Rock Mech. Min. Sc.*, 16, 1979, 77–92.
6. Walgraet, D. and Aifantis, E.C., "On Theories in Formulation and Stability of Dislocation 429. Patterns, I-III," *Int. J. Eng. Science*, 12, 1985, 1351–1372.
7. Kratochvil, J., "Dislocation Pattern Formulation in Metals," *Revue Phys. Appl.*, 23, 1988, 419–429.
8. Benallal, A., Billarden, R. and Geymonat, G., "Localization Phenomena at the Boundaries and Interfaces of Solids," *Proc., Third Int. Conf. on Constitutive Laws for Engineering Materials: Theory and Application*, Desai, C.S. et al. (Editors), Univ. of Arizona, Tucson, AZ, 1991, 387–390.
9. Valanais, K.C., "A Global Damage Theory and Hyperbolicity of the Wave Problems," *J. of Appl. Mech., ASME*, 58, 1991, 311–316.
10. de Borst, R., Sluys, L.T., Mühlhaus, H.B. and Pamin, J., "Fundamental Issues in Finite Element Analyses of Localization of Deformation," *Int. J. Eng. Computations*, 10, 1993, 99–121.
11. Bazant, Z.P. and Cedolin, L., *Stability of Structures*, Oxford Univ. Press, New York, 1991.
12. Mühlhaus, H.B. (Editor), *Continuum Models for Materials with Microstructure*, John Wiley, Chichester, U.K., 1995.
13. Bazant, Z.P., "Nonlocal Damage Theory Based on Micromechanics of Crack Interactions," *J. Eng. Mech., ASCE*, 120, 3, 1996, 593–617.
14. Desai, C.S., "Constitutive Modelling Using the Disturbed State as Microstructure Self-Adjustment Concept," Chapter 8, in *Continuum Models for Materials with Microstructure*, H.B. Mühlhaus (Editor), John Wiley, Chichester, U.K., 1995.
15. Desai, C.S. and Toth, J., "Disturbed State Constitutive Modelling Based on Stress-Strain and Nondestructive Behavior," *Int. J. Solids Struct.*, 33, 11, 1996, 1619–1650.
16. Schreyer, H.L. and Cheng, Z., "One-Dimensional Softening with Localization," *J. Appl. Mech.*, s53, 1986, 791–799.
17. Lasry, D. and Belytchko, T., "Localization Limiters in Transient Problems," *Int. J. Solids Struct.*, 24, 1988, 581–597.
18. Needleman, A., "Material Rate Dependence and Mesh Sensitivity in Localization Problems," *Int. J. Comp. Meth. Appl. Mech. Eng.*, 67, 1988, 69–86.
19. Simo, J.C., "Strain Softening and Dissipation: A Unification of Approaches," in *Cracking and Damage, Strain Localization and Size Effect*, Mazars, J. and Bazant, Z.P. (Editors), Elsevier, London, 1989, 440–461.
20. Desai, C.S., Kundu, T. and Wang, G., "Size Effect on Damage Parameters for Softening in Simulated Rock," *Int. J. Num. Analyt. Meth. Geomechanics*, 14, 1990, 509–517.
21. Loret, B. and Prevost, J.H., "Dynamic Strain Localization in Fluid-Saturated Media," *J. of Eng. Mech., ASCE*, 117, 1991, 907–922.
22. Zienkiewicz, O.C. and Huang, M., "Localization Problems in Plasticity Using Finite Elements and Adaptive Remeshing," *Int. J. Numer. Analy. Meth. Geomech.*, 19, 1995, 127–148.
23. Sulen, J., Vardoulakis, I., and Papamichos, E., "Microstructure and Scale Effect in Granular Rocks," Chapter 7 in in *Continuum Models for Materials with Microstructure*, Mühlhaus, H.B. (Editor), John Wiley, U.K., 1995, 201–237.
24. van Mier, J.G.M., "Mode-I Fracture of Concrete: Discontinuous Crack Growth and Crack Interface Grain Bridging," *J. of Cement Concrete Res.*, 21, 1991, 1–15.
25. van Mier, J.G.M., *Fracture Processes of Concrete*, CRC Press, Boca Raton, FL, 1997.

26. Gutierrez, M.A. and de Borst, R., "Studies in Material Parameter Sensitivity of Softening Solids," *Comp. Meth. in Appl. Mech. and Eng.*, 162, 1998, 337–350.
27. Bazant, Z.P. and Pijaudier-Cabot, G., "Measurement of Characteristic Length of Nonlocal Continuum," *J. of Eng. Mech., ASCE*, 115, 4, 1989, 775–767.
28. Mühlhaus, H.B. and Vardoulakis, I., "The Thickness of Shear Bands in Granular Materials," *Geotechnique*, 37, 1987, 271–283.
29. Desai, C.S., Basaran, C. and Zhang, W., "Numerical Algorithms and Mesh Dependence in the Disturbed State Concept," *Int. J. Num. Meth. Eng.*, 40, 1997, 3059–3083.
30. Desai, C.S. and Zhang, W., "Computational Aspects of Disturbed State Constitutive Models," *Int. J. Comp. Meth. Appl. Mech. Eng.*, 151, 1998, 361–376.
31. Oden, J.T. and Pires, E.B., "Nonlocal and Nonlinear Friction Laws and Variational Principles for Contact Problems in Elasticity," *J. of Applied Mechanics*, 50, 1983, 67–75.
32. Pastor, M., Zienkiewicz, O.C., Vilotte, J.P. and Quecedo, M., "Mesh Dependence Problems in Viscoplastic Materials Under Quasi-static Loading," in *Computational Plasticity: Fundamentals and Applications*, Owen, D.R.J., et al. (Editors), Pineridge Press, Swansea, U.K., 1995.
33. Desai, C.S. and Abel, J.F., *Introduction to the Finite Element Method*, Van Nostrand Reinhold, New York, 1972.
34. Desai, C.S. and Kundu, T., *Introductory Finite Element Method*, CRC Press, Boca Raton, FL, 1999, under publication.
35. Sluys, L.J., "Wave Propagation, Localization and Dispersion in Softening Solids," *Doctoral Dissertation*, Delft University of Technology, The Netherlands, 1992.
36. Wang, W.W., "Stationary and Propagative Instabilities in Metals—A Computational Point of View," *Doctoral Dissertation*, Delft Univ. of Technology, The Netherlands, 1997.
37. Eringen, A.C., "A Unified Theory of Thermomechanical Materials," *Int. J. Eng. Sci.*, 4, 1966, 179–202.
38. Kachanov, L.M., *Introduction to Continuum Damage Mechanics*, Martinus Nijhoft Publishers, The Netherlands, 1986.
39. Pijaudier-Cabot, G. and Bazant, Z.P., "Nonlocal Damage Theory," *J. Eng. Mech., ASCE*, 113, 1987, 1512–1533.
40. Krajcinovic, D., "Damage Mechanics," *Mech. Materials*, 8, 1987, 117–197.
41. Bazant, Z.P. and Lin, F.B., "Non-local Yield Limit Degradation," *Int. J. Num. Meth. Eng.*, 26, 1988, 1805–1823.
42. Bazant, Z.P. and Pijaudier-Cabot, G., "Nonlocal Continuum Damage, Localization Instability and Convergence," *J. Appl. Mech.*, 55, 1988, 287–293.
43. Lamaitre, J. and Chaboche, J.L., *Mechanics of Solid Materials*, Cambridge Univ. Press, U.K., 1989, Dunod-Bordas, Paris, 1985.
44. Mühlhaus, H.B., de Borst, R., Sluys, L.J. and Pamin, J., "A Thermodynamic Theory for Inhomogeneous Damage Evolution," *Proc., 9th Int. Conf. on Computer Methods and Advances in Geomechanics*, Balkema, 1994, 635–640.
45. Pijaudier-Cabot, G., "Non Local Damage," Chapter 4 in *Continuum Models for Materials with Microstructure*, H.B. Mühlhaus (Editor), John Wiley, U.K., 1995, 105–143.
46. Voyiadjis, G.Z. and Kattan, I., *Advances in Damage Mechanics: Metals and Metal Matrix Composites*, Elsevier Science Ltd., Oxford, U.K., 1999.

47. Coleman, B.D. and Hodgden, M.L., "On Shear Bands in Ductile Materials," *Arch. Rational Mech. Analysis*, 90, 1985, 219–247.
48. Aifantis, E.C., "The Physics of Plastic Deformation," *Int. J. of Plasticity*, 3, 1987, 211–247.
49. Zbib, H.M. and Aifantis, E.C., "On the Localization and Postlocalization Behavior of Plastic Deformation, I, II, III," *Res. Mechanica*, 23, 1988, 261–277, 279–292, 293–305.
50. Mühlhaus, H.B. and Aifantis, E.C., "The Influence of Microstructure Induced Gradients on the Localization of Deformation in Viscoplastic Materials," *Acta Mechanica*, 89, 1991, 217–231.
51. Vardoulakis, I. and Aifantis, E.C., "A Gradient Flow Theory of Plasticity for Granular Materials," *Acta Mechanica*, 87, 1991, 197–217.
52. de Borst, R. and Mühlhaus, H.B., "Gradient-Dependent Plasticity: Formulation and Algorithmic Aspects," *Int. J. Num. Meth. Eng.*, 35, 1992, 521–539.
53. Oka, F., "A Gradient Dependent Elastic Model for Granular Materials and Strain Localization," Chapter 5 in *Continuum Models for Materials with Microstructure*, Mühlhaus, H.B. (Editor), John Wiley, U.K., 1995, 145–158.
54. Peerlings, R.H.J., de Borst, R., Brekelmans, W.A.M. and de Vree, J.H.P., "Gradient Enhanced Damage for Quasi-Brittle Materials," *Int. J. Numer. Meth. Eng.*, 39, 1996, 3391–3403.
55. Cosserat, E. and Cosserat, F., *Theorie des Corps Deformables*, Herman et Fils, Paris, France, 1909.
56. Eringen, A.C., "Theory of Micropolar Continuum," *Proc., 9th Midwestern Mech. Auf.*, Univ. of Wisconsin, Madison, WI, 1965, 23-40.
57. Eringen, A.C., "Theory of Micropolar Elasticity," in *Fracture, An Advanced Treatise*, H. Liebowitz (Editor), Academic Press, New York, 1968.
58. Mühlhaus, H.B., "Application of Cosserat Theory in Numerical Solutions of Limit Load Problems," *Ing.-Archiv*, 59, 1989, 124–137.
59. Willam, K., Dietsobe, A., Iordache, M.M. and Steinmann, P., "Localization in Micropolar Continua," Chapter 9 in *Continuum Models for Materials with Microstructure*, Mühlhaus, H.B. (Editor), John Wiley, U.K., 1995.
60. Kröner, E., "Elasticity Theory of Materials with Long Range Cohesive Forces," *Int. J. Solids Struct.*, 3, 1967, 731–742.
61. Kubin, L.P. and Lepinoux, J., "The Dynamic Organization of Dislocation Structures," *Proc., 8th Int. Conf. on Strength of Materials and Alloys*, Tampere, 1988, 35–39.
62. Frantziskonis, G. and Desai, C.S., "Constitutive Model with Strain Softening," *Int. J. Solids Struct.*, 23, 1987, 751–767.
63. Desai, C.S. and Hashmi, Q., "Analysis, Evaluation and Implementation of a Nonassociative Model for Geologic Materials," *Int. J. of Plasticity*, 5, 1989, 397–420.
64. Desai, C.S. and Woo, L., "Damage Model and Implementation in Nonlinear Dynamic Problems," *Int. J. Comput. Mech.*, 11, 1993, 189–206.
65. Desai, C.S., Basaran, C., Dishongh, T. and Prince, J., "Thermomechanical Analysis in Electronic Packaging with Unified Constitutive Model for Materials and Joints," *Components, Packaging and Manuf. Tech., Part B: Advanced Packaging, IEEE Trans.*, 21, 1, 1998, 87–97.
66. Davis, R.O., Private communication, 1998.
67. Van Mier, J.G.M., "Strain Softening of Concrete under Multiaxial Loading Conditions," *Ph.D. Dissertation*, Eindhoran Univ. of Tech., The Netherlands, 1984.

68. Whitenack, R., Private communication, 1999.
69. Larsson, R., Runesson, K., and Ottoson, N.S., "Discontinuous Displacement Approximation for Capturing Plastic Localization," *Int. J. Numer. Methods in Engng.*, 36, 1993, 2087–2105.
70. Zienkiewicz, O.C. and Huang, M., "Localization Problems in Plasticity Using Finite Elements with Adaptive Remeshing," *Int. J. Numer. Analy. Meth. Geomech.*, 19, 1995, 127–148.
71. Gyi, M.M., "Multiaxial Cyclic Testing of Saturated Ottawa Sands," *M.S. Thesis*, Dept. of Civil Eng. and Eng. Mechs., University of Arizona, Tucson, AZ, 1997.
72. Desai, C.S., Park, I.J., and Shao, C., "Fundamental Yet Simplified Model for Liquefaction Instability," *Int. J. Numer. Analyt. Meth. Geomechanics*, 22, 1998, 721–748.
73. Desai, C.S., "Evaluation of Liquefaction Using Disturbance and Energy Approaches," *J. of Geotech. and Geoenv. Eng.*, ASCE, 126, 7, 2000, 618–631.

13

Implementation of DSC in Computer Procedures

CONTENTS

- 13.1 Finite Element Formulation ... 530
  - 13.1.1 Incremental Equations ... 532
- 13.2 Solution Schemes ... 534
  - 13.2.1 FA Stress ... 536
  - 13.2.2 Observed Behavior ... 541
- 13.3 Algorithms for Creep Behavior ... 544
  - 13.3.1 Algorithms for Elastic, Plastic, Viscoplastic Behavior with Thermal Effects ... 545
  - 13.3.2 Elastic Behavior ... 545
  - 13.3.3 Plane Stress ... 545
  - 13.3.4 Plane Strain ... 545
  - 13.3.5 Axisymmetric ... 545
  - 13.3.6 Thermoplastic Behavior ... 546
  - 13.3.7 Thermoviscoplastic Behavior ... 547
  - 13.3.8 DSC Formation ... 549
- 13.4 Algorithms for Coupled Dynamic Behavior ... 550
  - 13.4.1 Fully Saturated Systems ... 550
  - 13.4.2 Dynamic Equations ... 550
  - 13.4.3 Finite Element Formulation ... 551
  - 13.4.4 Time Integration for Dynamic Problems ... 554
- 13.5 Implementation ... 557
- 13.6 Partially Saturated Systems ... 558
- 13.7 Cyclic and Repetitive Loading ... 559
  - 13.7.1 Unloading ... 561
  - 13.7.2 Reloading ... 563
  - 13.7.3 Cyclic-Repetitive Hardening ... 565
- 13.8 Initial Conditions ... 567
  - 13.8.1 Initial Stress ... 567
  - 13.8.2 Initial Disturbance ... 568
- 13.9 Hierarchical Capabilities and Options ... 570
- 13.10 Mesh Adaption Using DSC ... 572
  - 13.10.1 Prepeak Response ... 572

	13.10.2 Postpeak Response	573
13.11	Examples of Applications	574
	13.11.1 Laboratory Tests	581
	13.11.2 Field Validation	586
	13.11.3 Computer Results	594
13.12	Computer Codes	618

The incremental constitutive equations in the DSC and its hierarchical versions are presented in Chapters 4–12. In this Chapter, we consider the implementation of the models in computer (finite element) procedures. The finite element static, dynamic and creep formulations are presented, together with details of a number of schemes for the introduction of the DSC models in the computer procedures. Here, elastic, plastic, and creep models with disturbance are included together with thermal effects. Use of the disturbance as the criterion for mesh adaption is also described. A number of example problems are then presented. They include computational aspects such as the effect of (load) increments and a wide range of problems from civil and mechanical engineering, electronic packaging, dynamics and earthquake engineering, solid mechanics and pavements.

13.1 Finite Element Formulation

The following formulations and algorithms are presented in the context of the finite element (FE) methods. However, they can be adopted, with some modifications, for other numerical procedures such as the boundary element method and finite difference technique.

With the displacement based FE method, the stationary potential energy or the virtual work principle is often invoked to derive the element and assemblage questions for problems discretized into FE meshes [Fig. 13.1 (1–4)]. According to the virtual work principle, we can write

$$\int_V (\delta \underline{\varepsilon}^a)^T \underline{\sigma}^a \, dV = \int_V \delta \underline{u}^T \, \overline{\underline{X}} \, dV + \int_{S_1} \delta \underline{u}^T \, \overline{\underline{T}} \, dS \tag{13.1}$$

where $\underline{\sigma}^a$ and $\underline{\varepsilon}^a$ are the vectors of observed or actual stresses and strains components, respectively, δ denotes virtual quantity, V is the volume, $\underline{u}$ is the vector of displacement components, $\overline{X}$ is the vector of applied body forces, and $\overline{T}$ is the vector of applied surface tractions on a part of the surface, S_1. The displacement vector $(\underline{u}^a)^T = [u \ v \ w]$ at a point P (Fig. 13.1) is expressed as

$$\underline{u}^a = \underline{N} \underline{q}^a \tag{13.2}$$

where $\underline{N}$ is the matrix of interpolation functions (of various orders) and $\underline{q}^a$ is the vector of (observed) nodal displacements. The strain vector, $\underline{\varepsilon}^a$, is obtained from Eq. (13.2) as

$$\underline{\varepsilon}^a = \underline{N}' \underline{q} = \underline{B} \underline{q}^a \tag{13.3}$$

Implementation of DSC in Computer Procedures

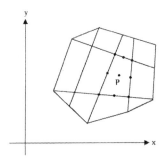

(a) Two-Dimensional

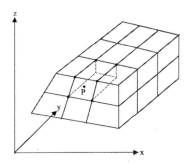

(b) Three-Dimensional

FIGURE 13.1
Schematic of finite element discretization.

where the prime denotes derivative and $\underline{B}$ is the strain-displacement transformation matrix. In general, $\underline{\varepsilon}^a$ is a 6 × 1 vector; however, for a two-dimensional (plane stress, plane strain, or axisymmetric) idealization, $\underline{\varepsilon}^a$ will have three to four components (1–4).

Substitution of Eqs. (13.2) and (13.3) in Eq. (13.1) gives

$$(\delta \underline{q}^a)^T \Big|_V \underline{B}^T \underline{\sigma}^a \, dV = (\delta \underline{q}^a)^T \left[\int_V \underline{N}^T \, \underline{\bar{X}} \, dV + \int_V \underline{N}^T \, \underline{\bar{T}} \, dS \right] \quad (13.4)$$

As the virtual displacement, $\delta \underline{q}^a$, is arbitrary, the FE equilibrium equations result as

$$\int \underline{B}^T \underline{\sigma}^a \, dV = \underline{Q} \quad (13.5a)$$

where $\underline{Q}$ is the applied load vector:

$$\underline{Q} = \int \underline{N}^T \underline{\bar{X}} \, dV + \int_{S_1} \underline{N}^T \underline{\bar{T}} \, dS \quad (13.5b)$$

If there is an initial stress, σ_o^a, the load vector, $\underset{\sim}{Q}$ [Eq. (13.5b)] will be augmented by

$$\underset{\sim}{Q}_o = -\int \underset{\sim}{B}^T \underset{\sim}{\sigma}_o^a \, dV \tag{13.5c}$$

13.1.1 Incremental Equations

For nonlinear analysis, we need to consider incremental form of Eq. (13.5). Consider equilibrium at the incremental time step, $(n + 1)$, Fig. 13.2; then Eq. (13.5a) becomes

$$\int_V \underset{\sim}{B}^T \underset{\sim}{\sigma}_{n+1}^a \, dV = \underset{\sim}{Q}_{n+1} \tag{13.6a}$$

where

$$\underset{\sim}{\sigma}_{n+1}^a = \underset{\sim}{\sigma}_n^a + d\underset{\sim}{\sigma}_{n+1}^a \tag{13.6b}$$

and d denotes an increment. Substitution of Eq. (13.6) into Eq. (13.5a) gives

$$\int_V \underset{\sim}{B}^T d\underset{\sim}{\sigma}_{n+1}^a \, dV = \underset{\sim}{Q}_{n+1} - \int_V \underset{\sim}{B}^T d\underset{\sim}{\sigma}_n^a \, dV$$
$$= \underset{\sim}{Q}_{n+1} - \underset{\sim}{Q}_n^b$$
$$= d\underset{\sim}{Q}_{n+1} \tag{13.7}$$

Here, the second term on the right-hand side, $\underset{\sim}{Q}_n^b$ represents the internal balanced load at step n, Fig. 13.2, and $d\underset{\sim}{Q}_{n+1}^b$ denotes the out-of-balance or residual load vector.

Now, from Eq. (4.1), Chapter 4, we have the DSC incremental constitutive equations

$$d\underset{\sim}{\sigma}_{n+1}^a = (1 - D)d\underset{\sim}{\sigma}_{n+1}^i + Dd\underset{\sim}{\sigma}_{n+1}^c + dD(\underset{\sim}{\sigma}_{n+1}^c - \underset{\sim}{\sigma}_{n+1}^i) \tag{13.8}$$

in which D can refer to step n or can be evaluated as an average value over step n to $n + 1$. Substitution of Eq. (13.8) in Eq. (13.7) gives

$$\int_V (1 - D_n)\underset{\sim}{B}^T d\underset{\sim}{\sigma}_{n+1}^i \, dV + \int_V D_n \underset{\sim}{B}^T d\underset{\sim}{\sigma}_{n+1}^c \, dV$$
$$+ \int_V \underset{\sim}{B}^T (\underset{\sim}{\sigma}_{n+1}^c - \underset{\sim}{\sigma}_{n+1}^i) dD_n \cdot dV = d\underset{\sim}{Q}_{n+1} \tag{13.9}$$

where

$$d\underset{\sim}{\sigma}_{n+1}^i = \underset{\sim}{C}_n^i d\underset{\sim}{\varepsilon}_{n+1}^i \tag{13.10a}$$

$$d\underset{\sim}{\sigma}_{n+1}^c = (1 + \alpha)\underset{\sim}{C}_{n+1}^c d\underset{\sim}{\varepsilon}_{n+1}^i \tag{13.10b}$$

Implementation of DSC in Computer Procedures

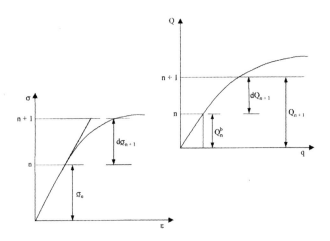

(a) Load-Displacement and Stress-Strain Responses

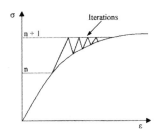

(b) Incremental-Iterative Scheme

FIGURE 13.2
Incremental analysis.

Here, we used Eq. (4.5), i.e., $d\varepsilon^c_{n+1} = (1 + \alpha) d\varepsilon^i_{n+1}$, where α is the relative strain (motion) parameter. Now, substitution of Eqs. (13.10) in Eq. (13.9) leads to

$$\int B^T \left[(1 - D_n)\underline{C}^i_n + D_n(1 + \alpha)\underline{C}^c_n + R^T \underline{\sigma}^r_n \right] d\underline{\varepsilon}^i_{n+1} = d\underline{Q}_{n+1} \quad (13.11)$$

Here $\underline{\sigma}^r_{n+1} = \underline{\sigma}^c_{n+1} - \underline{\sigma}^i_{n+1}$ is the vector of relative stresses in the FA and RI parts, and $dD_n = R^T d\underline{\varepsilon}^n_n$ from Eq. (4.14). Now, writing

$$\underline{C}^{DSC}_n = (1 - D_n)\underline{C}^i_n + D_n(1 + \alpha)\underline{C}^c_n + R^T \underline{\sigma}^r_n \quad (13.12)$$

and substitution of $d\underline{\varepsilon}^i_{n+1} = B d\underline{q}^i_{n+1}$ from Eq. (13.3), we have

$$\int_V B^T \underline{C}^{DSC}_n B \, dV \cdot d\underline{q}^i_{n+1} = d\underline{Q}_{n+1} \quad (13.13a)$$

or

$$k_n^{DSC} \, dq_{n+1}^i = dQ_{n+1} \tag{13.13b}$$

where

$$k_n^{DSC} = \int_V B^T C_n^{DSC} \cdot B \cdot dV \tag{13.13c}$$

is the stiffness matrix.

Note that in Eq. (13.13), the unknown displacement, dq_{n+1}^i is referred to the RI behavior, while the other quantities such as the out-of-the balance load refer to the observed behavior.

13.2 Solution Schemes

Equations (13.13) represent the incremental formulations of the DSC in which matrices C^{DSC} and k_n^{DSC} are nonsymmetric. During the incremental solution of Eq. (13.13) (Scheme 1 below), they become negative definite (after the peak) due to the third term in Eq. (13.12), depending upon its relative magnitude in comparison to those of the first two terms.

Simplified schemes can be developed such that the system matrix remains positive definite, by moving the contribution of the third term to the right-hand side as an equivalent load vector evaluated at the end of the previous step (n) (5–8). This is discussed in Scheme 2 below.

In another simplified Scheme 3, the solution is first obtained by solving the equations for the RI (elastic, elastoplastic hardening, etc.) response, and then evaluating the observed response from Eq. (13.8) through an iterative procedure in which the disturbance is modified based on the observed stress-strain behavior (9–11).

In the general procedure, the RI, observed and FA strains are different; this is discussed below as Scheme 4.

In the foregoing schemes described below, the iterative solution is obtained at constant values of applied or computed strains (increments) so that unique convergence is achieved for the strain softening behavior. They can provide satisfactory and consistent solutions, but may sometimes suffer from certain limitations.

Scheme 1
In this scheme, the incremental equations are expressed as

$$k_n^{DSC} \, dq_{n+1}^i = dQ_{n+1} = Q_{n+1} - Q_n^b \tag{13.13}$$

where k_n^{DSC} is evaluated at step n, Fig. 13.3. It is nonsymmetric and becomes negative definite after the peak. The load applied, Q_{n+1}, refers to the observed or actual condition, Fig. (13.3), and the balance load, Q_n^b is computed by using the observed stress, σ_n^a, at step n. Solution of Eq. (13.13)

Implementation of DSC in Computer Procedures

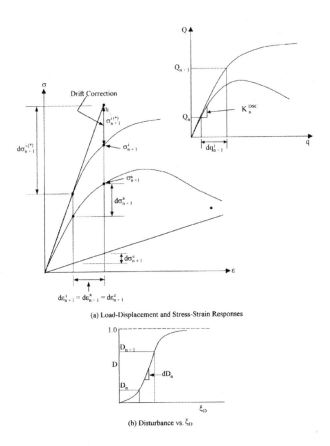

FIGURE 13.3
Incremental analysis at constant strain and disturbance.

provides the increments of RI displacement, dq^i_{n+1}, which are used to evaluate the RI strains, $d\varepsilon^i_{n+1}$, and stress, $d\sigma^{i(*)}_{n+1}$, using the following equations:

$$d\varepsilon^i_{n+1} = B\, dq^i_{n+1} \tag{13.14}$$

and

$$d\sigma^{i(*)}_{n+1} = C^i\, d\varepsilon^i_{n+1} \tag{13.15}$$

where C^i is the RI constitutive matrix which can be characterized by using elastic, elastoplastic, etc. idealizations.

In the case of the elastoplastic (δ_0-model), the increment of RI stress, $d\sigma^{i(*)}_{n+1}$ is corrected by using the drift correction procedure described in Chapter 7. Then the converged values, $d\sigma^i_{n+1}$ and the corresponding total stress, σ^i_{n+1}, are calculated. The iterative procedure is performed at constant strain (increment), $d\varepsilon^i = d\varepsilon^a = d\varepsilon^c$, Fig. (13.3). Such constant value can be due to applied strains (displacement) or, in general, due to the computed strains as a result of the applied load (Q_{n+1}) or stress. Such a procedure can lead to a unique converged value of stress for a given strain and avoid the duality of strains if convergence is sought based on the applied stress.

Now, we need the FA stress $(d\underline{\sigma}^c)$ and disturbance, D_n, for the calculation of the observed stress, $d\underline{\sigma}^a_{n+1}$, Eq. (13.8). Details of their calculations are given below.

13.2.1 FA Stress

The FA stress increment is expressed as

$$d\underline{\sigma}^c_{n+1} = d\underline{S}^c_n + \frac{1}{3}dJ^c_{1n}\underline{I} \qquad (13.16)$$

where $\underline{I}^T = [1\ 1\ 1\ 0\ 0\ 0]$. As discussed in Chapters 3 and 4, a number of simulations are possible to characterize the FA state; e.g., (a) "void", (b) hydrostatic strength, and (c) critical state. These are discussed below.

Void: If the FA part is treated like a "void", as in the classical damage model (12), it can carry no stress at all. Then

$$\underline{\sigma}^c_n = d\underline{\sigma}^c_{n+1} = \underline{0} \qquad (13.17)$$

This assumption is considered to be unrealistic because the microcracked or FA part, which is surrounded by the RI parts, can possess certain strength.

Hydrostatic strength: If the FA state is assumed to carry hydrostatic stress or mean pressure, it acts like a constrained liquid. Then it can carry no shear stress, but can carry hydrostatic stress, Eq. (4.27). Therefore,

$$d\underline{S}^c_{n+1} = \underline{0} \qquad (13.18a)$$

$$\frac{1}{3}dJ^c_{n+1} \neq \underline{0} \qquad (13.18b)$$

The FA response can be characterized based on the bulk modulus, K^c, for the FA material. As a simple approximation, it can be assumed that the hydrostatic stresses in the RI and FA parts are equal, i.e.,

$$dJ^c_{1(n+1)} = dJ^i_{1(n+1)} \qquad (13.18c)$$

Then Eq. (4.19) can be used. Alternatively, Eq. (13.19b) below can be used to define the FA response.

Critical state: As discussed in Chapter 4, the constitutive equations that govern the response of the FA material at the critical state are given by

$$\sqrt{J^c_{2D}} = \overline{m}J^c_1 \qquad (13.19a)$$

and

$$e^c = e_o^c - \lambda \ln(J_1^c / 3p_a) \tag{13.19b}$$

or

$$J_1^c = 3p_a \exp\left(\frac{e_o^c - e^c}{\lambda}\right) \tag{13.19c}$$

Then the matrix, $\underline{C}^c$, is defined by assuming that $J_1^i = J_1^c$, and the shear stress, J_{2D}^c, is defined based on Eq. (4.31). The constitutive equations with the assumption of critical state are derived in Chapter 4, Eqs. (4.35), in which the incremental critical stress is given by

$$d\sigma_{n+1}^c = \frac{J_1^i}{\lambda}(1 + e_o)\left(\frac{\bar{m}\underline{S}^i}{\sqrt{J_{2D}^i}} + \frac{1}{3}\underline{I}\right)\underline{I} d\varepsilon_{n+1}^c$$

$$+ \frac{J_1^c \bar{m}}{\sqrt{J_{2D}^i}}\left[\underline{C}^i - \frac{1}{3}\underline{I}\bar{C}^i - \frac{\underline{S}^{iT}\underline{S}^i}{2J_2^i}\bar{C}^i\right]d\varepsilon_{n+1}^i \tag{13.20a}$$

$$= \underline{M}\, d\varepsilon_{n+1}^c + \underline{N}\, d\varepsilon_{n+1}^i \tag{13.20b}$$

$$\underline{M} = \frac{J_{1(n+1)}^c}{\lambda}(1 + e_o)\left(\frac{\bar{m}\underline{S}^i}{\sqrt{J_{2D(n+1)}^i}} + \frac{1}{3}\underline{I}\right)$$

where

$$\underline{N} = \frac{J_{1(n+1)}^c \bar{m}}{\sqrt{J_{2D(n+1)}^i}}\left[\underline{C}^i - \frac{1}{3}\underline{I}\bar{C}^i - \frac{\underline{S}^{iT}\underline{S}^i}{2\sqrt{J_{2D}^i}}\bar{C}^i\right]$$

Here, the RI stresses are computed at step $(n + 1)$ and J_c^1 is computed using Eq. (13.19c) and dJ_1^c as

$$dJ_{1(n+1)}^c = \frac{J_{1(n+1)}^c}{\lambda}(1 + e_o)\varepsilon_{ii(n+1)}^c \tag{13.21}$$

where e_o is the initial void ratio and $e = -(1 + e_o)\varepsilon_{ii}$.

Disturbance: The disturbance, D, at steps n and $n + 1$ is computed as

$$D_n = D_u\left(1 - e^{-A\xi_{D_n}^z}\right) \tag{13.22a}$$

and

$$D_{n+1} = D_u\left(1 - e^{-A\xi_{D(n+1)}^Z}\right) \tag{13.22b}$$

where, as an approximation, ξ_D can be evaluated based on the RI response as

$$\xi_D^i = \int\left[(d\underline{E}^{pi})^T d\underline{E}^{pi}\right]^{1/2} \tag{13.22c}$$

Disturbance based on the observed response can be used in an iterative procedure; this is described later, in the context of the simplified procedure (Scheme 3).

The increment, dD_n can be found approximately as

$$dD_n = D_{n+1} - D_n \tag{13.23a}$$

or by using Eq. (4.14b) as

$$dD_n = \underline{R}_{n+1}^T d\underline{\varepsilon}^i \tag{13.23b}$$

The computations are performed at constant strain, $d\underline{\varepsilon}^i = d\underline{\varepsilon}^a = d\underline{\varepsilon}^c$, and the foregoing procedure yields $d\underline{\sigma}_{n+1}^a$ and $d\underline{\varepsilon}_{n+1}^a$, which in turn, can be used to evaluate total values as

$$\underline{\varepsilon}_{n+1}^a = \underline{\varepsilon}_n^a + d\underline{\varepsilon}_{n+1}^a \tag{13.24a}$$

$$\underline{\sigma}_{n+1}^a = \underline{\sigma}_n^a + d\underline{\sigma}_{n+1}^a \tag{13.24b}$$

$$\underline{\varepsilon}_{n+1}^c = \underline{\varepsilon}_n^c + d\underline{\varepsilon}_{n+1}^c \tag{13.24c}$$

$$\underline{\sigma}_{n+1}^c = \underline{\sigma}_n^c + \underline{\sigma}_{n+1}^c \tag{13.24d}$$

The balanced load, $\underline{Q}_{n+1}^b$ is found as

$$\underline{Q}_{n+1}^b = \int \underline{B}^T \underline{\sigma}_{n+1}^a dV \tag{13.26}$$

With the above quantities at $(n + 1)$, we revise the constitutive matrix at $n + 1$ as

$$\underline{C}_{n+1}^{DSC} = (1 - D_{n+1})\underline{C}_{n+1}^i + D_n \underline{C}_{n+1}^c + \underline{R}^T \underline{\sigma}_{n+1}^r \tag{13.27}$$

Here, as $d\underline{\varepsilon}^i = d\underline{\varepsilon}^a = d\underline{\varepsilon}^c$, $\alpha = 0$ in Eq. (13.12) are assumed.

Now, we apply the next load increment, $\underline{Q}_{n+2}$, solve Eq. (13.13) and repeat the foregoing procedure.

Implementation of DSC in Computer Procedures

Iterations. The drift correct procedure (Chapter 7) can entail iterations during the incremental analysis during which the stress converges from $\sigma_{n+1}^{i(*)}$ to σ_{n+1}^{i}, Fig. 13.3. Iterations can be performed in the computation of $d\sigma_{n+1}^{a}$, Eq. (13.8), during which the disturbance is revised; this is described later (Scheme 3).

Scheme 2

The term, $\underset{\sim}{R}^T \sigma_n^r$ or $dD(\sigma^c - \sigma^i)$ in Eq. (13.11) is moved to the right-hand side as

$$\int_V \underset{\sim}{B}^T \left[(1 - D_n)\underset{\sim}{C}_n^i + D_n(1+\alpha)\underset{\sim}{C}_n^c \right] \underset{\sim}{B}\, dq_{n+}^i = d\underset{\sim}{Q}_{n+1} - \int_V \underset{\sim}{R}^T \sigma_n^r \underset{\sim}{B}\, dV \quad (13.28a)$$

or

$$\bar{k}_n^{DSC} dq_{n+1}^i = d\underset{\sim}{Q}_{n+1} - \underset{\sim}{Q}_n^r \quad (13.28b)$$

where

$$\bar{k}_n^{DSC} = \int_V \underset{\sim}{B}^T \bar{\underset{\sim}{C}}^{DSC} \underset{\sim}{B}\, V$$

$$\underset{\sim}{Q}_n^r = \int \underset{\sim}{R}^T \sigma_n^r \underset{\sim}{B}\, dV$$

The matrices $\bar{\underset{\sim}{C}}^{DSC}$ and $\bar{k}_n^{DSC}$ are positive definite, and $\underset{\sim}{Q}_n^r$ is evaluated at the end of the previous step n, based on the known values of the terms involved. The solution of Eq. (13.28) gives the incremental RI displacements, strains and stresses, and the procedure for the evaluation of the observed stress, $d\sigma_{n+1}^a$, is similar to that in Scheme 1.

It may be noted that the modification of equations in this scheme that yields positive definite stiffness matrix does not change the nature of the basic problem. Also, depending upon the relative magnitude of $\underset{\sim}{Q}_n^r$ in comparison to $d\sigma_{n+1}^a$, convergence and stability problems may arise for some situations (6).

Scheme 3

This scheme is conceptually similar to Scheme 2 in which the stiffness matrix is positive definite (5–7). The incremental nonlinear analysis is first performed to evaluate the RI response in which the stiffness matrix is positive definite because it relates to the continuum characterization such as nonlinear elastic and elastoplastic (9,11). The FA response is defined based on the assumptions described under Scheme 1. Then the observed stress is computed by using Eq. (13.8) and an iterative procedure during which the disturbance is revised.

Figure (13.4) shows a schematic of the procedure. The RI response is first evaluated by applying a load increment, $d\underset{\sim}{Q}^i$, relevant to the RI response, in

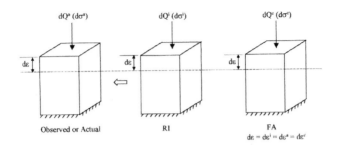

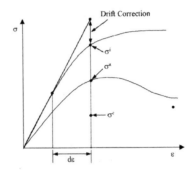

FIGURE 13.4
RI-based incremental procedure for scheme 3.

which the RI stress, $\underline{\sigma}^i$, is evaluated using the drift correction procedure. The observed stress, $\underline{\sigma}^a$, is evaluated at constant value of the computed RI strain, $d\underline{\varepsilon}^i = d\underline{\varepsilon}^a = d\underline{\varepsilon}^c = d\underline{\varepsilon}$.

Details of Scheme 3: The incremental FE equations for the RI response (elastic, elastoplastic hardening, etc.) are given by

$$\underline{k}_n^i \, d\underline{q}_{n+1}^i = d\underline{Q}_{n+1}^i = \underline{Q}_{n+1}^i - \underline{Q}_n^{b(i)} \tag{13.29a}$$

where $d\underline{q}_{n+1}^i$ is the incremental vector of RI displacements:

$$\underline{k}_n^i = \int \underline{B}^T \underline{C}_n^i \underline{B} \, dV \tag{13.29b}$$

and

$$\underline{Q}_n^{b(i)} = \int \underline{B}^T \underline{\sigma}_n^{b(i)} \, dV \tag{13.29c}$$

The matrices $\underline{C}_n^i$ and $\underline{k}_n^i$ are positive definite, as the RI response is characterized as nonlinear elastic or elastoplastic hardening (δ_0-model), etc.

Solution of Eq. (13.29) yields dq^i_{n+1} and then incremental strains and stresses as

$$d\varepsilon^i_{n+1} = \underline{B}\, dq^i_{n+1} \tag{13.30a}$$

$$d\sigma^{i(*)}_{n+1} = \underline{C}^i_n d\varepsilon^i_{n+1} \tag{13.30b}$$

The drift correction procedure leads to $d\sigma^i_{n+1}$, Fig. 13.3(a). The total quantities are evaluated as

$$\varepsilon^i_{n+1} = \varepsilon^i_n + d\varepsilon^i_{n+1} \tag{13.31a}$$

and

$$\sigma^i_{n+1} = \sigma^i_n + d\sigma^i_{n+1} \tag{13.31b}$$

13.2.2 Observed Behavior

The observed incremental stress, $d\sigma^a_{n+1}$, Eq. (13.8) is now found, with appropriate model for the FA behavior, as

$$d\sigma^{a(1)}_{n+1} = (1 - D^{(1)}_{n+1})d\sigma^i_{n+1} + D^{(1)}_{n+1}d\sigma^c_{n+1} + dD^{(1)}_{n+1}(\sigma^c_{n+1} - \sigma^i_{n+1}) \tag{13.32}$$

where the superscript (1) denotes the first iteration. As $D^{(1)}_{n+1}$ and $dD^{(1)}_{n+1}$ are based on the deviatoric plastic strains for the observed response, their values are not available for the current step $(n + 1)$. Hence, they are computed by assuming that for the first iteration:

$$\xi^{a(1)}_{D(n+1)} = \xi^{i(1)}_{D(n+1)} \tag{13.33}$$

and hence,

$$D^{(1)}_{n+1} = D_u(1 - e^{-A\xi^z_D}) \tag{13.34a}$$

and

$$dD^{(1)}_{n+1} = \underline{R}^{(1)T}_{n+1} d\varepsilon^i_{n+1} \tag{13.34b}$$

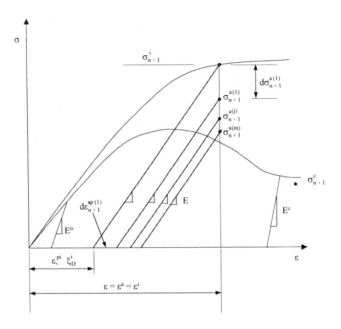

FIGURE 13.5
Iterative method for scheme 3.

where $\xi_D = \xi_{D(n+1)}^{a(1)}$. The term, $dD_{n+1}^{(1)}$ can also be computed by using Eq. (13.23a), which for the first iteration will be zero. Now, the total observed stress, $\sigma_{n+1}^{a(1)}$, (Fig. 13.5) is found as

$$\sigma_{n+1}^{a(1)} = (1 - D_{n+1}^{(1)})\sigma_{n+1}^i + D_{n+1}^{(1)}\sigma_{n+1}^c \qquad (13.35)$$

which would usually not be the "correct" value as the disturbance is found on the basis of the RI strains.

In order to improve the computed values in Eqs. (13.32) and (13.35), an iterative procedure can be used, in which the modified value of the observed plastic strain increment is found approximately as (Fig. 13.5)

$$d\varepsilon_{n+1}^{ap(1)} = (\underset{\sim}{C}^e)^{-1} d\sigma_{n+1}^{a(1)} \qquad (13.36)$$

where

$$d\sigma_{n+1}^{a(1)} = \sigma_{n+1}^i - \sigma_{n+1}^{a(1)}$$

and $\underset{\sim}{C}^e$ is the elastic constitutive matrix computed by assuming that the unloading elastic moduli in the RI and FA states are equal. Alternatively, modified

Implementation of DSC in Computer Procedures

value of the observed elastic moduli can be adopted by using the following equation (13)

$$E^a_{n+1} = (1 - D^{(1)}_{n+1})E^o + D^{(1)}_{n+1}E^c \tag{13.37}$$

where E^o is the initial elastic modulus and E^c is the modulus in the FA region, Fig. 13.5. The Poisson's ratio in C^c can be assumed to be constant.

The modified (increased) observed plastic strains, $\varepsilon^{ap(2)}_{n+1}$, for the second iteration are now given by

$$\varepsilon^{ap(2)}_{n+1} = \varepsilon^{ap(1)}_{n+1} + d\varepsilon^{ap(1)}_{n+1} \tag{13.38}$$

Then, the revised deviatoric plastic strain trajectory for the second iteration is evaluated as

$$\xi^{a(2)}_{D(n+1)} = \xi^{a(1)}_{D(n+1)} + d\xi^{a(2)}_{D(n+1)} \tag{13.39}$$

in which the second term on the right-hand side is computed by using Eq. (13.22c); the deviatoric plastic strains, dE^p, relate to the observed strains, Eq. (13.36). The disturbance is revised by using Eq. (13.22), and the observed stresses are computed for the second and the subsequent iterations (j) as

$$\sigma^{a(j)}_{n+1} = (1 - D^{(j)}_{n+1})\sigma^i_{n+1} + D^j_{n+1}\sigma^c_{n+1} \tag{13.40}$$

The iterations are continued until convergence at $j = m$, which can be defined as

$$\frac{d\xi^{a(j-1)}_D}{d\xi^{a(j)}_D} \leq \delta(\approx 10^{-6}) \tag{13.41}$$

At the end of iterations, m, various observed quantities are found for the load increment, Q^i_{n+1}, as

$$d\sigma^{a(m)}_{n+1} = (1 - D^{(m)}_{n+1})d\sigma^i_{n+1} + D^{(m)}_{n+1} d\sigma^c_{n+1} + dD^{(m)}_{n+1}(\sigma^c_{n+1} - \sigma^i_{n+1}) \tag{13.42a}$$

$$d\varepsilon^{a(m)}_{n+1} = d\varepsilon^{i(m)}_{n+1} \tag{13.42b}$$

$$\varepsilon^{a(m)}_{n+1} = \varepsilon^i_{n+1} \tag{13.42c}$$

and the total stresses using Eq. (13.35). The observed or actual load increments, dQ^a_{n+1}, can be found as

$$dQ^{(a)}_{n+1} = \int B^T d\sigma^{a(m)}_{n+1} dV \tag{13.43}$$

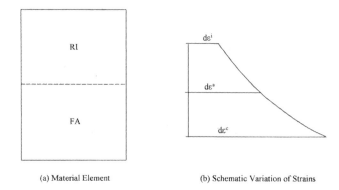

FIGURE 13.6
Schematic of RI, FA and observed strains in material element.

Then, we proceed to the next load increment, $n + 2$, and so on.

The foregoing procedure involves the basic solution for the RI response, in which the system matrix is positive definite. However, the coupling between the RI and FA responses, which affects the observed response, is incorporated through the iterative procedure in which the disturbance refers to the observed behavior, defined on the basis of the laboratory test data.

Scheme 4

In the foregoing three schemes, the FE equations, Eq. (13.13), involve solutions for the RI displacements, strains, etc. As described in Chapter 4, it is possible to formulate the equations in terms of the observed strains and displacements by postulating relations between the FA and RI, and RI and observed strains, Eqs. (4.49); a schematic is shown in Fig. 13.6. The procedure is described in Chapter 4, and would lead to different strains in the RI, observed and FA states. Further investigation would be required for this scheme, particularly with respect to convergence and computational effort. Hence, at this time, we recommend use of Schemes 1 to 3, and in particular Scheme 3, which is found to provide convergent and satisfactory results with considerable computational efficiency for many problems solved, including those presented later.

13.3 Algorithms for Creep Behavior

Description of the models based on the multicomponent DSC and overlay approach for viscoelastic, viscoplastic and elastoviscoplastic behavior is presented in Chapter 8. Now, we present algorithms for implementing typical creep models in computer (FE) procedures.

13.3.1 Algorithms for Elastic, Plastic, Viscoplastic Behavior with Thermal Effects

Among the factors to be considered for thermal response are (1) the existence of "initial" strains due to the (known) temperature, and (2) temperature dependence of material parameters during thermomechanical (cyclic) loading. First, we present algorithmic aspects for initial strains.

13.3.2 Elastic Behavior

In the case of elastic behavior, for isotropic material, the effect of known temperature change causing initial strains are expressed as follows:

$$\underline{\varepsilon}_o = \underline{\alpha}_T \, dT \tag{13.44}$$

where $\underline{\varepsilon}_o^T = [\varepsilon_x \; \varepsilon_y \; \varepsilon_z \; 0 \; 0 \; 0]$, $\underline{\alpha}_T = [\alpha_T \; \alpha_T \; \alpha_T \; 0 \; 0 \; 0]$, and dT is the temperature change $= T - T_0$, where T_0 is the initial (previous) temperature, α_T is the co-efficient of thermal expansion, and T is the current temperature. For two-dimensional idealizations, Eq. (13.44) is specialized as follows:

13.3.3 Plane Stress

$$\begin{aligned} \varepsilon_x(T) &= \alpha_T \, dT \\ \varepsilon_y(T) &= \alpha_T \, dT \\ \gamma_{xy}(T) &= 0.0 \end{aligned} \tag{13.45}$$

13.3.4 Plane Strain

$$\begin{aligned} \varepsilon_x(T) &= \alpha_T \, dT(1 + \nu) \\ \varepsilon_y(T) &= \alpha_T \, dT(1 + \nu) \\ \gamma_{xy}(T) &= 0.0 \\ \sigma_z(T) &= -E\alpha_T \, dT \end{aligned} \tag{13.46}$$

where E and ν are the elastic parameters.

13.3.5 Axisymmetric

$$\begin{aligned} \varepsilon_r(T) &= \alpha_T \, dT \\ \varepsilon_z(T) &= \alpha_T \, dT \\ \varepsilon_\theta(T) &= \alpha_T \, dT \\ \gamma_{rz}(T) &= 0.0 \end{aligned} \tag{13.47}$$

Then the incremental elastic constitutive relation is given by

$$d\sigma = \underline{C}^e\, d\underline{\varepsilon}^e$$
$$= \underline{C}^e\left[d\underline{\varepsilon} - d\underline{\varepsilon}(T)\right] \qquad (13.48)$$

where $\underline{C}^e$ is the elastic (tangent) constitutive matrix, and $d\underline{\varepsilon}$, $d\underline{\varepsilon}^e$, and $d\underline{\varepsilon}(T)$ are the vectors of total, elastic and thermal strains, respectively.

If the parameters E and v vary with temperature, they can be expressed in terms of temperature as (10, 14):

$$E = E_r\left(\frac{T}{T_r}\right)^{C_T} \qquad (13.49a)$$

$$v = v_r\left(\frac{T}{T_r}\right)^{C_v} \qquad (13.49b)$$

where E_r and v_r are values at reference temperature, T_r (e.g., room temperature 23°C or 300 K), and c_T and c_v are parameters found from laboratory tests.

13.3.6 Thermoplastic Behavior

The normality rule gives the increment of plastic strain vector $d\underline{\varepsilon}(T)$ as (11,14,15)

$$d\underline{\varepsilon}^p(T) = \lambda\frac{\partial Q(\sigma, \xi, T)}{\partial \underline{\sigma}} \qquad (13.50)$$

where Q is the plastic potential function; for associative rule, $Q \equiv F$, where F is the yield function (Chapters 6 and 7). Now, the total incremental strain vector $d\underline{\varepsilon}^t$ is given by

$$d\underline{\varepsilon}^t(T) = d\underline{\varepsilon}^e(T) + d\underline{\varepsilon}^p(T) + d\underline{\varepsilon}(T) \qquad (13.51a)$$

where $d\underline{\varepsilon}(T)$ is the strain vector due to the temperature change. Hence,

$$d\underline{\varepsilon}^e(T) = d\underline{\varepsilon} - \lambda\frac{\partial Q}{\partial \underline{\sigma}} - d\underline{\varepsilon}(T) \qquad (13.51b)$$

and

$$d\underline{\sigma} = \underline{C}^e(T)d\underline{\varepsilon}^e$$
$$= \underline{C}^e(T)\left(d\underline{\varepsilon} - \lambda\frac{\partial Q}{\partial \underline{\sigma}} - \alpha_T I_1^0\, dT\right) \qquad (13.51c)$$

Implementation of DSC in Computer Procedures

where $\underline{I}_1^0 = [1\ 1\ 0]$ for the two-dimensional case, and $[1\ 1\ 1\ 0\ 0\ 0]$ for the three-dimensional case.

Now, the consistency condition gives

$$dF(\underline{\sigma}, \xi, T) = 0 \tag{13.52a}$$

Therefore,

$$dF = \frac{\partial F^T}{\partial \underline{\sigma}} d\underline{\sigma} + \frac{\partial F}{\partial \xi} \cdot d\xi + \frac{\partial F}{\partial T} \cdot dT = 0 \tag{13.52b}$$

Then, use of Eqs. (13.51) and (13.52) gives

$$\lambda = \frac{\dfrac{\partial F^T}{\partial \underline{\sigma}} \underline{C}^e(T) d\underline{\varepsilon} + \dfrac{\partial F}{\partial T} \cdot dT - \alpha_T \dfrac{\partial F^T}{\partial \underline{\sigma}} \underline{C}^e(T) \underline{I}^0 dT}{\dfrac{\partial F^T}{\partial \underline{\sigma}} \underline{C}^e(T) \dfrac{\partial Q}{\partial \underline{\sigma}} - \dfrac{\partial F}{\partial \xi} \left(\dfrac{\partial Q^T}{\partial \underline{\sigma}} \dfrac{\partial Q}{\partial \underline{\sigma}} \right)^{1/2}} \tag{13.53a}$$

Therefore,

$$d\underline{\sigma} = \underline{C}^e(T)\left[\underline{I} - \frac{\dfrac{\partial Q^T}{\partial \underline{\sigma}} \dfrac{\partial F}{\partial \underline{\sigma}} \underline{C}^e(T)}{\dfrac{\partial F^T}{\partial \underline{\sigma}} \underline{C}^e(T) \dfrac{\partial Q}{\partial \underline{\sigma}} - \dfrac{\partial F}{\partial \xi}\left(\dfrac{\partial Q^T}{\partial \underline{\sigma}} \cdot \dfrac{\partial Q}{\partial \underline{\sigma}}\right)^{1/2}} \right] d\underline{\varepsilon}$$

$$- \underline{C}^e(T)\left[\alpha_T \underline{I}^0 - \frac{\alpha_T \dfrac{\partial Q}{\partial \underline{\sigma}} \dfrac{\partial F^T}{\partial \underline{\sigma}} \underline{C}^e(T) \underline{I}^0 - \dfrac{\partial F}{\partial T}\dfrac{\partial Q}{\partial \underline{\sigma}}}{\dfrac{\partial F^T}{\partial \underline{\sigma}} \underline{C}^e(T) \dfrac{\partial Q}{\partial \underline{\sigma}} - \dfrac{\partial F}{\partial \xi}\left(\dfrac{\partial Q^T}{\partial \underline{\sigma}} \cdot \dfrac{\partial Q}{\partial \underline{\sigma}}\right)^{1/2}} \right] dT \tag{13.53b}$$

The parameters in the elastoplastic model, e.g., HISS-δ_0, can be expressed as a function of temperature as

$$P(T) = P_r \left(\frac{T}{T_r} \right)^c \tag{13.54}$$

where P is any parameter such as E, ν, Eq. (13.49); γ, β, R, n, Eq. (7.1); a_1, η_1, Eq. (7.11); P_r is its value at reference temperature T_r, and c is parameter found from laboratory tests.

13.3.7 Thermoviscoplastic Behavior

The total temperature dependent strain rate vector, $\underline{\dot{\varepsilon}}$, is assumed to be the sum of the thermoelastic strain rate, $\underline{\dot{\varepsilon}}^e(T)$, thermoviscoplastic strain rate, $\underline{\dot{\varepsilon}}^{vp}(T)$, and the thermal strain rate due to temperature change dT, $\underline{\dot{\varepsilon}}(T)$, as

$$\underline{\dot{\varepsilon}} = \underline{\dot{\varepsilon}}^e(T) + \underline{\dot{\varepsilon}}^{vp}(T) + \underline{\dot{\varepsilon}}(T) \tag{13.55}$$

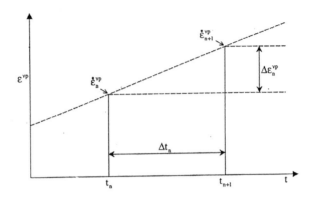

FIGURE 13.7
Time integration for viscoplastic strains.

Here, the thermoviscoplastic strain is contributed by creep and temperature effects.

With Perzyna's (16) viscoplastic theory, Eq. (8.2d) (Chapter 8), Eq. (13.55) can be written as

$$\dot{\underline{\varepsilon}} = \dot{\underline{\varepsilon}}^e(T) + \Gamma(T)\left\langle \phi\left(\frac{F(T)}{F_o}\right)\right\rangle \frac{\partial F}{\partial \underline{\sigma}} + \dot{\underline{\varepsilon}}^c(T) \tag{13.56}$$

where Γ and ϕ are temperature dependent fluidity parameter and flow function, respectively. Then the constitutive equations are given by

$$\dot{\underline{\sigma}} = \underline{C}^e(T)\left(\dot{\underline{\varepsilon}} - \Gamma(T)\left\langle \phi\left(\frac{F(T)}{F_o}\right)\right\rangle \frac{\partial F}{\partial \underline{\sigma}} + \dot{\underline{\varepsilon}}^c(T)\right) \tag{13.57}$$

Viscous or creep behavior requires integration in time. The thermoviscoplastic strain rate (Chapter 8) is evaluated at time step n, Fig. 13.7. Then the strain rate at step $(n + 1)$ can be expressed by using Taylor series expansion as (14, 17)

$$\dot{\underline{\varepsilon}}^{vp}(T)^{n+1} = \dot{\underline{\varepsilon}}^{vp}(T)_n + \frac{\partial \dot{\underline{\varepsilon}}^{vp}(T)_n}{\partial \underline{\sigma}} \partial \underline{\sigma}_n + \frac{\partial \dot{\underline{\varepsilon}}^{vp}(T)_n}{\partial T} dT_n \underline{I}$$

$$= \dot{\underline{\varepsilon}}^{vp}(T)^n + \underline{G}_{1n}\, d\underline{\sigma}_n + \underline{G}_{2n}\, dT_n \cdot \underline{I} \tag{13.58}$$

where $\partial \underline{\sigma}^n$ is the stress increment, dT^n is the temperature increment, and $\underline{G}_1^n$, $\underline{G}_2^n$ denote gradient matrices at time step, n; their details are given in Eq. (8.45), Chapter 8.

Implementation of DSC in Computer Procedures

The increment of viscoplastic strain, $\{d\varepsilon^{vp}(T)\}^n$, can be found during the time interval $\Delta t_n = t_{n+1} - t_n$, Fig. 13.7, as

$$d\varepsilon^{vp}(T)_n = \Delta t_n[(1-\theta)\dot{\varepsilon}^{vp}(T)_n + \theta\dot{\underline{\varepsilon}}^{vp}(T)_{n+1}] \qquad (13.59)$$

where $0 \le \theta \le 1$. For $\theta = 0$, Eq. (13.59) gives the Euler scheme, for $\theta = 0.5$ the Crank-Nicolson scheme, and so on.

Now, Eq. (13.57) can be written in the incremental form as

$$d\underline{\sigma}_n = \underline{C}^e(T)(d\underline{\varepsilon}_n - d\underline{\varepsilon}^{vp}(T)_n - d\underline{\varepsilon}(T)) \qquad (13.60)$$

Use of Eqs. (13.58) and (13.59) in (13.60) leads to

$$d\underline{\sigma}_n = \underline{C}^{evp}(T)(d\underline{\varepsilon}_n - \dot{\underline{\varepsilon}}^{vp}(T)_n \cdot \Delta t_n - \theta \underline{G}_{2n} \Delta t_n dT_n \underline{I} - \underline{\varepsilon}(T)) \qquad (13.61)$$

where

$$\bar{\underline{C}}(T) = \underline{C}^{evp}(T) = \underline{C}^e(T)\left[\underline{I} + \theta \underline{C}^e(T)\underline{G}_{1n} \cdot \Delta t_n\right]^{-1}$$

The finite element equations at step $(n+1)$ can now be written as

$$\underline{k}_n d\underline{q}_{n+1} = d\bar{\underline{Q}}_{n+1} \qquad (13.62)$$

where

$$\underline{k}_n = \int \underline{B}^T \bar{\underline{C}}_n(T)\underline{B} \, dV$$

and

$$d\bar{\underline{Q}} = \int_V \underline{B}^T \bar{\underline{C}}^n(T) \cdot \dot{\underline{\varepsilon}}^{vp} \Delta t_n \cdot dV - d\underline{Q}_{n+1}$$

Solution of Eq. (13.62) for $d\underline{q}_{n+1}$ allows calculation of $d\dot{\varepsilon}^{vp}_{n+1}(T), d\varepsilon^{vp}_{n+1}(T)$, and $d\underline{q}_{n+1}$ required for the next load increment $(n+2)$, and so on.

13.3.8 DSC Formation

In the case of the DSC model, Eq. (13.12), the RI response can be simulated as elastic, elastoplastic, or viscoelastic or viscoplastic (as in the multicomponent DSC or overlay model, Chapter 8). The disturbance parameters, D_u, A and Z, can be expressed as functions of temperature using Eq. (13.54). Then the DSC model can be implemented in the finite element procedure; details are given in Chapter 8.

13.4 Algorithms for Coupled Dynamic Behavior

The previous algorithms were developed for dry materials; hence, the stresses were effective. We now present algorithms for fully and partially saturated systems discussed in Chapter 9.

13.4.1 Fully Saturated Systems

The finite element procedure is based on the coupled theory of dynamics of porous saturated media proposed by Biot (18,19). The present procedure allows for various additional factors such as nonlinear behavior of porous (soil) media and interfaces, characterized by using the disturbed state concept (DSC) constitutive models. The following procedure is presented for dynamic analysis; it can be specialized to time dependent deformation (consolidation) and static behavior also.

13.4.2 Dynamic Equations

The coupled deformation and flow equations are given by (18–25)

$$\sigma_{ij,j} + (1 + \bar{n})\rho_s b_i + \bar{n} \cdot \rho_f b_i - (1 - \bar{n})\ddot{u}_i - n\rho_f \ddot{u}_{fi} = 0 \quad (13\text{-}63a)$$

or

$$\sigma_{ij,j} + \rho b_i - \rho \ddot{u}_i - \rho_f \ddot{w}_i = 0$$

and

$$p_i + \rho_f b_i - \rho_f \ddot{u}_i - \frac{\rho_f}{\bar{n}}\ddot{w}_i - k_{ij}^{-1}\dot{w}_j = 0 \quad (13\text{-}63b)$$

where σ_{ij} is the stress tensor, $\bar{n} = V_v/(V_v + V_s)$ is the porosity, V_v is the volume of voids, V_s is the volume of solids, Fig. 13.8, ρ_s is the solid density, ρ_f is the fluid density, ρ is the density of the solid-fluid mixture, b_i is the tensor of body force per unit mass, u_i and u_{fi} are the displacement components in solid skeleton and fluid, respectively, Fig. 13.9, the overdots denote time derivative, k_{ij} is the permeability tensor, and w_i denotes the displacement of the fluid relative to that of solid skeleton, Fig. 13.9, given by

$$w_i = \frac{Q_i}{A_i} = \bar{n}(u_{fi} - u_i) \quad (13\text{-}64)$$

Implementation of DSC in Computer Procedures

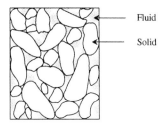

a). Element of Solid Skeleton

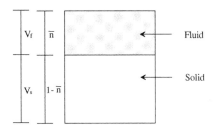

b). Porosity Diagram

FIGURE 13.8
Porous saturated material element and porosity (24).

where Q_i is the volume of fluid moving through an area of the skeleton normal to the ith direction, and A_i is the area normal to the ith direction ($i = 1, 2, 3$).

Equations (13.63) represent coupled differential equations with two variables u and w. Formulation based on these equations is general, and is referred to as $u - w$. If the relative acceleration, $\ddot{w}$, between the solid and fluid displacement is ignored, Eqs. (13.63) lead to the $u - p$ formulation, in which p is the (excess) fluid (pore water) pressure (22). For quasistatic problems, such as consolidation, $\ddot{u} = \ddot{w} = 0$.

13.4.3 Finite Element Formulation

The approximation functions for u and w are expressed as (Fig. 13.10) (24)

$$u_i = N_u^a U_i^a \tag{13.65a}$$

and

$$w_i = N_w^b W_i^b \tag{13.65b}$$

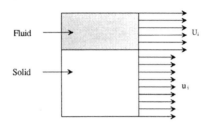

(a) Displacements in Two Phases

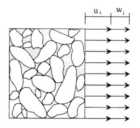

(b) Relative Displacements in Fluid

FIGURE 13.9
Deformations in two phase element.

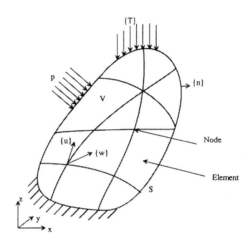

FIGURE 13.10
Finite element discretization and approximation function (24).

Implementation of DSC in Computer Procedures 553

where u_i and w_i ($i = 1, 2, 3$) are the solid displacements and relative displacements at a point, respectively; N_u and N_w are the interpolation functions corresponding to u and w, respectively; $a = 1, 2, \ldots N_{ue}$, $b = 1, 2, \ldots N_{we}$; N_{ue}, N_{we} = node numbers per element for u and w, respectively; and U_i and W_i denote the nodal solid and relative displacements, respectively.

Use of the virtual work principle and substitution of various quantities and derivatives, Eq. (13.65), lead to the following finite element equations (24):

$$\begin{bmatrix} M^{ac}_{uuij} & M^{ad}_{uwij} \\ M^{bc}_{wuij} & M^{bd}_{wwij} \end{bmatrix} \begin{Bmatrix} \ddot{U}^c_j \\ \ddot{W}^d_j \end{Bmatrix} + \begin{bmatrix} 0 & 0 \\ 0 & C^{bd}_{wwij} \end{bmatrix} \begin{Bmatrix} \dot{U}^c_j \\ \dot{W}^d_j \end{Bmatrix} + \begin{Bmatrix} \int_V N^a_{u,i}\sigma_{ij}\, dV \\ \int_V N^b_{w,i} p\, dV \end{Bmatrix} = \begin{Bmatrix} f^a_{ui} \\ f^b_{wi} \end{Bmatrix}$$

(13.66)

where $a, c = (1 \text{ to } NN)$; $b, d = (1 \text{ to } MM)$ with $NN = MM$ = number of nodes in the whole domain, and $i, j = (1, 2, 3)$, and

$$M^{ac}_{uuij} = \delta_{ij} \int_V \rho N^a_u N^c_u\, dV \tag{13.67a}$$

$$M^{ad}_{uwij} = \delta_{ij} \int_V \rho_f N^a_u N^d_w\, dV \tag{13.67b}$$

$$M^{bc}_{wuij} = \delta_{ij} \int_V \rho_f N^b_w N^c_u\, dV \tag{13.67c}$$

$$M^{bd}_{wwij} = \delta_{ij} \int_V \frac{\rho_f}{\bar{n}} N^b_w N^d_w\, dV \tag{13.67d}$$

$$C^{bd}_{wwij} = \int_V k_{ij}^{-1} N^b_w N^d_w\, dV \tag{13.67e}$$

$$f^a_{ui} = \int_S N^a_u T_i\, dS + \int_V \rho b_i N^a_u\, dV \tag{13.67f}$$

$$f^b_{wi} = \int_S p N^b_w n_i\, dS + \int_V \rho_f b_i N^b_w\, dV \tag{13.67g}$$

In Eqs. (13.66) and (13.67), the integration domain V is the whole volume and S the whole boundary, as sums of element volumes and surfaces. Equation (13.66) has c, d, and j as dummy indices (summation indices), and a, b, and i as free indices. Therefore, there are $(NN + MM) \times 3$ equations in Eq. (13.66). σ_{ij} and p in the third term on the left side in Eq. (13.66) are expressed in the incremental form as

$$d\sigma_{ij} = C^{ep}_{ijk\ell} d\varepsilon_{k\ell} + adp\delta_{ij} \tag{13.68a}$$

and

$$dp = M(a\, d\varepsilon_{kk} + d\zeta) \qquad (13.68b)$$

where ε_{ij} is the strain tensor for the solid skeleton, p is the fluid (pore water) pressure, $C_{ijk\ell}^{ep}$ is the elastoplastic constitutive tensor for the relative intact behavior (see later), $\zeta = w_{i,i}$ is the change in fluid volume in a unit volume of skeleton and

$$M = \frac{\bar{n}}{K_f} - \frac{a - \bar{n}}{K_s} \qquad (13.69)$$

K_f and K_s are the bulk moduli of fluid and solids, respectively, and a is given by (19, 21–24)

$$a = 1 - \frac{\delta_{ij} C_{ijk\ell}^{ep} \delta_{k\ell}}{9 K_s} \qquad (13.70a)$$

which for elastic porous materials reduces to

$$a = 1 - \frac{K}{K_s} \qquad (13.70b)$$

where K is the bulk modulus of the soil skeleton. For elastoplastic materials,

$$K = \frac{\delta_{ij} C_{ijk\ell}^{ep} \delta_{k\ell}}{9} \qquad (13.71)$$

13.4.4 Time Integration for Dynamic Problems

Equation (13.66) may be written in general as (24)

$$M_{ij} \ddot{x}_j + C_{ij} \dot{x}_j + K_{ij} x_j = f_i \qquad (13.72)$$

where $i, j = 1, 2 \ldots N_D$, N_D is the number of degrees-of-freedom in the system; M_{ij}, C_{ij}, and K_{ij} denote the components of mass, damping, and stiffness matrices of the system, respectively; f_i represents the force function or loads on the system; x_j, $\dot{x}_j$, and $\ddot{x}_j$ are displacements, velocities and accelerations at each degree-of-freedom, and include both U_j and W_j components. To solve the initial value problem is to find $x_j = x_j(t)$ satisfying Eq. (13.72), given initial conditions $x(0)$ and $\dot{x}_j(0)$, and force function $f_i(t)$. A number of algorithms are

Implementation of DSC in Computer Procedures

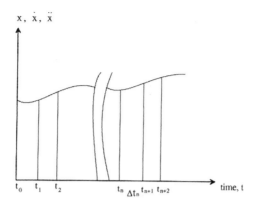

FIGURE 13.11
Dynamic time integration.

available for this purpose; the Newmark (β) method is used here and is described below.

The time domain is divided into time steps as shown in Fig. (13.11). Time integration is employed to find values x_j, $\dot{x}_j$ and $\ddot{x}_j$ at time t_{n+1} when their values at time t_n are known; here

$$t_{n+1} = t_n + \Delta t \tag{13.73}$$

where Δt is the time increment from t_n to t_{n+1}. By using the Taylor's series expansion, $x_i(t_{n+1})$ is expressed as

$$x_i(t_{n+1}) = x_i(t_n) + \Delta t \dot{x}_i(t_n) + \frac{\Delta t^2}{2} \ddot{x}_i(t_{n+\alpha_1}); \quad t_n \leq t_{n+\alpha_1} \leq t_{n+1} \tag{13.74}$$

In the Newmark's method (1, 3, 26), $\ddot{x}_i(t_{n+\alpha_1})$ is approximated as

$$\ddot{x}_i(t_{n+\alpha_1}) = (1 - 2\beta)\ddot{x}_i(t_n) + 2\beta\ddot{x}_i(t_{n+1}) \tag{13.75}$$

where β is a parameter, and $t_{n+\alpha_1}$ denotes time between t_n and t_{n+1}, Fig. (13.11). From Eqs. (13.74) and (13.75), $\ddot{x}_i(t_{n+1})$ can be expressed as

$$\ddot{x}_i(t_{n+1}) = \frac{1}{\beta \Delta t^2}\{x(t_{n+1}) - x(t_n) - \Delta t \dot{x}_i(t_n)\} - \frac{1 - 2\beta}{2\beta}\ddot{x}_i(t_n) \tag{13.76}$$

Similarly, $\dot{x}_i(t_{n+1})$ can be expressed as

$$\dot{x}_i(t_{n+1}) = \dot{x}_i(t_n) + \Delta t \ddot{x}_i(t_{n+\alpha_2}); \quad t_n \leq t_{n+\alpha_2} \leq t_{n+1} \tag{13.77}$$

and $\ddot{x}_i(t_{n+\alpha_2})$ in Eq. (13.77) is approximated as

$$\ddot{x}_i(t_{n+\alpha_2}) = (1-\gamma)\ddot{x}_i(t_n) + \gamma \ddot{x}_i(t_{n+1}) \tag{13.78}$$

where γ is another parameter, and $t_{n+\alpha_2}$ is the intermediate time, Fig. (13.11). From Eqs. (13.77) and (13.78), $\dot{x}_i(t_{n+1})$ can be expressed as

$$\dot{x}_i(t_{n+1}) = \dot{x}_i(t_n) + \Delta t\{(1-\gamma)\ddot{x}_i(t_n) + \gamma \ddot{x}_i(t_{n+1})\} \tag{13.79}$$

Substitution of Eqs. (13.76) and (13.79) into Eq. (13.72) yields

$$\overset{*}{K}_{ij} x_j(t_{n+1}) = f_i^* \tag{13.80}$$

where

$$\overset{*}{K}_{ij} = \frac{1}{\beta \Delta t^2} M_{ij} + \frac{\gamma}{\beta \Delta t} C_{ij} + K_{ij} \tag{13.81a}$$

$$f_i^* = f_i(t_{n+1}) + M_{ij}\left\{\frac{x_j(t_n)}{\beta \Delta t^2} + \frac{\dot{x}_j(t_n)}{\beta \Delta t} + \left(\frac{1}{2\beta} - 1\right)\ddot{x}_j(t_n)\right\} \tag{13.81b}$$

$$+ C_{ij}\left\{\frac{\gamma}{\beta \Delta t} x_j(t_n) + \left(\frac{\gamma}{\beta} - 1\right)\dot{x}_j(t_n) + \left(\frac{\gamma}{2\beta} - 1\right)\Delta t\, \ddot{x}_j(t_n)\right\} \tag{13.81c}$$

In the above equations, the mass matrix M_{ij} and damping matrix C_{ij} are constant. For linear problems, stiffness K_{ij} is constant and $x_i(t_{n+1})$ can be obtained by solving Eq. (13.80). For nonlinear problems, the stiffness matrix K_{ij} depends on $x_i(t_{n+1})$, and iterative techniques such as Newton-Raphson method have to be used to solve for $x_i(t_{n+1})$. After $x_i(t_{n+1})$ are found from Eq. (13.80), $\ddot{x}_i(t_{n+1})$ and $\dot{x}_i(t_{n+1})$ may be found from Eqs. (13.76) and (13.79).

The stability of the Newmark's scheme for linear systems has been investigated by many researchers and is expressed as (3, 26)

$2\beta \geq \gamma \geq 0.5$ ----- unconditionally stable
$\gamma \geq 0.5$ ----- conditionally stable with the restriction on time step given by

$$\omega \Delta t \leq \Omega_{crit} = \xi\left(\gamma - \frac{1}{2}\right) + \frac{\left[\frac{\gamma}{2} - \beta + \xi^2\left(\gamma - \frac{1}{2}\right)^2\right]^{\frac{1}{2}}}{\frac{\gamma}{2} - \beta} \tag{13.82}$$

where ω is the maximum natural frequency and ξ is the damping ratio of the system. This condition has to be satisfied for the maximum natural frequency of the system. The recommended values of α and β are 0.5 and 0.25, which are found to yield stable results for the dynamic problems.

13.5 Implementation

The dynamic Eqs. (13.66) can be written in matrix notation as (24, 25)

$$\begin{bmatrix} M^{ac} & M^{ad} \\ M^{bc} & M^{bd} \end{bmatrix} \begin{Bmatrix} \ddot{U}^c \\ \ddot{W}^d \end{Bmatrix} + \begin{bmatrix} 0 & 0 \\ 0 & C^{bd} \end{bmatrix} \begin{Bmatrix} \dot{U}^c \\ \dot{W}^d \end{Bmatrix}$$

$$+ \begin{Bmatrix} \int_V B_u^T \cdot \sigma^a \, dV \\ \int_V B_w^T \cdot pI \, dV \end{Bmatrix} = \begin{Bmatrix} Q_u^a \\ Q_w \end{Bmatrix} \quad (13.83)$$

by denoting $\ddot{x}^T = [\ddot{U}^c \; \ddot{W}^d]$, $\dot{x}^T = [\dot{U}^c \; \dot{W}^d]$ and $x^T = [U^c \; W^c]$, Eq. (13.83) can be written as

$$M\ddot{x} + C\dot{x} + Kx = Q \quad (13.84)$$

Time integration with Eq. (13.84) using the Newmark β-method then leads to

$$K^* x(t_{n+1}) = Q^*(t_{n+1}) \quad (13.85)$$

It usually requires iterative-incremental analysis to solve Eq. (13.85). Then, at iteration r, Eq. (13.85) can be written as

$$K^*(x^r(t_{n+1}) + dx^r(t_{n+1})) = Q^*(t_{n+1}) \quad (13.86)$$

where $dx^r(t_{n+1})$ is the increment of displacement during iteration, r, for increment of load (displacement), $n + 1$. Now, Eq. (13.85) is written as

$$K^* dx^r(t_{n+1}) = Q^*(t_{n+1}) - K^* x^r(t_{n+1}) \quad (13.87a)$$

where

$$K^* = \frac{1}{\beta \Delta t^2} M + \frac{\gamma}{\beta \Delta t} \cdot C^* + K \quad (13.87b)$$

and

$$Q^*(t_{n+1}) = Q(t_{n+1}) + \underset{\sim}{M}\left[\frac{\underset{\sim}{x}(t_n) - \underset{\sim}{x}^r(t_{n+1})}{\beta \Delta t^2} + \frac{\underset{\sim}{\dot{x}}(t_n)}{\beta \Delta t} + \left(\frac{1}{2\beta} - 1\right)\underset{\sim}{\ddot{x}}(t_n)\right]$$

$$+ \underset{\sim}{C}^*\left[\frac{\gamma}{\beta \Delta t} \cdot (x(t_n) - \underset{\sim}{x}^r(t_{nH})) + \left(\frac{\gamma}{\beta} - 1\right)\underset{\sim}{\dot{x}}(t_n) + \left(\frac{\gamma}{2\beta} - 1\right)\Delta t\, \underset{\sim}{\ddot{x}}(t_n)\right]$$

(13.87c)

In Eq. (13.87), the mass ($\underset{\sim}{M}$) and damping matrices ($\underset{\sim}{C}^*$) are assumed to remain constant, and for nonlinear material behavior, the stiffness matrix $\underset{\sim}{K}$ varies and depends on displacement and pore pressure $\underset{\sim}{x}$, through the constitutive matrix, $\underset{\sim}{C}^{DSC}$, Eq. (13.12), as

$$\int \underset{\sim}{B}^T d\underset{\sim}{\sigma}_n^a \, dV =$$

$$= \int \underset{\sim}{B}^T \underset{\sim}{C}^{DSC} dq^i dV \qquad (13.88)$$

$$= \underset{\sim}{K}^{DSC} dq^i$$

Here we assumed that the strains in the RI, observed and FA parts are compatible, and

$$\underset{\sim}{C}^{DSC} = (1 - D)\underset{\sim}{C}^i + D\underset{\sim}{C}^c + \underset{\sim}{R}^T(\underset{\sim}{\sigma}^c - \underset{\sim}{\sigma}^i) \qquad (13.89)$$

where $\underset{\sim}{C}^i$ is the RI matrix, $\underset{\sim}{C}^c$ is the FA matrix and dD, the increment or rate of D, expressed as

$$dD = \underset{\sim}{R}^T d\underset{\sim}{\varepsilon}^i \qquad (13.90)$$

The vector $\underset{\sim}{R}$ can be derived based on the given yield and disturbance functions. In the computer procedure, dD can also be evaluated as $dD = D(t_{i+1}) - D(t_n)$.

Thus, $\underset{\sim}{M}$ and $\underset{\sim}{C}^*$ are constant and the tangent stiffness matrix, $\underset{\sim}{K}^*$, involves variations during the nonlinear behavior because $\underset{\sim}{K}^{DSC}$ is evaluated as the tangent quantity.

13.6 Partially Saturated Systems

Computer (FE) procedures for partially saturated materials have been proposed by many investigators, based on constitutive models discussed in Chapter 9. In a general and coupled formulation, displacements, fluid pressures, and suction are considered as independent variables. (See References in Chapter 9, and 27–29.) Such coupled procedures can, however, be complex and involve significant computational effort. It is possible to develop

13.7 Cyclic and Repetitive Loading

Cyclic and repetitive loading, involving loading, unloading and reloading, occur in many problems such as dynamics and earthquakes, thermomechanical response such as in electronic packaging and semiconductor systems, and pavements. If the simulated behavior involves continuous increase in stress along the same loading path, without unloading and reloading, Fig. 13.12, it is often referred to as monotonic or *virgin* loading. The unloading and reloading are often referred to as *nonvirgin* loading. Loading in the opposite side, i.e., negative side of the (stress) response, is sometimes referred to as reverse (reloading) loading. Cyclic loading without stress reversal is often referred to as *one-way*, while with stress reversals, it is referred to as *two-way*. In the case of degradation or softening, decrease in stress beyond the peak occurs, but it is considered different from unloading. For the virgin loading, the constitutive equations, Eq. (13.12), apply. For nonvirgin loading, it is required to consider additional and separate, often approximate, simulations.

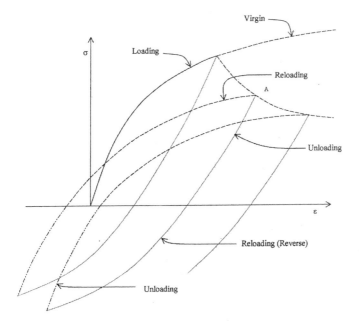

FIGURE 13.12
Schematic of loading, unloading, and reloading.

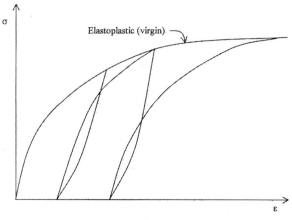

(a) Elastoplastic Response with Unloading and Reloading

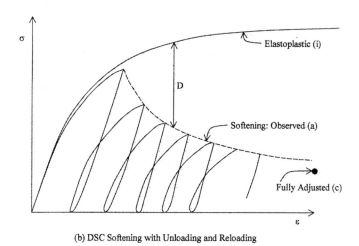

(b) DSC Softening with Unloading and Reloading

FIGURE 13.13
Schematic of elastoplastic and softening (DSC) response.

In the case of an elastoplastic model (e.g., HISS-δ_0), the simulated virgin response allows for the effect of plastic strains and plastic hardening or yielding, Fig. 13.13(a). In the case of the softening behavior, the plasticity model can simulate the RI behavior, and the use of DSC allows for the degradation, Fig. 13.13(b).

Plastic deformations can occur during unloading and reloading, and can influence the overall response, Fig. 13.13. Although models to allow for such behavior have been proposed in the context of kinematic harding plasticity (31, 32), they are often relatively complex and may involve computational difficulties. Hence, approximate schemes that are simple but can provide satisfactory simulation have often been used; one such method is described below.

13.7.1 Unloading

As indicated in Fig. 13.12, the unloading response is often nonlinear. However, as a simplification, it is often treated as linear elastic. Here, both linear and nonlinear elastic simulations are included. For the nonlinear case, of which the linear simulation is a special case, the following procedure is used (24, 25). During unloading, the incremental stress-strain equation is given by

$$d\underset{\sim}{\sigma} = \underset{\sim}{C}^{UL} d\underset{\sim}{\varepsilon} \qquad (13.91)$$

where $\underset{\sim}{C}^{UL}$ is the elastic constitutive matrix with variable elastic unloading modulus, E^u, Fig. 13.14, and the Poisson's ratio, v, is assumed to be constant. The modulus E^u is given by

$$\frac{1}{E^u} = \frac{1}{E^b} + \frac{1}{E^p} \qquad (13.92)$$

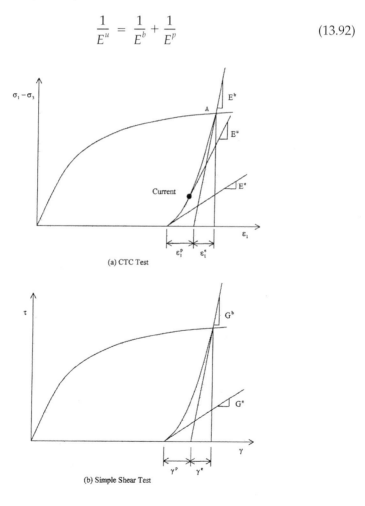

FIGURE 13.14
Unloading interpolation function for CTC and SS tests (25).

where E^b is the slope of the unloading curve (response) at the point (A) of unloading, Fig. 13.14(a), and E^p is the "plastic" modulus, which is evaluated by using the following equation:

$$E^p = p_a K_1 \left[\frac{p_a}{\sqrt{J_{2D}^b} - \sqrt{J_{2D}}} \right]^{K_2} \tag{13.93}$$

where K_1 and K_2 are constants, p_a is the atmospheric pressure (used for nondimensionalization), and J_{2D}^b and J_{2D} are the second invariants of the deviatoric stress tensor, S_{ij}, at the start of unloading (point A), and at the current state during unloading, respectively. The values of K_1 and K_2 are found from laboratory tests. For triaxial compression (CTC: $\sigma_1 > \sigma_2 = \sigma_3$) and simple shear (SS) tests (Chapter 7, Fig. 7.13), their values are derived as follows:

Triaxial Compression (CTC) Test

$$K_2 = \frac{\sqrt{3}(\sqrt{J_{2D}^b} - \sqrt{J_{2D}})}{\varepsilon_1^p} \left(\frac{1}{E^e} - \frac{1}{E^b} \right) - 1.0 \tag{13.94a}$$

$$K_1 = \frac{\sqrt{3}}{(K_2 + 1)\varepsilon_1^p} \cdot \left[\frac{\sqrt{J_{2D}^b} - \sqrt{J_{2D}}}{p_a} \right]^{K_2 + 1} \tag{13.94b}$$

where E^e is the elastic modulus (slope) at the end of unloading and ε_1^p is the "plastic" strain, Fig. 13.14(a).

Simple Shear (SS) Test, Fig. 13.14(b)

The relation between the elastic (Young's) and shear moduli (G) is given by

$$E^b = 2(1 + v)G^b \tag{13.95a}$$

$$E^e = 2(1 + v)G^e \tag{13.95b}$$

Substitution of Eq. (13.95) into Eq. (13.94) and replacing $\sqrt{J_{2D}}$ by τ (shear stress) and ε_1^p by $\sqrt{3}\gamma^p/2(1 + v)$, where γ^p is the "plastic" shear strain, Fig. 13.14(b), leads to

$$K_2 = \frac{\tau^b - \tau}{\gamma^p} \left(\frac{1}{G^e} - \frac{1}{G} \right) - 1.0 \tag{13.96a}$$

$$K_1 = \frac{1 + v}{\gamma^p (K_2 + 1)} \left[\frac{\tau^b - \tau}{p_a} \right]^{k_2 + 1} \tag{13.96b}$$

where τ^b and τ are the shear stresses at the point of unloading and during unloading, respectively.

The values of ε_1^p and γ^p are evaluated by using the following equations:

$$\varepsilon_1^p = \frac{\sqrt{3}\sqrt{J_{2D}^b}}{2}\left(\frac{1}{E^e} - \frac{1}{E^b}\right) \tag{13.97a}$$

and

$$\gamma^p = \frac{\tau^b}{2}\left(\frac{1}{G^e} - \frac{1}{G^b}\right) \tag{13.97b}$$

13.7.2 Reloading

Figure 13.15 shows two cases of reloading, for the one-way and two-way. In both cases, the following constitutive equation is used:

$$d\underline{\sigma}^a = R\underline{C}^{DSC}d\underline{\varepsilon} + (1-R)\underline{C}^e d\underline{\varepsilon} \tag{13.98}$$

where R is the interpolation parameter such that $0 \le R \le 1$; $R = 0$ for the beginning of reloading and $R = 1$ at the end of reloading. Thus, at the beginning of reloading, the behavior is elastic, given by

$$d\underline{\sigma}^a = \underline{C}^e d\underline{\varepsilon} \tag{13.99a}$$

At the end of reloading, virgin response resumes:

$$d\underline{\sigma}^a = \underline{C}^{DSC} d\underline{\varepsilon} \tag{13.99b}$$

The reloading elastic modulus, E^R, for the two cases, Fig. 13.15, is different. For case 1, the elastic modulus at the start of reloading, E^{br}, is given by

$$E^{br} = E^b \tag{13.100a}$$

where E^b is the unloading slope at the beginning of unloading, Fig. 13.15(a). For case 2,

$$E^{br} = E^e \tag{13.100b}$$

where E^e is the slope at the end of unloading, Fig. 13.15(b).

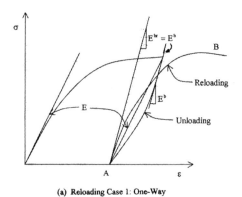

(a) Reloading Case 1: One-Way

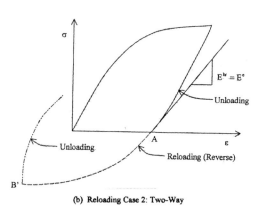

(b) Reloading Case 2: Two-Way

FIGURE 13.15
Two reloading cases (25).

The interpolation parameter, R, for both cases is found as

$$R = \frac{\sqrt{J_{2D}}}{\sqrt{J_{2D}^b}} \tag{13.101}$$

where J_{2D}^b and J_{2D} are the second invariants of the stress tensor at the beginning of the last unloading and current level, respectively.

In computer (finite element) analysis, the reloading stress path may be between the above two cases. Then, a parameter, S, is defined as an indicator of the direction of reloading:

$$S = \frac{(\underline{\sigma}^b - \underline{\sigma})^T d\underline{\sigma}}{\|(\underline{\sigma}^b - \underline{\sigma})^T\| \|d\underline{\sigma}\|} \tag{13.102}$$

where $-1 \leq S \leq 1$, $\underline{\sigma}^b$, $\underline{\sigma}$, and $d\underline{\sigma}$ are the stress vectors before unloading, for the current stress and for the next stress increment, respectively; $S = -1$ indicates case 1 reloading, while $S = 1$ indicates case 2 reloading. Now, E^{br} is interpolated between E^b and E as

$$\frac{1}{E^{br}} = \frac{1-S}{2E^b} + \frac{1+S}{2E^e} \qquad (13.103a)$$

Then, the modulus for reloading, E^R, is found as

$$\frac{1}{E^R} = \frac{1-R}{E^{br}} + \frac{R}{E} \qquad (13.103b)$$

where E is the elastic modulus of the material, which is often found as the (average) slope of the line joining the unloading and end of unloading points or the initial slope, Fig. 13.15(a). Then at the beginning of reloading when $R = 0$, $E^R = E^{br}$, which ensures smooth transition from unloading to reloading, Fig. 13.15(b). At the end of reloading ($R = 1$), $E^R = E$, which ensures smooth transition from reloading to the virgin loading.

13.7.3 Cyclic-Repetitive Hardening

In the case of elastoplastic behavior, there exists an *in situ* yield surface (F_0) corresponding to the initial or past state of stress experienced by the material before the present cyclic or repetitive load is applied, Fig. 13.16. When unloading occurs, the plastic strains can change (increase or decrease), and hence, for the reloading after the unloading, the yield surface that defines the elastic limit usually expands from F_0 to the initial surface, F_i, corresponding to each cycle $N (= 1, 2, ...)$. As a result, the magnitudes of plastic strains decrease from one cycle to the next, which is often referred to as *cyclic hardening*.

For a given load or stress (increment), the final or bounding surface, F_b, can be defined by solving for the incremental equations, Eq. (13.13). In the case of repetitive loading under constant amplitude of load (stress), Fig. 13.16(b), the maximum load (P_{max}) will be the amplitude of the load (stress). In the case of cyclic (one-way) loading, Fig. 13.16(c), the bounding surface, F_b, would change for each stress increase. Note that in this repetitive load analysis, the time effects are not included.

Mroz et al. (31) proposed a model for cyclic hardening, which was adopted by Bonaquist and Witczak (33) for materials in pavement structures. A similar and approximate (modified) method for cyclic hardening is presented below.

For the given load or stress increment, two bounding surfaces are defined, F_0 and F_b, Fig. 13.16, and the corresponding hardening functions and parameters are α_0 and α_b, Eq. (7.11), and ξ_0 and ξ_b, respectively. Here, ξ denotes the accumulated plastic strains or trajectory:

$$\xi = \int [(d\underline{\varepsilon}^p)^T \cdot (d\underline{\varepsilon}^p)]^{1/2} \qquad (13.104)$$

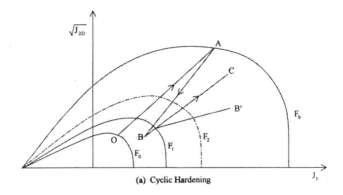

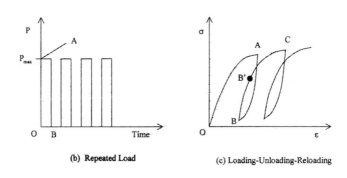

FIGURE 13.16
Cyclic hardening under repeated loading.

where $d\underline{\varepsilon}^p$ is the vector of incremental plastic strains. Then the initial yield surface parameter, ξ_i, for a given cycle, i, is expressed as

$$\xi_i = \xi_0 + \left(1 - \frac{1}{N^{h_c}}\right)(\xi_b - \xi_0) \qquad (13.105a)$$

where h_c is the cyclic hardening parameter, determined from laboratory repetitive load tests. It controls the rate of expansion of the initial yield surface, F_i, at the end of unloading for a given cycle, N. If $h_c = 0$, no cyclic hardening occurs. Bonaquist and Witczak (33) considered repeated tests involving the same stress (amplitude) to an initially unstrained material specimen with $\xi_0 = 0$. Then, Eq. (13.105a) becomes

$$\xi = \xi_b - \xi_i = \frac{1}{N^{h_c}}\xi_b \qquad (13.105b)$$

Implementation of DSC in Computer Procedures

or

$$\frac{\xi}{\xi_b} = \frac{1}{N^{h_c}}$$

where ξ is the plastic strain trajectory up to cycle N. Plots of normalized trajectory ξ/ξ_b vs. number of cycles are used to find h_c using a least square procedure. For a granular material, $h_c = 1.06$ was found (33).

With the above formulation, the value of ξ_i, Eq. (13.105) is used to evaluate the hardening function, α_i, Eq. (7.11). It is used to define the elastoplastic constitutive matrix $[C^{ep}] = [C^i]$, Eq. (7.51), and the general DSC matrix $[C^{DSC}]$, Eq. (13.12), when reloading occurs.

13.8 Initial Conditions

In the foregoing algorithmic procedures, it is necessary to establish the initial conditions, which usually involve the location of the initial state (I) on the yield surface, F, Fig. 13.17a, for the RI response, and the initial disturbance, D_0, Fig. 13.17b.

13.8.1 Initial Stress

The initial state of stress, $\underline{\sigma}^a$, is often known from the *in situ* or residual stresses. We assume that the observed and RI stresses are initially equal, i.e., $\sigma_0^i = \sigma_0^a$. The value of the initial hardening function, α_0, is found from Eq. (7.11) as (34)

$$\alpha_0 = \left[\gamma - \frac{J_{2D0}^i}{J_{10}^{i2}(1 - \beta S_{ro})^{-0.50}} \right] \left(\frac{J_{10}^i}{p_a} \right)^{2-n} \quad (13.106)$$

where the subscript 0 denotes initial quantity. Thus, the initial state I (σ_0^i, α_0) on the yield surface, Fig. 13.17(a), is defined. The value of the initial plastic strain trajectory, ξ_0, can be found as

$$\xi_0 = \left(\frac{a_1}{\alpha_0} \right)^{1/\eta_1} \quad (13.107a)$$

If the initial state of stress is isotropic or hydrostatic, the foregoing equations are simplified as

$$\alpha_0 = \gamma \left(\frac{J_{10}}{p_a} \right)^{2-n} \quad (13.108a)$$

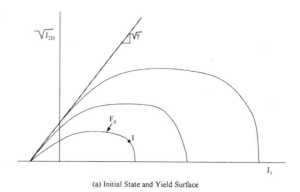

FIGURE 13.17
Initial state: hardening and disturbance.

and

$$\xi_0 = \xi_{v0} = (a_0/\alpha_0)^{1/\eta_1} \tag{13.108b}$$

and

$$\xi_D = 0 \tag{13.108c}$$

13.8.2 Initial Disturbance

It is usually difficult to define the initial disturbance, D_0, due to effects such as existing microcracking and anisotropy, and manufacturing defects. However, it can be possible to quantify it on the basis of nondestructive measurements and/or mechanical testing.

Nondestructive ultrasonic velocities and attenuation can be measured in different directions in a laboratory test specimen. Such measurements for

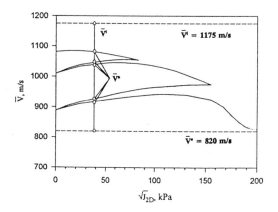

FIGURE 13.18
Measured ultrasonic P-wave average velocity: CTC 30 test ($\sigma_3 = -207$ kPa).

directional velocities (V_x, V_y, and V_z) were obtained; the average velocity ($\overline{V}$) for cubical specimens of a cemented sand are shown in Fig. 13.18 (13, 35). The velocities before the application of the load can be used to define D_0 as

$$D_0 = \frac{V_0^i - V_0^a}{V_0^i - V_0^c} \qquad (13.109)$$

where in Fig. 13.18, $V_0^i = 1175$ m/s, $V_0^c = 820$ m/s, and V_0^a is the average observed velocity.

Mechanical tests on specimens in the initial condition, involving application of small values of hydrostatic stress, $\sigma_1 = \sigma_2 = \sigma_3 = \sigma_0$, can provide values of the corresponding strains, ε_{10}, ε_{20}, and ε_{30}. By using elastic moduli (e.g., E and ν or K and G), the plastic strains can be found from

$$d\varepsilon_0^p = d\varepsilon_0 - d\varepsilon_0^e \qquad (13.110)$$

which in turn can be used to find the deviatoric plastic strains, dE_0^p, and the trajectory, ξ_0. Then the initial disturbance, D_0, is found from

$$D_0 = D_u\left(1 - A^{-A\xi_{D_0}^z}\right) \qquad (13.111)$$

The value of ξ_{D_0} can be found approximately as

$$\xi_{D_0} = \sqrt{\xi_0^2 - \xi_{v_0}^2} \qquad (13.112a)$$

where

$$\xi_{V_0} = \frac{\xi_0 \frac{\partial Q}{\partial \sigma}}{\sqrt{3}\left[\left(\frac{\partial Q}{\partial \sigma}\right)^T \cdot \frac{\partial Q}{\partial \sigma}\right]^{1/2}} \quad (13.112b)$$

The values of σ_0^i, α_0 and D_0 are then used to define the constitutive matrix, $\underline{C}_0^{DSC}$, for the first load increment in Eq. (13.12). Then various quantities during the incremental loading are found as

$$\underline{\sigma}_{n+1} = \underline{\sigma}_0 + \sum_{1}^{N} d\underline{\sigma}_{n+1} \quad (13.113a)$$

$$\alpha_{n+1} = \alpha_0 + \sum_{1}^{N} d\alpha_{n+1} \quad (13.113b)$$

$$\xi_{n+1} = \xi_0 + \sum_{1}^{N} d\xi_{n+1} \quad (13.113c)$$

$$D_{n+1} = D_0 + \sum_{1}^{N} dD_{n+1} \quad (13.113d)$$

where N denotes the number of load increments.

13.9 Hierarchical Capabilities and Options

The DSC allows integrated and hierarchical capabilities to adopt, depending upon specific need(s) and materials, models for various features such as elastic, plastic and creep responses, microcracking leading to fracture, degradation and softening and stiffening or healing. Thus, if the general constitutive matrix, $\underline{C}^{DSC}$, Eq. (13.12), is developed, one can choose various hierarchical options, depicted in Table 13.1.

The computer (FE) procedures with the DSC model can permit solutions of problems with options for materials that are characterized as elastic, plastic, and creep, with inclusion of degradation and stiffening. For instance, Table 13.2 shows twelve such options available in the codes (8, 10, 25). It may be noted that the user needs to provide input for only those parameters relevant for the specific hierarchical model chosen.

TABLE 13.1
Hierarchical Options in DSC

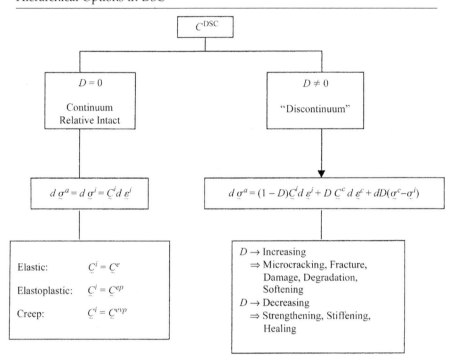

TABLE 13.2
Various Options in the DSC and Parameters

Model	Parameters				Thermal Effects (T)*
	Elastic	Plastic	Creep	Disturbance	
1. Elastic (E)	E, v				2. E(T)
3. Elastoplastic classical (ep/c) e.g., von Mises	E, v	σ_y			4. ep/c(T)
5. Elastoplastic hardening (ep/h)	E, v	$\gamma, \beta; n; a_1, \eta_1$			6. ep/h(T)
7. ep/h with disturbance	E, v	$\gamma, \beta; n; a_1, \eta_1$		A, Z, D_u	6. ep/h(T) with disturbance
9. Elastoviscoplastic**	E, v	$\gamma, \beta; n; a_1, \eta_1$	Γ, N		10. evp(T)
11. Elastoviscoplastic (evp) with disturbance**	E, v	$\gamma, \beta; n; a_1, \eta_1$	Γ, N	A, Z, D_u	12. evp(T) with disturbance

*Thermal effects are included by expressing parameters as in Eq. (13.54).
**Other creep overlay models (e.g., ve, vevp) (Chapter 8) are also available in the codes.
σ_y = yield stress, T = temperature.

13.10 Mesh Adaption Using DSC

Adaption of the finite element mesh during loading, and depending on the physical state of the deforming material, has been one of the active areas of recent investigations. For the case of linear or nonlinear elastic and plastic hardening materials when the problem is self-adjoint elliptic and the system matrices are positive definite, adaptive strategies such as those based on energy norm are well established. However, for problems involving microcracking, localization, and softening or degradation, there does not appear to exist a global error norm similar to the energy norm for the self-adjoint elliptic problem (36). Hence, it becomes necessary to consider and develop criteria for adaptive strategies based on other scalar variables (36–38). Zienkiewicz et al. (36) suggested definition of error (e) in terms of a scalar variable, ϕ, as

$$e = \phi - \phi^h \tag{13.114}$$

where ϕ is the exact solution and ϕ^h is the finite element solution; this definition may, however, be difficult to implement, as the value of ϕ is not known. However, we can now use the scalar disturbance, D, in the DSC, which is computed and available as a part of incremental nonlinear analysis, to develop criteria for mesh adaptivity. In the following, we describe an adaptive mesh procedure in which the traditional approach is used in the prepeak region, while a disturbance-based procedure is used for the postpeak region in which localization occurs and grows (39).

13.10.1 Prepeak Response

The uniform degree of error distribution (UDED) scheme proposed in (40) is used. Here, the error, $e(i)$, is expressed as

$$e(i) = \text{abs}\left(\frac{\underline{\sigma}^* - \underline{\sigma}}{\underline{\sigma}}\right) \tag{13.115}$$

where $\underline{\sigma}$ is the vector of computed stresses (from the displacement FE formulation), $\underline{\sigma}^*$ is the vector of "exact" stresses, and i denotes an element. The exact stress, $\underline{\sigma}^*$, is computed by using a hybrid FE procedure; details of the procedure are given in (39, 41).

The strategy for generating a nearly optional mesh is based on the following criterion:

$$e \leq \bar{e} \tag{13.116}$$

where e is the computed error, Eq. (13.115), and $\bar{e}$ is the permissible error, say, 5%. The approximation error, e_m, in each element is given by

$$e_m \approx \bar{e}\|h\|m^{1/2} \tag{13.117}$$

where m is the number of elements in the mesh, and $\|h\| = h_{max}$ is the norm or maximum size of the element. The local mesh enrichment indicator, ζ, is given by

$$\zeta = \frac{\|E\|_i}{e_m} \tag{13.118}$$

where $\|E\|_i$ is the energy norm for element i:

$$\|E\|_i = \int_{V_1} \underline{e}(i)^T \underline{C}^{-1} \underline{e}(i) dV \tag{13.119}$$

The element size (h) for the optimal mesh is evaluated as (40)

$$h = \frac{h_i}{\zeta^{1/p}} \tag{13.120a}$$

where h_i is the original maximum size of elements, and p is the order of interpolation functions. For elements near singularities of order, k, the new element size is estimated from

$$h = \frac{h_i}{\zeta^{1/k}} \tag{13.120b}$$

where k represents the strength of the singularly, which usually is adopted to be equal to 0.50.

13.10.2 Postpeak Response

As stated earlier, the foregoing UDED scheme alone may not apply or is not sufficient for the postpeak region with localization. Here, the disturbance, D, Eq. (13.22) is used as the "error" indicator for mesh adaption. For example, critical values of disturbance, $D^* = 0.5, 0.75$, or 0.90 can be specified as the indicator; that is, during the incremental analysis when the critical disturbance, D^*, is reached and exceeded, mesh refinement is initiated. As the disturbance is expressed in terms of plastic strain trajectory or dissipated energy, it represents the intensity of (strain) localization. Then, use of disturbance as the remeshing indicator is consistent with the physical state of the structure undergoing microcracking and localization.

The error indicator in the postpeak region is given by

$$D \geq D^* \quad (13.121)$$

The error ratio for stress is given by

$$\zeta = \frac{\|e_0\|}{\|e_i\|} \quad (13.122)$$

where e_0 is the lowest error and e_i is the error in any element. The mesh is refined by using the following strategy:

$$h_i = \zeta^t h_0 = \left(\frac{\|e_0\|}{\|e_i\|}\right)^t h_0 \quad (13.123)$$

where h_0 is the size of the element with the lowest error, and t is the exponent; usually, $t = 2.0$ provides satisfactory results. Examples of adaptive meshing using the above procedure are given in (39).

13.11 Examples of Applications

Typical problems involving computer analysis of factors such as localization, spurious mesh dependence, and stability are included in Chapter 12. Now, we describe solutions of a wide range of problems (civil, mechanical, electronic packaging, pavements) obtained by using two- and three-dimensional computer procedures with the DSC and its hierarchical versions (10, 25, 42). The computer predictions are compared with analytical, and from other numerical procedures, and/or laboratory and field behavior of practical engineering problems.

Example 13.1: Material Block: DSC Model, Effect of Load Increments

Figure 13.19(a) shows a one-element material (concrete) block, idealized as axisymmetric, and subjected to the total vertical load of 80 MPa. The load is applied in different incremental steps equal to 20, 50, and 100 (43). The material properties are given below (6, 44).

Elastic: $E = 3.7 \times 10^4$ MPa, $v = 0.25$;
Plasticity (HISS δ_0-Model): $\gamma = 0.06784$, $\beta = 0.75526$, $3R = 47.5$ MPa, $n = 5.23697$, $a_1 = 0.46 \times 10^{-10}$, $\eta_1 = 0.862$;
Disturbance: $A = 668$, $Z = 1.50277$, $D_u = 0.875$

Implementation of DSC in Computer Procedures

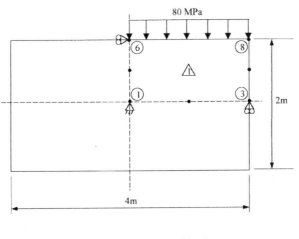

(a) One Element Mesh and Load

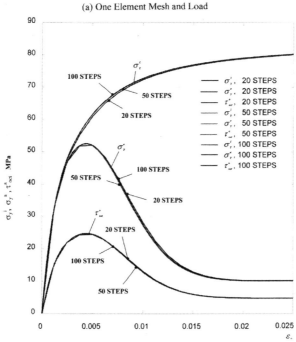

(b) Comparison of Stress-Strain Responses for Different Load Steps

FIGURE 13.19
Effect of load steps or increments.

Figure 13.19(b) shows computed vertical stress (σ_y^i = relative intact, and σ_y^a = observed or actual), and octahedral shear stress (τ_{oct}^a) versus the axial strain (ε_y) responses for the analyses with three different load steps. The simplified DSC scheme 3 was used for these calculations. It can be seen that the computed results are essentially the same from the analyses with three sizes of load steps, indicating the validity and robustness of the simplified solution scheme 3.

Example 13.2: Load (Footing) on Half Space (Soil): Elastoplastic (HISS-δ_1) Model

Computer algorithms for the finite element method and routines for the implementation of the HISS-δ_0 and δ_1 models have been presented in (5, 34). The FE procedure was used to predict the laboratory-measured stress and deformation behavior of a circular footing on cohesionless material (Leighton Buzzard sand).

The material parameters for the sand were obtained from a comprehensive series of laboratory multiaxial and triaxial tests on cubical and cylindrical specimens under different initial confining pressures ($\sigma_0 = \sigma_3$) and densities (5, 34). The elastic modulus, E, was considered to vary with depth (or σ_3) as (34)

$$E = K p_a \left(\frac{\sigma_3}{p_a}\right)^{\bar{n}} \tag{13.124}$$

where K is the modulus, $\bar{n}$ is the modulus exponent and p_a is the atmospheric pressure. The material parameters for the sand with relative density $D_r = 95\%$ were found by using procedures described in Chapters 4 to 7; they are given below.*

Footing (steel)
Elastic: $E = 30 \times 10^6$ psi (207 GPa), $\nu = 0.30$
Soil:
Elastic: $K = 1256$, $\bar{n} = 0.70$, $\nu = 0.29$;
Plasticity (HISS δ_1-Model): $\gamma = 0.10212$, $\beta = 0.36242$, $n = 2.5$;
                                                         $a_1 = 0.134575$, $\eta_1 = 450.0$, $a_2 = 0.0047$,
                                                         $\eta_2 = 1.02$;
Nonassociative: $\kappa = 0.29$
Atmospheric Pressure: $p_a = 14.7$ psi (101.35 kPa)
In situ Stress Ratio: $K_o = 0.5$
Unit Weight of Sand: $\bar{\gamma} = 0.10$ lb/in^3 (27.14 KN/m^3)

Here, the hardening (α) function is given by

$$\alpha = a_1 e^{-\eta_1 \xi} \left(1 - \frac{\xi_D}{a_2 + \eta_2 \xi_D}\right) \tag{13.125}$$

The finite element mesh for the soil and circular footing (steel) is shown in Fig. 13.20. It consists of 207 nodes and 58 eight-noded elements; 2 × 2 Gauss integration was used. The total footing load of 25.0 psi (172 kPa) was applied in steps of 2.50 psi (17 kPa), as it was done in the laboratory footing test (5).

---
* ©John Wiley & Sons Ltd. Reproduced with permission.

Implementation of DSC in Computer Procedures 577

Figures 13.21 and 13.22 show comparisons between predicted and measured displacements at the center node (No. 15), and at a distance of 12.0 inches (30.0 cm) from the center of the footing (node 138), respectively. Figure 13.23 shows computed and measured vertical stress (σ_y) in the element under the footing. The predictions and measurements show satisfactory correlations.

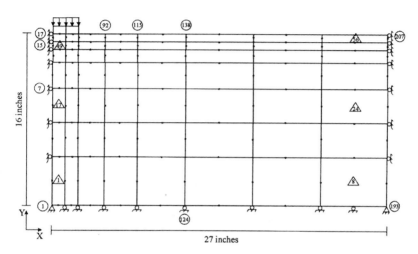

FIGURE 13.20
Finite elements mesh of soil-footing (1 in = 2.54 cm) (5, 34) (with permission from Elsevier Science and ©John Wiley & Sons Ltd; reproduced with permission).

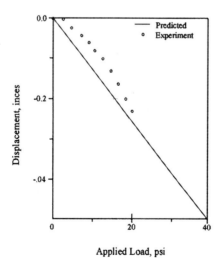

1 psi = 6.89 kPa, 1 in = 2.54 cm

FIGURE 13.21
Load displacement response at center of footing (Node 15) (34). ©John Wiley & Sons Ltd. Reproduced with permission.

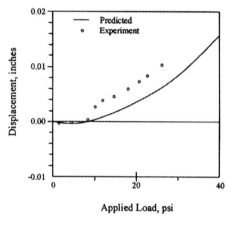

FIGURE 13.22
Load displacement response at 12 inches from center of footing (Node 138) (34). ©John Wiley & Sons Ltd. Reproduced with permission.

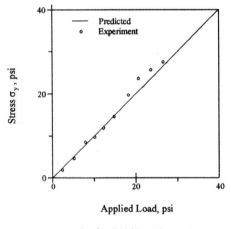

FIGURE 13.23
Normal stress, σ_y under the footing (34). ©John Wiley & Sons Ltd. Reproduced with permission.

Example 13.3: Layered (Pavement) System: Elasticity and Elastoplastic (HISS-δ_0) Models

Permanent deformation leading to rutting is one of the important distresses in the analysis and design of highway and airport pavements (45, 46). The finite element procedure with various characterizations involving elastic and elastoplastic (HISS-δ_0) models for different layers was used to analyze the development of deformations and stresses in a four-layered pavement system; the finite element mesh, idealized as axisymmetric, is shown in Fig. 13.24.

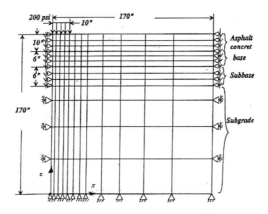

FIGURE 13.24
FE mesh for layered system: 1 in = 2.54 cm, 1 psi = 6.89 kPa.

Laboratory triaxial test data for asphalt (47) and for unbound materials (base, subbase, and subgrade) (33) were used to determine the elasticity and plasticity parameters for the δ_0-model; they are shown below.

Material Parameters for Asphalt Pavement System

Parameters	Asphalt concrete[1]	Base[2]	Subbase[2]	Subgrade[2]
E	500000.0 psi	56532.85	17098.56	10013.17
v	0.30	0.33	0.24	0.24
γ	0.1294	0.0633	0.0383	0.0296
β	0.00	0.70	0.70	0.70
n	2.40	5.24	4.63	5.26
α_1	1.23×10^{-6}	2×10^{-8}	3.6×10^{-6}	1.2×10^{-6}
η_1	1.944	1.231	0.532	0.778

[1]Based on tests by Witczak (47).
[2]Based on tests by Bonaquist (33).

The maximum load of 200 psi (1380 kPa) was applied in increments so as to simulate the wheel load over a radius of 10.0 inch (25.0 cm). A parametric study was performed to evaluate the influence of characterizing material behavior as nonlinear elastoplastic vs. linear elastic. Accordingly, the following combinations were considered (10, 43, 45, 46):

EEEE   All four layers linear elastic,
HEEE   Pavement nonlinear (HISS-δ_0), other layers linear elastic,
EHHH   Pavement linear elastic, other layers nonlinear (HISS-δ_0),
HHHH   All layers nonlinear (HISS-δ_0).

Figures 13.25(a) and (b) show vertical, v, and horizontal (radial), u, displacements of the top surface for the final load (= 200 psi) for the four combinations. The results show satisfactory trends. They also indicate the influence of nonlinear (elastoplastic) material response on the predictions. For example, the

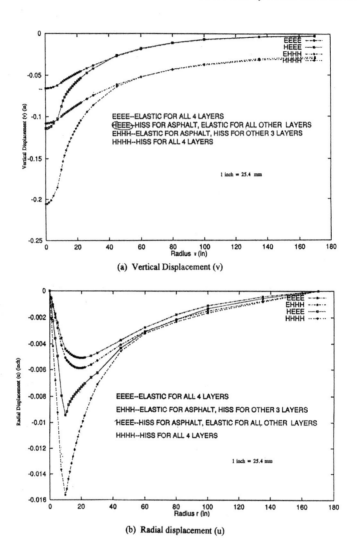

FIGURE 13.25
Displacements at top surface (43).

maximum displacements from the plasticity (HHHH) case are about three times those from the elasticity (EEEE) case. It is believed that the elastoplastic (HISS) model can provide for realistic evaluation of permanent displacements (rutting) in pavement systems.

Example 13.4: Reinforced Earth: DSC and Elastoplastic (HISS-d1) Model

Various types of reinforcements such as fibers, nails and geogrids are used to strengthen compacted soils for different construction systems. They represent composite systems discussed in Chapter 10. The DSC model with the

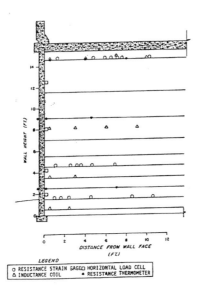

FIGURE 13.26
Tensar reinforced retaining wall no. 26-30 (48).

HISS (δ_0/δ_1) plasticity models were used to characterize the behavior of soil backfill and interfaces between soil and Tensar (geogrid) reinforcement for an earth reinforced retaining wall constructed in a highway project in Tucson, Arizona, USA; a typical wall panel with various measuring instruments is shown in Fig. 13.26 (48).

The instrumented wall panels had a height of 4.72 m and the wall facing consisted of 15.24 cm (6.0 in) thick by 3.05 m (10.0 ft) wide precast concrete panels. The geogrids were laid on compacted backfill to various depths and were connected to the concrete facing panels at different elevations and extended to a length of 3.66 m (12 ft). A pavement was installed on the top of the backfill and consisted of 10.16 cm (4.0 in) base course covered by 24.13 cm (9.50 in) Portland cement concrete.

Measurements were made over a period of about five years and involved strains in the geogrid (resistance strain gages), displacement in soil backfill (inductance coils), forces near the wall (horizontal load cells) and temperature on the wall (resistance thermometers).

13.11.1 Laboratory Tests

Triaxial (cylindrical: 7.0 × 14.0 cm) and multiaxial (cubical: 10 × 10 × 10 cm) specimens of the backfill soil were tested under HC and CTC stress paths (Chapter 7, Fig. 7.13) with different confining pressures σ_3 = 7.5, 17.5, 35.0, 52.0, 70.0, 140.0, 210.0, 345.0, and 420.0 kPa (49, 50). The index properties of the backfill soil are: Specific gravity, G_s = 2.64; D_{30} = 1.00 mm, uniformity coefficient, C_u = 3.64, maximum and minimum void ratios, e_{max} = 0.71 and

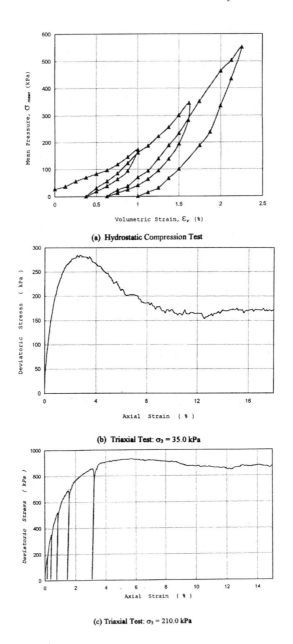

FIGURE 13.27
Typical stress-strain responses for backfill soil (49, 50).

$e_{min} = 0.37$. The initial density of soil was about 17.3 kN/m^3. Figures 13.27 show typical test data for dry backfill under HC and CTC tests.

Interface shear tests between backfill and Tensar geogrid (16.00 cm diam) were conducted by using the CYMDOF device (Chapter 11) under different

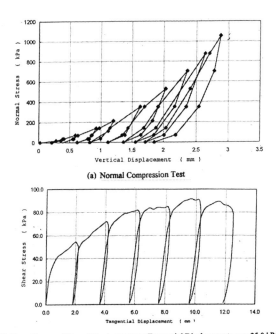

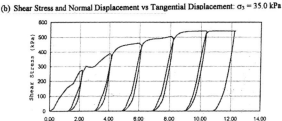

FIGURE 13.28
Typical stress-strain responses for dry soil — Tensar geogrid interface.

normal stresses, σ_n = 17.5, 35.0, 70.0, 140.0, 210.0, 350.0, 525.0, 700.0, 875.0, and 1050.0 kPa. Figures 13.28 show typical results for the normal loading, and shear loading (σ_n = 35.0 and 875.0 kPa) under displacement controlled condition.

The test data was used to evaluate parameters for the DSC model for the soil and interface by using procedures described in Chapters 7 and 11. The parameters are shown below (49, 50).

Soil

Elastic:	$61600 \times \sigma_3^{0.28}$, and $\nu = 0.3$;
Plasticity:	$\gamma = 0.12, \beta = 0.45, n = 2.56$;
Hardening:	$a_1 = 3.0 \times 10^{-5}, \eta_1 = 0.98, \kappa = 0.20$;

Disturbance: $D_u = 0.93$, $A = 0.368$, $Z = 1.60$;
Unit Weight = 18.84 KN/m³, co-efficient of earth pressure at rest, $K_0 = 0.40$.

Interface

Elastic: $K_n = 29100 \times \sigma_n^{0.28}$, $K_s = 18350 \times \sigma_n^{0.29}$;
Plasticity: $\gamma = 2.3$, $\beta = 0.0$, $n = 2.8$;
Hardening: $a_1 = 0.03$, $\eta_1 = 1.0$, $\kappa = 0.40$

As shown above, the elastic parameters E for soil, and normal and shear stiffness for interface, were found to be dependent on the confining pressure, σ_3, and normal stress, σ_n, respectively. The DSC model was used for the soil, while the HISS (δ_1) nonassociative model was used for interfaces.

The material parameters for the soil above were obtained from the triaxial tests with cylindrical specimens. The volume change behavior was not available from these tests; however, such behavior was available from the multiaxial tests for the same soil. Hence, the multiaxial tests were back predicted as *independent* tests. Figure 13.29 shows a typical comparisons between prediction and test data in terms of τ_{oct} and volumetric strain vs. axial strain for test with $\sigma_3 = 7.5$ psi (52.0 kPa).

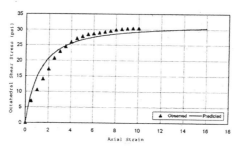

(a) τ_{oct} vs Axial Strain

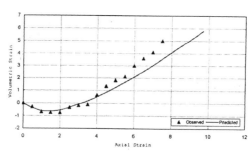

(b) ε_v vs Axial Strain

FIGURE 13.29
Comparisons between predictions and multiaxial test data: $\sigma_3 = 52.0$ kPa (49, 50).

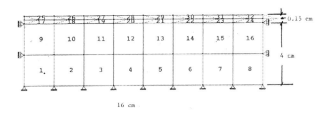

FIGURE 13.30
FEM mesh for simulation of interface test.

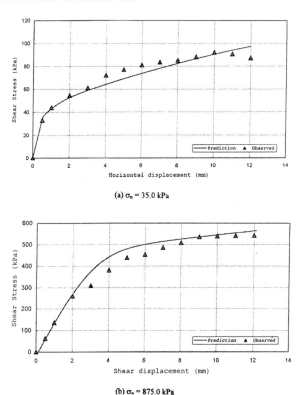

FIGURE 13.31
Comparisons between FE predictions and interface test data.

The back predictions for the interface test were obtained by using the finite element code, DSC-SST2D (10). Figure 13.30 shows the finite element mesh for the test specimen: soil with 16.0 cm diam. and 4.00 cm height, and interface with 16.0 cm diam. and 0.15 height. The bottom nodes for soil were restrained in the horizontal and vertical directions, while the side boundaries were restrained in the horizontal direction. The nodes of the interface zone were free to move. Typical comparisons between back predictions and test data for $\sigma_n = 35.0$ and 875.0 kPa are shown in Fig. 13.31.

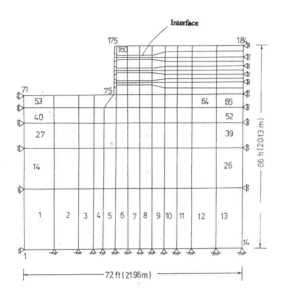

(a) Case 1: Coarse Mesh

FIGURE 13.32
Coarse and finer FE meshes for field Tensar reinforced wall (49, 50).

13.11.2 Field Validation

The DSC-SST2D code (10) was used to predict the behavior of the Tensar-reinforced field wall, Fig. 13.26. The analyses were performed for two cases: Case 1, Fig. 13.32(a), involved a rather coarse mesh with three layers of Tensar reinforcement with interface elements, while for Case 2, a finer mesh with ten layers of reinforcement was used; here, bar elements were used for the reinforcement, and interface elements were provided on the top and bottom of the reinforcement, Fig. 13.32(b). In addition, interface elements were also provided between the soil and wall for both cases.

The material parameters for the soil and interface were given previously. Parameters for the concrete wall facing and Tensar reinforcement, assumed to be linear elastic, are given below:

Wall Facing: $\qquad E = 2.1 \times 10^7$ kPa, $\nu = 0.15$

Reinforcement: $\qquad E = 1.5 \times 10^6$ kPa.

The analysis involved simulation of excavation and embankment sequences in which the Tensar reinforcement was placed sequentially on compacted backfill layers. The traffic load was simulated by applying a load = 20 kPa on the top surface, after the end of construction. Typical results are given below.

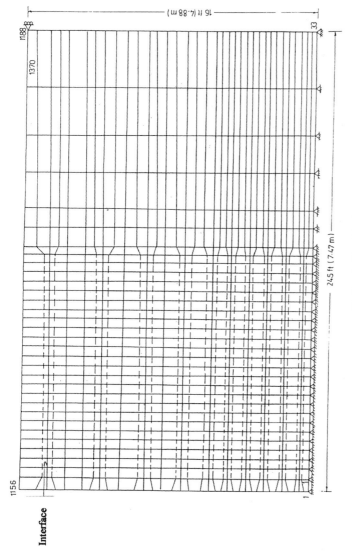

(b) Case 2: Finer Mesh

FIGURE 13.32
(continued).

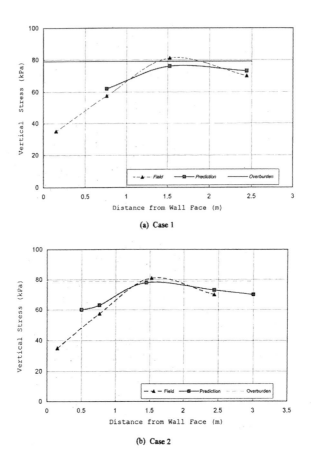

FIGURE 13.33
Comparisons between predicted and observed vertical soil stress at elevation 1.53 m after opening to traffic.

Figure 13.33 shows comparisons between back predictions and field data for vertical soil stress at the elevation 1.53 m (Fig. 13.26) after opening to traffic. These figures also show (the maximum) overburden pressure, $\sigma_v = \gamma h$, where γ is the density and h is the height. It can be seen that both Case 1 and 2 results show good agreement with the field data. The vertical stress near the wall is lower, partly due to the soil-structure interaction and relative motions at the interface. At a distance of about 1.52 m from the wall, the stress reaches maximum value equal to the overburden stress.

Figure 13.34 shows comparisons between the predictions from Case 2 for the horizontal soil stress in the elements near the inside wall face. It shows good correlation, while the horizontal stress distribution using the classical Rankine theory exhibits much higher values compared to the observed

Implementation of DSC in Computer Procedures

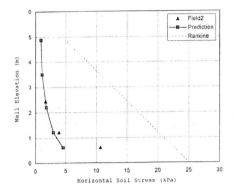

FIGURE 13.34
Comparisons between predicted and observed horizontal soil stress after opening to traffic: Case 2.

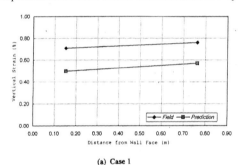

(a) Case 1

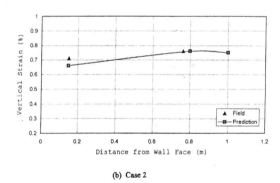

(b) Case 2

FIGURE 13.35
Comparisons between predicted and observed vertical soil strains at elevation 1.08 m after opening to traffic.

results. In other words, the provision of reinforcement reduces the horizontal stresses (pressure) significantly compared to those from the Rankine theory.

Figures 13.35 (a) and (b) show comparisons between vertical soil strains at elevation 1.08 after opening to traffic. Figure 13.36 shows (axial) strains in

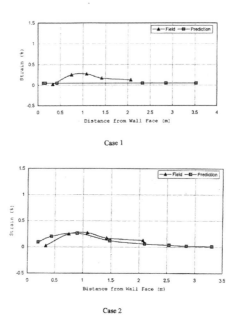

FIGURE 13.36
Comparisons between predicted and observed geogrid strains at elevation 1.37 m.

the Tensar geogrid at elevation of 1.37 m. It can be seen that the results from Case 2 with finer mesh, Fig. 13.32(b), show improved agreement with the observations.

This example illustrates that the DSC/HISS model can provide very good simulation of the behavior of reinforced (composite) systems, including nonlinear, interface (relative motions) and interaction effects.

Example 13.5: Borehole Stability: DSC-Elastoplastic (HISS-δ0) Model

Stability of boreholes (drilled) in soils and rocks constitutes an important problem that requires consideration of various factors such as initial state of stress, anisotropic character of materials, nonlinear response of geologic materials and interface/joints, coupled (displacement and fluid pressures) effects in saturated or partially saturated materials, and multidimensional geometrics.

One of the specific situations occurs in petroleum engineering when boreholes are drilled to great depths in jointed, anisotropic and saturated rock masses (51-53). Very often, a borehole is kept open during drilling by bentonite slurry under pressure. Under increased and high borehole pressure, the rock mass can experience microcracking leading to hydraulic fracturing. If for some reason the borehole pressure drops, cracking and failure at the borehole wall can lead to caving-in or breakout of the borehole. These phenomena are influenced by elastic, plastic and creep responses, and microcracking and fracture due to strain softening or degradation response of the anisotropic and jointed rock mass.

The unified DSC model is considered to be ideally suitable for the characterization of the mechanical behavior of rocks and joints. It has been implemented in a three-dimensional finite element procedure (42) for coupled response of saturated media based on Biot's generalized theory (18,19).

It is difficult to simulate the entire borehole problem involving large depth and surrounding region by using a computer (finite element) procedure. Very often, laboratory model tests are performed on finite-sized rock masses subjected to simulated fluid pressure and loading conditions.

The finite element procedure (42) with the DSC model can be used to predict the behavior of boreholes simulated in the laboratory and field. Typical three-dimensional finite element mesh for a borehole is shown in Fig. 13.37.

The DSC model parameters are obtained from laboratory triaxial tests on cylindrical specimens of the rock under different confining pressures (53). The rock was considered to be initially radially anisotropic or transversely

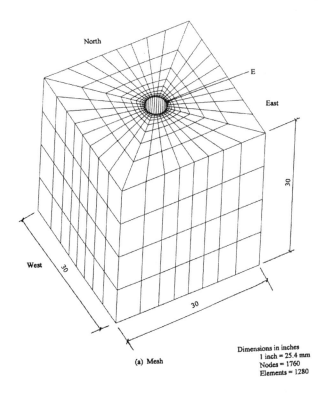

FIGURE 13.37
Finite element mesh for bore hole.

isotropic; the elastic stress-strain relation is given by (53, 54)

$$\begin{Bmatrix} \varepsilon_1 \\ \varepsilon_2 \\ \varepsilon_3 \\ \gamma_{12} \\ \gamma_{23} \\ \gamma_{31} \end{Bmatrix} = \begin{pmatrix} \frac{1}{E_1} & -\frac{\nu_{21}}{E_2} & -\frac{\nu_{21}}{E_2} & 0 & 0 & 0 \\ -\frac{\nu_{12}}{E_1} & \frac{1}{E_2} & -\frac{\nu_{23}}{E_2} & 0 & 0 & 0 \\ -\frac{\nu_{12}}{E_1} & -\frac{\nu_{23}}{E_2} & \frac{1}{E_2} & 0 & 0 & 0 \\ 0 & 0 & 0 & \frac{1}{G_{12}} & 0 & 0 \\ 0 & 0 & 0 & 0 & \frac{1}{G_{23}} & 0 \\ 0 & 0 & 0 & 0 & 0 & \frac{1}{G_{12}} \end{pmatrix} \times \begin{Bmatrix} \sigma_1 \\ \sigma_2 \\ \sigma_3 \\ \tau_{12} \\ \tau_{23} \\ \tau_{31} \end{Bmatrix}$$

(13.126)

where E_1 and E_2 are elastic moduli, G_{12} is the shear modulus, and ν_{12} and ν_{23} are the Poisson's ratios. These elastic parameters were expressed as functions of the mean pressure, $p = \sigma_3 = J_1/3$, where $\sigma_3 =$ confining stress and $J_1 = \sigma_1 + \sigma_2 + \sigma_3$:

$$E_1 = E_{10} + \frac{(E_{1S} - E_{10})k_{E1}J_1}{k_{E1}J_1 + E_{1S} - E_{10}}$$

$$E_2 = E_{20} + \frac{(E_{2S} - E_{20})k_{E2}J_1}{k_{E2}J_1 + E_{2S} - E_{20}} \qquad E_{10} \text{ at } J_1 = 0$$

(13.127)

$$\nu_1 = \nu_{10} + \frac{(\nu_{1S} - \nu_{10})k_{\nu 1}J_1}{k_{\nu 1}J_1 + \nu_{1S} - \nu_{10}} \qquad E_{1s} \text{ at } J_1 = \infty$$

$$\nu_2 = \nu_{20} + \frac{(\nu_{2S} - \nu_{20})k_{\nu 2}J_1}{k_{\nu 2}J_1 + \nu_{2S} - \nu_{20}}$$

The parameters for the HISS-δ_0 plasticity model used are given below.

$$\gamma = 5.80, \ \beta = 0.60, \ n = 4.84,$$
$$R = 67.0 \text{ psi (460 kPa)},$$
$$a_1 = 1.14 \times 10^{-17}, \ \eta_1 = 2.56.$$

In order to allow for failure condition at low mean pressures, the yield function in the HISS-δ_0 model (Chapter 7) was modified as

$$F = J_{2D} - (-\alpha J_1^n + \gamma J_1^q)(1 - \beta S_r)^{-0.5} = 0 \qquad (13.128)$$

where the exponent, q, allows for the nonlinearity of the ultimate yield function (55); the value of (q) was found to be 1.4. Figure 13.38(a) shows the curved ultimate envelopes, Eq. (13.128) for tests with loading in the parallel and normal directions. Figure 13.38(b) shows the plot of D vs. ξ_D.

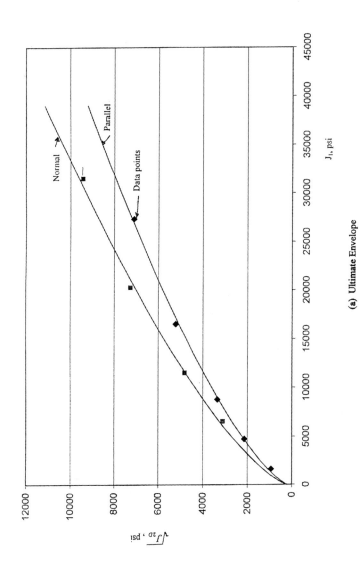

FIGURE 13.38
Nonlinear ultimate envelope and disturbance.

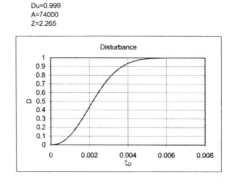

(b) Disturbance vs ξ_D

FIGURE 13.38
(continued).

In the DSC model, the response of the fully adjusted part was characterized by using the following equation:

$$\sqrt{J_{2D}^c} = m J_1^c \qquad (13.129)$$

where c denotes FA (critical) state. The parameter m was found to be 0.28.

The disturbance parameters, Eq. (3.16), used are as follows:

$$A = 74 \times 10^4$$
$$Z = 2.265$$
$$D_u = 0.999.$$

Figure 13.38b shows a plot of disturbance vs. ξ_D.

13.11.3 Computer Results

The 3-D finite element mesh is shown in Fig. 13.37. Figure 13.39 shows predicted E-W and N-S displacements with the pressure applied on the borehole. These results show consistent trends as observed in laboratory tests. The predictions of the pressure at the breakout, when large displacements occur, are considered to be realistic.

The growth of disturbance in element E, Fig. 13.37, in the E-W direction is also plotted in Fig. 13.39. It shows that the critical disturbance, $D_c \approx 0.80$, is reached between pressures of 3000 to 3500 psi (21 to 24 MPa). Thus, the instability and breakout can be identified in the finite element analysis based on the critical disturbance derived from laboratory tests.

Implementation of DSC in Computer Procedures

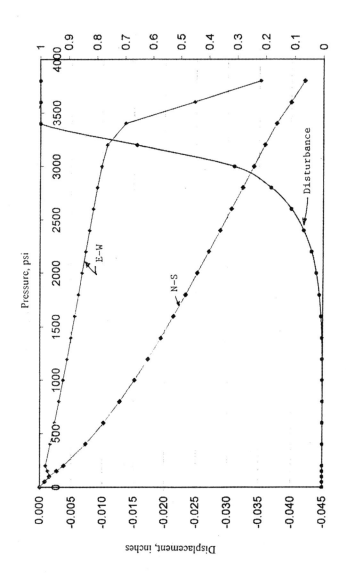

FIGURE 13.39
Predicted displacements and disturbance for borehole: 1 inch = 2.54 cm, 1 psi = 6.89 kPa.

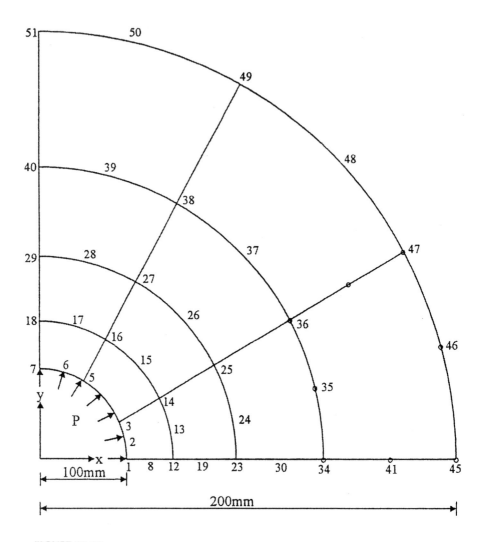

FIGURE 13.40
Mesh for elastoviscoplastic problem (17) (with permission).

Example 13.6: Viscoplastic Model: Cylinder (Tunnel) with Internal Pressure

Figure 13.40 shows the finite element mesh for a thick cylinder (tunnel), idealized as plane strain; because of the symmetry, only a quarter of the cylinder is discretized (56). The internal diameter of the cylinder is 100 mm, with wall thickness = 100 mm. It is subjected to internal pressure, P, equal to 14 N/mm². The material parameters for the viscoplastic (δ_0-HISS) model (10) and those used by Owen and Hinton (17) in their viscoplastic (von Mises) model are given below:

Elastic: $\quad E = 2.10 \times 10^3$ N/mm², $\nu = 0.3$;

von Mises: $\quad \sigma_y$ (yield stress) = 2.40 N/mm²,

HISS-δ_0:   H' (strain hardening parameter) = 0.00;
   $\gamma = 0.05$, $\beta = 0.62$, $n = 3.9$,
   $a_1 = 10^{-9}$, $\eta_1 = 0.47$, $R = 2.0$ N/mm^2;
Creep: $\Gamma = 0.01$ 1/day, $N = 1.00$.

Time Integration (Chapter 8):

   Factor for graded time = 0.05
   Steppping time increment factor = 1.50
   Time integration parameter, $\theta = 0.0$.

Figure 13.41(a) shows radial (x) displacement with time for node 1, obtained by using the viscoplastic (von Mises) model. The displacement converges to the value of 0.13997, which compares well with that of 0.13959 reported by Owen and Hinton (17). Figure 13.41(b) shows the radial (x) displacement for node 1 with time obtained by using the viscoplastic (HISS-δ_0) model. The converged value of displacement of 0.160 mm is somewhat higher than that from the von Mises model, Fig. 13.41(a). The difference can be due to the fact that in the HISS-δ_0 model, plastic yielding is incorporated from the beginning of loading.

Viscoplastic Model: Creeping Natural Slope. Natural slopes often exhibit continuous movements under gravity load, which are influenced by creep or viscous response of (geologic) materials in the slopes and the interface zone between the slope (solid body), and parent rock mass. Finite element analysis and comparisons between predictions and field observation in a natural slope at Villarbeney, Switzerland, are presented in (57–59).

Example 13.7: Layered Pavement System: DSC Model, Microcracking and Fracture

Figure 13.42 shows the finite element mesh for a two-layered system with 4.0 in (10.0 cm) thickness of pavement over a 42 in (107 cm) thick base (60). It was idealized as axisymmetric. Two analyses were performed: (1) without initial crack and (2) with an initial vertical crack (starting from the surface) of depth 0.5 inch (1.27 cm) and thickness of 0.05 inch (0.127 cm) at the end of the circular loaded area of 12.0 inch (30 cm) radius (10). The load was applied in (variable) increments [4 at 50 psi (350 kPa), 10 at 20 psi (140 kPa) and the remainder at 10 psi (70 kPa)] until failure, when the displacements became excessive (10, 45, 46).

The material in the pavement was characterized by using the DSC model that allows for elastoplastic response with disturbance (damage) leading to microcracking and fracture. The base material was characterized by using the elastoplastic HISS-δ_0 model. The material parameters are shown below. The crack region was assumed to be linear elastic with a very low value (100

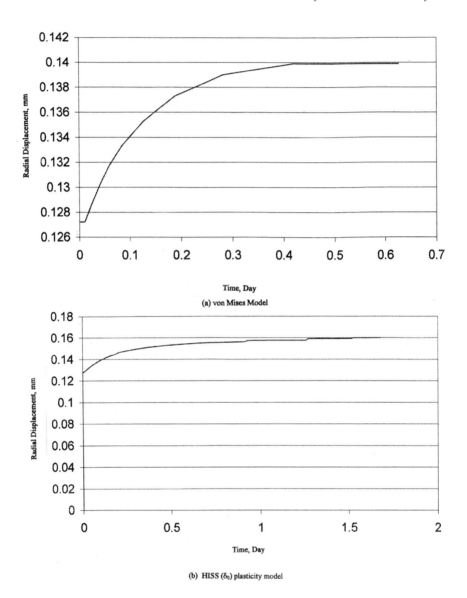

FIGURE 13.41
Radial displacement vs. time: viscoplastic model.

psi = 0.69 MPa) of the elastic modulus (E). Before the load was applied, *in situ* stresses due to the (overburden) weight were computed and included in the analysis using the unit weights and K_0 values shown below.

Figure 13.43 shows the computed load displacement curves from the no-crack and crack analyses; it shows that the system with the crack is less stiff compared to that without a crack, and that the initiation of yielding and microcracking occurred at lower load in the case of the crack.

Implementation of DSC in Computer Procedures

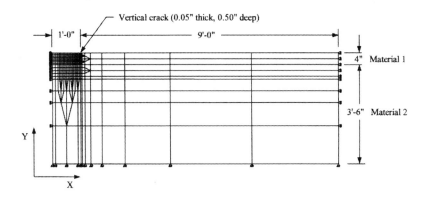

FIGURE 13.42
Typical mesh for layered pavement (1.0 inch = 2.54 cm).

Material Parameters.

Parameter	Material 1: Pavement (Concrete)	Material 2: Base	Crack
Elastic			
E psi (MPa)	3×10^6 (20684)	3×10^5 (2068)	100 (0.69)
v	0.25	0.24	0.24
Plasticity (HISS-δ_0) model			
γ	0.0678	0.030	
β	0.755	0.700	
n	5.24	5.26	
a_1	4.61×10^{-11}	1.20×10^{-6}	
η_1	0.826	0.778	
R, psi (MPa)	8122 (56)	28.88 (0.20)	
A	668		
Z	1.502		
D_u	0.875		
In situ stress			
Unit weight lb/in.³ (kg/m³)	0.087 (2325)	0.064 (1771)	
Co-efficient of lateral Earth pressure, K_o	0.20	0.70	

Computed distributions and contours of disturbance (in a part of the pavement, near the applied load) at different load levels for the two cases (no crack and with crack) were analyzed. For the case without an initial crack, critical disturbance $D_c = 0.80$ indicating initiation of fracture occurs at the interface between the pavement and the base (under the load), at the load

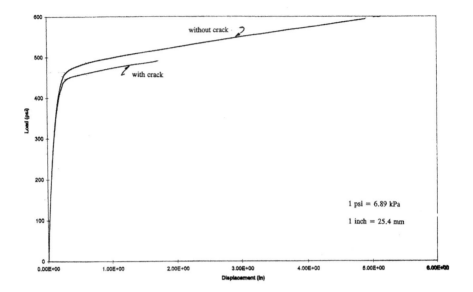

FIGURE 13.43
Computed load displacement curves.

level of about 480 psi (3.30 MPa). At about 550 psi (3.80 MPa) load, $D_c = 0.80$ is exceeded in almost all of the pavement under the load. In the case of the initial crack, the critical disturbance $D_c = 0.80$ is indicated at the interface, and in the upper part (near the surface) of the cracked region at about 450 psi (3.10 MPa) load. As the load increases, the extent of zones with $D \geq 0.8$ increases, and at about 490 psi (3.38 MPa) load, the entire pavement (below the load) experiences microcracking and fracture.

The above results show the capability of the DSC model for predicting microcracking and growth of fracture (in a layered system). Note that the nonlinear FE with the DSC allows for the identification of the initiation and growth of cracking, without the need for the introduction of linear fracture mechanics based crack intensity factors, and it avoids the need for the *a priori* assumption of the location and geometry of the crack. In other words, the DSC identifies cracking and its propagation at locations dictated by the nature of the problem (geometry, loading, boundary conditions, etc.)

Example 13.8: Chip Substrate 313-PIN PBGA: DSC Model, Thermomechanical Cyclic Fatigue Failure

The DSC model has been used to characterize the thermomechanical behavior of joining materials (e.g., solders) in chip-substrate systems in electronic packaging. The computer finite element codes with the DSC model have been used to analyze thermomechanical response of a number of chip-substrate systems: leadless ceramic chip carrier (LCCC) (7, 8, 14), ceramic ball

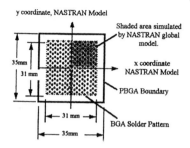

FIGURE 13.44
313-PBGA package (63, 67).

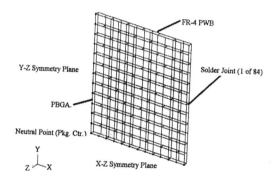

FIGURE 13.45
PBGA PWB solder joint NASTRAN model (63).

grid array (CBGA) (61, 62), plastic ball grid array (PBGA) (63–66) and thin small outline package (TSOP) (9). The computer predictions for these problems compared very well with laboratory test responses in terms of disturbance (damage), microcracking and fracture, and cycles to failure. Two typical problems are presented here; 313-Pin PBGA in this example and the CBGA in the next example.

Figure 13.44 shows the 313-PIN plastic ball grid array (PBGA) manufactured by AMKOR–ANAM, which is 35.0 mm square and 2.33 mm high. The solder ball diameter varies from 0.60 to 0.90 mm with the spacing of 1.27 mm. The PBGA package was tested in the laboratory under thermal cycles of loading at the Jet Propulsion Lab (JPL) (67). The PBGA module consisted of bimaleimide triazine (BT) epoxy glass laminate, mold compound, silicon chip, bond wires and micro solder balls. It is mounted on a printed wire board (PWB) composed of FR-4 glass/epoxy laminated with copper traces or planes (63).

The analysis was performed in two stages: macro and micro. In the macro model, the package was idealized as three-dimensional with linear elastic properties, Fig. 13.45; the PBGA and PWB were divided in 144, 4-node plane strain plate elements each, and were connected by 84-bar elements to represent the solder joints in the quarter of the system. The thicknesses used correspond to the photomicrographs of the test hardware (63, 64). Other details such as boundary conditions and loading are given by Zwick and Desai (63, 64). The macro model provided displacements at each nodal point and stresses in each element. The highest displacements occurred in the solder joint farthest from the neutral point (package center) at temperature = $-55°C$.

Figure 13.46(a) shows the temperature and displacement cycles with the highest magnitude applied for the analysis of a single ball joint, Fig. 13.46(b), in the micro model. The material parameters for the Pb-Sn solder are given below (14, 63, 64). They were determined on the basis of test data on Pb/Sn (40/60) solders reported in (68–71). At this time, the analysis involved elasto-plastic disturbance model, i.e., creep effects, are not considered.

Parameters for 63 Sn/37 Pb Solder (63)

Parameter	P_{300}	c(*)
Elasticity		
E	15.7 GPa	−4.1
ν	0.40	0.0
Plasticity		
γ	0.00081	−0.1579
a_1	0.78E-05	0.00
η_1	0.46	0.23
n	2.10	0
R	208.0 MPa	−5.34
Disturbance		
A	0.102	1.55
Z	0.676	0
D_u	0.85	—

(*)Temperature effects are included by expressing any parameter, P, as in Eq. (13.54).

The finite element analysis was performed for about 4000 cycles. Figure 13.47(a) shows the computed values of stresses σ_x, σ_y and τ_{xy} in the top left corner element, up to 2000 cycles. Cyclic degradation is evident in this figure. Figure 13.47(b) shows the growth of disturbance with cycles (N). Figure 13.48 shows the contours of disturbance at typical cycles N = 2500 and 2700 cycles.

Different values of design disturbance (D_{cm}), when microcracking is considered to initiate, can be specified for specific design needs. For instance, if D_{cm} = 0.50, the critical value (D_{cm} = 0.5) occurred in the solder after about 500 cycles, Fig. 13.49. Then it grew with cycles, and after about 3000 cycles (to failure, N_f), about 90% of the solder volume had experienced $D \geq 0.50$, Fig. 13.49, which can be considered to represent fatigue failure. Alternatively, $D_{cm} = D_c = 0.85$, which represents the critical value from laboratory tests (8, 14, 63), when significant fracture occurs as a consequence of the coalescence of microcracks, can be adopted as the design criterion. It occurred for the first time after about 2000 cycles; then about 10% of the solder volume reached $D \geq 0.85$ after about 2800 cycles (N_f),

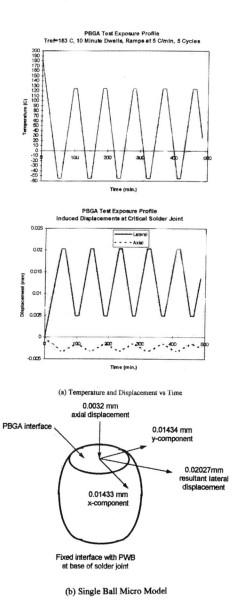

FIGURE 13.46
Displacement and temperature loading cycles and single ball micro model (63).

Fig. 13.49, which can be considered to be failure. Thus, cycles to failure (N_f) can be found based on a given design criterion on the disturbance, D. Figure 13.49 shows that the foregoing values of cycles to failure compare well with those observed in the laboratory tests on the four PBGA 313-PIN packages (67).

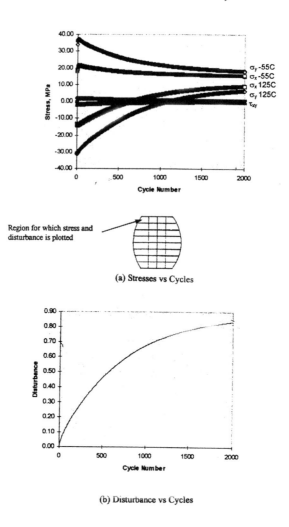

FIGURE 13.47
Stress variation and disturbance with cycles (63).

Inclusion of creep behavior with elastoplastic disturbance model can provide improved results. Also, the computer code can be used to perform parametric studies in which the material properties, geometry, ball spacing, etc., can be varied, leading to computation of cycles to failure and reliability, and design optimization (72).

Example 13.9: Chip Substrate CBGA Package: DSC Model, Thermomechanical Cyclic Fatigue Failure

Figure 13.50 shows a *ceramic ball grid* array (CBGA) package tested in the laboratory by Guo et al. (73). It is used in the IBM 604 power PC chip and is composed of either alumina or aluminum nitride (ALN), which encases a

Implementation of DSC in Computer Procedures

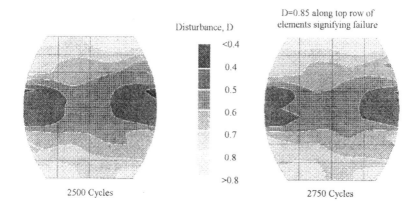

FIGURE 13.48
Contours of disturbance at cycles N = 2500 and 2750 (63).

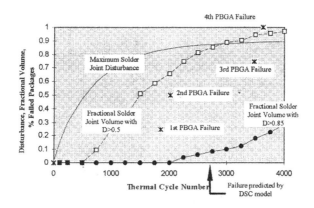

FIGURE 13.49
Disturbance and fractional volume vs. cycles and comparison with test data (63).

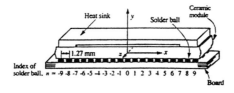

FIGURE 13.50
Test vehicle for CBGA in simulated laboratory test (61).

controlled collapsible chip connection (C-4) silicon chip. The solder balls in the CBGA-SBC (solder ball connect) package were composed of 90/10 (Pb/Sn) with 63/37 (Pb/Sn) solder fillers. Details are given in (73, 74).

The material properties for the chip, printed wire board (PWB) and the joining solder with 63% Pb-37% Sn and 90% Pb-10% Sn were determined on the basis of laboratory tests and other data available in (75–77). These parameters are given below (61).

Material Parameters

Parameter	63% Pb-37% Sn		90% Pb-10% Sn	
	P_{300}	c	P_{300}	c
Elasticity				
E	15.7 GPa	−4.1	9.29 GPa	−1.92
ν	0.40	0.0	0.40	0.0
Plasticity				
γ	0.00081	−0.158	0.000822	−0.16
a_1	0.78×10^{-5}	0.00	1.10×10^{-5}	−0.61
η_1	0.46	0.23	0.44	0.24
n	2.10	0.00	2.10	0.00
R	208.0 MPA	−5.34	122.0 MPA	−1.67
Disturbance				
A	0.102	1.55	4.07	0.00
Z	0.676	0.0	1.95	0.00
D_u	0.90	0.0	0.90	0.00
Ceramic Module				
E	318 GPa			
ν	0.23			
α_r	7.0×10^{-6} /°C			
PWB (FR-4)				
E	11.0 GPa			
ν	0.28			
α_r	20×10^{-6} /°C			

Here, P_{300} and c are parameters in Eq. (13.54)

The solder materials were characterized by using the DSC with elastoplastic δ_0-model for the RI behavior; the FA response was assumed to be hydrostatic. The chip and PWB material were assumed to be linear elastic. A series of tension and compression tests were performed on ALN so as to determine its elastic constants (8, 61).

The finite element analyses were performed by considering three different ball spacings, 1.00, 1.27, and 1.50 mm; they involved two steps: macro and micro. In the macro level, half of the structure was idealized as plane strain; Figure 13.51 shows the FE mesh with 4-node isoparametric elements for the case of 1.27 mm spacing. The mesh consisted of 360 nodes and 304 elements. In the macro model, a thermal cycle, Fig. 13.52(a), was applied and the critically strained solder ball was identified. Figure 13.52(b) shows displacement cycles with computed (critical) amplitudes of axial and shear displacements that were applied in the analysis of the single ball for the micro level calculations. Figure 13.53 depicts the finite element mesh for the single ball.

Figure 13.54 shows the computed axial (normal) and shear displacements for different ball spacings, together with the experimental data (73) and another finite element analysis for the same package reported by Corbin (74).

Implementation of DSC in Computer Procedures

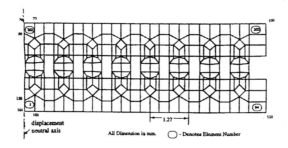

FIGURE 13.51
Typical finite element mesh for ball spacing = 1.27 mm (61).

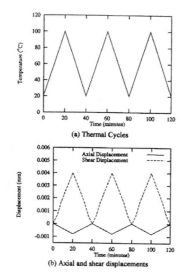

FIGURE 13.52
Thermal cycles and applied displacements: from macromodel applied to micromodel for ball spacing = 1.27 mm (8,61). ©1998 IEEE.

The present results show satisfactory correlation with the test data and Corbin's analysis. A part of the difference can be attributed to the fact that the size of the module analyzed here is smaller than that used by Corbin (74) and in the testing. Also, in the present analysis, the structure analyzed was 21 × 21 mm, while Corbin considered a 25 × 25 mm SBC module (8, 61). In the present analysis, a two-dimensional plain strain idealization was used, while Corbin considered a three-dimensional idealization. Furthermore, the molybdenum and copper pads at the top and bottom of the ball were not included in the present analysis; this was done as a simplification, and because failure usually initiates in the filler, not in the pads.

Disturbance (microcracking leading to fracture) growth and its localization with thermal cycles were computed for the three ball spacings. Figure 13.53

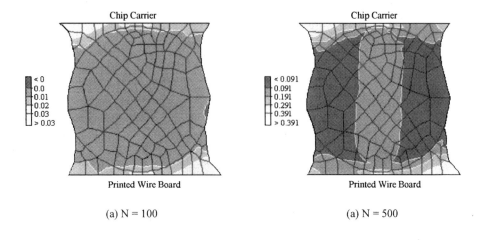
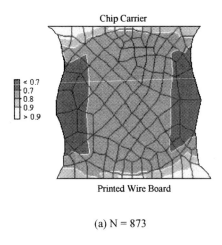

FIGURE 13.53
Finite element mesh and distribution of disturbance in solder ball for 1.27 at different thermal cycles, N (8, 61). ©1998 IEEE.

shows typical distributions of computed disturbance for the ball spacing of 1.27 mm at cycles $N = 100$, 500 and $N_f = 873$; here, N_f denotes failure cycle, which corresponded to the critical disturbance, $D_c = 0.90$. Figure 13.53 shows that at $N_f = 873$, the growth of $D \geq 0.90$ has extended to the major part of the package, for about 80 to 90% of the width of the ball. This can be considered to represent failure, as the finite crack for about 80 to 90% of the width could lead to failure of the electrical connection. The distributions of disturbance, which represent accumulated plastic strains in Fig. 13.53, compare well with those reported from interferometry results (73) and the finite element analysis (74).

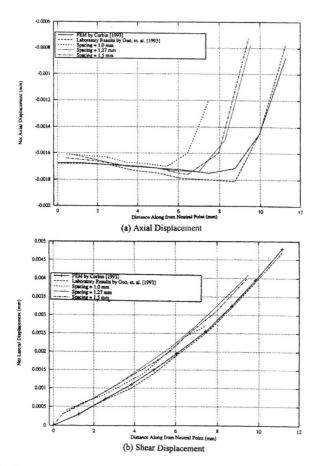

FIGURE 13.54
Computed displacements and comparisons with previous FE analysis (74) and test data (73). ©1993 by IBM Corp., reprinted with permission.

Figure 13.55 shows a plot of the number of cycles to failure, N_f, vs. the via spacing. It can be seen that the values of N_f decrease as the spacing increases. In other words, a package with a smaller ball spacing can sustain a greater number of thermal and mechanical cycles. The DSC model and computer procedure can thus also allow design optimization and evaluation of reliability under variations of factors such as ball spacing, material properties, geometry and loading cycles (rate, dwell, etc.) (72).

Example 13.10: Cyclic Hardening, Repetitive Loading, HISS Plasticity (δ_0) Model

Figure 13.56 shows the finite element mesh with four 8-noded elements. The top surface was subjected to cyclic or repetitive pressure loading with the amplitude of 50.0 MPa, increased in twenty increments and then decreased

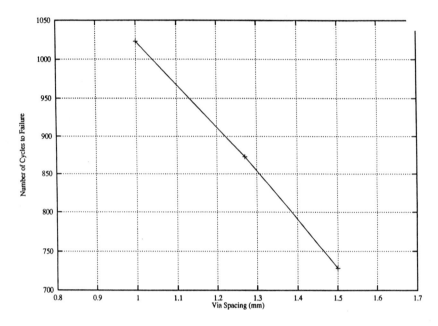

FIGURE 13.55
Plot of the number of cycles to failure vs. via spacing for three ball spacings (8, 61). ©1998 IEEE.

to zero in twenty decrements. Five such cycles were applied in the present analysis (10, 56). The material parameters used are given below:

Elastic: $\quad\quad\quad\quad\quad E = 37000$ MPa, $\nu = 0.25$;
Plasticity (HISS-δ_0) Model: $\gamma = 0.06784$, $\beta = 0.755$, $n = 5.24$;
$$a_1 = 0.4614 \times 10^{-10}, \eta_1 = 0.826,$$
$$3R = 56.64 \text{ MPa}$$

For the nonlinear unloading, Fig. 13.14, $E^b = 50,000$ MPa and $E^e = 35,000$ MPa were used. The values of the cyclic hardening parameter, h_c, Eq. (13.105a), were adopted as 0.00 and 1.00.

Three cases were analyzed: (1) $h_c = 0.00$ indicating no cyclic hardening and with linear unloading, (2) $h_c = 1.00$ indicating cyclic hardening and with linear unloading, and (3) $hc = 1.0$ indicating cyclic hardening and with linear unloading, Fig. 13.14.

Figure 13.57 shows computed results for the three cases in terms of axial stress (σ_y) versus axial strain (ε_y) for the five cycles. Figure 13.57(a) shows that there is no cyclic hardening, as the plastic strains for all cycles are the same. The results for cases 2 and 3, Fig. 13.57(b) and (c), show that the plastic strains decrease with increasing cycles. However, Case 3 can allow the effect of nonlinear unloading on the observed behavior of some materials.

Implementation of DSC in Computer Procedures

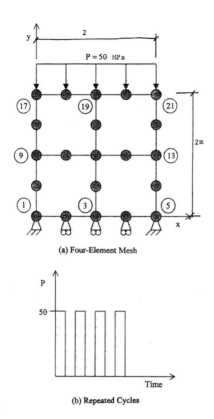

FIGURE 13.56
Four-element mesh and repeated loading.

Example 13.11: Pile in Clay under Cyclic Loading: DSC Model

Interaction and coupled response between structures and geologic materials play an important role in the behavior of structure-foundation systems. The behavior is complex and affected by many factors such as nonlinear response of soils and interfaces, type of loading and sequences of construction. The finite element procedure with the DSC model was used to predict the field behavior of an instrumented axially loaded pile in saturated marine clay (23–25, 78–82).

The pile load tests were performed by Earth Technology Corporation (78) at Sabine, Texas, with pile segments of diameter 4.37 cm and 7.62 cm. Typical analysis for the 7.62 cm pile is presented below. The field conditions including *in situ* stresses and sequences—pile driving, consolidation, tension (one-way) tests and cyclic two-way loading—were simulated. The *in situ* stresses, before the pile was driven, were computed by using the coefficient of lateral earth pressure, $K_o = 0.786$ (23). The driving stresses and pore water pressures were evaluated by using the strain path method (SPM) (23, 79, 83). The driving stresses and pore water pressures were introduced as "initial" conditions for

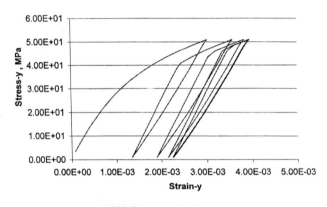
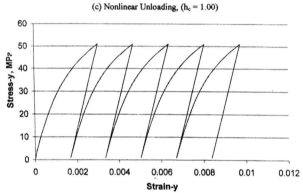
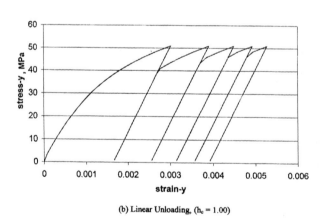

FIGURE 13.57
Computed cycles stress, σ_y vs. strain, ε_y.

Implementation of DSC in Computer Procedures

the next stage. Time-dependent deformations (consolidation) after the driving were evaluated by using the FE procedure (25). The cyclic two-way loading was then simulated as it was done in the field.

Laboratory triaxial (cylindrical specimens, 7.0 × 15.0 cm) and multiaxial (cubical specimens, 10.0 × 10.0 cm) tests were performed on the clay specimens obtained from the field (80). Interface tests between the clay and pile (steel) were conducted by using the CYMDOF device described in Chapter 11 (81). These tests were used to find the DSC model parameters for loading, unloading and reloading; they are listed below.

Material Parameters Used for Finite Element Analyses (24, 82)

	Clay	Interface
Relative intact state		
E	10350 kPa	4300 kPa
ν	0.35	0.42
γ	0.047	0.08
β	0.0	0
n	2.8	2.6
$3R$	0.0	0.0
h_1 Eq. (7.11b)	0.0001	0.000408
h_2	0.78	2.95
h_3	0.0	0.02
h_4	0.0	0.08
Critical state		
λ	0.1692	0.3
e_0^c	0.9	1.36
m	0.07	0.12
Disturbance function		
D_u	0.75	1.0
A	1.73	0.82
Z	0.31	0.42
Unloading and reloading:		
E^b	34500 kPa	4300 kPa
E^e	3450 kPa	400 kPa
ε_1^p	0.005	0.03
Other parameters		
Permeability, k	2.39×10^{-10} m/sec	2.39×10^{-10} m/sec
Density of soils, ρ_s	2.65 mg/m^3	2.65 mg/m^3
Bulk modulus of soil grain, K_s	10^9 kPa	10^9 kPa
Bulk modulus of water, K_f	10^8 kPa	10^8 kPa
Density of water, ρ_f	1.0 mg/m^3	1.0 mg/m^3

©John Wiley & Sons Ltd. Reproduced with permission.

Figure 13.58(a) shows the finite element mesh for the pile segment and the surrounding soil; Fig. 13.58(b) shows the applied displacement history at the pile nodes (78). Figure 13.59 shows the comparison for consolidation behavior and the growth of disturbance during the consolidation. Here, the HISS

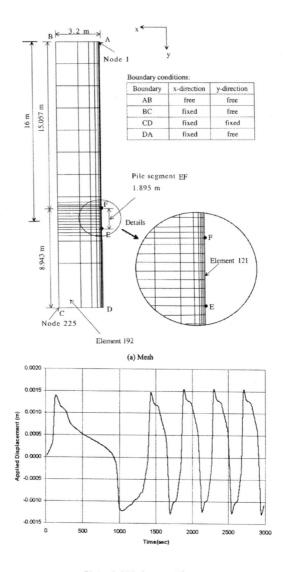

FIGURE 13.58
Finite element mesh of field pile test and displacement history (24, 82). ©John Wiley & Sons Ltd. Reproduced with permission.

(δ_0^*) indicates calculations based on previous anisotropic hardening HISS plasticity model (23, 79). Figures 13.60(a) and (b) show comparisons between the predicted responses using the DSC and HISS (δ_0^*) model and the field data for shear transfer at the pile interface vs. displacement and time during the one-way cyclic loading, respectively. The pore water pressure and the growth of disturbance vs. time during the one-way cyclic loading are shown in Figs. 13.60(c) and (d), respectively. The results for the cyclic two-way load-

Implementation of DSC in Computer Procedures 615

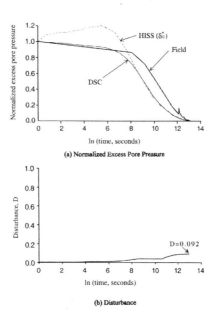

FIGURE 13.59
Comparisons between field consolidation behavior and predicted results from HISS and DSC models (24, 82). ©John Wiley & Sons Ltd. Reproduced with permission.

ings in terms of shear transfer vs. displacement, shear transfer vs. time, pore water pressures and growth of disturbance vs. time are shown in Figs. 13.61(a)–(d), respectively.

It can be seen that the DSC predictions correlate very well with field observations. Also, the DSC model, which allows for cyclic (disturbance) degradation, provides improved predictions compared to those from the HISS-plasticity model.

Example 13.12: Dynamic Analysis and Liquefaction in Shake Table Test: DSC Model

Behavior of structure foundation systems subjected to dynamic (earthquake) loading is affected, among other factors, by interaction effects due to relative interface motions. In the case of saturated granular (sandy) materials with certain initial densities, dynamic loading can cause microstructural instability or liquefaction (Chapter 9). Applications of computer procedures with the DSC model are given in (84–88). The DSC parameters for the Ottawa sand and sand-concrete (steel) interfaces were obtained from cyclic triaxial and multiaxial (84, 85, 88, 89) and shear tests (90), respectively. Here, we present dynamic and liquefaction analysis of a shake table test (91).

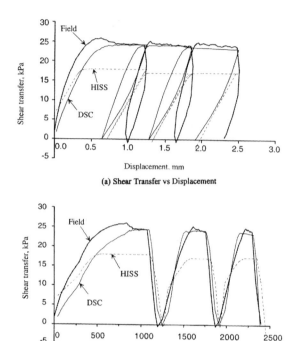

FIGURE 13.60
Comparisons between field measurements and predictions from DSC and HISS models: one-way cyclic load test (24, 82). ©John Wiley & Sons Ltd. Reproduced with permission.

The finite element procedure with the DSC model was used to predict the dynamic behavior in shake table tests reported by Akiyoshi et al. (91). Figure 13.62(a) shows the shake table test set-up with location of instruments. The saturated sand was Fuji River sand with total density, $\gamma_t = 19.8$ KN/m³. Here, the properties of the Ottawa sand with $\gamma_t = 19.63$ KN/m³ were used as an approximation; the rationale for the approximation is given in (85).

The finite element mesh used is shown in Fig. 13.62(b). The mesh contained 160 elements and 190 nodes with 120 elements for the soil and 40 elements for the steel box. The concept of repeating side boundaries was used (21, 92); accordingly, the displacement of the side boundary nodes on the same horizontal plane was assumed to be the same. The bottom boundary was restrained in the vertical direction, while it was free to move in the horizontal direction.

The applied loading involved horizontal displacement, X, at the bottom nodes, given by the following function:

$$X = \bar{x} \sin(2\pi f\, t) \qquad (13.130)$$

where $\bar{x}$ is amplitude ($= 0.0013$ m), f is the frequency ($= 5.0$ HZ) and t is the time. The analysis was carried out for 50 cycles with time step $\Delta t = 0.001$

Implementation of DSC in Computer Procedures

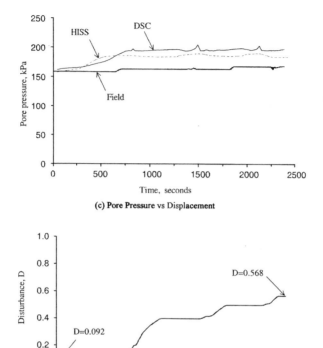

FIGURE 13.60
(continued).

sec (from time 0.0 to 2.0 secs), and $\Delta t = 0.05$ sec (from time 2.0 to 10.0 secs).

Figure 13.63 shows comparisons between the measured and predicted excess pore water pressures with time at the (dark) point in the box at depth = 300 mm from the top, Fig. 13.62(b). The test results indicate liquefaction after about 2.0 secs when the pore water pressure equaled the initial effective stress. The finite element predictions also show liquefaction after about 2.0 secs.

Figures 13.64(a) to (d) show contours and growth of disturbance in the mesh at typical times = 0.50, 1.0, 2.0, and 10.0 secs. The computed plot of the growth of disturbance at depth = 300 mm is shown in Fig. 13.64(e). Laboratory tests on the Ottawa sand showed that liquefaction initiated at average critical disturbance, $D_c = 0.84$ (87). At time = 0.50 sec, the disturbance was below the critical value in all elements. At $t = 1.0$ sec, the disturbance had grown and its value was greater than 0.70 in the middle zone, including depth = 300 mm. At time of 2.0 secs, the disturbance had reached the value of about 0.80 in the zone above depth = 300 mm, while its value

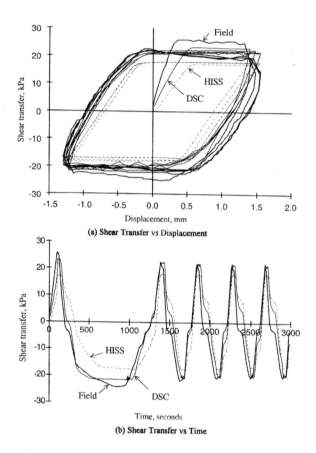

FIGURE 13.61
Comparisons between field measurements and predictions from DSC and HISS models: two-way cyclic load test (24, 82). ©John Wiley & Sons Ltd. Reproduced with permission.

was greater than 0.80 in the lower middle zone, indicating that liquefaction had initiated at about 2.0 secs. At time = 10.0 secs, the disturbance has grown to the value of about 0.90 in about 80% of the test box, indicating that the soil in the test box has liquefied and failed. These trends are considered to be consistent with the observed behavior.

13.12 Computer Codes

A number of computer codes with the DSC model have been developed and are used to solve the problems in this chapter. Some of the codes are described in various publications, e.g., DSC-SST2D (10), DSC-DYN2D (25) and DSC-SST3D (42). These codes can be acquired by writing to the author.

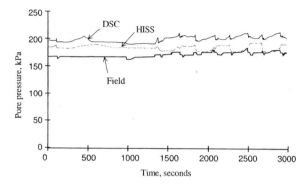

(c) Pore Pressure vs Displacement

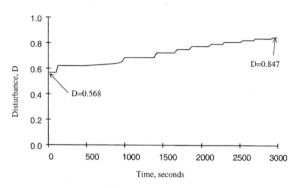

(d) Disturbance vs Displacement

FIGURE 13.61
(continued)

References

1. Desai, C. S. and Abel, J. F., *Introduction to the Finite Element Method*, Van Nostrand Reinhold, New York, 1972.
2. Zienkiewicz, O. C., *The Finite Element Method*, 3rd Edition, McGraw Hill, London, UK, 1997.
3. Bathe, K. J., *Finite Element Procedures*, Prentice-Hall, Englewood Cliffs, NJ, 1996.
4. Desai, C. S. and Kundu, T., *Introductory Finite Element Method*, CRC Press, Boca Raton, FL, 1999, under publication.
5. Desai, C. S. and Hashmi, Q., "Analysis, Evaluation and Implementation of a Nonassociative Model for Geologic Materials," *Int. J. Plasticity*, 5, 1989, 397–420.
6. Desai, C. S. and Woo, L., "Damage Model and Implementation in Nonlinear Dynamic Problems," *Int. J. Comp. Mechanics*, 11, 2/3, 1993, 189–206.

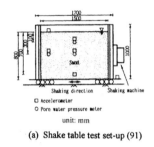

(a) Shake table test set-up (91)

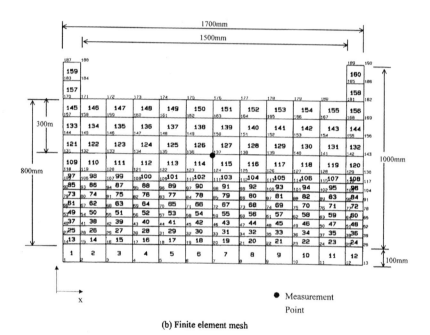

(b) Finite element mesh

FIGURE 13.62
Shake table and finite element mesh (85). (a) ©John Wiley & Sons Ltd. Reproduced with permission.

7. Basaran, C., Desai, C. S. and Kundu, T., "Thermomechanical Response of Materials and Interfaces in Electronic Packaging: Parts I and II," *J. of Electron. Packaging,* 120, 1, 1998, 41–47, 48–53.

8. Desai, C.S., Basaran, C., Dishongh, T. and Prince, J. L., "Thermomechanical Analysis in Electronic Packaging with Unified Constitutive Model for Materials and Joints," *IEEE Trans.: Components, Packaging and Manuf. Tech.,* Part B, 21, 1, 1998, 87–97.

9. Desai, C. S., Wang, Z. and Whitenack, R., "Unified Disturbed State Constitutive Models for Materials and Computer Implementation," *Keynote Paper, 4th Int. Conf. on Constitutive Laws for Engineering Materials,* Rensselaer Polytechnic Inst., Troy, NY, 1999.

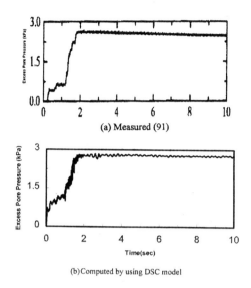

FIGURE 13.63
Excess pore presure at depth of 300 mm (85).

10. C. Desai, "DSC-SST2D-Computer Code for Static, Dynamic, Creep and Thermal Analysis: Solid, Structure and Soil-Structure Problems," Reports and Manuals, Tucson, AZ, 2000.
11. Wang, Z. and Desai, C. S., "Modelling and Computer Thermomechanical Analysis of Problems Using the Disturbed State Concept," *Report,* under publication, 1999.
12. Kachanov, L. M., *Introduction to Continuum Damage Mechanics,* Martinus Nijhoft, Dorchrecht, The Netherlands, 1986.
13. Desai, C. S. and Toth, J., "Disturbed State Constitutive Modeling Based on Stress-Strain and Nondestructive Behavior," *Int. J. Solids Struct.,* 33, 11, 1996, 1619–1650.
14. Desai, C. S., Chia, J., Kundu, T. and Prince, J., "Thermomechanical Response of Materials and Interfaces in Electronic Packaging: Parts I and II," *J. of Electronic Packaging,* 119, 1997, 294–305.
15. Prager, W., "Nonisothermal Plastic Deformations," *Bol. Koninke Nederl. Acad. Wet,* 8, 6113, 1958, 176–182.
16. Perzyna, P., "Fundamental Problems in Viscoplasticity," in *Advances in Applied Mechanics,* Academic Press, New York, 1996, 244–368.
17. Owen, D. R. and Hinton, E., *Finite Elements in Plasticity: Theory and Practice,* Pineridge Press, Swansea, UK, 1980.
18. Biot, M. A., "General Solutions of the Equations of Elasticity and Consolidation for a Porous Material," *J. Appl. Mech.,* 23, 1956, 91–96.
19. Biot, M. A., "Mechanics of Deformation and Acoustic Propagation in Porous Media," *J. Appl. Physics,* 33, 4, 1962, 1482–1498.
20. Sandhu, R. S. and Wilson, E. L., "Finite Element Analysis of Flow of Saturated Porous Elastic Media," *J. of Eng. Mech. Div., ASCE,* 95, 3, 1969, 641–642.

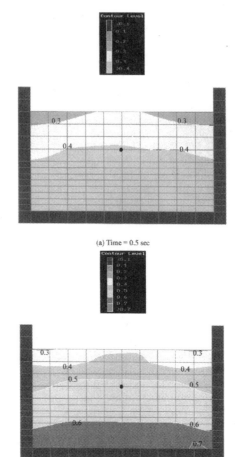

FIGURE 13.64
Growth of disturbance in sand at depth = 300 mm (85).

21. Desai, C. S. and Galagoda, H. M., "Earthquake Analysis with Generalized Plasticity Model for Saturated Soil," *Earthquake Eng. and Struct. Dyn.*, 18, 1989, 901–919.
22. Zienkiewicz, O. C. and Shiomi, T., "Dynamic Behavior of Saturated Porous Media: The Generalized Biot Formulation and Its Numerical Solution," *Int. J. Num. Analyt. Methods in Geomech.*, 8, 1984, 71–96.
23. Wathugala, G. W. and Desai, C. S., "Constitutive Model for Cyclic Behavior of Clays-Theory, Part I," *J. of Geotech. Eng.*, 119, 4, 1993, 714–729.
24. Shao, C. and Desai, C. S., "Implementation of DSC Model for Dynamic Analysis of Soil-Structure Interaction Problems," *Report*, Dept. of Civil Eng. and Eng. Mechanics, University of Arizona, Tucson, AZ, 1998.
25. "DSC-DYN2D-Dynamic and Static Analysis, Dry and Coupled Porous Saturated Materials," Reports and Manual, C. Desai, Tucson, AZ, 1999.

Implementation of DSC in Computer Procedures 623

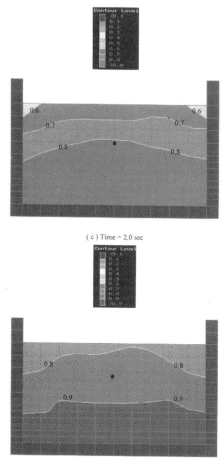

(c) Time = 2.0 sec

(d) Time = 10.0 sec

FIGURE 13.64
(continued)

26. Hughes, T. J. R., "Analysis of Transient Algorithms with Particular Reference to Stability Behavior," Chapter in *Computational Methods for Transient Analysis*, Belytschko, T. and Hughes, T. J. R. (Editors), North-Holland Publ. Co., Amsterdam, 1983.
27. Thomas, H. R. and He, Y., "Analysis of Coupled Heat, Moisture and Air Transfer in a Deformable Unsaturated Soil," *Geotechnique*, 45, 4, 1995, 677–689.
28. Thomas, H. R. and Li, C. L. W., "Modelling Transient Heat and Moisture Transfer in Unsaturated Soil Using a Parallel Computing Approach," *Int. J. Num. and Analyt. Methods in Geomech.*, 19, 1996, 345–366.
29. Yang, D. Q., Rahardjo, H., Leong, E.C. and Choa, V., "Coupled Model for Heat, Mositure, Air Flow and Deformation Problems in Unsaturated Soils," *J. of Eng. Mech., ASCE*, Vol. 124, No. 12, 1998, 1331–1338.
30. Ng, A. K. L. and Small, J. C., "Coupled Finite Element Analysis for Unsaturated Soils," *Int. J. Num. Analyt. Meth. in Geomech.*, in press, 1999.

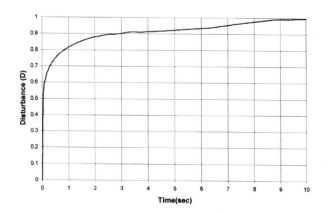

(e) Disturbance vs. time at depth = 300mm

FIGURE 13.64
(continued)

31. Mroz, Z., Norris, V. A. and Zienkiewicz, O. C., "An Anisotropic Hardening Model for Soils and Its Application to Cyclic Loading," *Int. J. Num. Analyt. Methods in Geomechanics*, 2, 1978, 203–221.
32. Somasundaram, S. and Desai, C. S., "Modeling and Testing for Anisotropic Behavior of Soils," *J. of Eng. Mech.*, ASCE, 114, 9, 1988, 1473–1497.
33. Bonaquist, R. F. and Witczak, M. W., "A Comprehensive Constitutive Model for Granular Materials in Flexible Pavement Structures," *Proc., 8th Int. Conf. on Asphalt Pavements*, Seattle, WA, 1997, 783–802.
34. Desai, C. S., Sharma, K. G., Wathugala, G. W. and Rigby, D. B., "Implementation of Hierarchical Single Surface δ_0 and δ_1 Models in Finite Element Procedure," *Int. J. Num. Analyt. Meth. in Geomech.*, 15, 1991, 649–680.
35. Desai, C. S., Jagannath, S. V. and Kundu, T., "Mechanical and Ultrasonic Anisotropic Response of Soil," *J. of Eng. Mech.*, ASCE, 121, 6, 1995, 744–752.
36. Zienkiewicz, O. C., Huang, M. and Pastor, M., "Localization Problems in Plasticity Using Finite Elements with Adaptive Remeshing," *Int. J. Num. and Analyt. Meth. Geomechanics*, 19, 2, 1995, 127–148.
37. Yu, J., Peric, D. and Owen, D. R. J., "Adaptive Finite Element Analysis of a Strain Localization Problem for the Elasto-Plastic Cosserat Continuum," in *Computational Plasticity III: Models, Software and Applications*, Owen, D. R. J. et al. (Editors), Pineridge Press, Swansea, 1992, 551–566.
38. Belytschko, T. and Tabbara, M., "H-Adaptive Finite Element Methods for Dynamic Problems with Emphasis on Localization," *Int. J. Numer. Methods Eng.*, 36, 1993, 4245–4265.
39. Desai, C. S. and Zhang, W., "Computational Aspects of Disturbed State Constitutive Models," *Int. J. Comput. Methods in Applied Mechanics and Engineering*, 151, 1998, 361–376.
40. Zienkiewicz, O. C. and Zhu, J. Z., "A Simple Error Estimator and Adaptive Procedure for Practical Engineering Problems," *Int. J. Numer. Meth. Eng.*, 24, 2, 1987, 337–357.

41. Zhang, W. and Chen, D. P., "The Patch Test Conditions and Some Multivariant Finite Element Formulations," *Int. J. Numer. Methods Eng.*, 40, 16, 1997, 3015–3032.
42. Desai, C., "DSC-SST3D-Computer Code for Static and Coupled Dynamic Analysis: Solid (Porous) Structure and Soil-Structure Problems," Reports and Manual, Tucson, AZ, 1999.
43. Wang, Z., Private communication, 1999.
44. Frantziskonis, G. and Desai, C. S., "Constitutive Model with Strain Softening," *Int. J. Solids Struct.*, 23, 6, 1987, 733–750.
45. Desai, C. S., "Review and Evaluation of Models for Pavement Materials," Report, Tucson, AZ, 1999.
46. Desai, C. S., "Unified Disturbed State Constitutive Model for Asphalt Concrete," *Proc., 14th Eng. Mech. Conference, ASCE*, Austin, TX, May 2000.
47. Witczak, M. W., "The Universal Airport Pavement Design System: Report II. Asphalt Mixture Material Characterization," The University of Maryland, College Park, MD, 1989.
48. Fishman, K. L., Desai, C. S. and Sogge, R. L., "Field-Behavior of Instrumented Geogrid Soil Reinforced Wall," *J. of Geotech. Eng., ASCE*, 119, 8, 1993, 1293–1307.
49. El-Hoseiny, K. and Desai, C. S., "Material and Interface Modeling and Computer Analysis of Instrumented Geogrid Soil Reinforced Wall," Interim Report, University of Arizona, Tucson, AZ, 1999.
50. El-Hoseiny, K., "Soil Stabilization Using Geosynthetics," *Ph.D. Thesis*, Univ. of Menoufia, Menoufia, Egypt, 1999.
51. Desai, C. S. and Reese, L. C., "Stress Deformation and Stability Analysis of Deep Boreholes," *Proc., 2nd Congress, Int. Soc. for Rock Mech.*, Belgrade, Yugoslavia, 1970.
52. Desai, C. S. and Johnson, L. D., "Influence of Bedding Planes on Stability of Boreholes," *Proc., 3rd Congress, Int. Soc. of Rock Mechs.*, Denver, CO, 1974.
53. Desai, C., "Finite Element 3-D Procedure with the Disturbed State Model for Wellbore Stability," *Report*, Tucson, AZ, 1999.
54. Niandou, H., Shao, J. F., Henry, J. P. and Fourmaintraux, D., "Laboratory Investigation of the Mechanical Behaviour of Tournemire State," *Int. J. Rock Mech. Min. Sc.*, 34, 1, 1997, 3–16.
55. Desai, C. S. and Ma, Y., "Modelling of Joints and Interfaces Using the Disturbed State Concept," *Int. J. Solids and Struct.*, 16, 1992, 623–653.
56. Sharma, K. G., Private communication, 1998.
57. Desai, C. S., Samtani, N. C. and Vulliet, L., "Constitutive Modeling and Analysis of Creeping Slopes," *J. of Geotech. Eng., ASCE*, 121, 1, 1995, 43–56.
58. Samtani, N. C., Desai, C. S. and Vulliet, L., "An Interface Model to Describe Viscoplastic Behavior," *Int. J. Num. Analyt. Meth. Geomech.*, 20, 1996, 231–252.
59. Vulliet, L., "Modelisation des Pentes Naturelles en Mouvement," *Thesis No. 635*, Ecole Polytechnique Federale de Lausanne, Lausanne, Switzerland, 1986.
60. Cohen, D., Private communication, 1998.
61. Dishongh, T. J. and Desai, C. S., "Disturbed State Concept of Materials and Interfaces with Applications in Electronic Packaging," *Report to NSF*, Dept. of Civil Engng. and Engng. Mechs., University of Arizona, Tucson, AZ, 1996.
62. Dishongh, T. J. and Desai, C. S., "Calibration of the Disturbed State Model for Thermomechanical Behavior of Various Solders," *Proc., Interpack 99*, Hawaii, 1999.

63. Zwick, J. W., "Thermostructural Analysis of a 313-PIN Plastic Ball Grid Array (PBGA)," Report for M.E., Dept. of Civil Eng. and Eng. Mechanics, University of Arizona, Tucson, AZ, 1998.
64. Zwick, J. W. and Desai, C. S., "Structural Reliability of PBGA Solder Joints with the Disturbed State Concept," *Proc., Interpack 99*, Hawaii, 1999.
65. Desai, C. S., "Review and Evaluation of Approaches for Thermomechanical Analysis for Stress, Failure and Reliability," *Proc., Interpack 99*, Hawaii, 1999.
66. Desai, C. S. and Whitenack, R., "Review of Models and the Disturbed State Concept for Thermomechanical Analysis in Electronic Packaging," *J. of Electronic Packaging, ASME,* 123, 1–15, 2001.
67. Ghaffarian, R., "Ball Grid Array Packaging RTOP: Interim Environmental Test Results," Jet Propulsion Lab, Cal Tech, Pasadena, CA, 1997.
68. Riemer, H. S., "Prediction of Temperature Cycling Life for SMT Solder Joints on TCE-Mismatched Substrates," *Proc., IEEE Electron. Comp.,* 1990, 418–423.
69. Skipor, A., Harren, S. and Bostis, J., "Constitutive Characterization of 63/37 Sn/Pb Eutectic Solder Using the Bodner-Parton Unified Creep-Plasticity Model," *Proc., Joint ASME/JSME Conf. Electron. Packag.,* Chen, W.T. and Abe, H. (Eds), 2, 1992, 661–672.
70. Solomon, H. D., "Isothermal Fatigue of LCCC/PWB Interconnections," *J. Electron. Packag., ASME,* 114, 1992, 161–168.
71. Pan, T. Y., "Thermal Cyclic Induced Plastic Deformations in Solder Joints – Part I: Accumulated Deformation in Surface Mount Joints," *Trans., ASME,* 113, 1991, 8–15.
72. Whitenack, R. and Desai, C. S., "Computer Procedure with Disturbed State Model for Cyclic Fatigue, Design and Reliability," *Report,* under preparation, 1999.
73. Guo, Y., Kim, C. K., Chen, W. T. and Woychik, C. G., "Solder Ball Connect (SBC) Assemblies under Thermal Loading: I. Deformation Measurements via Moire Interferometry and Its Interpretation," *IBM J. Res., Develop.,* 37, 5, 1993, 635–647.
74. Corbin, J. S., "Finite Element Analysis for Solder Ball Connect (SBC) Structural Design Optimization," *IBM J. Res. Develop.,* 37, 5, 1993, 585–596.
75. Cole, M., Caufield, T., Banks, D., Winton, M., Walsh, A. and Gonya, S., "Constant Strain Rate Tensile Properties of Various Load Based Solder Alloys at 0, 50 and 100°C," *Proc., Microelectron. Packag. Conf.,* Montreal, P.Q., Canada, 1991.
76. Savage, E. and Getzan, G., "Mechanical Behavior of 60 Tin/40 Lead Solder at Various Strain Rates and Temperatures," *Proc., Int. Soc. Hybrid Microelectron.* (ISHM), Nashville, TN, 1990.
77. Nir, N., Duddear, T. D., Wang, C. C. and Storm, A. R., "Fatigue Properties of Microelectronic Solder Joints," *J. Electron. Packag., ASME,* 113, 1999.
78. Earth Technology Corporation, "Pile Segment Tests, Sabine Pass: Some Aspects of the Fundamental Behavior of Axially Loaded Piles in Clay Soils," *ETC Report No. 85-007,* Houston, Long Beach, CA, 1986.
79. Desai, C. S., Wathugala, G. W. and Matlock, H., "Constitutive Model for Cyclic Behavior of Clay, II: Application," *J. of Geotech. Eng., ASCE,* 119, 4, 1993, 730–748.
80. Katti, D. R. and Desai, C. S., "Modeling and Testing of Cohesive Soils Using the Disturbed State Concept," *J. of Eng. Mech., ASCE,* 21, 1995, 648–658.
81. Desai, C. S. and Rigby, D. B., "Cyclic Interface and Joint Shear Decising Including Pore Pressure Effects," *J. of Geotech. & Geoenv. Eng., ASCE,* 123, 1997, 568–579.

82. Shao, C. and Desai, C. S., "Implementation of DSC Model and Application for Analysis of Field Pile Tests Under Cyclic Loading," *Int. J. Num. Analyt. Meth. Geomech.,* 24, 6, 2000, 601–624.
83. Baligh, M. M., "Undrained Deep Penetration, I: Shear Stresses; Part II: Pore Pressures," *Geotechnique,* 36, 1986, 471–485, 487–501.
84. Park, I. J. and Desai, C. S., "Disturbed State Modeling for Dynamic and Liquefaction Analysis," *Report to NSF,* Dept. of Civil Engng. and Engng. Mechs., University of Arizona, Tucson, AZ, 1997.
85. Park, I. J. and Desai, C. S., "Cyclic Behavior and Liquefaction of Sand Using Disturbed State Concept," *J. of Geotech. and Geoenv. Eng., ASCE,* 126, 9, 2000.
86. Desai, C. S., "Soil-Structure Analysis of Foundations with Unified Constitutive Models for Soils and Interfaces," *Proc., Analysis, Design and Construction, and Testing of Deep Foundations, OTRC Conference,* University of Texas, Austin, TX, ASCE, April 1999.
87. Desai, C. S., Park, I. J. and Shao, C., "Fundamental Yet Simplified Model for Liquefaction Instability," *Int. J. Num. Analyt. Meth. Geomech.,* 22, 1998, 721–748.
88. Desai, C. S., "Evaluation of Liquefaction Using Disturbed State and Energy Approaches," *J. of Geotech. and Geoenv. Eng.,* 126, 7, 2000, 618–631.
89. Gyi, M. M. and Desai, C. S., "Multiaxial Cyclic Testing of Saturated Ottawa Sand," *Report,* Dept. of Civil Engng. and Engng. Mechs., University of Arizona, Tucson, AZ, 1996.
90. Alanazy, A. S., "Testing and Modeling of Sand-Steel Interfaces Under Static and Cyclic Loading," *Ph.D. Dissertation,* Dept. of Civil Engng. and Engng. Mechs., University of Arizona, Tucson, AZ, 1996.
91. Akiyoshi, T., Fang, H. L., Fuchida, K. and Matsumoto, H., "A Nonlinear Seismic Response Analysis Method for Saturated Soil-Structure System with Absorbing Boundary," *Int. J. Num. Analyt. Meth. Geomech.,* 20, 5, 1996, 307–329.
92. Zienkiewicz, O. C., Leung, K. H. and Hinton, E., "Earthquake Response Behavior of Soils with Damage," *Proc., 4th Int. Conf. on Num. Methods in Geomech.,* Z. Eisenstein (Editor), Edmonton, Canada, 1982.

14

Conclusions and Future Trends

The disturbed state concept (DSC) is considered to be a unified and powerful approach for constitutive modeling of engineering materials, and interfaces and joints. It can be applied to characterize the thermomechanical behavior of a wide range of materials, geologic, concrete, metals, ceramics, alloys (solder), silicon, and reinforced or structured systems. The inherent idea in the DSC is based on a *holistic* consideration of the material's response to external influences (loads); the concept has been shown to have roots in philosophical and mechanistic viewpoints of the behavior of matter.

One of the important attributes of the DSC is that it provides a hierarchical framework, which permits the user to adopt a specialized version(s) for a specific material. Once a version is chosen, it is necessary to determine and use only the parameters relevant to that version. As a result, the DSC models are simpler, involve lower number of parameters compared to other available models of comparable capabilities, and can be implemented easily in solution (computer) procedures.

We have presented details of the theoretical and mathematical aspects of the DSC, including comparisons with other available models. The DSC and its specialized versions have been calibrated with respect to laboratory test data for a number of materials, interfaces and joints. We have presented parameters for many materials and interfaces, and have validated the models with respect to tests used to find them, and also independent tests *not* used in the calibration.

Details of the implementation of the DSC models in nonlinear, two- and three-dimensional computer procedures are presented. The computer codes are used to predict the actual behavior of simulated laboratory and field (practical) boundary value problems. The DSC-models are found to provide highly satisfactory predictions of the actual behavior.

Overall, we can conclude that the DSC provides a new and alternative approach for characterizing the mechanical behavior of materials and interfaces. It goes a step beyond the conventional continuum theories and models available within mechanics of materials.

The DSC represents a direction toward a unified theory for mechanics of materials and interfaces. Indeed, future work will be required for many issues such as additional analysis for mathematical characteristics, application to materials other than those included in the book, development of appropriate and advanced test devices and their use toward determination of parameters and validation of the actual behavior of problems in the field.

Appropriate and rational characterization of materials are vital for the solution of existing and emerging technological problems. This need will become more and more important for new and complex materials, particularly with the increased miniaturization of engineering systems. Hence, the pursuit of the development of new, alternative and unified concepts will continue.

APPENDIX I: Disturbed State, Critical State and Self Organized Criticality Concepts

We present a review, analysis and comparison between the disturbed state concept (DSC), critical state concept (CSC), and self-organized criticality (SOC) in the context of the mechanical behavior of engineering materials and interfaces/joints.

The DSC is based on the idea that a deforming material (interface) experiences *natural self-adjustment* (NSA) in its internal microstructure; *natural* implies that the material possesses the potential to adjust itself to external influences. During this process, the particles in the microstructure experience relative motions, e.g., sliding, rotation, and separation. The nature of deformation and particle motions will depend on factors such as initial stress and flaws, homogeneity, heterogeneity, density (loose or dense), and loading.

We arbitrarily divide materials in two categories or types. A normally fabricated or deposited material that is subjected to monotonically increasing loads is termed as Type I material and involves relatively homogeneous composition without significant flaws or preferred particle orientations. A normally consolidated (NC) soil, a granular material in loose state and a metal with relatively homogeneous composition are examples of Type I material. Materials (Type II) that have experienced prior histories such as unloading, and thermal and chemical loading, may exhibit relatively heterogeneous composition and brittle response. A material with preferred particle orientations and flaws, and an over-consolidated (OC) soil are examples of Type II material.

Both material types, in general, exhibit nonlinear response with irreversible (plastic) deformations and dissipation of energy. However, the mechanisms of deformation in the two would be different. Figure I.1 shows schematics of their behavior. In the case of Type I material, relative particle motions would cause continuous deformations and gradual breaking of weak bonds leading to breakdown under (infinitesimal) increase in strain or stress. The response would tend toward the FA or critical state; however, before it is reached, failure or breakdown will occur in the engineering sense. It is possible that the particle motions would cause "cracks" or "dislocations" with sizes below the microlevel; however, such cracks would coalesce to form major cracks or shear bonds essentially in the vicinity of the maximum or peak stress when catastrophic failure would take place.

In the case of Type II materials, the behavior may be initially similar to that for the Type I material. After a certain stress level (disturbance D_m), however,

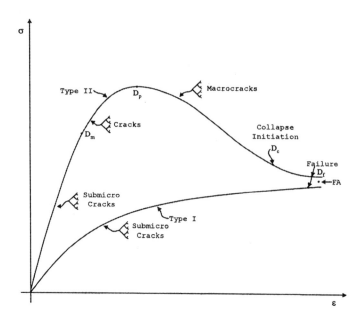

FIGURE I.1
Schematic of material responses.

microcracks would initiate. They could grow and become macrocracks, and after the peak (D_p) and at D_c, initiation of collapse or instability would occur. Continuing deformations beyond D_c would lead to "finite"-sized cracks and subsequent collapse or failure at D_f.

As the cracks form and grow with bifurcations, Fig. I.1, bursts of energy loss would occur. In the case of Type I material, such energy bursts and corresponding acoustic emissions (AE) or noise may be difficult to identify and measure at crack sizes below the microlevel. However, in the case of Type II material, it would be possible, with available engineering equipment, to measure AE at the crack initiation and growth, and define corresponding dissipated energy. The growth and coalescence of cracks would also cause measurable (AE) signals or noise.

A "homogeneous" or Type I material at relatively low (initial) density or in loose condition would compact continuously under compressive loading. Under tensile loading, continuous and gradual breaking of weak bonds would lead to the final breakdown. As a result, the motions of particles, Fig. I.2, may not involve any "sudden" changes or noise. Only at a high (critical) loading, failure or collapse will be caused due to sliding along shear planes. At this state, there can occur "sudden" and significant noise, which can be defined by measuring quantities such as acoustic emissions (AE) in the material (specimen). This situation would imply that before and after the peak or failure stress, the number of AE will be small without noticeable noise in the acoustic signals. At the same time, there can occur plastic or irreversible deformations

APPENDIX I

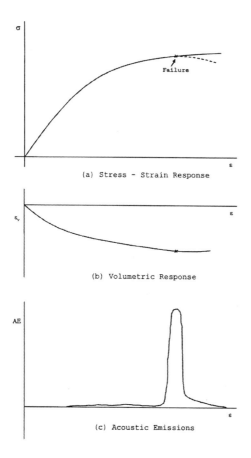

FIGURE I.2
Schematic of compactive and AE responses for relatively homogeneous (Type I) materials.

and dissipation of energy during the deformation process. One of the consequences of "continuous" irreversible deformations is the phenomenon of continuous yielding, which can lead to the critical state (Chapter 7).

A deforming heterogeneous or brittle material of Type II, on the other hand, develops microcracks or microfractures, which can nucleate and grow into clusters, and coalesce to form macrofractures. During this process, the material experiences irreversible deformations, and dissipation or release of seismic (acoustic) energy. As a consequence, redistribution of stresses that cannot be carried by the fractures (or failed parts) to the material parts in the intact state can take place. The process of microcracking can initiate somewhere before or at the peak stress, Fig. I.3. The occurrence of the release of seismic energy or seismic events can be measured as AE counts, and can exhibit noisy signals with different frequencies, Fig. I.3.

The NSA in the DSC implies bringing material particles or parts to a more effective relative configuration so as to optimally respond to external forces

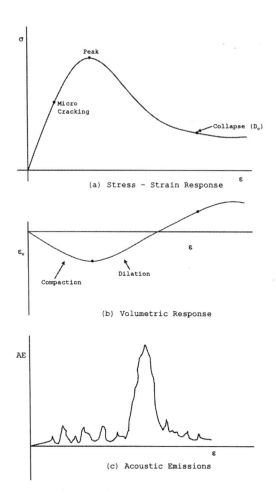

FIGURE I.3
Schematic of compactive-dilative, microfracturing and AE responses for relatively heterogeneous brittle (Type II) materials.

or influences. The process of self-adjustment results in the evolution or growth of the material's microstructure involving plastic yielding and microfracturing, or a combination of both, depending upon the nature of the composition and type of the material. The plastic yielding may not entail a process that causes sudden bursts of microcracks and energy, which can be termed as a relatively *quiet* phenomenon. On the other hand, microcracking (in a brittle material) may entail bursts of microcracks and energy, which can be termed as a *noisy* or *turbulent* phenomenon.

The behavior of both the above types of materials, homogeneous, and heterogeneous, involving growth and evolution of yielding or microfracturing, can be explained on the basis of the NSA approach. As it will be discussed later, the SOC can explain the response of brittle materials involving

APPENDIX I

microfracturing and noisy signals (e.g., flicker or 1/f noise). CSC and SOC mainly explain the phenomenon at the (final) critical state when avalanches and collapse (of the material's microstructure) initiate, e.g., D_c, Fig. I.3. On the other hand, the DSC can explain the entire deformation behavior including prepeak, postpeak, and critical or collapse. In other words, the DSC is considered to be general, as it can include the states defined by the CSC and SOC as special cases.

Self-Organized Criticality

Bak and coworkers (1–4) have proposed the self-organized criticality (SOC) concept. It is based on the idea that a large complex system such as a granular material containing many particles (components), under external influences, can evolve into a *poised* or *critical* state; its response is affected by past history and events. At the critical state (increasing) minor disturbances can lead to catastrophic events similar to those at the collapse or limiting critical state, called avalanches, of different sizes. Evolution of such a system to the critical state results as a consequence of the interactions between the constituents of the system. It was shown that a wide range of phenomena can be described on the basis of the SOC, e.g., behavior of a sand pile, occurrence of earthquakes, price variation of commodities, evolution and extinction of biological species, growth of cites, light from quasars and fluctuation of flow in rivers (4).

The SOC is often illustrated by using the behavior of a sand pile, Fig. I.4 (4). To start with, the sand pile is relatively flat and stable, and when sand grains are added, they remain at the locations where they land. As sand grains are

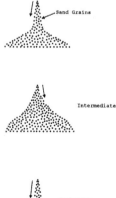

FIGURE I.4
Sandpile behavior (SOC).

further added, the pile becomes steeper with an increase in the sliding of grains over the steep surfaces. At the critical (stage) slope, sliding can occur over almost all of the pile, when the behavior of the sand pile is *out of balance*, and catastrophic sliding events or avalanches can take place. Such events cannot be understood in terms of the behavior of individual grains (components). In other words, in the behavior of avalanches, the coupling or iteration between the response of grains plays a vital role; thus, the behavior can be understood or defined only on the basis of *unified* or *holistic* considerations. Bak and Tang (2) have noted that the behavior of a complex material system may not be explained based on the response of individual particles, as in the micromechanics approach (5, 6).

In the SOC, the evolution of events (microfracturing, fluctuations in river levels, change in market prices, etc.) is characterized by long periods of stability and short periods of rapid changes during which new events or forms appear. Such a phenomenon is often referred to as *punctuated equilibrium* (4), with periods of *stasis* (state of static balance of equilibrium or stable states) of different durations separated by bursts of activity, which gives rise to 1/f or flicker noise and power law behavior. In the case of microfracturing in brittle materials (Type II), the occurrence of microcracks and fractures involving stable material states can be considered to exhibit punctuated equilibrium and stasis response.

It may be mentioned that the SOC (in the case of the sand pile) explains the phenomenon of collapse or avalanche formation. However, it dos not include the mechanical response of the growing sand pile before it reaches the critical state. In other words, in the case of the behavior of engineering materials, the SOC defines the initiation of collapse state (D_c, Fig. I.1), but may not define the entire pre- and postpeak response.

In engineering material systems, evolution of the critical states, in general, entails irreversible deformations, i.e., dissipation or release of energy. The earth's crust subjected to tectonic forces and motions of plates can lead to the critical states or earthquakes. If we consider the relation between the flow of sand (during sliding) with time, or the gradual developments of motions in the earth's crust leading to an earthquake, we would observe erratic signals of different durations due to sudden crack formation and energy bursts. Such signals are called flicker noise or 1/f noise (pronounced as "one over ef" noise). The name 1/f noise indicates that the strength of a signal is inversely proportional to its frequency (f).

The flicker or 1/f noise can be related to the systems that have evolved into the critical state. Such systems exhibit a power law behavior given by

$$N(f) \sim f^{-\beta} \tag{I.1a}$$

where N is any quantity such as the number of earthquakes (or acoustic emissions) with given energy, for which Eq. (I.1a) can be expressed as

$$N(s) \sim s^{-\tau} \tag{I.1b}$$

where s is the energy released by the earthquake and τ is a parameter. The well-known empirical Gutenberg-Richter law (7, 8) for earthquakes follows Eq. (I.1) and is given by

$$\log_{10} N = a - bm \tag{I.2a}$$

where parameters a and b depend on the (regional) location, and N is the number of earthquakes of size (or magnitude) greater than m. The energy, E, released during the earthquake increases exponentially with its size, and is given by

$$\log_{10} E = c - dm \tag{I.2b}$$

where c and d are parameters. Based on Eq. (I.2), the Gutenberg-Richter law represents the power law, Eq. (I.1), that connects the released energy (E) with the frequency distribution:

$$\frac{dN}{dE} = m^{-\tau} \tag{I.3a}$$

where $1.25 < \tau < 1.5$ (2). The relation between N and m can also be expressed as

$$\log N \sim -\tau \log m \tag{I.3b}$$

which represents a straight line on the log N vs. log m plot, Fig. I.5. In this figure, N represents the number of earthquakes (per year), and magnitude m is proportional to the released energy in the New Madrid seismic zone in the southeastern United States during the period 1974–83 (8).

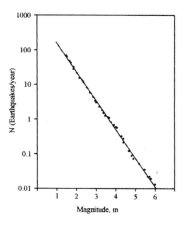

FIGURE I.5
Distribution of earthquake magnitudes in New Madrid zone (8). ©American Geophysical Union.

In the critical state concept (CSC) (9), a deforming material with given initial pressure approaches the critical state at which it continues to deform in shear under the constant shear stress with constant density or void ratio (Chapter 3). The initiation of the critical state can be considered to imply initiation of collapse of the material. In this sense, the basic idea in the SOC that defines the behavior of a system that evolves to collapse can be considered to be similar to the CSC. This has been shown by Evesque (10) on the basis of the stability response of glass spheres in a rotating drum, by using Eqs. (3.2a) and (3.2b), Chapter 3.

The foregoing brief descriptions suggest that a natural system (like an engineering material) composed of interacting particles or clusters of particles under external influences (forces, temperature, chemicals, etc.) can evolve into catastrophic states, affected by the past history, due to the self-organization of the internal microstructure of the system. As discussed in various chapters in this book, a deforming material evolves from the RI state to the FA state. During this process, the material's microstructure can experience irreversible deformations and energy release for homogeneous (Type I) materials, while it may experience both the energy release and microfracturing for brittle materials (Type II). Both may exhibit a power law type behavior; however, the occurrence of the 1/f noise may usually be observed and measurable only in the case of microfracturing. In other words, it is believed that the SOC that is based on the power law behavior, steady states and 1/f noise, at this time may be defined mainly in the case of microfracturing response.

The DSC, with the NSA idea, can explain both the foregoing responses and can allow the characterization of the entire (prepeak, postpeak, and critical or catastrophic) behavior, Figs. (I.1)–(I.3). The overall behavior can experience a number of *threshold transitions* in its microstructure such as the transition from compactive to dilative volume, peak stress, and the critical condition, at D_c (11–13), Fig. I.3.

Acoustic Emissions

One of the ways to ascertain if the deforming material exhibits noisy (1/f type) behavior is to perform indirect (nondestructive) measurements such as acoustic emissions. Although a number of studies have considered the microfracturing behavior, including measurements of AE, there are not many results available for the measurement of AE in homogeneous (Type I) materials. Figure I.6 shows computer simulations of stress-strain, AE and dissipated energy responses of a relatively (or more) heterogeneous rock specimen (with homogeneity index $m = 1.5$), and relatively (or more) homogeneous rock specimen ($m = 3$) reported by Tang and Kaiser (14,15); details of the computer simulation are given later in Example I.3. It can be seen that the response of the heterogeneous rock involves relatively "noisy" AE distribution. On the other hand, the homogeneous rock does not involve significant AE in the prepeak and postpeak regions; the major AE occur near the peak stress where failure due to sliding is indicated. Thus, the response of the

APPENDIX I

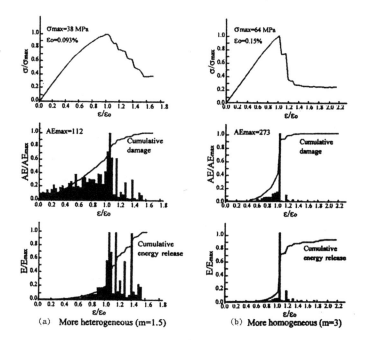

FIGURE I.6
Numerical simulation of tests for rock specimens (14). (Reprinted with permission from Elsevier Science.)

relatively heterogeneous rock indicates dispersed microcracking with AE that exhibit sudden bursts and noise, while that for the relatively homogeneous rock shows sudden bursts only near the peak. However, both may exhibit power law type behavior; this is shown in Fig. I.7 for the relatively heterogeneous rock. Here, the number of acoustic events for given AE count were taken from Fig. I.6; because the details of the laboratory data are not available, the values are essentially approximate. The rock exhibits power law behavior given by

$$N = \alpha(AE)^{-\tau} \qquad (I.4)$$

where the value of $\alpha = 166.15$ and $\tau = 1.012$. Since sufficient data is not available from Fig. I.6 for the homogeneous rock, it is not possible to obtain a similar plot.

In the following, we consider examples of the behavior of a number of materials including correlations and comparisons between the DSC and SOC.

Example I.1: Simulated Microfracturing Phenomenon in Elastoplastic Materials

Zapperi et al. (16), performed numerical tests by considering the "fuse" model constituting a resistor network system on a tilted square lattice. The disorder (or heterogeneity) was introduced in the model by assigning a random failure

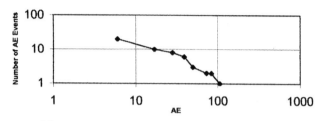
(a) AE vs Number of AE Events: Heterogeneous Rock Specimen

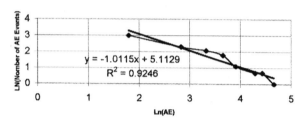
(b) Ln(AE) vs Ln(Number of AE Events): Heterogeneous Rock Specimen

FIGURE I.7
Number of AE events vs. AE event and determination of τ and α.

threshold, I_c, to each resistor, where I_c is the critical current analogous to stress. In the classic fuse model, failure is identified when the current flowing in a resistor exceeds the failure threshold, I_c; then the bond is removed from the lattice, and the electrical conductivity (c) drops to zero. Thus, the (resistor) model develops a macroscopic crack and eventually the lattice breaks. The simulated test (16) introduced permanent "damage" to the bond by decreasing its conductivity by a factor $a = (1 - \omega)$, where ω is the damage parameter, [Eq. (4.23), Chapter 4]. The analogy to linear elastic behavior can be used by expressing the observed or actual stress, σ^a, as

$$\underset{\sim}{\sigma}^a = (1 - D)\underset{\sim}{\sigma}^i = (1 - D)\underset{\sim}{C}^i \underset{\sim}{\varepsilon}^i \tag{I.5a}$$

where $\underset{\sim}{C}$ is the elastic constitutive matrix. Thus, the reduction in the conductivity, a, denotes the reduction in the initial stiffness, $\underset{\sim}{C}$, which is equivalent to the conductivity during the application of current, I; the voltage, V, is analogous to the strain. For the resistor model, Eq. (I.5a) can be written as

$$I^a = (1 - D)I^i = (1 - D)c^i \cdot V^i \tag{I.5b}$$

It may be noted that Eq. (I.5) represents the classical damage model [Eq. (4.24), Chapter 4]. In the context of the DSC, the general form of Eq. (I.5) will be given by

$$I^a = (1-D)c^i V^i + Dc^c \cdot V^c \tag{I.6}$$

where c^c and V^c are relevant to the FA material parts. It is believed that Eq. (I.6) can lead to more realistic representation of the behavior of the resistor model because it allows for the coupling between the undamaged and damaged parts (Chapter 4). Zapperi et al. (16) used the simplified Eq. (I.5). Such approximations (14–16) do not allow for the nonlinear response, which many materials exhibit.

In the resistor model test, an external voltage (strain) difference is applied to the lattice with periodic boundary conditions in the other direction. The current in each bond or resistor node is computed by solving numerically the Kirchoff equation using a multigrid relaxation algorithm with precision $\varepsilon = 10^{-12}$. As the voltage is increased, the current (stress) reaches the threshold value in (number of) bonds. For such damaged bonds, the disorder is changed according to the reduction in the conductivity. The process is repeated under new currents until no unstable bonds are present. The "redistribution" of the disorder can cause an avalanche of additional bonds breakage after the failure of a single bond. The network system was studied in the limit of slow driving, implying that the time scale over which microfractures form and propagate is much faster than the time scale of the external driving (voltage). This condition is one of the characteristics of systems that exhibit SOC (4).

During the application of increasing voltage, initially there occurs proportional increase in the total current (stress); i.e., the system behavior is linearly elastic, Fig. I.8(a). With further increase in the voltage and development of damage and broken bonds, the system deviates from linear elasticity, becomes macroscopically plastic, and approaches the steady state. In this state, an increase in the voltage (strain) is balanced by the damage such that the current (stress) remains approximately constant, and it is accompanied by high fluctuations (in the current), with avalanches of different sizes.

Zapperi et al. (16) studied the statistical properties of the bond breakage, fracture, or rupture events in time and the magnitude of fracturing, Fig. I.8(b). The avalanche or rupture event size probability distribution function, $P(s)$, was found to be given by

$$P(s) \sim s^{-\tau} \tag{I.7}$$

where s is the number of damaged bonds for a given voltage increment, Fig. I.8(b), and $\tau = 1.19 \pm 0.01$. Figure I.8(c) shows the plot of $P(s)$ vs. s for different sizes of the model ($L = 16, 32, 64$). It can be seen that the response of the system experiencing microfracture and rupture events displays the power-law behavior. The distribution of energy bursts during the testing was found to be related to the acoustic emissions recorded, and also exhibited the power-law behavior (16–18).

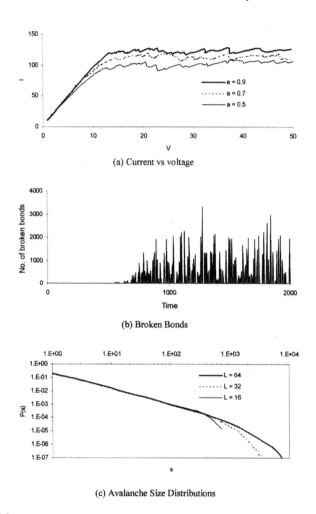

FIGURE I.8
Current vs. voltage, broken bonds, and avalanche size distributions (16). (Reprinted with permission from *Nature*, 1997, McMillan Magazines Ltd.)

Based on the above-simulated response of elastoplastic materials, it can be stated that under the strain or displacement (voltage) controlled loading, the system may exhibit SOC. However, it was stated that under constant stress (or current) controlled loading, the system may tend to instability corresponding to the critical stress (current) in the strain (voltage) controlled experiment (16). A system like a hardening (compacting) material may exhibit non-stationary distribution of critical state (avalanches) with a power law; however, it may not exhibit the SOC response. Thus, there are controversies and differences of opinion regarding whether all materials under different loading conditions are SOC (16, 19). Further investigations will be needed to resolve these and other issues. In the meantime, it is believed that the DSC, with the idea of natural self-adjustment, can provide a general framework for the behavior of engineering materials.

APPENDIX I

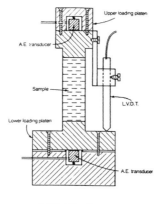

(a) Test Specimen

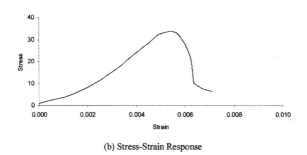

(b) Stress-Strain Response

FIGURE I.9
Rock specimen and stress-strain response: strain rate = 0.495 μm/sec ($\approx 7 \times 10^{-6}$/sec) (20). (Reprinted with permission from Elsevier Science.)

Example I.2: Microfracturing and Acoustic Emissions in Laboratory Tests for Rock

Cox and Meredith (20) performed unconfined compression tests on cylindrical specimens (25 mm dia. and 72 mm long), Fig. I.9(a), of Gosford sandstone from NSW, Australia. The rock has porosity of 13% and contains mostly sub-angular to sub-rounded quartz and quartzite grains with diameters in the range of 0.1–1.0 mm, with about 20% fine-grained clay and polycrystalline mica matrix material. The specimens are tested under strain-controlled loading at a constant rate of axial shortening. Acoustic emissions are measured by using PZT AE sensors located in cavities within the end platens, Fig. I.9(a).

The stress-strain behavior of one of the specimens, shortened at the rate of 0.495 μm/sec (approximate strain rate = 7×10^{-6}/sec) is shown in Fig. I.9(b). The stress vs. time (which bears approximately linear relation to strain) and AE or hit rate vs. time are shown in Figs. I.10(a) and (b), respectively. It can be seen from Figs. I.9(b) and I.10(a) that during the linear elastic phase of the stress-strain response, very few AE occurred because the active deformation

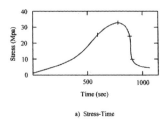

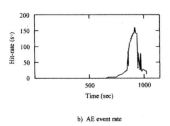

FIGURE I.10
Stress vs. time and hit (AE) rate vs. time for rock specimen, Figure I.9 (20). (Reprinted with permission from Elsevier Science.)

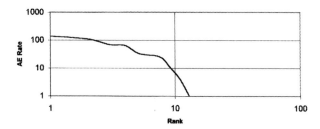

FIGURE I.11
AE rate vs. rank: Test on rock specimen.

mechanisms (pore-crack closure and elastic shortening) are *continuous*. In the zone beyond the linear elastic in which irreversible deformations occur, the rate of AE increases because of discontinuities due to microfractures and cracks. The maximum rate occurs near the peak and stress drop region, and then decreases in the post-drop region.

The AE rates at different times were computed (approximately) based on Fig. I.10(b). They were then ranked in descending order, Table I.1. Figure I.11 shows a plot of AE rate vs. rank on log-log scale. This plot is similar to that proposed in the Zipf's law (21). The (average) value of τ (Fig. I.11) is found to be 1.558 with α = 316.65, indicating the power-law behavior as in the SOC.

APPENDIX I

TABLE I.1
Acoustic Emission Rates and Rank for Microfracturing in Rock (20)

Time, Secs Origin + 500 secs	AE Rate	Rank	Order AE
62.5	1	1	140
125	2	2	110
187.5	4	3	70
219.5	7	4	65
250	12	5	35
282.5	28	6	30
312.5	65	7	28
344.75	110	8	22
375	140	9	12
389.25	70	10	7
437.5	35	11	4
469.75	30	12	2
500	22	13	1

In the following example, we analyze computer-simulated behavior of a rock, which provides comprehensive results in terms of stress-strain response, AE and dissipated energy.

Example I.3: Simulated Microfracturing and Acoustic Emission in Brittle Rock

Tang and Kaiser (14) and Kaiser and Tang (15) performed computer simulation of the behavior of rocks with different levels of heterogeneities. In addition to the response of rock itself (Fig. I.6), they also considered the behavior of rock (pillar) surrounded by host rocks [roof and floor, Fig. I.12(a)]. Two cases were considered, stiffer, and softer host rocks; we consider here analysis of the former. For the stiff host rock problem, the following material properties were adopted (15):

Property	Pillar	Roof and Floor
Elastic Modulus	60 GPa	300 GPa
Poisson's Ratio	0.25	0.25
Shear Strength	200 MPa	200 MPa
Homogeneity Index		
Elasticity	20	20
Strength	3	3

The simulations are performed by using a finite element procedure in which the rock was assumed to be linear elastic. A finite element is considered to have failed when the peak stress, σ_f, is reached. The seismic events or acoustic emissions are assumed to be proportional to the number of failed elements under the brittle failure. Then, the cumulative AE event rate, Ω, is

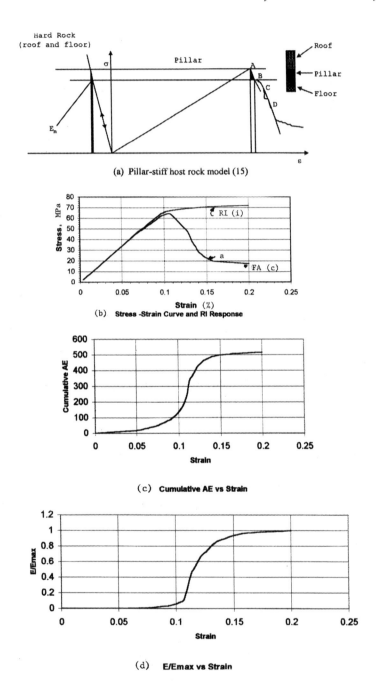

FIGURE I.12
Pillar-stiff host rock model and simulated stress-strain, AE, and energy responses.

considered to be proportioned approximately to damage or disturbance (D) as

$$\Omega = D = \frac{1}{N}\sum_{i=1}^{s} n_i \qquad (\text{I}.8\text{a})$$

where D is the damage given by

$$D = \frac{\Sigma v_f}{V} \qquad (\text{I}.8\text{b})$$

v_f is the volume of failed element, V is the total volume, n_i is the number of failed elements, s is the number of calculation steps, and N is the total number of elements in the (FE) mesh. The cumulative seismic energy (E), assuming brittle failure at σ_f, is given by (15)

$$E = \Sigma E_f \frac{v_e}{2} \Sigma \frac{\sigma_f^2}{C_f} \qquad (\text{I}.9)$$

where C_f is the elastic modulus, and E_f is the released energy for a given element. The magnitude, m_f, of an individual microseismic event, and the magnitude, M_f, of the seismic event cluster are obtained from (15)

$$m_f = \log(E_f) + C \qquad (\text{I}.10\text{a})$$

and

$$M_f = \log\left(\sum_{i=1}^{n} E_{fi}\right) + C \qquad (\text{I}.10\text{b})$$

where i is the number of events in an event cluster and C is the constant.

It may be noted that the above model is based on simple linear elastic characterization in which the "damaged" elements have no strength, as in the classical damage model. In actuality, the rock behavior is usually nonlinear, and the FA (or damaged) elements can possess certain strength and interact with the RI elements as in the DSC. If such interactions were allowed, the computed simulations can be different.

Figure I.12(b)–(d) (22) show the stress, cumulative AE and E/E_{max} vs. strain for the pillar in the stiff host rock model, Fig. I.12(a); details of plots showing AE counts at different loading steps and formation of shear planes leading to failure are given in (15).

Figures I.12(b)–(d) show that the highest number of AE and magnitude of energy dissipation occur around the first stress drop, Fig. I.12(a), after the peak stress. Then a number of stress drops occur in the descending part of the curve, and each drop is associated with elevated AE and energy release. For the pillar-stiff host rock model, the energy stored in the host rock that can be

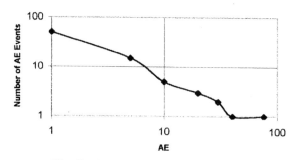

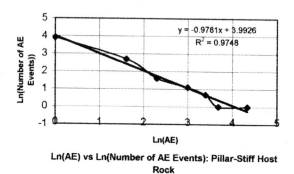

FIGURE I.13
Number of AE events vs. AE event and determination of τ and α.

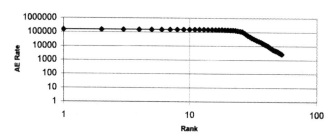

FIGURE I.14
AE rate with respect to strain vs. rank.

released before equilibrium, or steady state, is very small such that stability can be achieved at various stages in the post-peak response, e.g., points B and D, Fig. I.12(a).

Figure I.13 shows the logarithmic plot of number of times an AE occurred vs. AE, measured (approximately) from Fig. I.12(c). It shows that the slope, τ, is about 0.978 with $\alpha = 54.196$, showing power-law behavior.

Figure I.14 shows the plot of AE rate with respect to strain $AE/d\varepsilon\,(=AE_r)$ vs. rank, as in the Zipf's law (21). The AE rate is computed as average AE over a

APPENDIX I

given strain range divided by the corresponding strain increment. This is similar to the analysis by Petri (18) in which the number of broken bonds between two consecutive stretches (strains) are considered. The relation between $(AE)_r$ and rank (R) is given by

$$(AE)_r = \alpha R^{-\tau} \tag{I.11}$$

The value of τ, Fig. I.14, is found to be about 1.42 and $\alpha = 32 \times 10^5$.

Disturbance, AE, and Energy

The disturbance, D, can be expressed based on the stress-strain response, Fig. I.12(b), where the RI response is simulated as nonlinear elastic or elasto-plastic:

$$D_\sigma = \frac{\sigma^i - \sigma^a}{\sigma^i - \sigma^c} \tag{I.12a}$$

in which $\sigma^c \approx 17.0$ MPa. In terms of plastic strains, ε, disturbance is expressed as

$$D_\sigma = 1 - e^{-A_1 \varepsilon^{Z_1}} \tag{I.12b}$$

where $D_u = 1$ is assumed. The plastic strains, ε, are computed as

$$\varepsilon = \varepsilon^t - \varepsilon^e \tag{I.12c}$$

where ε^t denotes the total strain, Fig. I.12(b), and ε^e is the elastic strain computed by using elastic modulus, $E \approx 6.67 \times 10^4$ MPa. Figure I.15 shows computed, Eq. (I.12b), and measured, Eq. (I.12a), disturbance, with respect to ε. The values of A_1 and Z_1 were found by plotting $\ln[-\ln(1 - D_\sigma)]$ vs. $\ln(\varepsilon)$; their values were found to be $A = 48.584$ and 1.522, respectively.

The variations of the norm of AE_n ($AE/515$), Fig. I.12(c), and of E/E_m, Fig. I.12(d), can also be expressed in terms of plastic strains as

$$AE_n = 1 - e^{-A_2 \varepsilon^{Z_2}} \tag{I.13a}$$

and

$$E/E_m = 1 - e^{-A_3 \varepsilon^{Z_3}} \tag{I.13b}$$

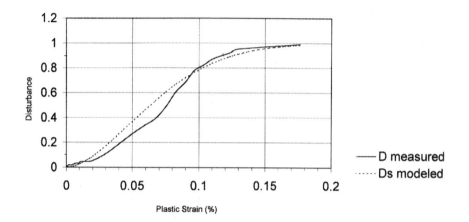

FIGURE I.15
Measured and modeled disturbance vs. plastic strain.

The values of $A_2 = 45.792$ and $Z_2 = 1.192$, and $A_3 = 168.629$ and $Z_3 = 1.912$ were found corresponding to Eqs. (I.13a) and (I.13b), respectively. Figures I.16(a) and (b) show computed variations of AE_n and E/E_m vs. plastic strains, Eq. (I.13). It can be seen from Figs. (I.15) and (I.16) that the disturbance, AE_n and E/E_m, show similar relationships with respect to plastic strains. Also, they can be correlated; Figs. I.17(a) and (b) show such relations between disturbance and AE and E/E_m. In fact, the relations can be expressed as

$$D_{AE} = 1 - e^{-A_4(AE_n)^{Z_4}} \tag{I.14a}$$

$$D_E = 1 - e^{-A_5(E/E_m)^{Z_5}} \tag{I.14b}$$

The foregoing analyses show that there exist correlations between disturbance, acoustic emissions and energy dissipation. Hence, the behavior of the rock can be explained by using disturbance as the measure of microstructural changes. This is further illustrated in Fig. I.18, where the disturbance rate (D_r) with respect to strains is plotted with rank; in Fig. (I.14), a similar plot for energy rate is presented. The behavior in Fig. I.18 exhibits the power law with $\tau = 1.027$ and $\alpha = 1228.68$.

In Fig. I.19(a)–(c) are plotted the variations of derivatives of disturbance, Eq. (I.12b), AE_n, Eq. (I.13a), and E/E_m, Eq. (I.13b), with plastic strains; they show behavior similar to the AE rate measured in the laboratory test on rock, Fig. I.10b. At the initiation of the collapse or residual condition, the plastic strain is about 0.12% and occurs at the critical value, $D_c \approx 0.90$, $(AE_n)_c \approx 0.95$, and $(E/E_m)_c \approx 0.97$.

APPENDIX I

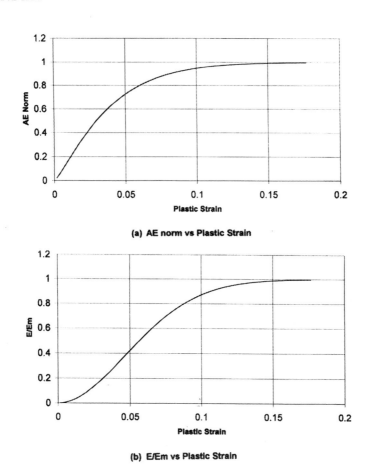

FIGURE I.16
Computed AE norm and E/E_m vs. plastic strain.

Example I.4: Instability and Liquefaction During Earthquake

We now present analysis of the Port Island, Japan earthquake in 1995, during which data for measured shear wave velocities, stress-strain behavior and dissipated energy were available (23). These results were used to identify liquefaction based on the DSC and energy approaches (12).

Figures I.20(a) and (b) show soil details and instruments, and typical measurements of shear wave velocities with time, respectively. Figures I.21(a) and (b) show dissipated energy vs. time and typical shear stress-strain curves, respectively. Figure I.22 shows disturbance (D) with time at various depths. Disturbance was computed on the basis of shear wave velocities [Eq. 3.9, Chapter 3] in which the RI shear wave velocity, $V^i = 250$ m/s, and the FA velocity, $V^c = 25$ m/s, were adopted; V^i relates to average velocity in the upper saturated soil layer, Fig. I.20(b) before the earthquake, and V^c represents the fully adjusted or residual velocity after the earthquake shaking. It

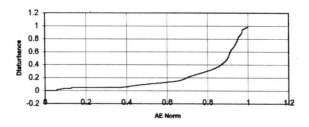

(a) AE norm vs Disturbance

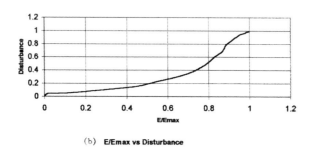

(b) E/Emax vs Disturbance

FIGURE I.17
Relations between AE norm and E/Em, and disturbance.

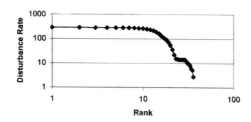

FIGURE I.18
Disturbance rate vs. rank.

was found that the soil experienced instability and liquefaction after about 14.6 seconds during the earthquake when the shear wave velocity was about 42 m/s and the critical disturbance, $D_c \approx 0.905$, Fig. I.22.

The dissipated energy and disturbance rates, E_r and D_r, were computed with respect to time. Here, average values of energy or disturbance, Figs. I.21 and I.22, over a given time range were divided by the corresponding time increment. Figures I.23(a) and (b) show E_r and D_r versus Rank (R) on the logarithmic scale. The values of the exponent, τ, were found to be about 1.182 and 1.1416, and $\alpha = 27.27$ and 6.45 for E_r and D_r, respectively, showing the

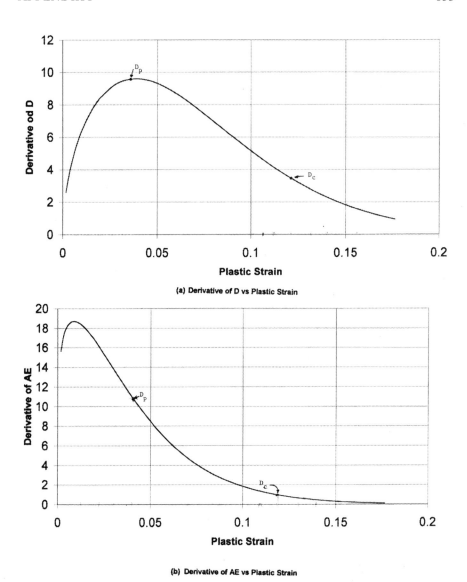

FIGURE I.19
Plots of derivatives of D, AE norm, and E/E_m vs. plastic strain.

power-law behavior. However, the existence of 1/f noise cannot be verified for this case. This is because the saturated soil under earthquake loading experiences continuous compaction, which causes increase in the pore water pressure. When the pore water pressure approaches the critical value (critical disturbance, D_c), microstructural instability or liquefaction occurs (11–13). This compaction phenomenon essentially involves particle motions and may not involve microfracturing that may usually be responsible for 1/f-type noise.

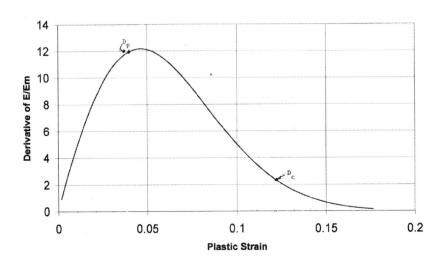

(c) Derivative of E/Em vs Plastic Strain

FIGURE I.19
(continued)

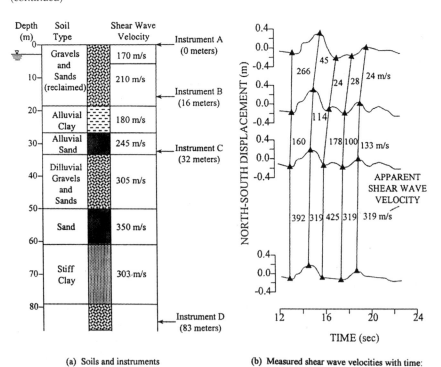

(a) Soils and instruments

(b) Measured shear wave velocities with time: north-south direction

FIGURE I.20
Soils, instruments, and measured shear wave velocities, Port Island, Kobe site (23).

APPENDIX I

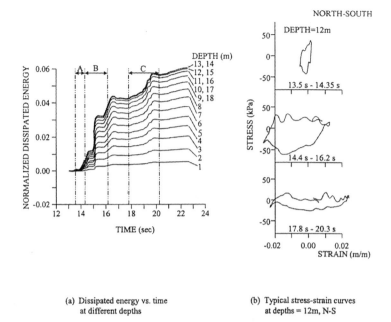

(a) Dissipated energy vs. time at different depths

(b) Typical stress-strain curves at depths = 12m, N-S

FIGURE I.21
Dissipated energy and typical stress-strain curves (23).

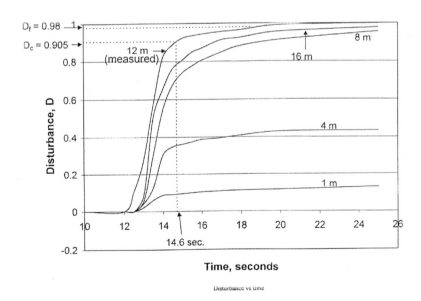

Disturbance vs time

FIGURE I.22
Disturbance vs. time at different depths, energy vs. depth, and growth of liquefaction (12). With permission.

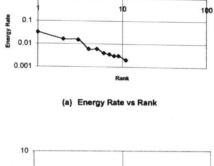

(a) Energy Rate vs Rank

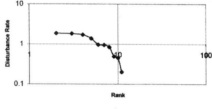

(b) Disturbance Rate vs Rank

FIGURE I.23
Energy rate and disturbance with respect to time vs. rank: Port Island earthquake.

Example I.5: Cyclic Behavior of Saturated Sands

We consider laboratory behavior of two saturated sands, Reid Bedford (24) and Ottawa (11–13, 25), tested by using torsional shear and multiaxial devices, respectively.

Figure I.24 shows the shear stress vs. strain and time (10 points/sec), and pore water pressure vs. time responses for the sand specimen with relative density $D_r = 60\%$, initial mean pressure $p_o = 124.10$ kPa tested under cyclic strain-controlled condition (frequency = 0.10 Hz) with amplitude of shear strain = 47%. Figure I.25 shows plots of disturbance (D) and accumulated energy per unit volume with number of cycles (N). The values of D were found based on the stress-strain response, Fig. I.24(a). Figure I.26(a) shows energy rate E_r vs. Rank; E_r was computed as the ratio of average energy over a range of pore water pressure to the pore water pressure increment. Figure I.26(b) shows disturbance rate vs. rank. The value of $\tau = 0.341$ and 0.327, and $\alpha = 2155.98$ and 1.17, respectively, are indicated.

Cyclic stress-controlled (frequency = 0.10 HZ) tests were performed on cubical ($10 \times 10 \times 10$ cm) specimens of saturated Ottawa sand with $D_r = 60\%$ and initial effective confining pressures $\bar{\sigma}_o' = 69, 138$, and 207 kPa. The measurements were obtained in terms of stress, strain, and pore water pressure responses (25). The results for pore water pressure (U_e) vs. time for the three confining pressures are shown in Fig. I.27. The effective stress $\bar{\sigma}'$ was computed as the difference between the total stress, σ, and pore water pressure, U_e; Fig. I.28 shows plots of $\bar{\sigma}'$ vs. N. The disturbance was computed by

APPENDIX I

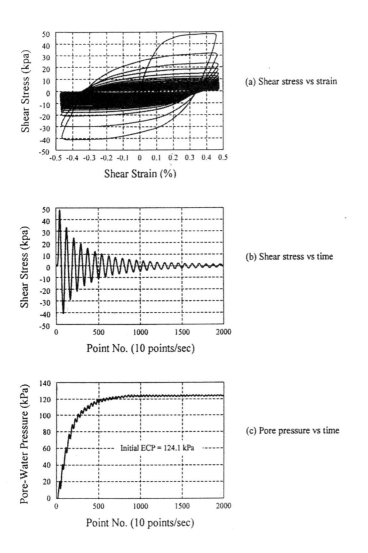

FIGURE I.24
Test results for Reid Bedford sand (24) (with permission).

assuming the RI behavior to be elastoplastic with the HISS δ_0-plasticity model (Chapter 7), given by

$$D = \frac{\bar{\sigma}'^{(i)} - \bar{\sigma}'^{(a)}}{\bar{\sigma}'^{(i)} - \bar{\sigma}'^{(c)}} \quad (I.13)$$

where $\bar{\sigma}'^{(c)}$ is the asymptotic effective stress, which can be adopted to be zero.

Figure I.29 shows the plot of the disturbance rate, D_r, with respect to pore water pressure vs. the rank; the values of $\tau = 1.886$ and $\alpha = 1.986$.

658   Mechanics of Materials and Interfaces

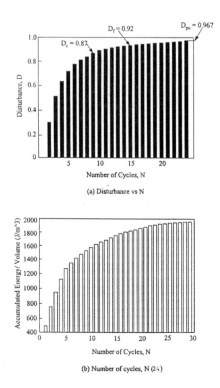

FIGURE 1.25
Disturbance and accumulated energy vs. time (N): Reid Bedford sand.

As indicated earlier, the saturated sand under shear loading (strain and stress controlled) experiences compaction and reaches unstable or collapse or liquefaction condition when the pore water pressure becomes approximately equal to the initial effective pressure. During the process, the material's microstructure can experience continuing compaction without (measurable) $1/f$-type noise. Hence, the behavior may not be considered to exhibit SOC-type response. However, the microstructural evolution can be explained in terms of self-adjustment in the DSC.

Summary

The foregoing review and analysis indicate that the DSC, with the microstructural self-adjustment approach, can explain mechanical behavior of a wide range of material systems. The disturbance, D, provides for evolution of both the continuous yielding and microfracturing responses, and can be correlated to dissipated energy and nondestructive measures such as acoustic emissions. The DSC provides characterization of the entire material

APPENDIX I

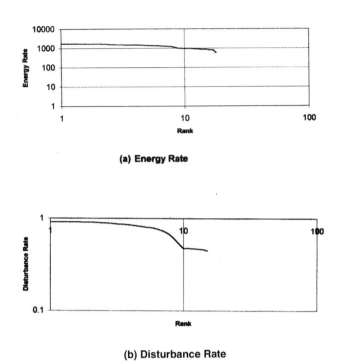

FIGURE I.26
Energy and disturbance rate with respect to cycles vs. rank: Reid Bedford sand.

response including prepeak, peak, postpeak, and collapse or failure. On the other hand, the CSC and SOC provide definition of collapse or critical state; in this context, the DSC is considered to be general, because it includes the collapse state defined by the CSC and SOC as a special case.

References

1. Bak, P., Tang, C., and Weisenfeld, K., "Self-Organized Criticality: An Explanation of 1/f Noise," *Physical Review Letters*, 59, 4, 1987, 381–384.
2. Bak, P. and Tang, C., "Earthquakes as a Self-Organized Critical Phenomenon," *J. of Geophysical Research*, 94, 311, 1989, 15635–15637.
3. Bak, P. and Chen, K., "Self-Organized Criticality," *Scientific American*, January 1991, 26–33.
4. Bak, P., *How Nature Works: The Science of Self-Organized Criticality*, Copernicus: Springer-Verlag, New York, 1996.
5. Nemat-Nasser, S. and Hori, M., *Micromechanics: Overall Properties of Heterogeneous Materials*, North-Holland, Amsterdam, 1993.

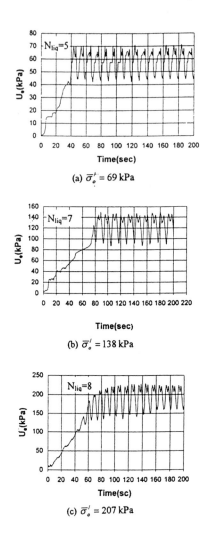

FIGURE I.27
Multiaxial test results for excess pore water pressure for Ottawa sand (11–13, 25).

6. Okai, Y. and Horii, H., "A Micromechanics-Based Continuum Theory for Microcracking Localization of Rocks under Compression," Chapter 2 in *Continuum Models for Materials with Microstructure*, H. B. Mühlhaus (Editor), John Wiley, Chichester, UK, 1995, 27–68.
7. Gutenberg, B. and Richter, C. F., *Seismicity of the Earth*, Princeton University Press, Princeton, NJ, 1949.
8. Johnston, A. C. and Nova, S., "Recurrence Rates and Probability Estimated for the New Madrid Seismic Zone," *J. of Geophysical Research*, 90, 1985, 6737.
9. Roscoe, K. H., Schofield, A. N., and Wroth, C. P., "On Yielding of Soils," *Geotechnique*, 8, 1958, 22–53.

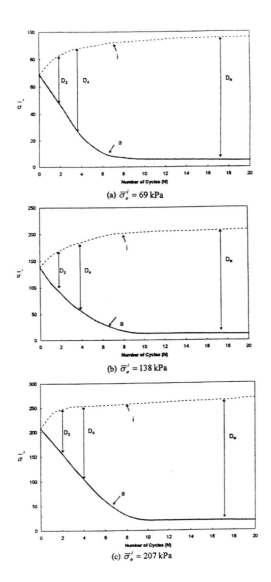

FIGURE I.28
Effective stress vs. number of cycles for Ottawa sand (13) (with permission).

10. Evesque, P., "Analysis of the Statistics of Sandpile Avalanches Using Soil-Mechanics Results and Concepts," *Physical Review A, The American Physical Society*, 43, 6, 1991, 2720–2740.
11. Desai, C. S., Park, I. J., and Shao, C., "Fundamental Yet Simplified Model for Liquefaction Instability," *Int. J. Num. and Analyt. Meth. Geomech.*, 22, 1998, 721–748.
12. Desai, C. S., "Evaluation of Liquefaction Using Disturbed State and Energy Approaches," *J. of Geotech. & Geoenv. Eng.*, ASCE, 126, 7, 2000, 618–631.

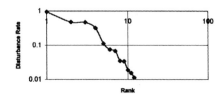

FIGURE I.29
Disturbance rate with respect to pressure vs. rank: Ottawa sand.

13. Park, I. J. and Desai, C. S., "Cyclic Behavior and Liquefaction of Sand Using Disturbed State Concept," *J. of Geotech. and Geoenv. Eng.*, ASCE, 126, 9, 2000.
14. Tang, C. A. and Kaiser, P. K., "Numerical Simulation of Cumulative Damage and Seismic Energy Release During Brittle Rock Failure—Part I: Fundamentals," *Int. J. Rock Mech. Min. Sc.*, 35, 2, 1998, 113–121.
15. Kaiser, P. K. and Tang, C. A., "Numerical Simulation of Damage Accumulation and Seismic Energy Release During Brittle Rock Failure—Part II: Rib Pillar Collapse," *Int. J. Rock Mech. Min. Sc.*, 35, 2, 1998, 123–134.
16. Zapperi, G., Vespignani, A., and Starkey, E. H., "Plasticity and Avalanche Behaviour in Microfracturing Phenomena," *Letter to Nature, Nature*, Macmillan Magazines Ltd., 388, 1997, 658–660.
17. Petri, A., Paparo, G., Vespignani, A., Alippi, A., and Costantini, M., "Experimental Evidence of Critical Dynamics in Microfracturing Processes," *Physics Review Letters*, 73, 1994, 3423–3426.
18. Petri, A., "Acoustic Emission and Microcrack Correlation," *Philosophical Magazine B*, 77, 2, 1998, 491–498.
19. Sornette, D., "Power Laws without Parameter Tuning: An Alternative to Self-Organized Criticality," *Physics Review Letters*, 72, 1994, 2306.
20. Cox, S. J. D. and Meredith, P. G., "Microcrack Formation and Material Softening in Rock Measured by Monitoring Acoustic Emissions," *Int. J. Rock Mech. Min. Sc.*, 30, 1, 1993, 11–24.
21. Zipf, G. K., *Human Behavior and the Principle of Least Effort*, Addison-Wesley, Cambridge, MA, 1949.
22. Desai, C. S. and Nickerson, M., "Analysis and Comparisons of DSC and SOC," under preparation, 2000.
23. Davis, R. O. and Berrill, J. B., "Energy Dissipation and Liquefaction at Port Island, Kobe," *Bulletin of New Zealand Nat. Soc. of Earthquake Eng.*, 31, 1998, 31–50.
24. Figueroa, J. L., Saada, A. S., Laing, L., and Dahisaria, M. N., "Evaluation of Soil Liquefaction by Energy Principles," *J. Geotech. Eng.*, ASCE, 120, 9, 1994, 1554–1569.
25. Gyi, M. M. and Desai, C. S., "Multiaxial Cyclic Testing of Saturated Ottawa Sand," *Report, Dept. of Civil Eng. and Eng. Mech.*, University of Arizona, Tucson, AZ, 1996.

APPENDIX II. DSC Parameters: Optimization and Sensitivity

In previous chapters, we discussed procedures for the determination of parameters in the DSC and its hierarchical versions, namely, elasticity, plasticity and elastoviscoplasticity, and disturbance by using essentially manual and conventional least-square procedures. In this appendix, we present details of the conventional least square, and general optimization procedures for the evaluation of parameters. We also consider the sensitivity of computed results to changes in the parameters.

The observed behavior of many materials may exhibit considerable scatter even for the test specimens reconstituted under the same initial conditions. Also, factors such as initial density, confining pressure, temperature, strain rates, and stress path need to be considered because the objective is to evolve the constitutive model that is valid for all such significant factors. Indeed, the averaging or optimization procedure may lead to parameters that may predict behavior under one factor better than that for the other. By assigning appropriate weight in the averaging or optimization procedure for a factor(s) that is more significant for a given problem, it is possible to obtain parameters that will lead to improved predictions for the given material.

Least Square Fit Procedure

We need to determine the parameters on the basis of laboratory test data such as stress-strain, volumetric and pore water pressure. In general, the measured quantities such as stress and strain can be denoted by symbols y and x, respectively. Then the function to fit the given set of data points $(x_i, y_i, i = 1, 2, \ldots, n)$, Fig. II.1, can be expressed as (1)

$$y = a_1 f_1(x) + a_1 f_2(x) + \cdots + a_m f_m(x) \qquad \text{(II.1)}$$

where a_i $(i = 1, 2, \ldots, m)$ are the constants to be determined from the least-square criterion described below, and x and y denote the experimental data. Hence,

$$y_i = \sum_{k=1}^{m} a_k f_k(x_i) \qquad \text{(II.2)}$$

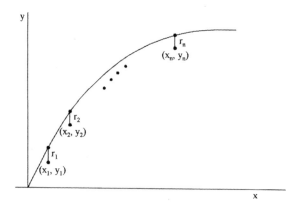

FIGURE II.1
Function and data points.

When the values of experimental data are substituted in Eq. (II.2), it will not be satisfied; i.e., there will be residuals (R_i). Hence,

$$\sum_{k=1}^{m} a_k f_k(x_1) - y_1 = R_1$$

$$\sum_{k=1}^{m} a_k f_k(x_2) - y_2 = R_2 \qquad \text{(II.3a)}$$

$$\vdots$$

$$\sum_{k=1}^{m} a_k f_k(x_n) - y_n = R_3$$

The best fit with the experimental data is obtained by making the residuals as small as possible, which is achieved by finding the values of a_i such that the sum of the squares of the residuals is a minimum, i.e.,

$$\frac{\partial}{\partial a_k} \sum_i (R_i)^2 = 0 \qquad \text{(II.3b)}$$

which can be written in matrix rotation as

$$\underset{\sim}{R} \underset{\sim}{F} = \underset{\sim}{O} \qquad \text{(II.4a)}$$

where $\underset{\sim}{R} = [R_1 \ R_2 \ ... \ R_n]$, and

$$\underset{\sim}{F} = \begin{bmatrix} f_1(x_1) & f_2(x_1) & ... & f_m(x_1) \\ f_1(x_2) & f_2(x_2) & ... & f_m(x_2) \\ & \vdots & & \\ f_1(x_n) & f_2(x_n) & ... & f_m(x_n) \end{bmatrix} \qquad \text{(II.4b)}$$

APPENDIX II

By using Eqs. (II.3) and (II.4), we can obtain

$$F^T F \underline{a} = F^T \underline{y} \tag{II.5}$$

where $\underline{a}^T = [a_1 \; a_2 \ldots a_m]$ and $\underline{y}^T = [y_1 \; y_2 \ldots y_n]$. Solution of the algebraic equations (II.5) leads to values of $\tilde{a}_i$ that provide the best fit to the observed data, Fig. II.1.

We can assign different weights (w_i) to different data points. In that case, the minimizing residual quantity is given by

$$w_1 R_1^2 + w_2 R_2^2 + \cdots + w_n R_n^2 = \text{minimum} \tag{II.6a}$$

Then, Eq. (II.5) is modified as

$$F^T \underline{w} F \underline{a} = F^T \underline{w} \underline{y} \tag{II.6b}$$

where $\underline{w}$ is the diagonal matrix with $w_1, w_2, \ldots$ as diagonal elements.

As an example, let us consider the evaluation of parameters a_1 and η_1 in the hardening function, α, Eq. (7.11a), Chapter 7, which can be written as

$$\ln \alpha = -\eta_1 \ln \xi + \ln a_1 \tag{II.7}$$

With a number of data points, $n \; (\alpha_i, \xi_i)$, this equation can be written in matrix notation as

$$\begin{bmatrix} 1 & \ln \xi_1 \\ 1 & \ln \xi_2 \\ \vdots & \\ 1 & \ln \xi_n \end{bmatrix} \begin{Bmatrix} \ln a_1 \\ -\eta_1 \end{Bmatrix} = \begin{Bmatrix} \ln \alpha_1 \\ \ln \alpha_2 \\ \vdots \\ \ln \alpha_n \end{Bmatrix} \tag{II.8}$$

which can be solved for a_1 and η_1 by using the least square procedure. Details of the procedures for finding parameters for the HISS and other plasticity and disturbance models are given elsewhere (2–4).

Optimization Procedure

In the least square procedure, which is termed as conventional, we obtain the parameters for a given test data, and then average them over a number of tests, say, under different loading conditions and stress paths. It can also allow evaluation of weighted average values by using data points from a number of tests, in which each test (stress path) is assigned an appropriate weight.

However, the quality of the parameters is also affected by the constraints arising from physical requirements. Hence, the quality of parameters can be improved by including the effects of various factors such as stress paths and density, and the constraining conditions in an optimization procedure. Such an optimization procedure developed for finding the parameters for the DSC model is described below (4).

Objective Function

In the optimization procedure, an objective function is defined (5). Here, the objective function, r, which consists of weighted basic objective functions (r_i) for different stress paths, is expressed as

$$r = \sum w_i r_i \qquad (II.9)$$

where w_i ($\Sigma w_i = 1$) denotes the weights corresponding to the objective functions, r_i.

For a given experimental response for a stress path, $r(x)$ denotes the total error given by

$$r(\underline{x}) = \frac{1}{x_1 - x_2} \int_{x_1}^{x_2} \left(\frac{f(\underline{x})}{f} - 1 \right)^2 dx \qquad (II.10)$$

where $f(\underline{x})$ denotes predictions by the (DSC) model, f denotes the test (stress-strain) data, x_1 and x_2 are the end points of the sampling data, Fig. II.2, and $\underline{x}$ is the vector of material parameters (see later). The ideal situation is that $r(\underline{x})$, Eq. (II.10), vanishes. However, usually, the optimization procedure is designed such that $r(\underline{x})$ is minimized, i.e., $r(\underline{x}) \to 0$.

The weights, w_i, in Eq. (II.9) are adopted on the basis of experience and engineering judgment. For example, for the stability analysis of a slope or dam, the extension behavior, Fig. 7.13 (Chapter 7), is more significant, and hence, a greater weight is assigned to the extension test data. For the footing load on a half space, the compressive behavior is predominant; hence, its data can be assigned a higher weight. Sometimes, as a simplification, data from all available stress path tests may be assigned equal weights such that $\Sigma w_i = 1$.

Constraints

In an optimization procedure, the analysis is carried out under constraints that influence the objective function. In the case of the DSC model, the following constraints are introduced (see Chapters 2–12) for various parameters. Here in the DSC model, the RI response is characterized by using the HISS-δ_0

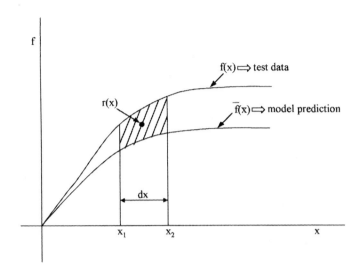

FIGURE II.2
Schematic of residual.

elastoplastic model.

Elastic

$$E > 0, \quad < \nu < 0.5; \quad G > 0, \quad K > 0 \quad (\text{II.11a})$$

Plasticity (HISS-δ_0 Model)

As the hardening function $\alpha \geq 0$, Eq. (7.11a), the constraining condition for the ultimate parameter, γ, is given by

$$\gamma \geq \frac{J_{2D}}{\overline{J}_1^2 (1 - \beta S_r)^{-0.5}} \quad (\text{II.11b})$$

The constraining condition for the parameter, β, for the shape of the yield surface is given by (Chapter 7).

$$0 \leq \beta < 0.756 \quad (\text{II.11c})$$

For the transition or phase change parameters, n, the condition is

$$n > 2.0 \quad (\text{II.11d})$$

The optimization problem is to evaluate $\underset{\sim}{x}$ given by

$$\underset{\sim}{x} = [E, \nu, n, \beta, \gamma, a_1, \eta_1, \overline{m}, \lambda, e_o^c, A, Z]^T \quad (\text{II.12})$$

such that $r(x)$ is minimized. Here, the elements of $\underset{\sim}{x}$ are the parameters in the DSC model that include elastic, elastoplastic (HISS-δ_0), critical state (for FA behavior) and disturbance; $D_u = 1.0$ is assumed.

Numerical Method for Optimization

A number of strategies are available for the optimization procedure. They include Newton's method, quasi-Newton method with constrained or unconstrained minimization (5). Here, we present and use the quasi-Newton method with unconstrained approach, which leads to a simplified procedure.

In the Newton's method, the function, $f(\underline{x})$, Eq. (II.10), is obtained by using truncated Taylor series expansion at about point $\underline{x}_k$, Fig. II.3, as

$$f(\underline{x}_k + \underline{\delta}) \approx q_k(\underline{\delta}) = f_k + g_k^T \underline{\delta} + \frac{1}{2} \underline{\delta}^T G_k \underline{\delta} \qquad (II.13)$$

where $\underline{x}$ denotes the vector of DSC parameters, Eq. (II.12), and $\underline{\delta} = \underline{x}_{k+1} - \underline{x}_k$, $g(\underline{x})$ is the vector of first partial derivatives or gradient vector given by

$$g(\underline{x}) = \nabla f(\underline{x}) \qquad (II.14)$$

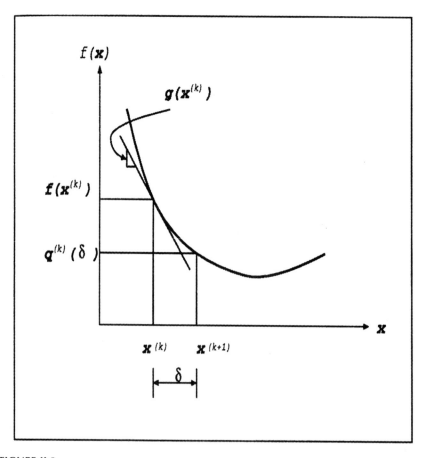

FIGURE II.3
Schematic Taylor series expansion of a function. ©John Wiley & Sons Ltd. Reproduced with permission.

APPENDIX II

and $G(\underline{x})$ is the matrix of second partial derivatives or Hessian matrix given by

$$G(\underline{x}) = \nabla^2 f(\underline{x}) \tag{II.15}$$

and q_k is the resulting quadratic function.

The iterative procedure involves evaluation of the parameters, which is equal to $\underline{x}_k + \underline{\delta}_k$ where the correction $\underline{\delta}_k$ minimizes $q_k(\underline{\delta}_k)$. In the Newton's method, we need f_k, $\underline{g}_k$, and G_k at different points (k) so as to evaluate $q_k(\underline{\delta}_k)$. The method gives well-defined solution if G_k is positive definite. A disadvantage of the Newton's method is that we need to compute the Hessian matrix, G, which can be time-consuming. Hence, the simplified quasi-Newton method given below is used here.

Quasi-Newton Method

Like the Newton's method, the quasi-Newton method involves the line search procedure (5). However, the matrix, G_k, is approximated by using the symmetric positive definite matrix, H_k, which is corrected during the iterative procedure. The following equations constitute the quasi-Newton method:

$$\underline{s}_k = -H_k \underline{g}_k \tag{II.16a}$$

$$\underline{x}_{k+1} = \underline{x}_k + \bar{\alpha}_k \underline{s}_k \tag{II.16b}$$

in which the matrix, H_k, is updated during the iterations as $H_{k+1}, H_{k+2}, \ldots, \underline{s}_k$ denotes the line search direction, and $\bar{\alpha}$ is the scalar line search step length, described below.

The initial matrix, H_1, can be adopted as any positive definite matrix, which can be chosen as $H_1 = \underline{I}$, where $\underline{I}$ is the identity matrix:

$$\underline{I} = \begin{bmatrix} 1 & 0 & 0 & \cdots & 0 \\ 0 & 1 & 0 & \cdots & 0 \\ \vdots & & & & \\ 0 & 0 & 0 & \cdots & 1 \end{bmatrix} \tag{II.17}$$

The BFGS algorithm (6–9), named after the first letter of the last names of the four authors, is often used to evaluate iterated values of H as

$$H_{k+1} = H_k + \left(1 + \frac{\underline{\gamma}^T H_k \underline{\gamma}}{\underline{\delta}^T \underline{\gamma}}\right) \frac{\underline{\delta}\underline{\delta}^T}{\underline{\delta}^T \underline{\gamma}} - \left(\frac{\underline{\delta}\underline{\gamma}^T H_k + H_k \underline{\gamma}\underline{\delta}^T}{\underline{\delta}^T \underline{\gamma}}\right) \tag{II.18}$$

where $\underline{\gamma}_k = \underline{g}_{k+1} - \underline{g}_k$.

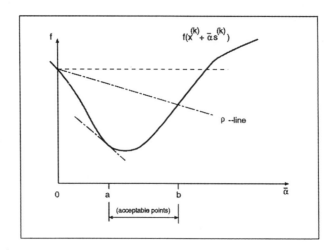

FIGURE II.4
Descent and stable conditions (5). ©John Wiley & Sons Ltd. Reproduced with permission.

Line Search Strategy

A number of line search strategies have been proposed (5). One such strategy is depicted in Fig. II.4, and is found to be efficient and stable. The following descent and stability conditions are embedded in this strategy for the computation of $\bar{\alpha}$ in Eq. (II.16b) (5):

$$f_k - f_{k+1} \geq -\rho g^T \underset{\sim}{\delta}_k \qquad \text{(II.19a)}$$

$$g_{k+1}^T \underset{\sim}{s}_k \geq \sigma g_k^T \underset{\sim}{s}_k \qquad \text{(II.19b)}$$

where ρ is a scalar in the range of 0.0 to 0.5, and σ is a scalar in the range of ρ to 1.0. The following equations are used for the evaluation of $\bar{\alpha}$

$$\hat{\bar{\alpha}} - \bar{\alpha} = 0.5 \frac{\bar{\alpha} - \bar{\alpha}_1}{\left(1 + \dfrac{f_1 - f}{(\bar{\alpha} - \bar{\alpha}_1) f_1'}\right)} \qquad \text{(II.19a)}$$

$$\hat{\bar{\alpha}} - \bar{\alpha} = \frac{(\bar{\alpha} - \bar{\alpha}_1) f'}{(f_1' - f')} \qquad \text{(II.19b)}$$

where the prime denotes derivative.

APPENDIX II

Unconstrained Optimization

The optimization procedure subject to the constraints, Eq. (II.11), can be simplified by using unconstrained minimization (5). The constraints are transformed by using trigonometric functions subject to the condition, $\ell_i \leq x_i \leq u_i$, where ℓ_i and u_i, are the lower and upper bounds for the ith variable (parameter), Eq. (II.11). The trigonometric transformation is given by

$$x_i = \ell_i + (u_i - \ell_i)\sin^2(y_i) \tag{II.20}$$

where y varies in the domain $-\infty \leq y_i \leq +\infty$. Such a transformation procedure works well and is reported by (10, 11).

The problem of unconstrained optimization is to

$$\text{minimize} \quad r(\underline{x}^*) \tag{II.21}$$

where $\underline{x}^* = [E^*, \nu^*, n^*, \beta^*, \gamma^*, a_1^*, \eta_1^*, \lambda^*, e_o^{c*}, A^*, Z^*]^T$ with the following transformations for various parameters and their initial values:

$$
\begin{array}{ll}
\textit{Transformation} & \textit{Initial Value: } \underline{x}_o^I \\[4pt]
\beta = \beta_{\max}\sin^2(\beta^*) & \beta_o^* = \sin^{-1}\sqrt{\dfrac{\beta_o}{\beta_{\max}}} \\[4pt]
\gamma = \gamma_{\min} + (\gamma^*)^2 & \gamma_o^* = \sqrt{\gamma_o - \gamma_{\min}} \\[4pt]
n = 2 + (n^*)^2 & n_o^* = \sqrt{n_o - 2} \\[4pt]
E = (E^*)^2 & E_o^* = \sqrt{E_o} \\[4pt]
\nu = 0.5\sin^2(\nu^*) & \nu_o^* = \sin^{-1}\sqrt{2\nu_o} \\[4pt]
a_1 = (a_1^*)^2 & a_{10}^* = \sqrt{a_{10}} \\[4pt]
\eta_1 = \eta_1^{*2} & \eta_{10}^* = \sqrt{\eta_{10}} \\[4pt]
\overline{m} = (\overline{m}^*)^2 & \overline{m}_o^* = \sqrt{\overline{m}_o} \\[4pt]
\lambda = (\lambda^*)^2 & \lambda_o^* = \sqrt{\lambda_o} \\[4pt]
e_o^c = (e_o^{c*})^2 & e_o^{c*} = \sqrt{(e_o^c)_o} \\[4pt]
A = (A^*)^2 & A_o^* = \sqrt{A_o} \\[4pt]
Z = (Z^*)^2 & Z_o^* = \sqrt{Z_o}
\end{array}
\tag{II.22}
$$

The initial values are found from the conventional averaging procedure.

Computer Code

A computer code (DSCOPT) for the unconstrained optimization was prepared and validated as described below (4). It has also been used to perform a sensitivity analysis to assess the relative influence of variations in parameters on the model predictions.

TABLE II.1

Laboratory Tests of Leighton Buzzard Sand (12)

Stress Path	Density (D_r)			Confining Pressure, kPa
	95%	65%	10%	
	*	*	*	89.6
CTC	*	*	*	275.6
(cylindrical)	*	*	*	826.8
	*	*		89.6
RTE				275.6
(cylindrical)			*	826.8
HC	*	*		
(cylindrical)				
			*	89.6
RTC				275.6
(cubical)				826.8
				89.6
TC (cubical)	*			275.6
				826.8
				89.6
TE (cubical)	*			275.6
				826.8

*denotes the test performed.
The void ratio: $e_{max} = 0.81$, $e_{min} = 0.53$.
   Accordingly,
   $D_r = 95\% \Rightarrow e_0 = 0.544$
   $D_r = 65\% \Rightarrow e_0 = 0.628$
   $D_r = 10\% \Rightarrow e_0 = 0.782$

Validations

The laboratory behavior of the Leighton Buzzard (LB) sand under different stress paths, Fig. 7.13, Chapter 7, is used as the basis of the validation. Table II.1 shows details of the multiaxial tests on cubical specimens (10 × 10 × 10 cm) of sand under different stress paths, confining pressure, (σ_3), and relative density, D_r (12).

For the optimization analysis, the material parameters were found by dividing the test data in various groups. Groups 1–5, Table II.2(a), refer to the tests for $D_r = 95\%$, while Table II.2(b) refers to the tests for $D_r = 10\%$. In each group, only selected stress path tests were used for finding the parameters. For example, in Group 1 three tests under HC, CTC (89.6 kPa), and CTC (275.6 kPa), were used, and so on. Equal weights, Eq. (II.9), were used for each test in a given group.

The parameters were found by using the conventional (least square) averaging and the constrained optimization procedures. Then the stress-strain-volumetric responses of tests under different stress paths were predicted by using the optimized parameters and those from the conventional procedures,

APPENDIX II

TABLE II.2(a)
Groups for $D_r = 95\%$

Stress Path	Confining Stress (kPa)	1	2	3	4	5	Group
HC	N/A	×	×	×	×	×	
CTC	89.6	×	×	×	×	×	
CTC	275.6	×					
CTC	826.8		×				
RTE	89.6			×			
TC	275.6				×		
TE	89.6					×	

TABLE II.2(b)
Group for $D_r = 10\%$

Stress Path	Confining Stress (kPa)	Data Set Used 6	Group
HC	N/A	×	
CTC	89.6	×	
CTC	275.6	×	
CTC	826.8	×	

TABLE II.3
Parameter Values Determined from Data Group 1

Parameter	Conventional	Optimized	Change (%)
E	101450 kPa	104507 kPa	+3.01
ν	0.41	0.32584	−20.53
n	2.537465	2.40913	−5.06
β	0.7	0.60156	−14.06
γ	0.08102	0.0894954	+10.46
a_1	0.00296	3.00302E-4	−89.85
η_1	0.2849	0.298106	+4.64
$\bar{m}$	0.00929	9.31397E-3	+0.26
e_0	0.544	0.544	0
e_0^c	0.683211	0.752341	+10.12
λ	0.0317	0.2940957	+827.75
A	0.4579	0.84379	+84.27
Z	0.3201	0.499684	+56.10

and were compared with the observed behavior. Typical results for only selected groups are given below.

Tables II.3–II.5 show parameters for Groups 1 and 5 for $D_r = 95\%$ and Group 6 for $D_r = 10\%$, respectively. The percentage changes are with respect to the parameters from the conventional procedure.

TABLE II.4
Parameter Values Determined from Data Group 5

Parameter	Conventional	Optimized	Change (%)
E	40567.9 kPa	44194.6 kPa	+8.94
ν	0.4	0.35904	−10.24
n	3	2.89034	−3.66
β	0.75	0.75213	−0.28
γ	0.0827565	0.07904	+4.49
a_1	0.00128265	0.001282	−0.05
η_1	0.307546	0.31001	+0.80
$\bar{m}$	0.102821	0.110981	+7.94
e_0	0.544	0.544	0
e_0^c	0.783012	0.679201	−13.26
λ	0.182442	0.2956	+62.02
A	0.965391	0.889101	−7.90
Z	0.29839	0.27092	−9.21

TABLE II.5
Parameter Determined from Data Group 6

Parameter	Conventional	Optimized	Change (%)
E	35456.1 kPa	42830.5 kPa	+20.80
ν	0.4	0.379	−5.25
n	2.6	2.758	+6.08
β	0.7	0.751456	+7.35
γ	0.049215225	0.05315372	+8.00
a_1	0.00122739	0.00102946	−16.13
η_1	0.304157	0.3375602	10.98
$\bar{m}$	0.268212	0.2459786	−8.29
e_0	0.544	0.544	0
e_0^c	0.0327233	0.02002876	−38.79
λ	0.0194252	0.0234572	+20.76
A	1.53464	1.300956	−15.23
Z	0.0800357	0.0689714	−13.82

Typical comparisons between predicted and observed stress-strain and volumetric responses for different groups are shown in various figures as follows:

Figure	Group	Stress Path	σ_3	Density
Fig. II.5	1	CTC	89.6 kPa	95%
Fig. II.6	5	TE	89.6	95%
Fig. II.7	6	CTC	826.8	10%

It can be seen from Figs. II.5–II.7 that the predictions by using the optimized parameters yield much closer correlations with the observed responses compared to those from the conventional procedures.

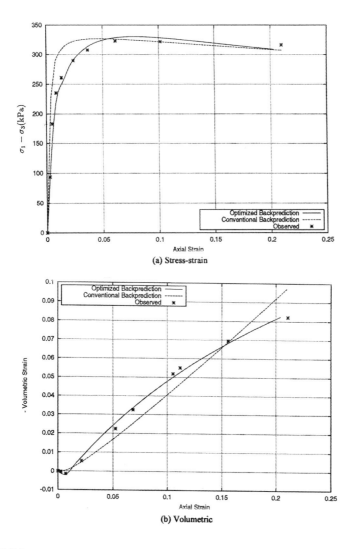

FIGURE II.5
Comparisons between predictions and CTC test: $D_r = 95\%$, $\sigma_o = 89.6$ kPa (4).

Effect of Weight

In the previous analysis, equal weights were assigned to each stress path. In order to assess the effect of weights on the predicted behavior, three special analyses were performed by using the CTC ($\sigma_3 = 89.6$ kPa) and RTE ($\sigma_3 = 89.6$ kPa) stress paths in Group 3, Table II.2(a):

(i) $w_{CTC} = w_{RTE} = 0.5$
(ii) $w_{CTC} = 0.65$, $w_{RTE} = 0.35$
(iii) $w_{CTC} = 0.35$; $w_{RTE} = 0.65$

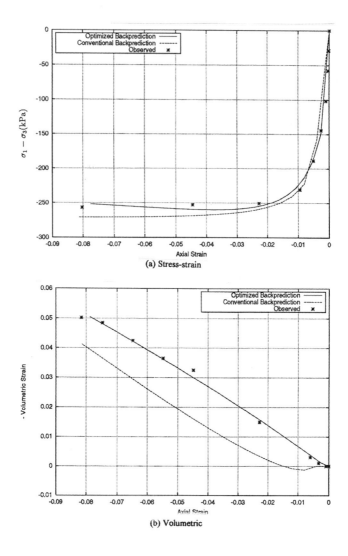

FIGURE II.6
Comparisons between predictions and TE test: $D_r = 95\%$, $\sigma_o = 89.6$ kPa (4).

Only results for the first two sets are presented here. Table II.6 shows parameters corresponding to the use of the sets of weights. The percentage error is with respect to the equal weights.

Figure II.8 shows comparisons between predictions using equal weights (= 0.5) and $w_{CTC} = 0.65$ and $w_{RTE} = 0.35$ for CTC test ($\sigma_3 = 89.6$ kPa, $D_r = 95\%$).

It can be seen that the predictions for higher weight (= 0.65) for the CTC test yield improved predictions compared to those from the equal weight (= 0.50).

Thus, the predictive quality of the model can be improved by assigning higher weights to the stress paths that are important in a given practical problem.

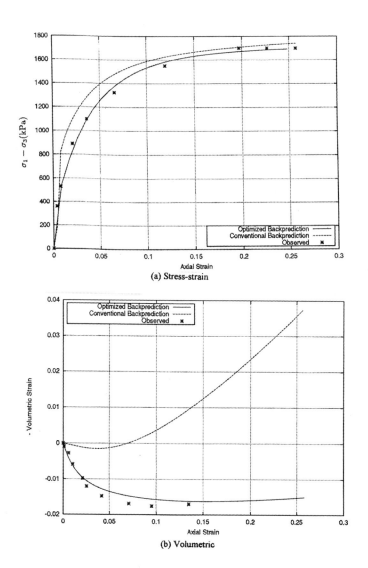

FIGURE II.7
Comparisons between predictions and CTC test: $D_r = 10\%$, $\sigma_o = 826.8$ kPa (4).

Sensitivity of Parameters

A special study was performed to establish the sensitivity of the DSC predictions to the change in parameters and relative importance of the parameters.

Table II.7 shows the set of parameters adopted for this analysis. Predictions for stress-strain responses for CTC ($\sigma_3 = 826.8$ kPa, $D_r = 10\%$) and RTE

TABLE II.6

Effect of Weight on Parameter Values of Test Group 3 ($w_{CTC} = 0.65; w_{RTE} = 0.35$)

Parameter	$w_{CTC} = w_{RTE} = 0.5$	$w_{CTC} = 0.65; w_{RTE} = 0.35$	Change (%)
E	354176.9 kPa	37334.2 kPa	+6.13
ν	0.3736	0.3549	−5.01
n	2.48745	2.56391	+3.07
β	0.755	0.6013	−20.36
γ	0.424063	0.498171	+17.48
a_1	0.0690234	0.0649979	−5.83
η_1	0.06012	0.05537	−7.90
$\bar{m}$	0.199052	0.190444	−4.32
e_0	0.544	0.544	0
e_0^c	0.400987	0.431145	+7.52
λ	0.209711	0.347845	+65.87
A	3.07893	2.84561	−7.58
Z	0.199761	0.198043	−0.86

TABLE II.7

Parameters for Sensitivity Analysis

No.	Parameter	Value
1	E	14345
2	ν	0.35
3	n	2.6
4	β	0.6
5	γ	0.0502571
6	a_1	0.000386232
7	η_1	0.7112
8	$\bar{m}$	0.0790039
9	e_0^c	0.0883205
10	λ	0.049583
11	A	1.01535
12	Z	0.4284032

($\sigma_3 = 826.8$ kPa, $D_r = 10\%$) tests were obtained for three values of each parameter; the value in Table II.7, and two other values as +20% and −20%. All other parameters were kept constant, while a given parameter was thus changed.

Figure II.9 shows averages of cumulative relative differences for the two (CTC and RTE) tests, which was found as

$$\text{Cumulative Relative Difference} = \sum \left(\frac{\bar{q} - q}{q}\right) \quad \text{(II.23)}$$

where $q (= \sigma_1 - \sigma_3)$ denotes the observed stress, and $\bar{q}$ denotes the computed value.

APPENDIX II

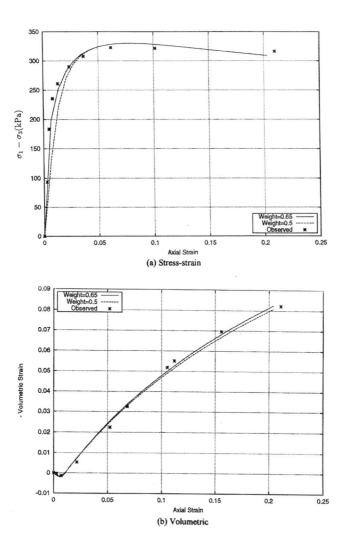

FIGURE II.8
Effect of weight on backprediction of CTC test: $D_r = 95\%$, $\sigma_o = 89.6$ kPa; $\omega_1 = \omega_{CTC} = 0.65$, $\omega_2 = \omega_{RTE} = 0.35$ (4).

Table II.8 shows the order of the parameter sensitivity to computed results. It can be seen that the changes in the phase change (n), ultimate slope of yield surface (γ), hardening (η_1) and the ultimate shape (β) parameters affect the predictions most. The other parameters in this analysis do not appear to influence the predictions significantly. Hence, it will be desirable to evaluate n, γ, η_1 and β as carefully as possible.

TABLE II.8
Order of Parameter Sensitivity

1	Parameter	Averaged Difference (%)
2	n	34.31
3	γ	14.11
4	η_1	6.48
5	β	5.99
6	a_1	4.22
7	A	3.43
8	Z	3.13
9	$\bar{m}$	2.73
10	e_0^c	2.36
11	λ	2.35
12	E	1.28
13	ν	0.05

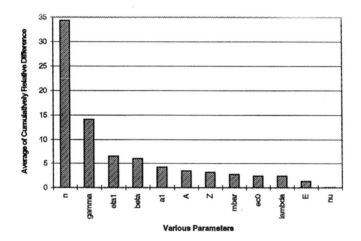

FIGURE II.9
Order of sensitivity of DSC model parameters (4).

References

1. James, M. L., Smith, G. M., and Wolford, J. C., "Applied Numerical Methods for Digital Computation with FORTRAN and CSMP," IEP-A Dyn. Donnelley Publisher, New York, 1977.
2. Zaman, M. M., Desai, C. S., and Faruque, M. O., "An Algorithm for Determining Parameters for Cap Model from Raw Laboratory Test Data," *Proc., 4th Int. Conf. on Num. Methods in Geomechanics*, Edmonton, Canada, 1982, 275–285.
3. Desai, C. S., "Procedures and Algorithms for Finding Parameters for HISS-δ_0 and δ_1 Models," *Report*, Dept. of Civil Eng. and Eng. Mechanics, The Univ. of Arizona, Tucson, AZ, 1990.

4. Chen, J. Y. and Desai, C. S., "Optimization in the Disturbed State Concept Constitutive Modeling and Application in Finite Element Analysis," *Report*, Dept. of Civil Eng. and Eng. Mechanics, The Univ. of Arizona, Tucson, AZ, 1997.
5. Fletcher, R., *Practical Methods of Optimization*, John Wiley & Sons, New York, 1987.
6. Broyden, C. G., "The Convergence of a Class of Double Rank Minimization Algorithms, Parts I and II," *J. Inst. Maths. Applications*, 6, 1970, 76–90.
7. Fletcher, R., "A New Approach to Variable Metric Algorithms," *Computer Journal*, 13, 1970, 317–322.
8. Goldfarb, D., "A Family of Variable Metric Methods Derived by Variational Means," *Maths. Comp.*, 24, 1970, 23–26.
9. Shanno, D. F., "Conditioning of Quasi-Newton Methods for Function Minimization," *Maths. Comp.*, 24, 1970, 149–160.
10. Gill, P. E. and Murray, W., *Practical Optimization*, Academic Press, New York, 1981.
11. DeNatale, J. S., Herrmann, L. R. and Dafalias, Y. F., "Calibration of the Bounding Surface Soil Plasticity Model by Multivariate Optimization," *Proc., Int. Conf. on Constitutive Laws for Engineering Materials*, Tucson, AZ, 1983.
12. Armaleh, S. H. and Desai, C. S., "Modeling Including Testing of Cohesionless Soils Using Disturbed State Concept," *Report*, Dept. of Civil Eng. and Eng. Mechanics, The Univ. of Arizona, Tucson, AZ, 1990.

Index

A

Acoustic emissions (AE), 632, 638, 643
Acoustic tensor, 510
Aerospace engineering, metal-to-metal contacts in, 422
Aluminum
  nitride (ALN), 604
  shaft–sand interfaces, 468
Amontons, friction law of, 421
Angle of friction, 191
Anisotropy, 64, 200
Approximate constitutive models, 4
Artificial mind, 8
Asphalt
  concrete
    creep response for, 328
    fracture energy for, 329
    hardening for, 328
    uniaxial compression test for, 326
    viscoplastic model for, 322
  pavement system, material parameters for, 579
Assembled equations, 211
Atmospheric pressure constant, 181
Avalanches, 642

B

Backfill soil
  stress-strain responses for, 582
  testing, 581
Back stress, 85
Behavior, see also Stress–strain behavior
  clay lumps, 26
  comparison of observed and predicted, 133
  complex interaction systems, 10
  compression, of structured soil, 398
  concrete, 174
  creep, 82, 344, 464
  critical state, 434, 356
  elastic, 545
  elastoplastic, 20
  engineering, 21
  first-cycle, 76
  fully adjusted, 68
  geologic material, 480
  granular material, 480
  loose and dense granular materials, 67
  metal, 174
  observed, 32, 379, 541
  piecewise linear, 117
  predicted, 379, 380
  rate-dependent, 289
  relatively intact, 67, 124, 427
  softening, applications of DSC model for, 403
  stiffening, 30
  tensile, of metallic materials, 194
  thermomechanical, in electronic chip-substrate systems, 470
  thermoplastic, 546
  thermoviscoplastic, 547
  yield, of natural and reinforced soil, 412
Bimaleimide triazine (BT) epoxy glass laminate, 601
Biot's generalized theory, 591
Boltzmann's constant, 85
Bonding
  material, 53, 55, 57
  stress, 193, 344
Bonds, breakage of, 53
Borehole wall, cracking and failure at, 590
Boundary value problem, wellposedness of, 482
Brittle rock, acoustic emission in, 645
BT epoxy glass laminate, see Bimaleimide triazine epoxy glass laminate
Bulk
  interface, 425
  modulus, 36, 117, 311

683

C

Cam clay model, 162, 163, 230
Cap model, 161, 162, 163
Cartesian coordinates, 224
Cauchy elastic model, 121, 124
Cemented sand, 142, 418
Ceramic
  composite, 139
  P-wave velocity for fiber-reinforced, 417
Chain rule of differentiation, 209
Characteristic dimension, 7, 32, 484
Chemical rebonding, 78
Chip substrate, 511, 600
CL, see Constrained liquid
Clay
  Leda, 400
  lumps, 26
  parameters for saturated, 239
  pile in under cyclic loading, 611
  plot for, 295
  saturated, 236, 370
CLS, see Constrained liquid–solid
Collapse, 363
Compaction, continuous, 78
Component-reference materials, material elements composed of, 39
Composite material systems, 408
Compression
  behavior, of structured soil, 398
  tests, 399
Computer
  codes, 618, 671
  implementation, 472
Computer procedures, implementation of DSC in, 529–624
  algorithms for coupled dynamic behavior, 550–556
    dynamic equations, 550–551
    finite element formulation, 551–554
    fully saturated systems, 550
    time integration for dynamic problems, 554–556
  algorithms for creep behavior, 544–550
    algorithms for elastic, plastic, viscoplastic behavior with thermal effect, 545
    axisymmetric, 545–546
    elastic behavior, 545
    plane strain, 545
    plane stress, 545
    thermoplastic behavior, 546–547
    thermoviscoplastic behavior, 547–550
  computer codes, 618
  cyclic and repetitive loading, 559–567
    cyclic-repetitive hardening, 565–567
    reloading, 563–565
    unloading, 561–563
  examples of applications, 574–618
    borehole stability, 590–594
    chip substrate CBGA package, 604–609
    chip substrate 313-PIN PBGA, 600–604,
    computer results, 594–595
    cyclic hardening, repetitive loading, HISS plasticity model, 609–610
    dynamic analysis and liquefaction in shake table test, 615–618
    field validation, 586–590
    laboratory tests, 581–585
    layered pavement system, 597–600
    layered system, 578–580
    load on half space, 576–577
    material block, 574–576
    pile in clay under cyclic loading, 611–615
    reinforced earth, 580–581
    viscoplastic model, 596–597
  finite element formulation, 530–534
  hierarchical capabilities and options, 570–571
  implementation, 557–558
  initial conditions, 567–570
    initial disturbance, 568–570
    initial stress, 567–568
  mesh adaption using DSC, 572–574
  mesh adaption using prepeak response, 572–573
  mesh adaption using postpeak response, 573–574
  partially saturated systems, 558–559
  solution schemes, 534–544
    FA stress, 536–541
    observed behavior, 541–544
Concrete
  behavior of, 174
  concrete interface, 451
  DSC model, 248
  DSC response for, 175
  laboratory test results for, 508
  plasticity model, 246
Constant normal stiffness test, 453
Constitutive equations, 11
Constitutive laws, 11
Constitutive matrix, 439, 499
Constitutive models, 11
Constrained liquid (CL), 70, 71, 232, 432
  FA as, 87, 101
  solid (CLS), 37, 69, 102, 232

Index

Constraints, in optimization procedure, 666
Contact(s)
  asperities, 431
  ductile, 423
  stress, variations of, 50
Continuous compaction, 78
Continuous yielding models, 165
Continuum damage model, 100, 439, 486
Continuum models, gradient enrichment of, 489
Conventional triaxial compression (CTC) loading, 154, 258
  parametric sensitivity analysis, 263
  tests, unloading interpolation function for, 561
Converged state, 218
Correction procedures, 215
Cosserat
  continuum, 491
  theory, 492
Coulomb, friction law of, 421, 422
Crack(s), 631
  density, 140
    during multiaxial loading, 146
    during uniaxial loading, 142
    plots of, 415
    variations of, 143
  finite, 487
  frozen, 485
  glued, 485
Cracking, at borehole wall, 590
Crank–Nicholson scheme, 279
Creep
  models, disturbance for, 287
  one-dimensional elastoviscoplastic model for, 333
  parameters, determination of from shear test for soil, 295
  predictions, 334
  response, for asphalt concrete, 328
  thermal, in restrained bar, 319
  viscoplastic, 303
Creep behavior, 82, 273–336, 464
  algorithms for, 344
  disturbance function, 286–289
  elastoviscoplastic model, 275–286
    elastoviscoplastic finite-element equations, 280–282
    mechanics of viscoplastic solution, 277–279
    one-dimensional formulations of Perzyna model, 282–285
    selection of time step, 286
    stress increment, 280
    theoretical details, 275–277
    viscoplastic strain increment, 279–280
  examples, 314–336
    closed-form solution of overlay model, 327–332
    elastoviscoplastic model for rock salt, 315–316
    elastoviscoplastic model for solders, 314–315
    one-dimensional elastoviscoplastic model for creep, 333
    one-dimensional elastoviscoplastic model for relaxation, 333–336
    stress relaxation, 316–319
    thermal creep in restrained bar, 319–321
    ve, evp, and vavp overlay models with von Mises plasticity criterion, 321
    ve, vp, and vevp models with HISS plasticity, 321–322
    viscoplastic model fro asphalt concrete, 322–327
  material parameters in overlay model, 303–314
    advantages of overlay models, 313–314
    elastoviscoplastic overlay model, 307
    finite element equations, 313
    other models, 313
    parameters for viscoelastic model, 309–312
    physical meanings of parameters, 312
    Poisson's ratio, 306–307
    viscoelastic overlay model, 303–305
    viscoelasticviscoplastic overlay model, 307–308
  multicomponent DSC and overlay model, 297–303
    disturbance due to viscoelastic and viscoplastic creep, 303
    multicomponent DSC, 297–302
  parameters for viscoplastic model, 290–297
    determination of parameters, 290–292
    disturbance parameters, 292
    laboratory tests and examples, 293–295
    temperature dependence, 296–297
  rate-dependent behavior, 289–290
Crescrograph, 7
Critical disturbance, 520
Critical parameters, determination of, 443
Critical state (CS), 15
  behavior, 356, 434
  characterization, for FA response, 102
  concept (CSC), 59, 165, 365, 631
  DSC equations with, 102
  FA as, 87, 432

line (CSL), 69, 158
model
   advantages and limitations of, 163
   parameters in, 161
representation, 159
soil mechanics, 25
strain equations using, 105
CTC loading, see Conventional triaxial compression loading
Cumulative relative difference, 678
Curved ultimate envelope, 190
Cycles
   effective stress vs., 375
   to fatigue failure, 396
Cyclic hardening, 255, 609, 610
Cyclic loading, 168, 559, 611
Cyclic-repetitive hardening, 565
Cyclic strain-controlled test, 76
Cycling multi-degree-of-freedom (CYMDOF) shear device, 449, 450
Cylinder with internal pressure, 596

D

Damage models, 12, 100
Damping ratio, 556
Data, stress–strain, 126
Deformation patterns, in imperfect bar, 515
Degradation, 363
Density clusters, 28
Derivatives, details of, 209
Deviatoric stress tensor, 209
Differentiation, chain rule of, 209
Dilation, 28
Dislocation(s), 64, 631
   density, 405
   locking of, 78, 394
Disordered material parts, 85
Displacement, virtual, 531
Dissipated energy, 83, 84
Disturbance, 46, 348
   based on disorder and free energy, 85
   based on observed response, 538
   creep models, 287
   critical, 520
   damage models and, 12
   definition of, 75, 392
   determination of, 375
   due to microcracking, 125
   due to structure, 393
   elastic moduli, 79
   expression of, 54
   function, 33, 73, 88, 266, 435

alternative form of, 466
schematic of, 81
structured soil, 397
global, 357
index, 398
initial, 233, 568
instability through, 502
irreversible strains vs., 131
models, 99, 262
neutral pressure decreasing with, 51
overall, 392
parameters, 89, 292, 446
plastic strain vs., 172, 346
rate, 652
representation of, 80
schematic of, 48
shear modulus, 79
test data, 74
ultimate, 496
Disturbed state, critical state and self-organized criticality concepts, 631–662
   acoustic emissions, 638–639
   cyclic behavior of saturated sands, 656–658
   disturbance, AE, and energy, 649–650
   instability and liquefaction during earthquake, 651–655
   microfracturing and acoustic emission in laboratory tests for rock, 643–645
   self-organized criticality, 635–638
   simulated microfracturing and acoustic emission in brittle rock, 645–649
   simulated microfracturing phenomenon in elastoplastic materials, 639–642
Disturbed state concept (DSC), 11–13, 17–60, 20, 340, 427, 550, see also DSC equations and specializations
   alternative formulations of DSC, 38–41
   approximate decoupled, 500
   averaging and weighting function in, 495
   bilinear elastic response, 138
   bonded materials, 53–58
      approach 1, 53
      approach 2, 53–55
      approach 3, 55–56
      approach 4, 56–57
      porous saturated bonded materials, 57–58
      structured materials, 58
   characteristics, 58–60
      comparisons and comments, 58–59
      self-organized criticality, 59–60
   disturbance and damage models, 12–13
   DSC and other models, 13
   engineering behavior, 21–23

Index   687

equations, 44
  alternative, 440
  critical state, 102
  general formulation of, 108
  incremental, 346
  one-dimensional specialization of, 129
  formulation of disturbed state concept,
    32–33
  hierarchical framework of, 59
  hierarchical versions in, 60
  incremental equations, 33–38
    effective or net stress, 38
    fully adjusted state, 36–38
    relatively intact state, 35–36
  instability in, 516
  mathematic and physical characteristics
    of, 447
  mechanism, 23–32
    additional considerations, 27–31
    characteristic dimension, 32
    fully adjusted state, 25–27
  mesh adaption using, 572
  model(s), 13, 21, 361, 376
    application of for softening and
      stiffening response of silicon, 266
    computer procedures with, 570
    formulation of, 402
    parameters, order of sensitivity of, 680
    stability condition for one-dimensional,
      503
  multicomponent DSC system, 41–53
    comments, 52–53
    disturbance, 46–47
    DSC equations, 44–46
    DSC for porous saturated media, 43–44
    example and analysis, 49–52
    Terzaghi's effective stress concept, 47–48
  nonassociative response with, 200
  observed behavior, 32
  options in, 571
  optimization procedure
    computer code, 671–672
    constraints, 666–667
    line search strategy, 670
    numerical method for optimization,
      668–669
    objective function, 666
    quasi-Newton method, 669
    unconstrained optimization, 671
  parameters, 663–680
    least square fit procedure, 663–665
    optimization procedure, 665–672
    sensitivity of parameters, 677–680
    validations, 672–677
  porous saturated media, 43, 45

predictions
  multiaxial compression and, 144
  multiaxial extension and, 145
  representations of, 4
  response
    for concrete, 175
    for metal, 175
  stability analysis of, 503
  system, multicomponent, 41
  thermodynamical analysis of, 522
Drift correction, 212, 213, 217
Drucker–Prager (D–P)
  model, 183, 513
  yield criterion, 154, 165, 189
Dry sand, 241
  concrete interface, 454
  steel interfaces, 458
DSC, see Disturbed state concept
DSC equations and specializations, 93–112
  alternative formulations of DSC, 110
  derivation of strain equations, 105–108
  examples, 110–112
  general formulation of DSC equations,
    108–110
  relatively intact response, 96–98
  specialization of DSC equations, 98–105
    classical continuum damage model, 100
    critical-state characterization for FA
      response, 102
    disturbance models, 99–100
    DSC equations with critical state,
      102–105
    DSC model without relative motions,
      101–102
    elastoviscoplastic, 99
    elstoplastic, 99
    linear elastic, 98–99
    thermal effects, 99
Ductile contacts, 423
Dynamic loading, 205
Dynamic problems, time integration for, 554,
  555
Dynamic yield surface, 278

E

Earth, reinforced, 409, 580
Earthquakes, 636, 637, 651
Earth Technology Corporation, 611
Effective stress, 37, 47, 50, 345
  concept, Terzaghi's, 47
  cycles vs., 375
  parameter, 358

variations of, 52
Elastic behavior, 545
Elasticity, theory of in DSC, 115–146
   disturbance function, 126–127
   examples, 129–146
      cemented sand, 142–143
      ceramic composite, 139–140
      correlation with crack density, 140–142
   fully adjusted behavior, 125–126
   linear elasticity, 115–118
   material parameters, 127–129
   relatively intact behavior, 124–125
   variable parameter models, 118–124
      first-order model, 121–123
      functional forms, 118–120
      hyperelastic models, 120–121
      second-order Cauchy elastic model, 142
Elastic moduli, 366
Elastic perfectly plastic material, 150
Elastic-plastic
   loading, 214
   models, 430
Elastic response, DSC bilinear, 138
Elastic stress–strain response, 135
Elastoplastic (ep), 20
   constitutive matrix, 208
   constitutive tensor, 99
   equations, derivation of, 205
   material, 150
   matrix, 98
   response, 112
Elastoviscoplastic (evp) model, 275
   Perzyna model, 301
   for rock salt, 315
   for solders, 314
Electronic chip-substrate systems, solder joint in, 470
Energy
   based models, 489
   dissipated, 83, 84
   fracture, 329
   free, 85
   Peirel's, 85
   responses, pillar-stiff host rock model, 646
Engineering
   behavior, 21
   failure condition, 21
   shear strain, 122
Engineering materials, 1, 8–11
   continuous or discontinuous, 8
   levels of understanding, 9–10
   matter and, 4
   role of material models in, 10–11

   transformation and self-adjustment, 9
ep, see Elastoplastic
Error
   indicator, in postpeak region, 574
   ratio, for stress, 574
evp model, see Elastoviscoplastic model
Extension, 232

F

FA, see Fully adjusted
Failed grout specimen, vertical reconstruction of under triaxial compression, 27
Fatigue failure, cycles to, 396
Fiber-reinforced ceramic composite, 139
Field
   pile test, 614
   testing, determination of parameters by, 448
   validation, 586
Final state, unique, 59
Finite element (FE), 530
   analysis, 509
   discretization, 552
   equations, 288
   formulation, 511, 530
   mesh, 577, 608
First-cycle behavior, 76
First-order model, 121
Flaws, 64
Fluidity parameter, 284
Force equilibrium, 341
Fracture energy, for asphalt concrete, 329
Free energy, 85
Friction
   angle of, 191
   laws, 421
Frozen cracks, 485
Fully adjusted (FA), 2
   behavior, 68
   constrained liquid, 87
   critical state, 87, 432
   displacement, 438
   part, 32, 427
      constrained liquid, 101
      transformation of parts in RI state to, 64
   response, 97, 102, 347, 356
   state(s), 11, 25, 36, 87, see also Fully adjusted states, relatively intact and, and disturbance
      asymptotic, 20

during shear, 433
material parts in, 93
metallic materials, 71
quasi-, 20
stresses and strains in, 94
strains, 109
stress, 536
Fully adjusted states, relatively intact and, and disturbance, 63–91
disturbance and function, 72–86
creep behavior, 82–84
disturbance based on disorder and free energy, 85–86
disturbance function, 73
laboratory tests, 73–78
rate dependence, 84–85
representation of disturbance, 80–81
stiffening effect, 82
stiffening or healing, 78–80
material parameters, 86–91
disturbance function, 88–91
fully adjusted state, 87–88
relatively intact and fully adjusted states, 65–71
characterization of material at critical state, 68–70
specialization, 70–71
Fully saturated materials, 340

G

Geologic material(s)
behavior of, 480
examples of, 236
in pavement, 249
Geotextile
nonwoven, 409
reinforcement, 410, 413
Global disturbance, 357
Global instability, 479
Global states, 7
Glued cracks, 485
Granite, comparisons between model predictions and test data for, 259
Granular materials
behavior of, 67, 480
dense, 196
loose, 196
Green elastic model, 124
Growth parameters, 197, 198

H

Hardening, 233
asphalt concrete, 328
cyclic, 255, 565
elastoplastic, 534
function, 187, 199, 323
models, 156
parameters, 197, 198, 218, 364, 444
response, linear, 334
Healing, 78, 416
Hessian matrix, 669
Hierarchical single-surface (HISS) plasticity models, 179–267
basic HISS model, 180–182
derivation of elastoplastic equations, 205–210
drift correction procedure, 217–220
examples, 222–227
examples of validation, 231–267
concrete, 246–248, 249
concrete, geological materials in pavements, 249–251
examples of geologic materials, 236–237
initial conditions, 233–234
metal alloys, 234–236
optimum tests and sensitivity of parameters, 255–260
repeated loading and permanent deformations, 251–252
rockfill material, 260–261
rocks, 244–245
rock under high pressure, 252–255
sand models, 237–239
saturated clay and dry sand, 239–243
saturated sand, 243
silicon crystal with dislocation, 262–267
HISS versions, 186
incremental iterative analysis, 210–217
correction procedures, 215–216
elastic-plastic loading, 214–215
possible stress states, 214
subincrementation procedure, 216–217
material parameters, 186–203
bonding stress, 193–194
curved ultimate envelope, 190
hardening or growth parameters, 197–199
nonassociative δ_1-model, 199–201
phase or state change parameter, 194–197

rate effects, 202–203
relation between ultimate parameters, cohesion, and angle of friction, 191–193
thermal effects on parameters, 201–202
repetitive loading, 204–205
specialization of HISS model, 183–186
stress path, 227–231
thermoplasticity, 220–222
HISS plasticity model, see Hierarchical single-surface plasticity model
Hooke's law, 99, 116, 121, 123
l'Hospital's rule, 284
Hyperbolic relation, 137
Hyperelastic models, 120

I

Imperfect bar, deformation patterns in, 515
Incremental equations, 164, 437, 532
Incremental iterative analysis, 210
Incremental loading, 212, 213, 281
Incremental nonlinear elastic model, 98
Incremental plasticity equations, 77
Independent test, prediction of, 369
Independent validation, 245, 453
Indiana limestone, 312
Interacting mechanisms, 12
Interfaces and joints, DSC for, 421–472
 computer implementation, 472
 determination of parameters, 441–448
  mathematical and physical characteristics of DSC, 447–448
  regularization and penalty, 448
 disturbance function, 435–437
 disturbed state concept, 427–435
  elasto plastic models, 430–432
  FA as critical state, 432–435
  fully adjusted state, 432
  relatively intact behavior, 427
  stress-displacement equations, 428–430
 examples, 449–472
  aluminum shaft–sand interfaces, 468–470
  concrete–concrete interface, 451
  dry sand–concrete interfaces, 454–458
  dry sand-steel interfaces, 458–460
  homework problems, 471–472
  rock joints, 451–452
  rock–pile interface, 452–454
  sand–geosynthetic interfaces, 467–468
  saturated clay–steel interfaces, 464–467
  soil–rock interface, 461–464
 solder joint in electronic chip-substrate systems, 470–471
 general problem, 422
 incremental equations, 437–441
  alternative DSC equations, 440–441
  specializations, 439–440
 review, 422–425
 testing, 448–449
 thin-layer interface model, 425–426
Internal characteristics dimension, 100
Interpolation
 functions, 530
 parameter, 564
Isotropic stress, 185

J

Joint(s), see also Interfaces and joints, DSC for
 examples of, 426
 rock, 451
 roughness coefficient (JRC), 451
 shear tests on, 434
Jointed rock, 409, 414
Jointed systems, 407

L

Laboratory
 curves, prediction of, 404
 stress–strain behavior
  rock salt, 244
  soap stone, 244
 tests, 73
  backfill soil, 581
  data, elastic constants from, 128
  determination of parameters, 448
  normally consolidated soils, 157
  prediction of, 210
  results, for concrete, 508
  rock, 643
  two-dimensional, 459
  viscoplastic model, 293
Leadless ceramic chip carrier (LCCC), 600
Least square fit procedure, 663
Leda clay, compression tests on, 400
Leighton Buzzard (LB) sand, 236, 241
Limestone, Indiana, 312
Linear elasticity, 115
Line search strategy, 670
Liquefaction, 14
 during earthquake, 651

Index

instability, 517
saturated sand, 395
Liquid–solid
assumption, critical-state concept for, 365
constrained, 37
Load
-carrying capacity, 483
on half space, 576
increments, 570
repetitive, 204
steps, effect of, 575
Loading
cyclic, 559, 566, 611
dynamic, 205
elastic-plastic, 214
hydrostatic, 30
incremental, 212, 213, 281
material stress during, 22
multiaxial, crack density and disturbance during, 146
nonvirgin, 559
proportional, 227
repetitive, 251, 559, 609
simulated, 371
uniaxial
crack density and disturbance during, 142
tension, 153
yield criterion in, 152
unloading–reloading cycles, 204
virgin, 559
Local instability, 479
Localization, cause of, 492
Local states, 7

M

Manufacturing defects, 64
Marquard–Levenberg (ML) method, 90
Mass density, 522
Material(s)
bonded, 53, 57
characteristics of deforming, 477
characterization of material at, 168
cohesive, 196
component-reference, 39
contacts between two, 421
elastic perfectly plastic, 150
elastoplastic, 150
element
composed of two materials, 40
discontinuities of, 18
equivalent constitutive matrix for, 299
force equilibrium and, 34
existence, levels of, 6
failed, 65
fully saturated, 340
geologic
examples of, 236
in pavement, 249
granular
dense, 196
loose, 196
matrix, deformation of, 478
metallic, 71, 194
models, role of in engineering, 10
parameters, 86, 363, 504
asphalt pavement system, 579
overlay model, 303, 306
parts
disordered, 85
ordered, 85
pavement, parameters for, 254
porous, 339
saturated, 551
void ratio for, 106
responses, schematic of, 632
rockfill, 264
stiffness of, 150
stress, during loading, 22
structured, 58
unified theory for mechanics of, 629
Matrix
constitutive, 499
definition of, 537
Hessian, 669
notation, 63, 97
constitutive tensor expressed in, 116
DSC equations given in, 98
rotation, 664
stiffness, 490, 534
suction, 342
Maxwell model, 301
Mechanical engineering, metal-to-metal contacts in, 422
Mechanical testing, 568
Mesh
adaption, using DSC, 572
dependence, 485, 494
localization and, 514
spurious, 507
finite element, 518
Metal
alloys, 234
behavior of, 174
DSC response for, 175
Microcracking, 18, 29, 53
deformations due to, 482

disturbance due to, 125
laboratory tests for rock, 643
simulated, 639
Microcrack interaction, 494
  damage model with, 100
  models, 485
Microstructure, 477–523
  deformation of, 479
  disturbed state concept, 494–507
    approximate decoupled DSC, 500–502
    instability through disturbance, 502–503
    stability analysis of DSC, 503
    stability condition for one-dimensional DSC model, 503–507
  examples, 507–522
    instability based on critical dissipated energy and disturbance, 522
    instability in DSC, 516–521
    localization, 510–514
    localization and mesh dependence, 514–516
    spurious mesh dependence, 507–510
  localization, 482–484
  nonlocal continuum, 488–492
    Cosserat continuum, 491–492
    gradient enrichment of continuum models, 489–491
    strain and energy based models, 489
  regularization and nonlocal models, 484–488
    continuum damage model, 486–487
    microcrack interaction models, 485–486
    models for nonlocal effects, 487–488
    rate-dependent models, 486
  self-adjustment of, 11
  stability, 492–494
  thermodynamical analysis of DSC, 522–523
  wellposedness, 482
ML method, see Marquard–Levenberg method
Model(s)
  approximate constitutive, 4
  cam clay, 162, 163, 230
  cap, 161, 162, 163, 167
  Cauchy elastic, 124
  classical continuum damage, 100
  concrete plasticity, 246
  constitutive, 11
  continuous yielding, 165
  continuum
    damage, 439, 486
    gradient enrichment of, 489
  Cosserat, 491
  creep, disturbance for, 287

CS
  advantages and limitations of, 163
  parameters in, 161
damage, 12, 100
disturbance, 99, 262
Drucker–Prager plasticity, 183
DSC, 13, 21, 361, 376
  application of for softening and stiffening response of silicon, 266
  computer procedures with, 570
  concrete, 248
  formulation of, 402
  without relative motions, 101
elasto plastic models, 430
elastoviscoplastic
  Perzyna, 301
  rock salt, 315
  solders, 314
energy based, 489
evp, responses of, 285
first-order, 121
Green elastic, 124
hardening, 156
HISS, 164, 361, 609
  derivations of, 222
  stress path analysis of, 228
  thermal effects in, 201
  yield function in, 220
hyperelastic, 120
incremental nonlinear elastic, 98
interface/joint, 449
linear elastic, 330
Maxwell, 301
microcrack interaction, 485
Mohr–Coulomb plasticity, 183
nonlocal
  effects, 487
  regularization and, 484
overlay, 297, 300
  advantages of, 313
  closed-form solution, 327
  material parameters in, 303
Perzyna, 275
  formulation of, 282
  viscoplasticity, 313
pillar-stiff host rock, 646
predictions
  comparisons between LB test data and, 242
  granite, 259
  sand, 261
rate-dependent, 486
review of various, 188
role of material, 10
sands, 237

Index

saturated sand, 243
stiff host rock, 647
thin-layer interface, 425
Tresca plasticity, 183
variable parameter, 118
viscoelasticviscoplastic overlay, 307
viscoplastic, 235, 290
    concrete, 322
    cylinder with internal pressure, 596
    soil–rock interface, 461
    theory based on the overstress, 313
Mohr–Coulomb
    criterion, for plastic response, 332
    plasticity model, 183
    stress spaces, 226
    yield criterion, 154, 155, 165, 286
Molding, of clay lumps, 26
Monistic idealism, 5
Motivation, 2–8
    engineering materials and matter, 4–7
    explanation of reference states, 3–4
    local and global states, 7–8
Multiaxial compression, comparisons between DSC predictions and, 144
Multiaxial loading, crack density and disturbance during, 146

N

Natural self-adjustment (NSA), 631
Net mean stress, 345
Net stress, 37
Newton's method, 668, 669
Noisy phenomenon, 634
Nonassociative response, 463
Nondestructive measurements, 568
Nonlocal continuum, 488
Nonlocal models, regularization and, 484
Nonvirgin loading, 559
Normality rule, 219
Normalized roughness, 456
Normally consolidated (NC) soil state, 391
Normally consolidated (NC) soil, 23, 157, 433

O

Observed behavior, 133, 541
Observed strain increment, 104
Optimization
    numerical method for, 668
    procedure, 665
    unconstrained, 671
Ordered material parts, 85
Overconsolidated (OC) soil, 23
Overconsolidated (OC) soil state, 391
Overconsolidation effect, 55
Overlay(s)
    model(s), 297, 300
        advantages of, 313
        closed-form solution, 327
        material parameters in, 303, 306
        viscoelasticviscoplastic, 307
    strain, 305, 329
Overstress, 278, 328

P

Parameter(s)
    creep, determination of from shear test for soil, 295
    determination of, 441, 448
    disturbance, 292, 446
    dry sands, 238
    effective stress, 358
    fluidity, 284
    growth, 197, 198
    hardening, 197, 198, 218, 364
    interpolation, 564
    material, for asphalt pavement system, 579
    pavement materials, 254
    phase change, 196, 364
    physical meanings of, 312
    rockfill materials, 264
    rock joint type, 445
    rocks, 240
    saturated clay, 239
    schematic of test for, 312
    sensitivity of, 677
    solder, 235
    state change, 196
    thermal effects on, 201
    ultimate, 191, 363
    viscoelastic model, 309
    viscoplastic model, 290
Parametric sensitivity analysis, CTC, 263
Partially saturated materials, computer procedures for, 558
Partially saturated soil, 374
Partially saturated systems, 558
Particle-to-particle contacts, 44
Pathological mesh dependence, 485
Pavement
    geologic materials in, 249
    materials, parameters for, 254

system, 578, 597
PBGA, see Plastic ball grid array
Peak stress, 21
Peirel's energy, 85
Perzyna
  model, 275, 282, 313
  theory, 275
Phase change (PC)
  envelope, 371
  line, slope of, 226
  parameter, 196, 364
  point, 457
Piecewise linear behavior, 117
Pile load tests, 611
Pillar-stiff host rock model, 646
PL, see Proportional loading
Plane
  strain, 154, 545
  stress, 545
Plastic ball grid array (PBGA), 601
Plastic displacement trajectory, deviatoric, 446
Plasticity
  constants, 257
  criterion, von Mises, 321
  equations, incremental, 77
  models, 362
Plasticity, theory of in DSC, 149–176
  advantages and limitations of CS and cap models, 163–164
  cap model, 161–163
  continuous yielding or hardening models, 156
  critical-state concept, 156–160
  examples, 168–176
  incremental equations, 164–168
    cyclic loading, 168
    parameters and determination from laboratory tests, 165–168
    thermoplasticity, 168
  mechanisms, 151–152
  Mohr–Coulomb yield criterion, 154–156
  parameters in CS model, 161
  theoretical development, 152
  yield criteria, 152–154
  yield surface, 160–161
Plastic relative displacement trajectory, 436
Plastic shear strain, 184
Plastic strain trajectory, 362
  deviatoric, 367
  revised deviatoric, 543
Poisson's ratio, 36, 140, 206, 306
Pore
  -collapse phase, 31

pressure measurements, CYMDOF shear device with, 449
  water pressure, 76, 343, 519, 653
Porous chalk, behavior of under hydrostatic loading, 30
Porous materials, 339
Porous media, nonlinear behavior of, 550
Porous saturated material element, 551
Porous saturated media, DSC for, 43, 45
Postliquefaction behavior, of saturated sand, 395
Postpeak
  region, error indicator in, 574
  response, 573
Predicted behavior, 133
Prediction, independent, 460
Predictor–correction procedures, 216
Premordial material, 7
Pressure, rock under high, 252
Printed wire board (PWB), 601, 606
Pristine material, 7
Projection surface, curvature of, 223
Proportional loading (PL), 227
P-wave velocity, 140, 141
  cemented sand, 418
  fiber-reinforced ceramic, 417
  measured ultrasonic, 569

Q

Quasi-Newton method, 668, 669

R

Rate effects, 202, 203
Reduced triaxial extension (RTE), 413
Reference states, analogies for, 20
Reid Bedford sand, 657
Reinforced earth, 409, 580
Reinforced soil
  sample, 411
  yield behavior of, 412
Reinforced specimen, 143
Reinforced systems, 407
Relatively intact (RI), 2
  behavior, 67, 124, 240, 427
  characterization, elastic model as, 125
  constitutive matrix for, 439
  response, 96, 347
  shear stress, 406
  simulation, linear, 131

Index

state(s), 3, 20
  during shear, 433
  material parts in, 93
  response of matrix in, 53
  stresses and strains in, 94
strains, 109
stress–strain curve, 135
Relative shear displacement, 429
Relaxation, one-dimensional
  elastoviscoplastic model for, 333
Reloading, 563
Repetitive loading, 251, 559, 609
Residual flow
  concept, 359
  procedure (RFP), 340, 359
Rest periods, 415
RFP, see Residual flow procedure
RI, see Relatively intact
Rock(s), 244
  acoustic emission
    brittle, 645
    laboratory tests for, 643
  joints, 409, 414, 424, 451
  parameters for, 240
  pile interface, 452
  salt
    elastoviscoplastic model for, 315
    laboratory stress–strain behavior of, 244
    problem, finite-element mesh for, 332
    testing of, 245
  specimen, test on, 644
  under high pressure, 252
Rockfill material, 260, 264
Roughness
  initial, 496
  normalized, 456
RTE, see Reduced triaxial extension

S

Sand(s)
  cemented, 142
  comparisons between model predictions and test data for, 261
  -concrete interfaces, 615
  dry, 241
  geosynthetic interfaces, 467
  Leighton Buzzard, 236
  liquefaction of saturated, 395
  models, 237
  observed behavior of, 368, 369
  parameters for dry, 238
  Reid Bedford, 657

saturated, 243, 366
steel interface, 454, 467
Sandy silt, saturated and partially saturated, 372
Saturated clay, 370
Saturated clay–steel interfaces, 464
Saturated sand(s), 366
  cyclic behavior of, 656
  liquefaction of, 395
  model, 243
Saturated soil, 373
Saturated and unsaturated materials, DSC for, 339–388
  disturbance, 348–361
    effective stress parameter, 358–359
    residual flow concept, 359–361
    stress-strain behavior, 349–356
    volumetric or void ratio, 356–357
  equations, 341–343
  examples, 365–380
    back predictions, 369–370
    DSC model, 376–380
    partially saturated soil, 374–376
    saturated clay, 370–372
    saturated and partially saturated sandy silt, 372–373
    saturated sand, 366–368
    saturated soil, 373
  fully saturated materials, 340–341
  HISS and DSC models, 361–363
  incremental DSC equations, 346–348
  material parameters, 363–365
  softening, degradation, and collapse, 363
  stress equations, 343–346
Saturation ratio, 347
Self-adjustment, 9
Self-organized criticality (SOC), 10
Shake table test, 615
Shear
  modulus, 36, 79
  RI and FA states during, 433
  stiffness, 428, 442
  strain(s)
    engineering, 122
    plastic, 184
  stress(es)
    relation between, 103
    results for, 377
    RI, 406
  tests, on joints, 434
  train components, nonzero, 122
  wave velocities, 654
Silicon
  application of DSC model for softening and stiffening response of, 266

chip, 605
crystal with dislocation, 262
dislocated, 401
Silt, saturated and partially saturated sandy, 372
Simple shear (SS)
   stress path, 192
   test, 562
Slippage, 18, 29
Soap stone, laboratory stress–strain behavior of, 244
SOC, see Self-organized criticality
Softening, 363
   behavior, applications of DSC model for, 403
   response, 394
   stress-strain response, 506
Soil(s)
   disturbance
      function for structured, 397
      in stiff, 24
   drained shear strength of at saturation, 358
   -footing, finite elements mesh of, 577
   mechanics, critical state in, 25
   normally consolidated, 23, 433
   overconsolidated, 23
   partially saturated, 374
   reinforced
      sample, 411
      yield behavior of, 412
   rock interface, 461
   saturated, 373
   shear test for, 295
   skeleton, measure of average stress carried by, 47
   strains, comparisons between predicted and observed vertical, 589
   stress, comparisons between predicted and observed vertical, 588
   structured, 396
   testing, backfill, 581
Solder(s), 194, 234
   elastoviscoplastic model for, 314
   joint, in electronic chip-substrate systems, 470
   parameters for, 235
Solids, DSC/HISS models for, 440
Solid skeleton, force in, 44
Solid stress, 345
Spurious mesh dependence, 485, 507
SS, see Simple shear
State change parameter, 196
Static yield surface, 278
Stiffening, 78
   behavior, 30

effect, 82
response, 394
stress-strain response, 506
Stiff host rock model, 647
Stiffness
   materials, 150
   matrix, 490, 493, 534
   normal, 442
   shear, 428, 442
Stiff soils, disturbance in, 24
Strain(s)
   computation of stresses for different, 137
   -displacement relations, 505
   disturbance vs. plastic, 172
   equations
      derivation of, 105
      using critical state, 105
   increment, 109, 231
   overlay, 306, 329
   plane, 154, 545
   plastic, 649
   rates, responses under different, 84
   shear component of, 116
   softening, 151
      response, 152
      -stiffening behavior, 151
   strain curve, uniaxial, 111
   tensor, 123, 291
   total, 283
   vector, 221, 530
   viscoplastic, 282
Stress
   acting on interface, 459
   approximate values of, 87
   back, 85
   bonding, 193, 344
   computed cycle, 611
   confining, 232
   contact, 50
   definitions of, 19
   -displacement equations, 428
   effective, 37, 47, 48, 50, 52
   equations, 343
   error ratio for, 574
   FA, 536
   increment, 280
   initial, 567
   isotropic, 185
   net, 37, 345
   path(s), 227
      analysis, of HISS model, 228
      explanations of, 229
      representation of, 88
      switching of, 230
      triaxial compression, 413

Index

peak, 21
plane, 545
relaxation, 316, 335
shear
  component of, 116
  relation between, 103
solid, 345
solution for, 336
spaces
  Mohr–Coulomb, 226
  plots of F in different, 182
states, possible, 214
tensor, 116, 209
time response vs., 320
total, at different depths, 51
yield, 169
Stress–strain
  behavior, 21, 66, 349
    elastic linear, strain hardening, 283
    observed, 173
  curve(s)
    predicted, 136
    RI, 135
    uniaxial, 130, 203
  data, 126
  response(s)
    backfill soil, 582
    elastic, 135
    functional representation of, 119
    prepeak region of, 402
    softening, 506
    stiffening, 506
  volume change behavior, prediction of triaxial, 413
Structural instability, 493
Structured soil(s), 396
  compression behavior of, 398
  disturbance in, 24
Structured and stiffened materials, DSC for, 391–418
  definition of disturbance, 392–396
  dislocations, softening, and stiffening, 401–407
    dislocated silicon with impurities, 401–402
    disturbance and dislocation density, 405–407
    formulation of DSC model, 402–403
    validation, 403–404
  reinforced and jointed systems, 407–415
    equivalent composite, 407–409
    individual solid and joint elements, 409
    jointed rock, 414
    reinforced earth and jointed rock, 409–413

    validation for test results with HISS model and FE method, 413
    rest periods, 415–417
  structured soils, 396–401
    compression behavior of structured soil, 398–399
    validations, 399–401
Subincrementation procedure, 216
Switch-on–switch-off operator, 277
Synergetics concepts, 59

T

Tangent stiffness matrix, 493
Taylor series expansion, 279
TCT, see Triaxial compression test
Technology, humanization of, 9
Tensile behavior, of metals and metallic materials, 194
Tension bar, with imperfection, 512
Terzaghi
  concept, inherent assumptions in, 43
  effective stress concept, 47
  equation, 344
Test(s)
  constant normal stiffness, 453
  cyclic strain-controlled, 76
  cyclic stress-controlled, 656
  data, disturbance from, 74
  laboratory stress–strain, 127
  mechanical, 568
  pile load, 611
  prediction of independent, 369
  shake table, 615
  simple shear, 562
  specimen, applied stress for, 378
  triaxial compression, 562
Theoretical maximum density (TMD), 20
Thermal creep, in restrained bar, 319
Thermodynamical analysis, of DSC, 522
Thermoelasticity, 127
Thermomechanical behavior, in electronic chip-substrate systems, 470
Thermoplastic behavior, 546
Thermoplasticity, 220
Thermoviscoplastic behavior, 547
Thin-layer interface model, 425
Threshold transitions, 157, 481
Time
  integration, for dynamic problems, 554, 555
  step, selection of, 286
TMD, see Theoretical maximum density

Total stresses, 51
Toyoura–concrete interface, 454
Tresca
    plasticity model, 183
    yield function, 286
Triaxial compression
    stress paths, 413
    test (TCT), 562
    vertical reconstruction of failed grout specimen under, 27
Tribology, 422
Turbulent phenomenon, 634

U

Ultimate disturbance, 496
Ultimate envelope, 224
Ultimate parameters, 191, 363, 443
Underground excavations, analysis and design of, 315
Uniaxial compression test, for concrete, 326
Uniaxial loading
    crack density and disturbance during, 142
    yield criterion in, 152
Uniaxial strain–strain curve, 111, 130
Uniaxial tension
    loading, 153
    stress–strain curves, 203
Uniform degree of error distribution (UDED), 572
Unique final state, 59
Unloading, 371, 415, 561
Unsaturated materials, see Saturated and unsaturated materials, DSC for

V

Validation(s), 672
    examples of, 231
    field, 586
    independent, 245, 453
Variable parameter models, 118
VBO models, see Viscoplastic theory based on the overstress models
vevp, see Viscoelasticviscoplastic
Virgin loading, 559
Virtual displacement, 531
Viscoelastic overlay model, Indiana limestone, 312
Viscoelasticviscoplastic (vevp), 274
    model, 302
    overlay model, 307

    response, 306
Viscoplastic creep, 303
Viscoplastic (vp) model, 235, 290
    creeping natural slope, 597
    cylinder with internal pressure, 596
    parameters for, 290
    soil–rock interface, 461
Viscoplastic solution, mechanics of, 277, 278
Viscoplastic strains, 83, 282
Viscoplastic theory based on the overstress (VBO) models, 313
Viscosity coefficient, 310
Void ratio, 47, 49, 66, 73, 348, 356
    interface material and linear relation, 440
    porous materials, 106
    relation between incremental volume and, 1-7
Voltage, current vs., 642
von Mises
    criterion, 112, 153, 154
    envelope, 165
    plasticity criterion, 321
    yield function, 286
vp model, see Viscoplastic model

W

Wall facing, 586
Wellposedness, of boundary value problem, 482

Y

Yield
    behavior, of natural and reinforced soil, 412
    criteria, 152
        Drucker–Prager, 154, 165, 189
        Mohr–Coulomb, 154, 155, 165
        uniaxial loading, 152
    function, 111, 170
    stress, 169
    surface, 160
        convex, 225
        dynamic, 278
        static, 278
Young's modulus, 79, 127

Z

Zipf's law, 648